Benchmark Papers in Geology

Series Editor: Rhodes W. Fairbridge
Columbia University

Volume

1 ENVIRONMENTAL GEOMORPHOLOGY AND LANDSCAPE CONSERVATION, Volume 1: Prior to 1900 / *Donald R. Coates*
2 RIVER MORPHOLOGY / *Stanley A. Schumm*
3 SPITS AND BARS / *Maurice L. Schwartz*
4 TEKTITES / *Virgil E. Barnes and Mildred A. Barnes*
5 GEOCHRONOLOGY: Radiometric Dating of Rocks and Minerals / *C. T. Harper*
6 SLOPE MORPHOLOGY / *Stanley A. Schumm and M. Paul Mosley*
7 MARINE EVAPORITES: Origin, Diagenesis, and Geochemistry / *Douglas W. Kirkland and Robert Evans*
8 ENVIRONMENTAL GEOMORPHOLOGY AND LANDSCAPE CONSERVATION, Volume III: Non-Urban / *Donald R. Coates*
9 BARRIER ISLANDS / *Maurice L. Schwartz*
10 GLACIAL ISOSTASY / *John T. Andrews*
11 GEOCHEMISTRY OF GERMANIUM / *Jon N. Weber*
12 ENVIRONMENTAL GEOMORPHOLOGY AND LANDSCAPE CONSERVATION, Volume II: Urban Areas / *Donald R. Coates*
13 PHILOSOPHY OF GEOHISTORY: 1785-1970 / *Claude C. Albritton, Jr.*
14 GEOCHEMISTRY AND THE ORIGIN OF LIFE / *Keith A. Kvenvolden*
15 SEDIMENTARY ROCKS: Concepts and History / *Albert V. Carozzi*
16 GEOCHEMISTRY OF WATER / *Yasushi Kitano*
17 METAMORPHISM AND PLATE TECTONIC REGIMES / *W. G. Ernst*
18 GEOCHEMISTRY OF IRON / *Henry Lepp*
19 SUBDUCTION ZONE METAMORPHISM / *W. G. Ernst*
20 PLAYAS AND DRIED LAKES: Occurrence and Development / *James T. Neal*
21 GLACIAL DEPOSITS / *Richard P. Goldthwait*
22 PLANATION SURFACES: Peneplains, Pediplains, and Etchplains / *George F. Adams*
23 GEOCHEMISTRY OF BORON / *C. T. Walker*
24 SUBMARINE CANYONS AND DEEP-SEA FANS: Modern and Ancient / *J. H. McD. Whitaker*
25 ENVIRONMENTAL GEOLOGY / *Frederick Betz, Jr.*
26 LOESS: Lithology and Genesis / *Ian J. Smalley*

27 PERIGLACIAL PROCESSES / *Cuchlaine A. M. King*
28 LANDFORMS AND GEOMORPHOLOGY: Concepts and History / *Cuchlaine A. M. King*
29 METALLOGENY AND GLOBAL TECTONICS / *Wilfred Walker*
30 HOLOCENE TIDAL SEDIMENTATION / *George deVries Klein*
31 PALEOBIOGEOGRAPHY / *Charles A. Ross*
32 MECHANICS OF THRUST FAULTS AND DÉCOLLEMENT / *Barry Voight*
33 WEST INDIES ISLAND ARCS / *Peter H. Mattson*
34 CRYSTAL FORM AND STRUCTURE / *Cecil J. Schneer*
35 OCEANOGRAPHY: Concepts and History / *Margaret B. Deacon*
36 METEORITE CRATERS / *G. J. H. McCall*
37 STATISTICAL ANALYSIS IN GEOLOGY / *John M. Cubitt and Stephen Henley*
38 AIR PHOTOGRAPHY AND COASTAL PROBLEMS / *Mohamed T. El-Ashry*
39 BEACH PROCESSES AND COASTAL HYDRODYNAMICS / *John S. Fisher and Robert Dolan*
40 DIAGENESIS OF DEEP-SEA BIOGENIC SEDIMENTS / *Gerrit J. van der Lingen*
41 DRAINAGE BASIN MORPHOLOGY / *Stanley A. Schumm*
42 COASTAL SEDIMENTATION / *Donald J. P. Swift and Harold D. Palmer*
43 ANCIENT CONTINENTAL DEPOSITS / *Franklyn B. Van Houten*
44 MINERAL DEPOSITS, CONTINENTAL DRIFT AND PLATE TECTONICS / *J. B. Wright*
45 SEA WATER: Cycles of the Major Elements / *James I. Drever*
46 PALYNOLOGY, PART I: Spores and Pollen / *Marjorie D. Muir and William A. S. Sarjeant*
47 PALYNOLOGY, PART II: Dinoflagellates, Acritarchs, and Other Microfossils / *Marjorie D. Muir and William A. S. Sarjeant*
48 GEOLOGY OF THE PLANET MARS / *Vivien Gornitz*
49 GEOCHEMISTRY OF BISMUTH / *Ernest E. Angino and David T. Long*
50 ASTROBLEMES—CRYPTOEXPLOSION STRUCTURES / *G. J. H. McCall*
51 NORTH AMERICAN GEOLOGY: Early Writings / *Robert Hazen*
52 GEOCHEMISTRY OF ORGANIC MOLECULES / *Keith A. Kvenvolden*
53 TETHYS: The Ancestral Mediterranean / *Peter Sonnenfeld*
54 MAGNETIC STRATIGRAPHY OF SEDIMENTS / *James P. Kennett*
55 CATASTROPHIC FLOODING: The Origin of the Channeled Scabland / *Victor R. Baker*

Benchmark Papers in Geology / 52

A BENCHMARK® Books Series

GEOCHEMISTRY OF ORGANIC MOLECULES

Edited by

KEITH A. KVENVOLDEN

U. S. Geological Survey
Menlo Park, California

Dowden, Hutchinson & Ross, Inc.

STROUDSBURG, PENNSYLVANIA

Benchmark Papers in Geology, Volume 52
Library of Congress Catalog Card Number: 79-18201
ISBN: 0-87933-353-7

82 81 80 1 2 3 4 5
Manufactured in the United States of America.

LIBRARY OF CONGRESS CATALOGING IN PUBLICATION DATA
Main entry under title:
Geochemistry of organic molecules.
(Benchmark papers in geology ; 52)
Includes bibliographical references and indexes.
1. Organic geochemistry—Addresses, essays, lectures. 2. Molecules—Addresses, essays, lectures. I. Kvenvolden, Keith A., 1930-
QE516.5.G46 551.9 79-18201
ISBN 0-87933-353-7

Distributed world wide by Academic Press, a subsidiary of Harcourt Brace Jovanovich, Publishers.

SERIES EDITOR'S FOREWORD

The philosophy behind the "Benchmark Papers in Geology" is one of collection, sifting, and rediffusion. Scientific literature today is so vast, so dispersed, and, in the case of old papers, so inaccessible for readers not in the immediate neighborhood of major libraries that much valuable information has been ignored by default. It has become just so difficult, or so time consuming, to search out the key papers in any basic area of research that one can hardly blame a busy person for skimping on some of his or her "homework."

This series of volumes has been devised, therefore, to make a practical contribution to this critical problem. The geologist, perhaps even more than any other scientist, often suffers from twin difficulties—isolation from central library resources and immensely diffused sources of material. New colleges and industrial libraries simply cannot afford to purchase complete runs of all the world's earth science literature. Specialists simply cannot locate reprints or copies of all their principal refence materials. So it is that we are now making a concerted effort to gather into single volumes the critical materials needed to reconstruct the background of any and every major topic of our discipline.

We are interpreting "geology" in its broadest sense: the fundamental science of the planet Earth, its materials, its history, and its dynamics. Because of training in "earthy" materials, we also take in astrogeology, the corresponding aspect of the planetary sciences. Besides the classical core disciplines such as mineralogy, petrology, structure, geomorphology, paleontology, and stratigraphy, we embrace the newer fields of geophysics and geochemistry, applied also to oceanography, geochronology, and paleoecology. We recognize the work of the mining geologists, the petroleum geologists, the hydrologists, and the engineering and environmental geologists. Each specialist needs a working library. We are endeavoring to make the task of compiling such a library a little easier.

Each volume in the series contains an introduction prepared by a specialist (the volume editor)—a "state of the art" opening or a summary of the object and content of the volume. The articles, usually some twenty to fifty reproduced either in their entirety or in significant extracts, are selected in an attempt to cover the field, from the key papers of the last century to fairly recent work. Where the original works are in foreign languages, we have endeavored to locate or commission translations. Geologists, because of their global subject, are often acutely aware of the oneness of our world. The selections cannot, therefore, be restricted to any one country, and whenever possible an attempt is made to scan the world literature.

To each article, or group of kindred articles, some sort of "highlight commentary" is usually supplied by the volume editor. This commentary should serve to bring that article into historical perspective and to emphasize its particular role in the growth of the field. References, or citations, wherever possible, will be reproduced in their entirety—for by this means the observant reader can assess the background material available to that particular author, or, if desired, he or she, too, can double check the earlier sources.

A "benchmark," in surveyor's terminology, is an established point on the ground, recorded on our maps. It is usually anything that is a vantage point, from a modest hill to a mountain peak. From the historical viewpoint, these benchmarks are the bricks of our scientific edifice.

RHODES W. FAIRBRIDGE

PREFACE

From the extensive literature that has been published since the early 1930s that is concerned with many aspects of the geochemistry of organic molecules, forty-two papers have been selected for this volume to provide the reader with a broad view and appreciation of the field. The papers attempt to show the developments, progress, and present applications of organic geochemistry. The book is divided into five parts. The first part includes papers that have given the field some of its substance and direction. The second part includes papers that are devoted to advances brought about by considerations of different kinds of organic molecules found in geological settings. The third section is concerned with the stereochemistry of organic molecules found in nature. In the first three parts, emphasis is placed on the geological occurrence of organic molecules, while in the fourth part the direction changes to laboratory experiments that attempt to explain natural occurrences. The last group of papers, the fifth part, show how organic geochemistry has been applied to problems in different scientific areas.

This book has been designed to serve both as a reference and to some extent as a textbook. Each part either cites references relevant to that group of papers or gives lists of additional suggested reading or both. By noting the reprinted papers and by consulting references or reading lists, the reader should be able to ascertain past stages of organic geochemical research and at which stage the research stands at present (1978).

The most difficult task in preparing a book of this sort is choosing those papers that will adequately represent the field from the thousands of available papers. A few guidelines were necessary to help narrow the choices. The decision was made that in all cases entire papers would be reprinted so that the reader would not have to look elsewhere for a complete document. Thus, a number of important, longer papers could not be included. Secondly, no paper that appears in book form was reprinted. The idea here was that the books, and the articles in them, are readily available in libraries. Articles scattered through various journals are more difficult to find. Thus, only individual journal articles are reprinted here. Undoubtedly, many important articles were missed in the selection process. Hopefully many of these articles are included in the references and additional reading lists. The reader should not expect an exhaustive treatment of the subject. At best, what is presented is a selective treatment that tries to provide a guide to the geochemistry of organic molecules.

Of course, this book would not be possible without the cooperation of the authors and publishers of the articles that have been reprinted here. I thank all of those who allowed me to include their work. I am also grateful to Ms. Jeanne Henning for her help in preparing the text, to George D. Redden for preparing the molecular illustrations, and to Reiner Protsch for the English translation of the first paper.

KEITH A. KVENVOLDEN

CONTENTS

PART III: MOLECULAR ARCHITECTURE

PART IV: MOLECULAR EXPERIMENTATION

CONTENTS BY AUTHOR

GEOCHEMISTRY OF ORGANIC MOLECULES

INTRODUCTION

Studies of the geochemistry of organic molecules, usually called organic geochemistry, borrow from the principles and procedures of the broader fields of geology, organic chemistry, and, to some extent, biology. A major goal of the resulting interdisciplinary subject is to determine the occurrence and fate of organic materials (both biogenic and abiogenic) that are found in geologically controlled contexts.

Organic geochemistry has been applied to both terrestrial and extraterrestrial substances. When applied to earth materials, organic geochemistry considers mainly organic matter and the processes to which this organic matter may be subjected, especially diagenetic changes in sediments and rocks. Moreover, organic geochemistry is also concerned with the precursors of this organic matter—most, if not all of which, come from the biosphere. Generally speaking, organic geochemistry is not concerned with what goes on inside of organisms because this is the realm of biochemistry. Rather, organic geochemistry focuses on the various contributions organisms make to the geochemical cycles in which these organisms are involved. Besides applications to terrestrial materials, the principles of organic geochemistry have also been used in the study of extraterrestrial substances. Meteorites and samples returned from the moon have been investigated in numerous laboratories. Samples from Mars have been examined only remotely by automated instrumentation.

This book deals mainly with organic geochemistry at the molecular level—that is, most of the papers discuss the geochemistry of certain organic molecules or classes of molecules. Aspects such as identifications, determination of structural configurations, occurrences, reactions, and implications are discussed. The title of this book and the part headings emphasize the molecular nature of the content. The first part, "Molecular Foundations," contains six fundamental and influential papers that form the foundations for much of the work in organic geochemistry that has followed. The second and longest part contains sixteen papers under the title "Molecular Advances." These papers describe early identifications of various organic molecules in geological materials, and they are arranged more or less in an order of increasing complexity of molecular structure and composition. Part II begins with papers on hydrocarbons of very simple structures and ends with papers that discuss very complex, polymeric materials such as humic acids and kerogen. The final paper describes the carbon isotopic composition of organic material. Part III, "Molecular Architecture," goes beyond molecular structure and composition and considers the geometric aspects of certain molecules. Six papers emphasize stereochemistry. These papers represent a natural step in the complete understanding of organic molecules and their geochemical significance. The fourth part entitled "Molecular Experimentation" includes six papers that outline laboratory experiments wherein attempts have been made to duplicate some of the chemical reactions that take place during the geochemical transformation of organic molecules. The last part, "Molecular Applications," contains eight examples of the application of organic geochemistry to geological problems. Because of the direct contributions organic geochemistry has made to the petroleum industry, three papers on petroleum geochemistry are included. The five remaining papers demonstrate the application of organic geochemistry to studies of microbiology, environmental chemistry, geochronology, paleothermometry, and cosmochemistry.

This book has been designed to provide an historical as well as practical view of the geochemistry of organic molecules. The papers selected for inclusion represent only a small fraction of the pertinent literature. No papers are included, for example, that are published as chapters or articles in other books, because these sources are usually available in libraries. Rather, the papers reprinted here are individual journal articles which, as time passes, are becoming less easy to obtain in their original form. The reprinted articles along with the lists of references and suggested additional readings serve as a guide to the field of molecular organic geochemistry during its first fifty years.

Part I

MOLECULAR FOUNDATIONS

Editor's Comments on Papers 1 Through 6

1 **TREIBS**
Chlorophyll- und Häminderivate in organischen Mineralstoffen
English Translation: *Chlorophyll and Hemin Derivatives in Organic-Mineral Substances*

2 **TRASK and WU**
Does Petroleum Form in Sediments at Time of Deposition?

3 **ROSSINI**
Hydrocarbons in Petroleum

4 **SMITH**
The Occurrence of Hydrocarbons in Recent Sediments from the Gulf of Mexico

5 **ABELSON**
Organic Constituents of Fossils

6 **BLUMER**
Organic Pigments: Their Long-Term Fate

Serious considerations of molecular organic geochemistry began in the 1930s. During that decade, two research efforts, one in Germany, led by Alfred Treibs, and the other in the United States, headed by Parker D. Trask, had tremendous impact and lasting consequences for the field. A general, well-founded consensus in the field is that the pioneering investigations of Treibs on porphyrin pigments represents the beginnings of molecular organic geochemistry, but the monumental work of Trask must not be overlooked. With support from the American Petroleum Institute and the U.S. Geological Survey, Trask and his coworkers were involved in extensive analyses of the organic content of recent and ancient sediments. Although their work considered only marginally specific organic molecules, these researchers were able to formulate broad generalizations about carbonaceous materials in sediments. The generalizations pertained mainly to petroleum origin and petroleum source beds. As one result of the petroleum oriented work, the petroleum industry began to recognize the importance of fundamental research in organic geochemistry and, consequently, provided both monetary and philosophical support for basic and applied research in organic geochemistry.

The discovery by Treibs in 1933 of porphyrin pigments in shale probably marks the first of modern studies on the geochemistry of organic molecules. By 1936, Treibs had identified free-base as well as metal-complexed porphyrins in shale, oil, and coal. Paper 1 reviews the state of knowledge of porphyrins and illustrates some of the significant contributions made by Treibs. The organic geochemical implications of this paper are remarkable. In 1936 the nature of the origin of petroleum was very poorly understood. It was uncertain whether petroleum was originally derived from biological materials or whether it was the product of strictly inorganic processes. Treibs showed that the porphyrins in petroleum could be degradation products from chlorophyll. His seven-step mechanism of porphyrin formation from chlorophyll has stood the test of time with only minor modifications. He devised thermal experiments to test the stability of porphyrins and determined a temperature ceiling for petroleum diagenesis processes. Treibs' observations and interpretations strongly supported the organic hypothesis for petroleum origin and cast doubt on the inorganic hypotheses. His paper ends with a prophetic statement about the future of organic geochemistry. Treibs said, in effect, that the preservation of porphyrin pigments in geological materials suggests that other kinds of organic molecules will also be found. Indeed, the prophecy has been amply fulfilled as partly demonstrated in Part II in which the geological occurrences of many organic molecules are illustrated. Treibs' broad approach has served as a model for many of the organic geochemical studies that have followed.

While Treibs established organic geochemistry at the molecular level, Trask at the same time demonstrated the wide geological applications of the field in his comprehensive surveys on the organic content of sediments and sedimentary rocks (Trask, 1932; Trask, 1939; Trask and Patnode, 1942). The surveys were designed for understanding the origin and environments of source sediments of petroleum. As a prelude to this work, Trask and Wu (Paper 2) wondered if petroleum actually forms in sediments at the time of deposition. To answer this question they extracted recent marine sediments that were rich in organic matter. They identified a number of classes of organic molecules including fatty acids, pigments, sterols, organic sulfur compounds, and solid paraffins, but they did not detect liquid hydrocarbons. They concluded that significant concentrations of liquid petroleum do not occur in recently deposited sediments. In 1952, twenty-two years later, Smith (Paper 4) showed that, contrary to the observations by Trask and Wu, recent marine sediments do contain liquid hydrocarbons. These hydrocarbon mixtures differ compositionally, however, from petroleum. Although Trask and Wu failed to detect liquid hydrocarbons, their basic conclusion regarding the lack of significant concentrations of liquid petroleum in recent sediments (except sediments that have been polluted with petroleum products) has been firmly established. Trask and Wu also concluded that much—if not most—petroleum comes from nonextractable organic matter (kerogen) in sediments. This conclusion corresponds with much of the current thinking of petroleum geochemists. See Paper 37.

During the same decade in which Treibs and Trask were establishing the chemical and geological foundations of organic geochemistry, a parallel research effort was underway that has had tremendous impact on studies of the geochemistry of organic molecules. In 1927 the American Petroleum Institute established Research Project 6 at the National Bureau of Standards. The purpose of the project was to investigate at the molecular level the hydrocarbon constitutents of petroleum. By means of precision distillation techniques and other separation methods, 175 individual hydrocarbons from a representative petroleum (Brett No. 6 well, Ponca City, Oklahoma Field) had been characterized by 1960. One of the significant results of this work has been the realization that, although it is a complex mixture, petroleum is composed of relatively few of the molecules that are theoretically possible. Paper 3 calls attention to the work of A.P.I. Project 6. The paper, a succinct review by F. D. Rossini, the director of the project from 1934 to 1960, provides a brief history of the project and inventories the current list of isolated and identified hydrocarbons.

Although the research of Treibs, Trask, and Rossini formed important foundations for molecular organic geochemistry, not until 1952 was a sustained interest in the field developed. This interest was initiated by the discovery by P. V. Smith, Jr. (Paper 4) of liquid hydrocarbons in ten recent sediment samples from the Gulf of Mexico. This discovery changed the course of organic geochemistry because until that time it had been assumed, based on the observations of Trask and Wu (Paper 2), that liquid hydrocarbons were not present in recent sediments. Smith clearly demonstrated that sediments from both oxidizing and reducing environments contain aliphatic and aromatic hydrocarbons. Smith concluded that the petroleumlike mixture in recent sediments "is a composite of the hydrocarbon remains of many forms of marine life." Many papers have followed since then that confirm the presence of liquid hydrocarbons in modern sediments. We should note that the hydrocarbons in recent sediments differ in chemical composition from hydrocarbons in petroleum. Therefore, the determination of possible relationships between the two hydrocarbon mixtures became the goal of much of the research that followed during the next ten years. For example, Smith (1954) and Meinschein (1959), although recognizing the differences between these mixtures, emphasized the similarities. They believed that petroleum could be accounted for by a selection and concentration of hydrocarbons in sediments. On the other hand, Forsman and Hunt (1958) and Stevens, Bray, and Evans (1958) emphasized the differences in the mixtures. These scientists suggested that the hydrocarbon mixtures in recent sediments cannot serve directly as precursors for petroleum hydrocarbons, but that sediment hydrocarbons must be altered and transformed in order to produce petroleum.

The papers considered thus far have focused on organic molecules as they relate to the origin of petroleum. By the 1950s the field of organic geochemistry was well established particularly within the petroleum industry. Then a new direction in molecular organic geochemistry was taken by P. H. Abelson (Paper 5) who announced the finding of amino acids in fossils that ranged in age from modern to Devonian (360×10^6 years). Besides the identification of various amino acids in fossil shells and bones, Abelson described experiments that showed that considerable variation exists in the rates of thermal degradation of different amino acids. Detailed studies of alanine demonstrated that this amino acid is sufficiently stable to resist degradation for at least a billion years under mild thermal conditions. This early work established amino acids as important chemical fossils (a term coined by Eglinton and Calvin [1967]). Further work reviewed by Abelson (1963), Hare (1969), and Kvenvolden (1975), has demonstrated the potential of amino acids for stratigraphic correlations, chemotaxonomy, paleothermometry, and geochronology.

During the late 1950s and 1960s, interest in organic geochemistry broadened considerably with research laboratories being established not only in petroleum companies but also in government agencies and academic institutions. Paper 6 by M. Blumer represents a model for academic research in organic geochemistry at the molecular level. He outlines geochemical occurrences of pigments (fringelites, porphyrins, and chlorins) and isoprenoid hydrocarbons. From the structural details of the molecules Blumer postulates possible geochemical reactions that account for the observed distributions of molecules. For example, reduction of carboxylic acid groups to methyl groups is required to account for the homologous series of porphyrins present in the Serpiano oil shale. Decarboxylation competes with reduction. Blumer's research is characterized by careful attention to structural details and clever assessment of the molecular evidence for reaction pathways.

BIOGRAPHIES

Brief biographies of six scientists follow to provide the reader with some idea about the people whose accomplishments have significantly influenced the beginning and development of organic geochemistry.

Alfred Treibs is considered by many to be the "father of organic geochemistry." He was born in Oberstein, Germany, in 1899 and studied organic chemistry in Aachen and Munich. In 1928 he began to work out the structure of chlorophyll and, in 1933, discovered and described porphyrins in shales and later in oils and coals. These discoveries, plus his laboratory experiments that were designed to account for geoporphyrins, mark the beginning of the first comprehensive work in organic geochemistry. He worked for eight years in the chemical industry and for most of his career has taught at the Institute of Technology in Munich. Besides establishing the foundations for organic geochemistry, Treibs also made major contributions to the much broader field of organic chemistry through his work on the chemistry of pyrrol.

Parker D. Trask was born in Springfield, Massachusetts, in the same year as Alfred Treibs. In contrast to Treibs, whose major contributions to organic geochemistry were chemical, Trask emphasized the geological aspects of the field. He studied at the University of Texas and at the University of California in Berkeley where he obtained the Ph.D. in geology. From 1926 to 1941 Trask was supported first by the American Petroleum Institute and later by the A.P.I. and the U.S. Geological Survey for a comprehensive study of source beds of petroleum. During this time Trask integrated geology and chemistry to develop clues for the recognition of source beds. His work established the importance of the geological framework for organic geochemical studies. Trask was a remarkably versatile geologist whose career was cut short by his death in 1961.

Frederick D. Rossini was also born in 1899, in the state of Pennsylvania, and was educated at the Carnegie Institute of Technology and received the Ph.D. in physical chemistry from the University of California at Berkeley. He has held a number of research and administrative positions at the National Bureau of Standards, Carnegie Institute of Technology, University of Notre Dame, and Rice University. Rossini's work in petroleum geochemistry has helped to provide a solid foundation for organic geochemistry. Through the American Petroleum Institute he and his associates were able to isolate many of the important constituents of a typical mid-continent petroleum, and much of his research in the physical chemistry of hydrocarbons and petroleum has been fundamental to understanding the origin and movement of hydrocarbons in the subsurface.

Philip H. Abelson is a physical chemist whose broad interests in science have included significant contributions to organic geochemistry. He was born in 1913 in Tacoma, Washington, was educated in that state, and received the Ph.D. in nuclear physics from the University of California. As director of the Geophysical Laboratory of the Carnegie Institution of Washington, D.C., he established one of the most respected laboratories for organic geochemistry. He was president of the Carnegie Institution and has served as editor of *Science* since 1962. Abelson has emphasized the biochemical aspects of organic geochemistry through his pioneering studies of fossil amino acids. His early thoughts on kerogen have been useful in explaining the origin of petroleum.

Max Blumer devoted almost his entire professional career to organic geochemistry. Born in Basel, Switzerland, in 1923, he received the Ph.D. in chemistry from the university there in 1949. After receiving his degree he did research at the University of Basel, the University of Minnesota, Scripps Institution of Oceanography, and Shell Oil Company. In 1959 he joined the staff at Woods Hole Oceanographic Institution where he was a senior scientist who investigated many aspects of organic geochemistry and comparative biochemistry. He was probably the first scientist to apply the principles of organic geochemistry to environmental chemistry. His early work on the fate and effects of petroleum in the marine environment served as a model for much of the work by others that has followed. His brilliant research in organic geochemistry was terminated by his untimely death in 1977.

Paul V. Smith is an organic chemist whose career with the petroleum industry has spanned more than thirty years. Smith was born in Lima, Ohio, in 1921 and attended Miami University and the University of Illinois where he received his Ph.D. degree in organic chemistry. During the time he was a research chemist and group leader for Esso Research and Engineering Company he made his fundamental observations concerning the occurrences of hydrocarbons in recent sediments. Since the time of his direct involvement with organic geochemistry he has held a number of different positions in Esso and Exxon and he is currently the manager of education and scientific relations in the public affairs office of Exxon Research and Engineering Company.

REFERENCES CITED

Abelson, P. H. 1963. Geochemistry of amino acids. In I. A. Breger, ed., *Organic Geochemistry*. Macmillan, N. Y., 431–455.

Eglinton, G., and M. Calvin. 1967. Chemical fossils. *Sci. American* **216**: 32–43.

Forsman, J. P., and J. M. Hunt. 1958. Insoluble organic matter (kerogen) in sedimentary rocks of marine origin. In L. G. Weeks, ed., *Habitat of Oil*, Am. Assoc. Petrol. Geol., Tulsa, Okla.:747–778.

Hare, P. E. 1969. Geochemistry of proteins, peptides and amino acids. In G. Eglinton and M. T. J. Murphy, eds., *Organic Geochemistry—Methods and Results*. Springer-Verlag, Berlin, 438–463.

Kvenvolden, K. A. 1975. Advances in the geochemistry of amino acids: *Ann. Rev. Earth and Planetary Sci.* **3**:183–212.

Meinschein, W. G. 1959. Origin of petroleum. *Am. Assoc. Petrol. Geol. Bull.* **43**:925–943.

Smith, P. V., Jr. 1954. Studies on origin of petroleum: Occurrence of hydrocarbons in recent sediments. *Am. Assoc. Petrol. Geol. Bull.* **38**:377–404.

Stevens, N. P., E. E. Bray, and E. D. Evans. 1956. Hydrocarbons in sediments of the Gulf of Mexico. *Am. Assoc. Petrol. Geol. Bull.* **40**:975–983.

Trask, P. D. 1932. Origin and environment of source sediments. The Gulf Publishing Co., Houston, Tex., 323 pp.

Trask, P. D. 1939. Organic content of recent marine sediments. In P. D. Trask, ed., *Recent marine sediments: A symposium*. Am. Assoc. Petrol. Geol., Tulsa, Okla., 428–453.

Trask, P. D., and H. W. Patnode. 1942. *Source beds of petroleum*. Am. Assoc. Petrol. Geol., Tulsa, Okla., 566 pp.

SUGGESTED ADDITIONAL READINGS

Advances in Organic Geochemistry Series—Proceedings of the International Meetings on Organic Geochemistry:

Colombo, U., and G. D. Hobson. 1964. *Advances in Organic Geochemistry 1962*. Macmillan, N. Y., 488 pp.

Hobson, G. D., and M. C. Louis. 1966. *Advances in Organic Geochemistry 1964*. Pergamon Press, Oxford, 330 pp.

Hobson, G. D., and G. C. Speers. 1970. *Advances in Organic Geochemistry 1966*. Pergamon Press, Oxford, 577 pp.
Schenck, P. A., and I. Havenaar. 1969. *Advances in Organic Geochemistry 1968*. Pergamon Press, Braunschweig, 617 pp.
Gaertner, H. R. v., and H. Wehner. 1972. *Advances in Organic Geochemistry 1971*. Pergamon Press, Braunschweig, 736 pp.
Tissot, B., and F. Bienner. 1974. *Advances in Organic Geochemistry 1973*. Editions Technip., Paris, 1080 pp.
Campos, R., and J. Goñi. 1977. *Advances in Organic Geochemistry 1975*. Empresa Nacional Adaro de Investigaciones Mineras, Madrid, 911 pp.

Albrecht, P., and G. Ourisson. 1971. Biogenic substances in sediments and fossils. *Angew. Chem. Internat. Ed.* **10**:209–286.
Baker, D. R. 1972. Organic geochemistry and geological interpretations. *Jour. Geol. Educ.* **20**:221–234.
Breger, I. A. 1963. *Organic Geochemistry*. Macmillan, N. Y., 658 pp.
Eglinton, G., and M. T. J. Murphy. 1969. *Organic Geochemistry—Methods and Results*. Springer-Verlag, Berlin, 828 pp.
Hunt, J. M. 1979. *Petroleum Geochemistry and Geology*. Freeman, San Francisco, 617 pp.
Manskaya, S. M., and T. V. Drozdova. 1968. *Geochemistry of Organic Substances*. Pergamon Press, Oxford, 345 pp.
Maxwell, J. R., C. T. Pillinger, and G. Eglinton. 1971. Organic geochemistry. *Quart. Rev., The Chem. Soc.* **25**:571–628.
Nagy, B., and U. Colombo. 1967. *Fundamental Aspects of Petroleum Geochemistry*. Elsevier Publishing Co., Amsterdam, 388 pp.
Swain, F. M. 1970. *Non-marine Organic Geochemistry*. Cambridge University Press, London, 445 pp.
Tissot, B. P., and D. H. Welte. 1978. *Petroleum Formation and Occurrence*. Springer-Verlag, Berlin, 538 pp.

1

Reprinted from *Angew. Chemie* **49**:682–686 (1936)

Chlorophyll- und Häminderivate in organischen Mineralstoffen.

Von Dr. ALFRED TREIBS.
Organisch-chemisches Institut der Technischen Hochschule München.
(Eingeg. 23. Juni 1936.)

Der Zufallsbefund eines organischen Minerals, des gut kristallisierten Farbstoffes Graebeit[1]), der mit großer Wahrscheinlichkeit als Anthrachinonfarbstoff anzusprechen ist, gab Veranlassung, definierte organische Substanzen in bituminösen Gesteinen aufzusuchen. Als Methode wurde die Destillation verworfen, da sie hier eine Zersetzungsreaktion ist, die aller Wahrscheinlichkeit nach fast ausschließlich Abbauprodukte vorhandener Stoffe liefern konnte. Die Extrakte eines bituminösen Schiefers mit organischen Lösungsmitteln hatten prächtig rote Farbe und das charakteristische Absorptionsspektrum der Metallkomplexe von Pyrrolfarbstoffen. Dieser Befund ergab sich an einem geologisch ziemlich alten Objekt, einem sog. Stinkschiefer aus dem Karwendelgebirge an der oberen Isar, der dem Hauptdolomit der oberen alpinen Trias angehört und zur Ichthyol-Erzeugung dient.

Dem Chlorophyll nahestehende grüne Farbstoffe hatte *Lubimenko*[2]) in rezenten und geologisch jungen Schlicksedimenten und Tonablagerungen des Schwarzen Meeres nachgewiesen, *Weigelt*[3]) hatte grüne Blätter in der Geiseltalbraunkohle des Tertiär aufgefunden, *Fikentscher*[4]) Porphyrin im fossilen Krokodilkot der gleichen Herkunft beobachtet.

Eine annähernde Gehaltsbestimmung des Porphyrinkomplexes zeigte eine Menge von 30—40 mg im Kilogramm Schiefer an. Durch Extraktion von 12 kg Schiefer mit heißem Eisessig und Spaltung des Komplexes mit Bromwasserstoff-Eisessig konnten 150 mg Porphyrin isoliert werden, das nach den Methoden der Porphyrinchemie untersucht, sich als ein Gemisch mehrerer Farbstoffe erwies, in der Hauptsache basische Porphyrine, wenig saure Porphyrine. Dies ist der erste Fall eines natürlichen Vorkommens von Ätioporphyrinen.

Die Hauptmenge der Basenfraktion besteht aus starkbasischem Ätioporphyrin, das mit Desoxophyllerythrin spektroskopische Übereinstimmung zeigte und durch Analyse als das zugehörige Desoxophyllerythro-ätioporphyrin I, die sauerstofffreie Grundsubstanz des Chlorophylls, erwiesen wurde. Später ist es dann durch *H. Fischer* und *H. J. Hofmann*[5]) synthetisiert worden. Unter den sauren Porphyrinen fand sich Desoxophyllerythrin II, was von besonderer Bedeutung für die sichere Identifizierung ist, denn die Ätioporphyrine besitzen keine Schmelzpunkte, während bei den sauren Porphyrinen die Esterschmelzpunkte und Mischschmelzpunkte mit Vergleichspräparaten herangezogen werden können, nach *H. Fischer* ein entscheidendes Kriterium der Identität. Desoxophyllerythrin ist das Endprodukt des sauren Chlorophyllabbaues und ist durch die Synthese von *H. Fischer* und *J. Riedmair*[6]) in der Konstitution klargestellt.

Schwachbasische, dem Desoxophyllerythro-ätioporphyrin sehr nahestehende Nebenprodukte sind noch nicht aufgeklärt, dagegen gelang es, sehr kleine Mengen eines Ätioporphyrins und eines sauren Porphyrins abzutrennen, die dem Spektraltyp des Mesoporphyrins der Blutfarbstoffreihe angehören, und für letzteres konnte die Identität mit Mesoporphyrin IV gezeigt werden, das Ätioporphyrin ist danach Mesoätioporphyrin III. Damit sind sowohl Porphyrine, die sich vom Chlorophyll, bzw. ihm nahestehenden Pigmenten, als auch vom Hämin herleiten, im Bitumen nachgewiesen. Diese Beobachtungen sind nun an einem sehr großen Material verallgemeinert worden[7]).

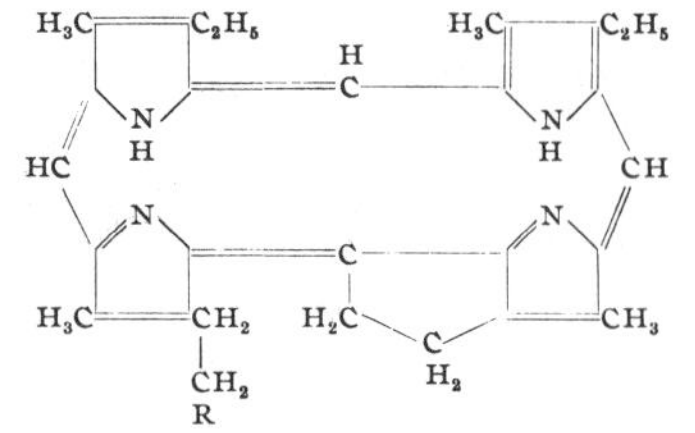

I R = H. Desoxophyllerythro-ätioporphyrin.
II R = COOH. Desoxophyllerythrin.

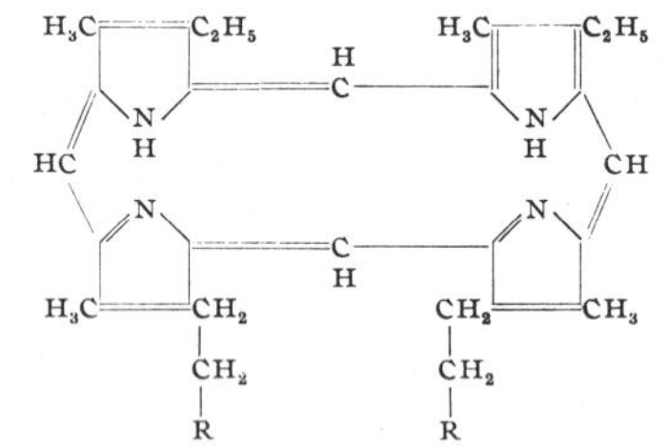

III R = H. Meso-ätioporphyrin
IV R = COOH. Mesoporphyrin.

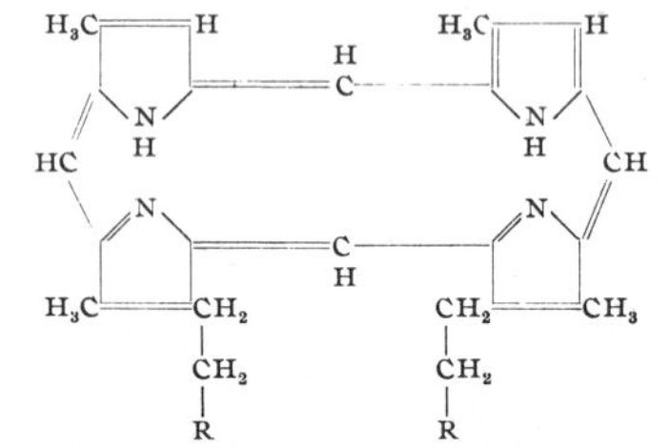

V R = H. Deutero-ätioporphyrin.
VI R = COOH. Deuteroporphyrin.

Die folgende Tabelle 1 soll, ohne auf die Einzelheiten weiter einzugehen, einen Überblick über das gesamte Versuchsmaterial geben.

Tabelle 1. Statistik der Befunde von Pyrrolfarbstoffen.

Erdöle	66	Porphyrinkomplexe
Asphalte	9	
Erdwachse	4	
Asphaltite	5	
Ölschiefer, bitum. Gest.	70	Porphyrinkomplexe, mehrfach Spuren grüner Farbstoffe
Phosphorite, Koprolithe	8	
Guano	1	
Gyttja, Cannel-Kohlen	7	
Steinkohlen	17	
Junge, rezente Gyttjen, Sapropele	14	Nur grüne Farbstoffe

1) *A. Treibs* u. *H. Steinmetz*, Liebigs Ann. Chem. **506**, 171 [1933].

2) C. r. hebd. Séances Acad. Sci. **190**, 813 [1930]; *Rauser-Cernoussowa*, Zbl. Mineral., Geol., Paläont., Abt. B, **1930**, 314; C. **1930** II, 2415.

3) Nova acta, Halle, Neue Folge **1**, 87 [1932].

4) Zoologischer Anzeiger **103**, 289 [1933].

5) Liebigs Ann. Chem. **517**, 274 [1935].

6) Ebenda **490**, 91 [1931].

7) Vgl. dazu ebenda **509**, 103 [1934]; **510**, 42 [1934]; **517**, 172 [1935]; **520**, 144 [1935].

Die Mengen an Porphyrin wechseln, in der Hälfte der Untersuchungsobjekte (die aus aller Welt stammen) finden sich nur Spuren, d. h. 0,5—10 γ in etwa 50 g Material, in den übrigen steigende Gehalte 0,2—2 mg, in vereinzelten Fällen mehr. Steinkohlen sind immer arm. Am meisten Porphyrin wurde gefunden in Ölschiefer von Meride (Schweiz), und zwar 0,34—0,4% Komplex, im Ölschiefer von Zavalje (Kroatien) 0,08%, weiter in Trinidad-Erdöl 0,04%. Diese ganz erheblichen Quantitäten kommen an die Gehalte von Blut, ungefähr 0,5% Hämin (auf Flüssigkeit berechnet), und den Chlorophyllgehalt grüner Blätter, 0,8% vom Trockengewicht, heran. In Sapropelen erreicht der grüne Farbstoff auch annähernd diese Größenordnung.

Dem geologischen Alter nach fallen die meisten Objekte ins Tertiär, zahlreiche sind aber älter: Die Steinkohlen entstammen dem Carbon, eine Braunkohle (Tula) dem Untercarbon, Erdöl Volkenroda dem Perm, amerikanische Öle dem Devon oder Silur, zahlreiche Ölschiefer der Trias, das älteste Objekt ist der Kuckersit aus dem unteren Silur.

Der Nachweis von Porphyrin-Metallkomplex kann vielfach in Extrakten bituminöser Gesteine direkt erfolgen, sie sind an ihren sehr charakteristischen 2-Banden-Spektren sehr leicht erkennbar. Zur Feststellung, welche Porphyrine vorkommen, ist jedoch die Zerlegung der Komplexe unerläßlich, die verschiedenen Porphyrine können dann an ihren fein differenzierten 4-Banden-Spektren und ihrer Basizität identifiziert werden. Dazu kommt dann noch der große Vorteil, daß die freien Porphyrine infolge ihrer basischen Eigenschaften sich von Begleitstoffen sehr leicht abtrennen lassen; letztere vereiteln infolge ihrer Eigenabsorption oft die Erkennung von Komplexen. Infolgedessen kann man so auch äußerst kleine Farbstoffmengen auffinden, da eine Anreicherung auf 100fache und größere Konzentration gelingt. Die untere Grenze der Erkennung und spektroskopischen Messung liegt bei etwa 0,0005 mg, und Farbstoffmengen dieser Größenordnung lassen sich bei einiger Erfahrung handhaben und aus großen Mengen von Begleitstoffen herauspräparieren. Ein Übersehen von kleinen, manchmal sogar beträchtlichen Mengen an Porphyrin ist zu befürchten, wenn beim Aufschluß mit Bromwasserstoff-Eisessig viel oberflächenaktive Verharzungsprodukte entstehen; außerdem muß mit der Möglichkeit von Oxydationswirkungen durch Schwermetallverbindungen gerechnet werden; weiter sind Porphyrine sehr empfindlich gegen Peroxyde des Äthers.

Der Geübte kann auch in Porphyringemischen Desoxophyllerythrin und Mesoporphyrin bzw. die entsprechenden Ätioporphyrine nebeneinander erkennen. Das ist wichtig, da diese Verbindungen, die Repräsentanten der Chlorophyll- und Häminreihe, oft im Bitumen nebeneinander vorkommen und sich in kleiner Menge nicht trennen lassen. Im großen gelingt die Trennung glatt mittels Chromatographie, wie *H. Fischer* und *J. Hofmann* gelegentlich der Synthese des Desoxophyllerythro-ätioporphyrins zeigten.

Die spektroskopische Messung, besser der direkte spektroskopische Vergleich mit Standardpräparaten gibt also Aufschluß über die Zugehörigkeit der aufgefundenen Porphyrine zur Chlorophyll- oder Häminreihe.

Außer bei zahlreichen weiteren Ölschiefern brachte auch die Untersuchung vieler Erdöle den Nachweis der gleichen Porphyrine, in einigen Fällen konnte auch die Isolierung und Analyse aus Öl durchgeführt werden. Weiterhin fanden sie sich in Erdwachsen, Asphalten und Phosphoriten. Damit ist ganz allgemein die geochemische Bedeutung der Porphyrine erwiesen. Die Untersuchung von Kohlen mußte vor allem die große Adsorptionskraft berücksichtigen, am besten werden mit organischen Lösungsmitteln erhaltene Extrakte verwendet. In einigen Braunkohlen und vor allen Dingen Cannel- sowie Bogheadkohlen, worin am ehesten Befunde zu erwarten waren, konnten auch tatsächlich Porphyrine nachgewiesen werden, und zwar die gleichen, die sich in Ölschiefern und Erdölen finden. Die genannten Kohlesorten enthalten sehr viel lösbare Bestandteile, woran im Gegensatz dazu Steinkohlen arm sind. Hier ist die Extraktion sehr viel schwieriger, und man kann nur bei feinster Mahlung und gründlichster Extraktion eine sichere Erfassung erwarten. Durch Anwendung μ-fein gemahlener Kohle, die nach *Peters* (KWI-Mülheim), dem ich dieses Material verdanke, den maximalen Extrakt liefert[8]), konnte in $^2/_3$ aller untersuchten Steinkohlen tatsächlich Porphyrinkomplex gefunden werden. Die freigelegten Porphyrine sind aber verschieden von den oben beschriebenen. Zur Isolierung in kristallisiertem Zustand war die bisher vorliegende Menge bei weitem unzureichend, aber nach dem Spektrum und dem gesamten Verhalten kann es sich in der Hauptsache nur um Deutero-ätioporphyrin V handeln. Deuterohämin entsteht bei der protahierten Fäulnis von Blut aus Hämin, dabei sind die Vinylgruppen eliminiert und durch H ersetzt, während sonst keine Veränderung erfolgt ist; präparativ kann dies durch Schmelzen von Hämin mit Resorcin erreicht werden. Das durch Eisenabspaltung daraus erhaltene Deuteroporphyrin VI läßt sich durch Brenzreaktion zu Deuteroätioporphyrin decarboxylieren.

Die Tabelle 2 der Porphyrin-Spektren läßt die Unterschiede der Absorption der verschiedenen Porphyrine erkennen, die Lage und die relative Intensität der Absorptionsmaxima zeigen sehr deutliche Abweichungen, die besonders beim Übereinanderprojizieren zweier Spektren im Vergleichsspektroskop hervortreten.

Tabelle 2. Porphyrin-Spektren.

Schwerpunkte der Hauptbanden in Äther.

	I	II	III	IV	Intensitäten
Desoxophyllerythrin ..	**621,6**	**567,5**	531,7	**497,5**	IV, I, II, III
Mesoporphyrin........	623,3	**567,6**	**528,2**	**494,9**	IV, III, II, I
Deuteroporphyrin	621,3	**565,2**	**525,3**	**492,8**	IV, III, II, I

Die Ätioporphyrine haben mit den zugehörigen Carbonsäuren völlig übereinstimmende Spektren.

Welche Metalle in den ursprünglichen Komplexen vorlagen, war zunächst nicht festzustellen, da die Spektren der Extrakte von den Komplexspektren der zugrunde liegenden Porphyrine mit allen bis dahin bekannten komplexbildenden Metallen verschieden waren. Außerdem erwies sich der vorwiegend vorhandene Komplex stabiler als alle bekannten Porphyrinmetallkomplexe, daneben zeigte das Spektrum einen zweiten Komplex an. Die Isolierung gelang mit Hilfe der chromatographischen Adsorptionsanalyse; durch gleichzeitige synthetische Versuche konnte der Hauptkomplex der Ölschiefer als Vanadinsalz erkannt werden, was durch die Analyse bestätigt wurde. Das seltene Vanadin findet sich häufig als auffälliger Bestandteil von Erdöl und Asphaltaschen und war daher besonders eingehend untersucht worden. Vanadinkomplexe entstehen viel schwerer als alle übrigen Porphyrinkomplexe, sie enthalten am Metall noch Sauerstoff gebunden und sind wahrscheinlich Vanadylverbindungen, die Analysen geben darüber nicht ganz sicher Auskunft, vielleicht ist das Vanadin auch 5wertig.

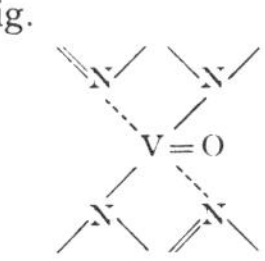

Tabelle 3. V-Salz-Spektren.

	I	II	Intensitäten
Desoxophyllerythrin	572,0	532,0	I, II
Mesoporphyrin...........	570,2	531,5	I, II

[8]) *A. Peters* u. *W. Cremer*, diese Ztschr. **47**, 576 [1934].

Die Vanadinsalze werden nur durch die stärksten Mineralsäuren gespalten und geben bemerkenswerterweise auch säurestabile Leukoverbindungen, wie sie bisher noch nicht bekannt waren.

Der zweite Komplex wurde auch von allen anderen Farbstoffen frei erhalten, konnte aber infolge zu geringer Menge nicht kristallisiert erhalten werden. Das Spektrum war unerwarteterweise ein reines Hämochromogenspektrum. Hämochromogene sind Molekularverbindungen 2wertiger Fe-Komplexe mit Basen, deren rotbraune Lösungen höchst charakteristische, äußerst intensive 2-Banden-Spektren aufweisen, die von denen der roten Metallkomplexe beträchlich abweichen. Die eingehende Untersuchung zeigte, daß tatsächlich Hämochromogen vorlag, und zwar das des Mesoätioporphyrins, aber mit abweichender Lage der Absorptionsmaxima und einer weiteren sehr bemerkenswerten Abweichung; die Lösungen zeigen nämlich völlige Stabilität gegen Luftsauerstoff, während normale Hämochromogene dagegen sehr empfindlich sind. Diese beiden Eigenschaften sind sonst nur beim Cytochrom C bekannt. Die Spaltung mit Mineralsäuren tritt jedoch leicht ein und ergibt 2wertiges Fe. Führt man jetzt in das Porphyrin wieder Fe ein und erzeugt das Hämochromogenspektrum durch Behandeln mit Pyridin-Hydrazin, so erhält man das normale Hämochromogen. Hier muß daher eine besondere Base am Fe^{II}-Komplex gebunden sein, die diese bemerkenswerte Eigenschaft erzeugt.

Tabelle 4.

	I	II
Hämochromogen aus Ölschiefer	552,7	517,0
Nach Spaltung, Fe-Einführung, Pyridin-Hydrazin	547,3	517,2
Mesoätiohämin, Pyridin-Hydrazin	547,3	517,4
Desoxophyllerythro-ätiohämin, Pyridin-Hydrazin	546,5	514,3

Leider fand sich bisher noch kein Objekt, das zur Reingewinnung dieses Komplexes genügende Farbstoffmengen enthielt, was zur Aufklärung der Base von größtem Interesse wäre. In den besonders zu Isolierungsversuchen verwendeten Ölschiefern lag in dem abgetrennten Hämochromogen nur das Häminderivat Mesoätioporphyrin vor, in vielen anderen Fällen, in denen die Messung der Extrakte nur Fe-Komplex anzeigte, findet sich auch Desoxophyllerythro-ätioporphyrin als Hämochromogen.

Komplexe mit anderen Metallen sind noch nicht beobachtet worden. In 3 bulgarischen Ölschiefern konnte jedoch 3wertiger Fe-Komplex aufgefunden werden; das verrät sich dadurch, daß der Pyridin-Extrakt normales Hämochromogenspektrum annimmt. Sehr verdünnte Lösungen von Häminen in Pyridin haben auch ohne Reduktionsmittel Hämochromogenspektrum[9]), hier kann auch durch organische Substanz noch Reduktionswirkung auftreten. In präparativer Hinsicht ist dieser Befund sehr unangenehm, da damit das beste Porphyrin-Lösungsmittel, Pyridin, bei der Extraktion derartiger Schiefer zu Extraktionen nicht verwendet werden darf, beim Kochen der Pyridinlösungen sind in kurzer Zeit die Komplexe völlig zerstört. (Die oben beschriebenen Komplexe vertragen tagelanges Kochen.)

Diese 3wertigen Komplexe scheinen sehr selten zu sein, doch mahnt ihr Auftreten zur größten Vorsicht in der Wahl der Lösungsmittel, am Platz sind Chloroform oder Eisessig und Benzol, Pyridin nur nach einer Vorprobe.

Die Komplexbildung der Bitumenporphyrine ist ein sekundärer Vorgang, wie sich schon aus dem Vorkommen von freien Porphyrinen und Fe-Komplexen der Chlorophyllderivate ergibt, nur die unter den obwaltenden Bedingungen begünstigten und stabilen Komplexe entstehen bzw. bleiben erhalten.

Einen V-haltigen Blutfarbstoff hat *Hentze* in niederen Meerestieren, Tunicaten, aufgefunden, die nähere Untersuchung dieses Farbstoffes zeigte, daß kein Porphyrinderivat vorliegt.

In einigen Ölschiefern kommt auch freies Porphyrin vor, außerdem sind häufig auch bei sehr alten Objekten noch langwellige Absorptionsbanden zu sehen, die dem Chlorophyll noch näherstehenden grünen Farbstoffen angehören. In einer Probe der bekannten Geiseltalbraunkohle aus dem Tertiär konnte nun die lang gesuchte Zwischenstufe der Chlorophyll-Umwandlung aufgefunden werden. Nebeneinander sind gleiche Mengen grüner Farbstoffe und Porphyrine vorhanden (Komplexe nur untergeordnet), darauf wird später noch zurückzukommen sein.

Tabelle 5 gibt eine Übersicht über die Spektren einer Reihe charakteristischer Objekte und die Zuordnung der Banden.

Ein brasilianischer Ölschiefer der Kreide oder des Tertiär, den *Ch. Dhéré* und *G. Hradil*, die auch einige porphyrinreiche Ölschiefer auffanden, zuerst untersuchten[10]), enthält nur grünen Farbstoff, der dem Chlorophyll noch nahesteht, in nichtkomplexer Bindung. Ähnliches findet sich in geologisch jungen und in rezenten Gyttjen und Sapropelen. (Gyttjen sind unter beschränktem Luftzutritt gebildete See-Ablagerungen, die aus zahlreichen Pflanzen und Tierresten bestehen, darunter vielen Algen; tierisches Leben ist noch möglich. In der obersten Zone der Schicht ist noch Sauerstoff, die organische Substanz wird durch Tiere stark koprogen umgeformt, in geringer Tiefe darunter beginnt die sauerstofffreie Reduktionszone. Derartige Ablagerungen enthalten viel organische Substanz. Sapropele lagern sich in sauerstofffreien, H_2S-vergifteten Gewässern ab und enthalten nur anaerob lebende Organismen; sie sind am reichsten an organischer Substanz[11]).) Nachgewiesen sind also alle Übergänge vom kaum veränderten

Tabelle 5. Spektren von Extrakten.

	Grüne Farbst.		Porph.		V-Kompl.	Fe-Kompl.	V-Kompl.	Fe-Kompl.	Porph.
Ölschiefer Wallgau	662,6	645,9	623,7	608,2	572,8	552,3	534,2		
Ölschiefer Meride		643	620,9	609,1	571,4		534,4		497
Ölschiefer Häring					572,7		534,0		
Ölschiefer Mandra						550,5			
Ölschiefer Mistelbach					575,1	547,6		516,0	
Asphalt Syrien					573,4	553,0	533,6		
Asphalt Batrak					574,0				
Asphalt Wurtzilite						550,5			
Erdöl Tegernsee (16 rumänische usw.)						551,1			
Erdöl Trinidad					572,9	553,4	534,6		
Phosphorit Gassa					572,0	552,5			
Kohle, Boghead Oelsnitz					571,5	552,3	534	518	
Steinkohle Ruhr						550			
Ölschiefer Marahou	660,9			603,8			536,65	515,5	

Im kurzwelligen Teil des Spektrums ab 540 $m\mu$ durchschnittlich stört die Eigenabsorption der Extrakte die Beobachtung sehr stark.

[9]) Vgl. *A. Treibs*, Hoppe-Seyler's Z. physiol. Chem. 168, 68 [1927].

[10]) Bull. Suisse de Min et Petr. 14, 291 [1934].

[11]) *K. Krejci-Graf:* Erdöl. Berlin 1936.

Chlorophyll bis zur Endstufe, V- oder Fe-Komplex des Desoxophyllerythro-ätioporphyrins.

Die Frage nach dem **Mechanismus der Porphyrinbildung aus Chlorophyll** gewinnt unmittelbare geochemische Bedeutung.

Die nötigen Schritte des Abbaues sind:

1. Mg-Abspaltung. Diese erfolgt sehr leicht, schon mit schwachen Säuren.
2. Verseifung der Estergruppen. In rezenten, aber auch in fossilen Ablagerungen sind die grünen Farbstoffe als Neutralstoffe vorhanden, die sich durch Laugen großenteils verseifen lassen. Die Verseifung ist Säuren zuzuschreiben, Alkalien spalten den isocyclischen Ring.
3. Hydrierung der Vinylgruppe zur Äthylgruppe. Gelingt recht leicht und ist als biochemische Reaktion nachgewiesen.
4. Dehydrierung des grünen Systems zum roten Porphyrinsystem. Kommt einer Stabilisierung gleich.
5. Reduktion der CO-Gruppe zu CH_2.
   Im Chlorophyll b, das in Algen nicht, oder nur ganz untergeordnet vorkommt, ist die der Vinylgruppe benachbarte Methylgruppe zur Aldehydgruppe umgewandelt, deren Reduktion natürlich ebensogut möglich ist, so daß aus der Komponente b grundsätzlich die gleichen Porphyrine entstehen können. Für Bakteriochlorophyll gilt ähnliches für die zur Acetylgruppe abgewandelte Vinylgruppe.
6. Decarboxylierung. Die am isocyclischen Ring gebundene COOH-Gruppe wird natürlich als Bestandteil einer β-Ketosäure nach der Verseifung des Esters sehr leicht abgestoßen. Die COOH-Gruppe des Propionsäurerestes ist dagegen sehr fest gebunden.
7. Komplexbildung. Erfolgt zum Teil wohl schon in der grünen Stufe.

Die Reaktionen 1—4 verlaufen relativ leicht, ebenso die Abspaltung der ringgebundenen Carboxylgruppe, das so entstehende Produkt ist Phylloerythrin VIII, ein verbreitetes biologisches Abbauprodukt des Chlorophylls, das im Kot, in Galle und Darm- und Gallensteinen vieler Säugetiere nachgewiesen ist. Im Bitumen ist es bisher noch nicht aufgetreten. Reaktion 5 läßt sich im Laboratorium mit energischen Reduktionsmitteln durchführen. Die Decarboxylierung der Propionsäuregruppe ist bei Porphyrinen ganz allgemein bisher nur durch Brenzreaktion bei sehr hoher Temperatur, 300—350°, im Dauerversuch 250°, möglich; in der grünen Stufe ist sie bisher überhaupt noch nicht gelungen.

Durch thermische Behandlung von Phäophytin, Erhitzen in rohem Erdöl gelang es den gesamten Abbau bis zum Desoxophyllerythro-ätioporphyrin nachzuahmen. Als Zwischenstufe trat dabei Phylloerythrin VIII auf. Die neuerdings aufgefundenen Fälle von Gehalten grüner Farbstoffe neben Porphyrin, und zwar auch bei Objekten, die wie die eocäne Geiseltal-Braunkohle mit Sicherheit keinen hohen Temperaturen ausgesetzt waren, führt zur Anschauung, daß die Decarboxylierungsreaktion (hier ist immer nur die Propionsäuregruppe gemeint) auch als biochemischer Prozeß möglich ist. Dies wird bewiesen durch die Auffindung von unverseifbaren grünen Farbstoffen (neben verseifbaren Anteilen) in zahlreichen rezenten und geologisch jungen Gyttja- und Sapropelablagerungen, die zum Teil aus dem Wasser geschöpft sind. Eine Isolierung in kristallisiertem Zustand ist bei den grünen Farbstoffen bisher noch nicht gelungen.

Damit ist für den Chlorophyllabbau außer der Möglichkeit einer thermischen Reaktion auch eine Bioreaktion erwiesen.

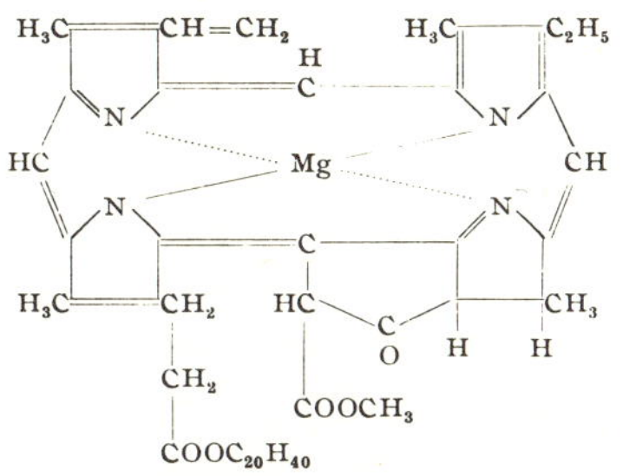

VII. Chlorophyll a.

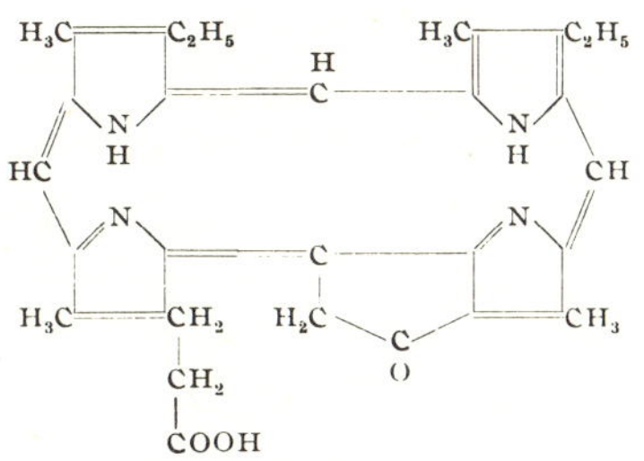

VIII. Phylloerythrin.

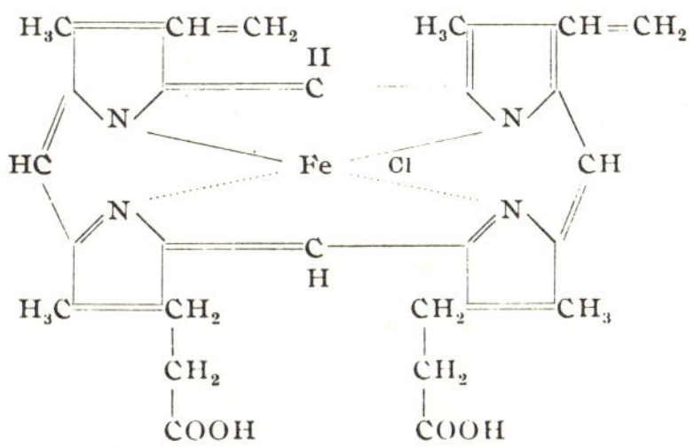

IX. Hämin.

Bezüglich der Einzelheiten der Struktur und der Reaktionen des Chlorophylls und Hämins sei auf die Arbeiten von *H. Fischer* verwiesen.

Von besonderem Interesse wäre es, die Stoffwechselprodukte niederer Tiere, vor allem von Wasserbewohnern auf Chlorophyllabbauprodukte zu untersuchen. Nach den Arbeiten von *E. Hecht*[12]) nimmt die koprogene Umformung organischer Substanz wichtigen Anteil am Bitumisierungsprozeß.

Die Porphyrinnebenprodukte des Chlorophyllabbaues stehen spektroskopisch dem Desoxophyllerythrin bzw. dessen Ätioporphyrin sehr nahe; hier ist an die Möglichkeit der Abspaltung der Vinylgruppe zu denken. Bei den grünen Farbstoffen wurden einigemal Spektren gemessen, die denen von Phäophorbid a sehr ähnlich waren, mehrfach bestand größere Ähnlichkeit mit Dihydrophäophorbid a (dessen Vinylgruppe hydriert ist). In den unverseifbaren Anteilen liegen wohl die hypothetischen Ätiophorbid oder Dihydro-ätiophorbid bzw. ganz nahe verwandte Farbstoffe vor.

Der Abbau des Hämins nimmt 2 grundsätzlich verschiedene Wege, die zum Meso-ätioporphyrin III oder zum Deutero-ätioporphyrin V führen.

[12]) Schrift. a. d. Gebiet d. Brennstoff-Geologie. *O. Stutzer*. 10. Erdölmuttersubstanz.

Meso-ätioporphyrin.

1. Reduktion der Vinylgruppen direkt zum Mesohämin, dem Fe-Komplex des Mesoporphyrins IV.
2. Decarboxylierung der beiden Propionsäuregruppen ergibt den Komplex des Mesoätioporphyrins. Hier gilt das gleiche wie oben ausgeführt.

Deutero-ätioporphyrin.

1. Abspaltung der beiden Vinylgruppen zum Fe-Salz des Deuteroporphyrins VI. Sie erfolgt, wenn man Blut im verschlossenen Gefäß langer Fäulnis überläßt, ist also der Wirkung bestimmter Bakterien zuzuschreiben.
2. Decarboxylierung der Propionsäuregruppen, wie oben.

Die Hämochromogenbildung ist eine Reaktion, die nur am Fe-Atom vor sich geht. An V gebundene Häminderivate sind bisher niemals beobachtet worden, Fe scheint also nicht abgespalten zu werden.

Welchen der beiden Wege der Häminabbau nimmt, hängt von den biologischen Bedingungen ab. Hier ist der biologische Primärprozeß entscheidend, und das bedeutet auch bereits, daß die Natur der zugrunde liegenden Organismen mitbestimmend ist, da deren Schicksal ja vom Lebensraum abhängt.

Es ist sicher kein Zufall, daß in Steinkohlen, die aus einer Baumflora entstanden sind, keine Chlorophyllderivate auffindbar sind. Beim Primärprozeß wurden Chlorophyll und mit ihm wohl auch andere Körperklassen zerstört, das Deutero-ätioporphyrin entstammt verwesenden Tieren, die vielleicht zeitlich später kamen. In allen andern Fällen sind Häminderivate in Mesoporphyrin und Mesoätioporphyrin übergegangen unter Bedingungen, die auch Chlorophyll konservierten. Das ist noch gerade bei Braunkohlen, die man von Mooren ableitet, der Fall; alle anderen organischen Mineralstoffe sind wohl Seebildungen; Gyttjakohlen z. B. sind mehrfach als Ölalgenabsätze erwiesen.

Die Porphyrinbefunde stehen somit in gutem Einklang mit gesicherten geologischen Anschauungen.

Besonders wichtig sind sie in bezug auf eine Beurteilung der Muttergesteine des Erdöls und der **Erdölentstehung**[13]). Das Vorkommen saurer Porphyrine schließt Entstehungstemperaturen über 200° aus. Der hier nachgewiesene biochemische Chlorophyllabbau zeigt die Möglichkeit der Mitwirkung biologischer Prozesse bei der Ölbildung auf, wie sie vor allem von russischen Forschern, *Archangelski*, *Vernadsky* u.a., angenommen wird. In allen Erdölen, ausgenommen solchen, die ganz hell und klar waren, konnten Porphyrine als Komplexe gefunden werden, und meist war Desoxophyllerythroätioporphyrin vorhanden, während Meso-ätioporphyrin meist ganz untergeordnet und selten allein vorkam. Man kann daraus entnehmen, daß im Ölmuttergestein immer ursprünglich Chlorophyll vorhanden war. Beim Suchen nach heutigen Ablagerungen zukünftiger Erdöllager muß diese Bedingung erfüllt sein. Wenn in hellen, gleichsam raffinierten Ölen kein Porphyrin mehr enthalten ist, so ist das durch Adsorption an die durchwanderten Schichten zu erklären. Desoxophyllerythro-ätioporphyrin wird leichter adsorbiert als Meso-ätioporphyrin, und V-Salze ganz allgemein leichter als Hämochromogene; danach muß also Mesoätiohämochromogen am längsten übrigbleiben. Die Carbonsäuren werden selbstverständlich noch wesentlich leichter adsorbiert.

[13]) Vgl. dazu *K. Krejci-Graf*, Naturwiss. **23**, 599 [1935].

Aus der Menge und Art der Porphyrinkomplexe können sehr wichtige Rückschlüsse auf die Natur des Öles gezogen werden. Man kann annehmen, daß hoher Komplexgehalt, vor allem von V-Komplex und erst recht Gehalt von saurem Porphyrin (Desoxophyllerythrin) nur bei einem Öl vorkommt, das wenig gefiltert ist und sich daher als mehr ursprüngliches Öl primärer Lagerstätte deuten läßt. Ein Öl geringen Porphyringehaltes kann nicht aus einem solchen hohen Gehaltes hervorgegangen sein (man müßte denn eine Auslaugung bituminöser Gesteine hohen Gehaltes beweisen), vielmehr ist das Gegenteil möglich. Wenn ein Asphalt viel Porphyrinkomplex enthält und dazu noch V-Salze von sauren Porphyrinen, was bei fast der Hälfte (der allerdings noch geringen Zahl) der Fall war, während nur ein kleiner Bruchteil der Erdöle saure Porphyrine enthielt, so spricht das durchaus gegen eine Entstehung dieser Asphalte aus einem stark gefilterten Erdöl. Hier ist die Polymerisation eines ursprünglichen oder eine Rückstandsbildung daraus anzunehmen. Hiermit sind die Grundsätze angedeutet, nach denen erfolgversprechende Untersuchungen spezieller Fragen der Erdölgeologie möglich sind. Nötig ist dazu die Porphyrinanalyse ganzer geologischer Horizonte in Verbindung mit der petrographischen Untersuchung. Diese Aufgabe ist in Gemeinschaft mit *K. Krejci-Graf* bereits begonnen. Aus Einzelbefunden sollte man auf alle Fälle keine allzu weit reichenden Schlüsse ziehen.

Ähnlich differenziertes Verhalten bei der Adsorption zeigen auch andere Erdölbestandteile, und die Sammlung und Wanderung von Erdöl ist gleichbedeutend mit einer chromatographischen Adsorptionsanalyse. Die Aufzeigung einer Erhaltung komplizierter Farbstoffe über geologische Epochen läßt dies auch für andere Körperklassen erhoffen.

[A. 91.]

1

CHLOROPHYLL AND HEMIN DERIVATIVES IN ORGANIC-MINERAL SUBSTANCES

Alfred Treibs

This translation was prepared expressly for this Benchmark volume by Reiner Protsch, Institut für Anthropologie, Frankfurt/Main, West Germany from "Chlorophyll- und Häminderivate in organischen Mineralstoffen," Angew. Chemie, **49***:682–686 (1936)*

The accidental discovery of an organic mineral of a well crystallized pigment named Graebeit,[1] which should, in all probability, be referred to as an anthraquinone pigment, prompted a search for specific organic substances in bituminous materials. Distillation methods were avoided because, in this case, distillation is a decomposition reaction which can probably produce a nearly completely decomposed product from the existing material. An organic solvent extract of a bituminous oil shale showed a magnificent red color and the characteristic absorption spectrum of the metal complexes of pyrrole pigments. This finding was made on a rather old object, geologically speaking, the so-called stink shale of the Carvendel Mountains of the upper Isar which belongs to the main dolomite formation of the upper Alpine Triassic and served in the production of ichthiol (fossil fish oil).

Leubimenko[2] observed green pigments similar to chlorophyll in recent and geologically young sediments and in clay sediments of the Black Sea. Similarily, Weigelt[3] found green leaves in the Tertiary Geiseltal lignite coal and Fikentscher[4] observed porphyrins in crocodile coprolites of the same origin.

An approximate figure for the concentration of porphyrin complexes is 30–40 milligrams per kilogram of shale. By extraction of 12 kilograms of shale with hot glacial acetic acid and splitting the complexes with bromoacetic acid, about 150 milligrams of porphyrin could be isolated. Through application of the methods of porphyrin chemistry this proved to be a mixture of many pigments—chiefly, basic porphyrins with a few acid porphyrins. This represents the first example of a natural occurrence of etioporphyrins.

The larger part of the basic fraction consists of a strongly basic etioporphyrin which appears to correspond spectroscopically with deoxophylloerythrin and by analysis is found to be a member of deoxophylloerythroetioporphyrin I, which is the acid-free basic substance of chlorophyll. It was then later synthesized by Fischer and Hofmann.[5] They found deoxophylloerythrin II among the acid porphyrins, which are of particular importance in providing a conclusive identification because the

[1]For all footnotes, see original article.

etioporphyrins have no melting point. On the other hand, the acid porphyrin, from which the ester melting point and mixed melting points could be compared with standard preparations, provided the final criterion of identity according to Hans Fischer. Deoxophylloerythrin is the end product of acid chlorophyll decomposition, and its structure was explained by the synthesis of Fischer and Riedmair.[6]

Weak basic by-products very intimately related to deoxophylloerythroetioporphyrins are not yet clarified. On the other hand, one was able to separate very small amounts of an etioporphyrin and an acid porphyrin which belong to the spectral types of mesoporphyrin—of blood pigment series. The latter could be proven identical to mesoporphyrin IV, the etioporphyrin being accordingly mesoetioporphyrin III. Thereby porphyrins, which are derived from chlorophyll or closely related pigments, also from hemin, could also be proven to exist in bitumen. These determinations were now generalized, based on a very large material inventory.[7] (Structures are shown on page **12** for deoxophylloerythroetioporphyrin, deoxophylloerythrin, mesoetioporphyrin, deuteroetioporphyrin, and deuteroporphyrin.)

In order to avoid going into further details, Table 1 (see original paper) provides a summary of the entire experimental material.

The amounts of porphyrin change; in half of the examined material only trace amounts of porphyrin show, specifically, 5–10 micrograms in about 50 grams of material. In the other half, increasing concentrations of 0.2–2 milligrams are evident. In isolated cases this increases even more. Soft or bituminous coal is always low in concentration. The greatest amounts of porphyrins are found in oil shales from Meride in Switzerland, which yielded 0.34–0.4%, also in oil shales from Croatia with 0.08% and in Trinidad petroleum with 0.04%. Such high quantities approach those quantities present in blood with about 0.5% hemin (calculated on the liquid content), and the chlorophyll quantities in green leaves, up to 0.8% dry weight. In sapropel, the green pigment material almost approaches this order of magnitude.

Generally, the geological age of most samples falls into the Tertiary; however, numerous ones are older. The bituminous coal was derived from the Carboniferous, one brown coal (Tula) from the Lower Carboniferous, the Volkenroda petroleum from the Permian, and American oil from the Devonian or Silurian and numerous oil shales from the Triassic. The oldest object is the Kuckersit, coming from the Lower Silurian.

The proof of porphyrin-metal complexes can in many cases be derived directly from extraction of bituminous rocks. They are very easily distinguishable through their characteristic two-band spectra. For determination of specific porphyrins, however, the decomposition products of the complexes are indispensable. The various porphyrins can then be differentiated by their specific four-band spectra and identified by their basicity. In this way a great advantage is derived since free porphyrins are easily separated from their accompanying substances based on their basic characteristics. The accompanying substances are often hindrances to the recognition of units due to their characteristic self absorption. Consequently, one is also able to identify very minute amounts of colored pigment because one can successfully enrich the concentration one hundredfold and higher. The lower limit of the identification and spectroscopic measurement is at about 0.0005 milligrams. Colored substances in such concentration do exist, and can be easily removed from accompanying substances. An oversight of small, sometimes even considerable, amounts in the porphyrins is suspected if many surface-active resinous products are present when removed by shaking, with hydrobromic acid, or acetic acid. Moreover, consideration must be given to the possibility of oxidation through heavy metal bonding. Further consideration must be given to the sensitivity of porphyrins to peroxides in ether.

The experienced person can also identify deoxophylloerythrin and mesoporphyrin respectively, and more specifically the related etioporphyrins. This is particularly important since these compounds of the chlorophyll and hemin series often occur simultaneously in bitumen, and cannot be separated in small quantities. Overall, success has easily been achieved in separation by chromatography, as illustrated by Fischer and Hofmann through the synthesis of deoxophylloerythroetioporphyrins.

Spectroscopic measurements, or better, direct spectroscopic comparisons to standard preparations, therefore provide a method for determining the possible relation of porphyrins to those of the chlorophyll and the hemin series.

The investigations of numerous other petroleums, not including many oil shales also revealed evidence for the same porphyrins. In several cases the isolation and analysis could be performed out of the oils as well. In addition, some were found in native paraffins, asphalts, and phosphorites. Very generally speaking, the geochemical significance of porphyrins is proven. In the investigation of coal, one must primarily consider the great absorptive power of coals. The best way to accomplish this is to use extracts derived through organic solvents. In some brown coals, most significantly cannel and boghead coals, from which one could expect the fastest results, one could actually recognize porphyrins, more specifically, the same porphyrins as those found in oil shales and petroleum. These coal types contain large amounts of soluble constituents in contrast to the few evident in soft coal. In the case of soft coal, the extraction is much more difficult, and only by fine crushing and grinding and very thorough extraction could a reliable result be found.[8] By using micro-fine ground coal, which according to Peters, from KWI-Mulheim to whom I am indebted for this research material, could yield the maximum amount of extract, one was actually able to find porphyrin complexes in two-thirds of all the examined soft coal. These free porphyrins are, however, different from those described above. Until now the existing amount had been largely insufficient for isolation in the crystalline state, but according to the spectrum and the overall reaction, one is dealing generally only with the deuteroetioporphyrin. Deuterohemin is derived from hemin by the advanced decay of blood; thereby the vinyl groups are eliminated and replaced by hydrogen; otherwise no change takes place. In a preparation this can be achieved through the melting of hemin with resorcinol. The deuteroporphyrin VI derived through the splitting of iron can be converted by decarboxylation through Brenz reaction to deuteroetioporphyrin.

Table 2 (see original paper) shows the variation in absorption of the different porphyrins. The location and the relative intensities of the absorption peaks show very distinct differences which are especially illuminated or emphasized by superimposed projection of two spectra in a comparison spectroscope.

The etioporphyrin spectra are wholly consistent with the spectra of accompanying carbonic acid.

The identity of the metal in the original complex was not establishable because the spectra of the extract from the complex spectrum of the basic porphyrins was different from all the until then unknown complex-forming metals. Furthermore, it was proved that the major available complex was more stable than all known porphyrin metal complexes—also this spectrum indicated a second complex. The isolation was accomplished with the help of chromatographic absorption analysis. Through simultaneous synthetic experiments the main complex of the oil shales could be recognized as vanadium salt, a fact verified through the analysis. This rare element, vanadium, is often found as an important ingredient of petroleum and asphalts and was thus investigated in great detail. Vanadium complexes are much more difficult to form than all other porphyrin complexes. They contain metal-bound oxygen and are probably vanadyl compounds. Analysis does not provide reliable information concerning this. The vanadium could also be in the +5-valence state.

The vanadium salts could only be demetallated by means of the strongest mineral acids and also manifest remarkable acid-stable leuco compounds which were never evidenced until now.

The second complex could also be kept free of all other pigments, but could not be maintained in a crystallized form due to insufficient quantities. The spectrum proved unexpectedly to be a pure hemochromogen spectrum. Hemochromogens are molecular compounds of a ferrous (Fe^{+2}) complex with bases, whose red-brown solution is highly characteristic, showing extremely intense two-band spectra which differ substantially from those of the red metal complexes. Exhaustive investigation showed that the hemochromogens did actually exist, more specifically in the form of mesoetioporphyrins, but with differing positions of the absorption maxima. The following are some further remarkable differences noted: the solutions showed complete stability to air oxygen to which normal hemochromogens are very sensitive. Both these properties are otherwise only attributable to cytochrome C. The splitting with mineral acid, however, occurs easily yielding ferrous (Fe^{+2}) iron. If more iron is again introduced into the porphyrins and if the hemochromogen spectrum is derived through treating it with pyridine-hydrazine, normal hemochromogen is produced. In this case a special base must be bound to the ferrous complex producing this remarkable phenomenon.

Unfortunately, nothing has yet been found to yield sufficient pigment in the purification of these complexes. This would be of great interest for clarification of the base. In those oil shales used especially for isolation experimentation studies, one could only find the hemin derivatives, mesoetioporphyrin, in the separated hemochromogen. In many other cases in which the measurement of the extract only showed iron complex, deoxophylloerythroetioporphyrin was found as a hemochromogen.

Complexes with other metals have not yet been observed. In three Bulgarian oil shales, however, the ferric (Fe^{+3}) complex could be found. This was illustrated by the fact that the pyridine extracts assume the normal hemochromogen spectrum. Very dilute solutions of hemin in pyridine, without reducing agents, show hemochromogen spectra.[9] In this instance a reducing action can also occur through organic substances. With respect to experimental work, this discovery is very unfortunate because it means that the best porphyrin solvent, pyridine, must not be used in the extraction of these shales. These complexes are completely destroyed within a short time by boiling pyridine solutions. (The above described complexes can remain stable after one day boiling.) These ferric (Fe^{+3}) complexes seem very rare. However, we urge the greatest caution in the choice of solvents. Chloroform, or acetic acid and benzene, or pyridine, are only acceptable after a trial run.

Complex formation of bitumen is a secondary process, as one can see by the occurrence of free porphyrins and iron complexes of chlorophyll derivatives. Only those complexes which are stable and favored are preserved under the existing conditions.

Hentze found vanadium containing blood pigment material in lower sea animals, tunicates. A closer investigation of this pigment showed that no porphyrin derivatives existed here.

In some oil shales, free porphyrins also occur. Furthermore, long wave absorption bands which resemble more closely the green pigmentation of chlorophyll are frequently evident even in the case of very old objects. In a sample of the famous Geiseltal brown coal deposit of the Tertiary, the long-sought after intermediates of chlorophyll transformation could be found. Here equal amounts of green pigment material and porphyrin exist next to one another (only secondary complexes). This will be elaborated on later.

Table 5 (see original paper) provides a synopsis of the spectra of a series of characteristic materials and the order of their absorption bands.

A Brazilian oil shale of the Cretaceous or Tertiary, which was first investigated[10] by Ch. Dhere and G. Hradil, who also found some porphyrin-rich oil shales, was found to contain only green pigment in a non-complex bond still resembling chlorophyll. Similar finds were reported in geologically recent gyttja and sapropels. (Gyttjas are lake deposits, formed under limited air-exposure, which consist of numerous plants and animals, including many algae. Animal life in these deposits is theoretically still possible. In the uppermost zone oxygen is still available. The organic substance is transformed to a coprolitic form as a result of animal action. In layers just below this one, the oxygen-free reduction zone begins. Such deposits contain substantial amounts of organic material. Sapropel deposits itself in oxygen-free, hydrogen sulfide-poisoned water, and contains only anaerobic living organisms. They are richest in organic substances.)[11] In other words, transition of the original state of chlorophyll up to the end step (vanadium or iron complexes of the deoxophylloerythroetioporphyrin) has been proven.

The question of the mechanism of the formation of porphyrin from chlorophyll takes on immediate geochemical significance. The necessary steps for the degradation are:

1. Magnesium demetallation. This occurs very readily even with weak acids.
2. Saponification of the ester group. In recent, but also in fossil deposits, the green pigment substances exist as neutral materials which are saponified mainly by alkali. The saponification is ascribed to acid. Alkalies open the iso-cyclic ring.
3. Hydrogenation of the vinyl group to an ethyl group. This is accomplished quite easily and is proven to be a biochemical reaction.
4. Dehydrogenation of the green system to the red porphyrin system, equivalent to stabilization.
5. Reduction of the carboxyl group to CH_2. In chlorophyll *b*, which does not occur in algae, or is only a less important occurrence in this case, the vinyl group related to the methyl group, is converted to the aldehyde group, whose reduction is naturally just as possible. This means that the same porphyrin always results from component *b*. In the case of bacterial chlorophyll, the situation is similar in that the vinyl group is replaced by an acetyl group.
6. Decarboxylation. The COOH bound on the isocyclic ring will naturally be removed easily as an ingredient of a β-keto acid, after saponification of the ester. The COOH on the propionic acid side chain is, on the contrary, bonded very tightly.
7. Complex formation. This already occurs partially in the green algae.

Steps 1 to 4 proceed easily, just as the splitting off of the ring-bound carboxyl group does. The product formed is phylloerythrin VIII, a widespread biological degradation product of chlorophyll, which has been proven to exist in excrement, gall and gut and gallstones of many animals. It has not been as yet observed in bitumen. Step 5 is accomplished in the laboratory with (or by) strong reducing materials. The decarboxylation of the propionic acid group is, in the case of porphyrins, generally possible only by Brenz reaction heating at very high temperatures, 300–350°C, in an endurance test at 250°C. In the green stage this has as yet not been successful.

Through thermal treatment of pheophytin, that is, heating of crude oil, it was possible to imitate the entire degradation scheme up to a deoxophylloerythroetioporphyrin stage. As an intermediate stage during this process, phylloerythrin VIII occurred. The recently discovered cases in which green pigment materials occur near porphyrin and specifically, even in objects such as Eocene Geiseltal brown coal, which were surely not exposed to high temperatures, leads one to conclude that the decarboxylation reaction (meaning in this case only the propionic acid group) is also a possible biochemical reaction. This is proven by the discovery of unsaponifiable green pigment material (along with saponifiable portions) in numerous recent and geologically young gyttja and sapropel deposits which are partially derived from the water. Isolation of green pigment matter in a crystalline stage, has not as yet been successful.

Thereby, the degradation of chlorophyll, aside from a possibility of a thermal reaction, is also proven to be a biochemical reaction.

It would be of particular interest to investigate the metabolic products of lower animals, particularly those living in water, for chlorophyll degradation products. According to the work of E. Hecht[12] the coprolitic transformation of organic substances is an important part of the bituminization process.

The porphyrin by-products of chlorophyll degradation are spectroscopically very similar to deoxophylloerythrin; more specifically to its etioporphyrin. Here, the possible splitting of the vinyl group must be considered. Spectral measurements were made of some of the green pigments which proved very similar to those of pheophorbide a. More tests frequently showed a greater similarity to dihydropheophorbide *a* whose vinyl group is hydrogenated. The hypothetical etiophorbide or dihydroetiophorbide, or more specifically, very closely related materials, probably remain in the unsaponifiable fraction.

The degradation of the hemins follow two fundamentally different courses leading either to mesoetioporphyrin III or to deuteroetioporphyrin V.

Mesoetioporphyrin:

1. Reduction of the vinyl group directly to mesohemin which is the iron complex of mesoporphyrin IV.
2. Decarboxylation of both propionic acid groups yields the complex of mesoetioporphyrin. Here the same is evident as described above.

Deuteroetioporphyrin:

1. The splitting off of both vinyl groups to the ferrous iron salt of deuteroporphyrin VI. If blood is left to ferment for a length of time in a tightly closed vessel this porphyrin can be attributed to the action of certain bacteria.
2. Decarboxylation of the propionic acid group as above. Hemochromogen formation is a reaction which only takes place in the presence of the iron atom. A vanadium-bound hemin derivative has as yet not been detected. Iron, in other words, seems not to split off.

Which of the two ways the hemin degradation takes depends on the biological conditions. Here the primary biological process is determinative, and this means also that the nature of the basic organisms is at the same time determined because their fate too depends upon their environment.

It is surely no accident that in soft coal, which originates from woody plants, no chlorophyll derivatives are found. In the primary process, chlorophyll along with other classes of compounds was destroyed. The deuteroetioporphyrins originated from decayed animals which perhaps came earlier in time. In all other cases, hemin derivatives are transformed to mesoporphyrins and mesoetioporphyrins under conditions which also conserve chlorophyll. This is particularly true of brown coal which one derives from moors. All other organic materials are formed in the sea. Gyttja coals, for example, are repeatedly proven to be oil algae deposits.

Porphyrin findings are in good agreement with known geological opinions. They are of special importance with respect to evaluation of the source rocks of petroleum and their petroleum origin.[13] The appearance of acidic porphyrins eliminates formational temperatures above 200°C. The hereby proven biochemical chlorophyll degradation provides the possibility that biological processes assist in oil formation, proposed by the Russian scientists Archangelski, Vernadsky et al. In all oils excepting those that were very bright and clear, porphyrins could be found as complexes and most deoxophylloerythroetioporphyrins were frequently present; mesoetioporphyrins on the other hand were subordinate and rarely occurred alone. On this basis one can conclude that in the source rock material from which the petroleum came, chlorophyll was always originally present. The sediments that create future oil deposits today must fulfill these conditions. If no porphyrins are contained in light oils refined the same way, it must have been removed by absorption by the layer through which it has migrated. Deoxophylloerythroetioporphyrin is more easily adsorbed than mesoetioporphyrin, and vanadium salts are generally more easily adsorbed than hemochromogen. It follows that mesoetiohemochromogens stay behind longest. Carbonic acids are still more easily adsorbed.

From the amount and type of porphyrin complexes, very important conclusions about the nature of oils can be made. One can assume that high complex content, especially of those containing vanadium, and particularly acid porphyrins (deoxophylloerythrin), occur only in an oil which is only partially filtered and, because of this small filtration, can be considered as a more original oil from a primary depositional site. An oil of high porphyrin content cannot have originated from one of a low content. (One would then have to prove a flushing out of bituminous rocks of high porphyrin content.) The opposite is much more plausible. When an asphalt contains both high porphyrin and vanadium salt of acid porphyrins, as was true in almost half the cases (still a small number), while only a small portion of the oil contains acid porphyrins, one must conclude that this asphalt could not have originated from a strongly filtered petroleum. Here is assumed polymerization of an original oil or a residue from it. Herewith the principles are outlined by which promising investigations of special problems in petroleum geology are made possible. Related to this it is necessary to investigate porphyrin analysis of whole geological horizons in connection with petrographic investigations. This problem has already been started in cooperation with K. Krejci-Graf. One should not come to too far-reaching conclusions on the basis of single or isolated findings.

Other petroleum components also show similar differential behavior in absorption, and the accumulation and migration of petroleum is of equal importance to an analysis by chromatographic adsorption. The appearance of a preservation of complicated pigment materials through geological eras permits hope that the same may also be true of other types of molecules.

Note: Later discoveries in the field found that hemochromogen in oil shale (for example, see Table 4) is in reality a Ni-complex. The misinterpretation occurred because the chemical tests for Ni at the time this paper was written were negative. Later Ni was indeed found in the original sample by emission spectroscopy. This sensitive method was not previously available. However, iron was also found and it is assumed to come from some contamination.

2

Reprinted from *Am. Assoc. Petrol. Geol. Bull.* **14**:1451–1463 (1930)

DOES PETROLEUM FORM IN SEDIMENTS AT TIME OF DEPOSITION?[1]

PARKER D. TRASK[2] and C. C. WU[3]
Stanford University, California

ABSTRACT

Chemical analyses of large quantities of recent marine sediments from the richest areas located during a survey of the distribution of organic matter in various types of modern deposits indicate that liquid petroleum probably does not occur in sediments at the time of their deposition, and that if it is present, it is in very small amounts, certainly less than 3 parts per 100,000. Therefore, petroleum present in newly deposited sediments is not an important factor in the genesis of oil. Presence of small quantities of solid paraffines suggests that some constituents of petroleum occur in fresh sediments. Fatty and oily substances constitute less than 1 per cent of the organic content of the marine sediments. Presumably much, if not most, of the petroleum comes from the remaining 99 per cent of the organic matter. Fatty acids ranging from cerotic ($C_{26}H_{52}O_2$) to melissic ($C_{30}H_{60}O_2$), comprise practically all the fatty acids observed and they constitute from 0.002 to 0.006 per cent of the sediment. Organic sulphur compounds form about 0.03 per cent of the deposits. Free sulphur is a common minor component of all the sediments, and ranges from 0.02 to 0.1 per cent in quantity.

INTRODUCTION

The question of whether or not petroleum forms in sediments at the time of their deposition is of great interest. In the writers' investigations on "The Origin and Environment of Source Sediments," conducted under the auspices of the American Petroleum Institute, they have recently obtained data which throw light on this problem. A preliminary survey of the distribution of organic matter on the sea bottom demonstrated the presence of a few regions in which deposits rich in organic matter are now forming. Large quantities of sediments from four of these rich areas were extracted with carbon tetrachloride and the extracts were examined in detail. As petroleum is soluble in CCl_4, in analyzing the extracts,

[1]This paper contains preliminary results of an investigation on "The Origin and Environment of Source Sediments," listed as Project 4 of the American Petroleum Institute research program. Financial assistance in this work has been received from a research fund of the American Petroleum Institute donated by John D. Rockefeller. This fund is being administered by the Institute with the coöperation of the Central Petroleum Committee of the National Research Council. Manuscript received by the editor, August 25, 1930.

[2]Research associate, American Petroleum Institute.

[3]Research fellow, American Petroleum Institute, now professor of chemistry, Chekiang College, Hangchow, China.

the writers gave special attention to ascertaining the distribution of liquid hydrocarbons in these, the richest marine organic deposits which have come to their attention.

No liquid hydrocarbons were detected. The maximum quantity of liquid hydrocarbons that might have been present without being observed is less than 3 parts per 100,000. This indicates that petroleum present in sediments at the time of their accumulation can not be a major factor in the formation of commercial oil pools. However, the presence of small amounts of solid paraffines suggests that some normal constituents of petroleum occur in newly deposited sediments.

The writers wish to express their appreciation of permission to use the chemical laboratory of Stanford University, in which these investigations were conducted, and their indebtedness to R. E. Swain and C. L. Alsberg, of that University, for advice and assistance. They are also grateful for suggestions and criticisms from their advisory committee, composed of Alex W. McCoy, *chairman*, Austin H. Clark, W. A. J. M. van Waterschoot van der Gracht, T. Wayland Vaughan, and S. A. Waksman; and from David White and K. C. Heald, of the Central Petroleum Committee of the National Research Council. Acknowledgment is due The Scripps Institution of Oceanography for aid in collecting sample 418 from the Channel Islands region, and to W. E. Cummer, Jacksonville, Florida, for sample 119 from the Florida algal lake.

This paper represents the joint work of the two writers, but the chemical analyses were made by Wu.

ANALYSES OF SEDIMENTS

GENERAL REMARKS

Description of sediments.—Each of the four sediments selected for analysis represents a distinct type of deposit. Three are marine and one is lacustrine. Sample 454 is a calcareous ooze from a mud-bank in Florida Bay, deposited in water less than a meter deep. Sample 418 is a greenish gray mud from the Santa Barbara Channel, California, in water 580 meters in depth. Sample 477 is a gray mud from Neuse River at the southwest end of Pamlico Sound, in 7 meters of water. Sample 119 is an algal deposit from Mud Lake, 30 kilometers northeast of Ocala, Florida, in water less than 20 centimeters deep.

The marine samples contain about 5 per cent of organic matter and on distillation produce about 2 gallons of oil per ton. Sample 119 from the algal lake probably is a mother rock of a rich oil shale. It is more than one-half organic matter and yields a distillate of 28 gallons of oil per ton. Sample 418 from the Channel Islands region of California contains remains of pelagic life and probably was deposited under conditions similar to those in which the source beds of the oil fields of the Los Angeles basin were formed. The organic matter in sample 477 from Pamlico Sound is probably strongly influenced by material draining from adjacent large swamps. Sample 454 is a limestone-forming deposit.

Extraction with CCl_4.—The first step in our analyses of the organic content of the sediments was extraction with an organic solvent in order to remove the fatty and oily constituents. Carbon tetrachloride, because of its uninflammability, its relative cheapness, and the high degree of purity in which it may be obtained, was selected as the solvent. Technical CCl_4 purified by one distillation was used. It contained only 0.02 per cent CS_2 and less than 1 part per 100,000 of volatile constituents.

Before the sediments could be extracted it was necessary to dry them. Oxidation was inhibited by drying the samples as rapidly and at as low a temperature as possible. Preliminary experiments showed that evaporation under reduced pressure at room temperature was too slow to be practicable. The procedure adopted was to dry the sediments in a long tunnel through which warm air at a temperature of about 60° C. was passed. The samples were spread on trays in a layer about 3 centimeters thick and were dried until they ceased to appear plastic. The quantity of water lost was enormous. Sample 119, evidently a colloidal gel, lost more than 99 per cent in weight; No. 477, from Pamlico Sound, lost more than 80 per cent; and No. 418, from the Channel Islands, more than 65 per cent. After drying, they were ground in a roll grinder until all particles passed a 20-mesh screen (maximum diameter about 1 millimeter).

In order to extract large quantities of sediment it was necessary to devise a special extractor. This consisted of a 5-gallon can fitted with a removable top. Suitable pyrex tubes and cork stoppers connected the can, at the bottom, with a 5-liter distilling flask, and at the top with a reflux condenser, which also was in connection with the 5-liter flask. A mixture of sodium silicate and barium sulphate was applied to the joints to prevent escape of solvent. Carbon tetrachloride was placed in the system and the sediments were extracted continuously until the extract coming from the sediment was colorless. This ordinarily required about 30 hours. The can was half filled with sediment and could accommodate 10 kilograms. About five runs could be made before the CCl_4 began to attack the cans.

Method of analysis.—The method of analysis adopted is the common procedure used for ether extracts. The material was saponified, the fatty acids were separated from the saponifiable fraction, and the pigments, sterols, and hydrocarbons from the unsaponifiable part. The presence of relatively large amounts of free sulphur in all the extracts necessitated preliminary crystallizations from benzene and chloroform to remove the sulphur. While removing the solvents, the writers tried not to overheat the mixtures or to continue the evaporation longer than necessary. In this manner loss of liquid parts of the extracts was minimized. Oxidation was inhibited by driving off the final traces of the solvent in an atmosphere of CO_2. Similarly the extracts, when not being studied, were stored in an atmosphere of CO_2.

SAMPLE 454 FROM FLORIDA BAY

Preliminary treatment.—The CCl_4 extract from 30.6 kilograms of dried sediment 454 from Florida Bay was filtered and distilled as much as possible to drive off the solvent, the temperature being kept below 80° C. While still warm the residue was dried with a slow stream of CO_2. The yield was 20.0 grams, or 65 parts per 100,000. The dried extract was treated with 95 per cent ethyl alcohol and decanted, the action being repeated until the alcoholic solution was practically colorless. The alcohol insoluble part was similarly treated repeatedly with ether. The residue left after the action of these two solvents was soluble in warm benzene. On concentration, yellow crystals of sulphur, melting at 115°-115.6° C., crystallized from the benzene solution. These crystals, most of which were monoclinic, were readily soluble in CS_2 and warm chloroform. Cooled to room temperature, the alcohol and ether solutions formed yellow amorphous precipitates. The solutions were combined, the solvents distilled off, and the residue taken up in warm chloroform, from which repeated crystallizations produced more crystalline sulphur. The total yield of sulphur was 6.76 grams, or 22 parts per 100,000.

The chloroform was distilled off and the residue saponified with alcoholic potash. After the resulting mixture had been concentrated and cooled, it was extracted repeatedly with ether to remove the unsaponifiables, and was saponified again to insure completeness of the action. However, it was found that the first saponification had been sufficient, as no additional saponifiable material was procured.

Examination of the saponifiables.—The saponifiable part was diluted with ether and acidified gradually in order to prevent the formation of H_2S. Analysis of the ether layer showed the presence of two fatty acids, each in 1 part per 100,000. One evidently was melissic acid, ($C_{30}H_{60}O_2$), as it melted at 90° C. and its impure lead salt at 115°. The other, characterized by a sweaty odor and melting at −6°, apparently was caproic acid ($C_6H_{12}O_2$). Addition of lead acetate to the ether-insoluble part brought down a heavy precipitate of lead sulphide, indicating the presence of organic sulphur compounds in the original extract. The amount of these compounds was 9.3 grams, or 30 parts per 100,000.

Examination of unsaponifiables.—The ether solution containing the unsaponifiables was washed free from alkali, dried with anhydrous sodium sulphate, filtered, and freed from ether. Treatment with warm chloroform yielded 0.88 grams of crystalline sulphur. This sulphur evidently came from the destruction of organic sulphur compounds during saponification, and is included in the estimate of 9.3 grams for those compounds. After the removal of the sulphur, the chloroform was driven off, the solution taken up with petroleum ether (boiling point 30°-60°), and refluxed with animal charcoal to take out the pigments. The petroleum ether was distilled off, and the residue, a reddish solid, was acetylated with acetic anhydride under reflux condenser in order to separate the higher alcohols from the hydrocarbons. While still warm, the acetylated mass was poured into boiling water. After the solution had cooled to room temperature, a dark solid floated on its surface, but no solids were present within it. This indicated the absence of higher aliphatic alcohols. The solid substance that floated on the mixture was washed free from acetic acid, taken up in petroleum ether, and extracted with absolute alcohol to separate it from the solution. This substance probably was phytosterol acetate, as it melted at 122° C. The amount obtained was 0.05 grams, or 2 parts per million.

Neutralization of the acetic solution with sodium hydroxide and evaporation to dryness gave sodium acetate, a further indication of the absence of higher alcohols. The residue was taken up in petroleum ether, from which repeated crystallizations produced an almost colorless waxy solid, melting at 62° C. This probably was a paraffine, possibly $C_{30}H_{62}$. Concentration of the petroleum ether solution to dryness produced a reddish yellow waxy solid, melting with decomposition at 62°-64°. It is probably composed largely of the same substance as was obtained from the first concentrations of the petroleum ether solution. The combined weight of the paraffinaceous materials was 1.78 grams, or 6 parts per 100,000. The pigments were recovered from the charcoal by refluxing successively with petroleum ether and carbon bisulphide. The petroleum ether was removed, the residue taken up in CS_2, combined with the other CS_2 part, and passed through a column of tightly packed calcium carbonate, which absorbed most of the color. This suggests that the pigments are of a carotinoid nature. By difference, the amount of pigments was estimated to be 1.38 grams, or 5 parts per 100,000. Table I presents a summary of the analysis.

TABLE I

SUMMARY OF CCl_4 EXTRACTS OF SEDIMENTS

SAMPLE 454—FLORIDA BAY

Substance	*Grams*	*Per Cent Extract*	*Parts per 100,000*
Weight of sediment	30,600.co		
Amount extracted	20.0		65
Free sulphur	6.76	33.8	22
Caproic acid ($C_6H_{12}O_2$)	.36	1.8	1
Melissic acid ($C_{30}H_{60}O_2$)	.38	1.9	1
Sulphur compounds	9.3	46.5	30
Pigments	1.37	6.9	5
Phytosterol	.05	.3	0.2
Paraffines	1.78	8.9	6

SAMPLE 418—CHANNEL ISLANDS, CALIFORNIA

Substance	*Grams*	*Per Cent Extract*	*Parts per 100,000*
Weight of sediment	6,440 00		
Amount extracted	5.6		87
Free sulphur	3.3	58.9	51
Montanic ($C_{28}H_{56}O_2$) and melissic ($C_{30}H_{60}O_2$) acids	.31	5.6	5
Sulphur compounds and pigments	.5	8.9	8
Cholesterol	Trace		
Paraffines	1.5	26.8	23

SAMPLE 418 FROM CHANNEL ISLANDS REGION, CALIFORNIA

Sample 418, from the Channel Islands region, was analyzed in essentially the same manner as was sample 454. Only 6.44 kilograms were available for extraction. This smallness of the quantity was due to the difficulty in collecting in deep water and to the high content of water in the fresh sediment. The yield was 5.6 grams, or 87 parts per 100,000. The amount of free sulphur was 3.3 grams, or 51 parts per 100,000. This is nearly 60 per cent of the extract. The fatty acid fraction melted at 86°-89° C. and evidently was a mixture of melissic ($C_{30}H_{60}O_2$) and montanic ($C_{28}H_{56}O_2$) acids, which melt at 91° and 83° respectively. The sulphur compounds were not separated and are included in the estimate of pigments, which was obtained by difference. The pigments probably contained chlorophyl, as magnesium was found in the ash resulting from ignition of the residue recovered from the animal charcoal.

The residue after removal of the pigments was a reddish brown solid. The washings from acetylation were not turbid, indicating the absence of higher aliphatic alcohols. A few reddish crystals, melting at 114°, probably consisting of cholesterol acetate, were obtained from the methyl alcohol solution of the residue from acetylation. Following the removal of the sterols, the methyl alcohol solution was cooled in water and chlorinated in sunlight. The excess of chlorine was removed by bubbling dried air through the solution, which caused hydrochloric acid fumes to be driven off. The chlorinated solution was concentrated by distilling off the alcohol, water was added, and the mixture heated. A reddish yellow solid, insoluble in water, melting at 93° C., and giving a positive test for chlorine, was obtained. This indicated the formation of substitution products, and demonstrated the presence of saturated hydrocarbons. The amount of paraffines present was 1.5 grams, or 23 parts per 100,000. A summary of the analysis is given in Table I.

SAMPLE 477 FROM PAMLICO SOUND

In analyzing sample 477 from Pamlico Sound, the writers modified the procedure. From 11.66 kilograms of sediment, 17.0 grams of extract, or 146 parts per 100,000 was obtained. After the removal of the free sulphur, which constituted 12.1 grams, or 71.2 per cent of the extract, the solution was distilled free of solvents and taken up in hot alcohol. On cooling, it formed a waxy mass, which melted at 73°-76° C. This mass was separated from the solution and saponified with alcoholic potash. The resulting soap, on being acidified with concentrated hydrochloric acid in dilute alcohol, formed a fatty acid having the same melting point, 73°-76°, as the mass which separated from the original extract. The identity of melting points before and after saponification seems to indicate that the fatty acid occurs in the free state rather than as a soluble soap or fat. The melting point approximates cerotic acid, which melts at 76°-82° C. The quantity of fatty acid recovered was 0.69 gram, or 6 parts per 100,000.

The alcoholic solution from which the fatty acid had been removed was distilled free of solvent, taken up in petroleum ether, and refluxed with animal charcoal. Dark green pigments to the extent of 0.55 gram, or 5 parts per 100,000, were recovered from the charcoal by successive treatment with warm alcohol, ether, acetone, and chloroform. After removal of the pigments, the petroleum ether was driven off, and the residue saponified with sodium ethylate. The residue from the ether-soluble part of the saponified mixture consisted of only a few milligrams of a green solid, which indicates that solid hydrocarbons were present in very small quantities. From the dark red ether-insoluble part, a dark, lustrous, pitch-like precipitate formed as the solution cooled. This pitch was fairly soluble in warm benzene. It softened at 60° C. and melted at about 80°. Approximately 3.5 grams, or 30 parts per 100,000 were recovered. As this residue represents the saponifiable part of the extract after the fatty acids had been removed, it probably consisted essentially of sulphur compounds. A summary of the analysis is shown in Table II.

TABLE II

SUMMARY OF CCl_4 EXTRACTS OF SEDIMENTS

SAMPLE 477

Substance	*Grams*	*Per Cent Extract*	*Parts per 100,000*
Weight of sediment	11,660.00		
Amount extracted	17.0		146
Free sulphur	12.1	71.2	104
Cerotic acid ($C_{26}H_{52}O_2$)	.69	4.1	6
Sulphur compounds	3.5	20.6	30
Pigments	.55	3.2	5
Sterols	0.00	0.0	0
Paraffines		Trace	
Loss in handling	.16	0.9	1

SAMPLE 119

Substance	*Grams*	*Per Cent Extract*	*Parts per 100,000*
Weight of sediment	10,260.00		
Amount extracted	24.4		238
Free sulphur	9.9	40.6	97
Cerotic acid	1.6	6.5	16
Sulphur compounds	3.7	15.2	36
Pigments	0.4	1.6	4
Phytosterol	1.6	6.5	16
Cholesterol	0.2	0.8	2
Paraffines	7.0	28.7	68

SAMPLE 119 FROM FLORIDA ALGAL LAKE

Sample 119, presumably a mother rock of an oil shale, yielded 24.4 grams, or 238 parts per 100,000, from 10.26 kilograms of sediment. The yield of free sulphur was 9.9 grams, or 97 parts per 100,000. The residue, after removal of the sulphur, was taken up in hot alcohol, and on cooling separated into a greenish yellow mass and red oily globules. Cooled to room temperature, the globules formed a brittle pitch, which when separated weighed 1.6 grams. The pigments were separated by the method of Willstaetter and Stoll.[1] The mucilaginous mass and the pitch were saponified separately with 30 per cent methyl alcoholic potash under reflux condenser for a short time. The resulting mixtures were cooled, washed into separatory funnels, acidified with 20 per cent hydrochloric acid, shaken with ether, neutralized with NH_4OH until the aqueous layer acquired a faint bluish tint, shaken with more ether, and washed with water to remove the methyl alcohol. Approximately 0.3 gram, or 3 parts per 100,000, of green pigment and 0.1 gram of brown were obtained.

After removal of the pigments the mixtures were combined, saponified with normal sodium ethylate, distilled free of alcohol, and extracted with ether. From the brown ether-insoluble part, 5.3 grams of solids were obtained. Of this, 1.6 grams, or 16 parts per 100,000, were cerotic acid, $C_{26}H_{52}O_2$. The remainder consisted essentially of sulphur compounds and contained a small amount of phosphorus. This suggested the presence of lecithins, but no nitrogen was found. The red ether-soluble part yielded 8.8 grams of brittle pitch, which, when dissolved in a large volume of alcohol and treated with 1 per cent digitonin, yielded a precipitate weighing 2.7 grams. As 0.9 gram of digitonin was used, 1.8 grams, or 18 parts per 100,000, of sterols were present. After the digitonin had been removed by boiling the precipitate in xylene and filtering, the sterols were taken up in CS_2, treated with bromine to form dibromides, and dissolved in a mixture of ether and glacial acetic acid to separate the cholesterol from phytosterol. Phytosterol was eight times more plentiful than cholesterol.

After the removal of the sterols from the unsaponifiable part, the mixture was distilled free of solvent, taken up in CCl_4, brominated, stoppered, and allowed to stand in the sun. Profuse fumes of HBr were liberated by blowing breath over the solution, thus indicating the formation of substitution products and demonstrating the presence of saturated hydrocarbons. After the brominated mixture had been concentrated, it formed a brownish red amorphous precipitate, representing the saturated hydrocarbons, and a small amount of red greasy fluid, probably consisting of brominated unsaturated hydrocarbons. A summary of the analysis is shown in Table II.

[1]Richard Willstaetter and Arthur Stoll, *Investigations on Chlorophyll*, translated by F. M. Schertz and A. R. Merz (Scientific Printing Company, 1928), p. 88.

INTERPRETATION OF ANALYSES

ABSENCE OF PETROLEUM IN RECENT SEDIMENTS

No liquid hydrocarbons were observed in any of the extracts of these, the sediments richest in organic matter, found in our survey of practically the entire world. If any petroleum had been present in these deposits it would have been extracted by carbon tetrachloride and should have been noticed in the extracts. Yet, all the material extracted was solid. One would expect low-boiling hydrocarbons to evaporate while the solvent was being driven off, but the higher-boiling liquid hydrocarbons, which are normal constituents of petroleum, if they had been present, would not have been completely driven off.

To test the reliability of this inference, two California crude oils, of 27° and 20° Bé. respectively, were dissolved in CCl_4 in the same concentration and treated in the same manner as the extracts of the four sediments described in this paper. The residue after removing the CCl_4 from these crude oils was liquid, although the lighter oil lost 20 per cent in weight and the heavier oil, 4 per cent.

Petroleum is a mixture of many members of the paraffine and olefine series, as well as of other hydrocarbon and related compounds. Consequently, one would not expect the CCl_4 extract of a petroleum to consist of a solid, most of which melted at a constant temperature. Yet, nearly all the hydrocarbon part of sample 454 melts at a temperature of 62° C. In fact, except for their content of paraffines, these extracts do not resemble petroleum.

It is conceivable that some liquid hydrocarbons could have been lost during removal of the solvent, but the quantity, if any, must have been very small. The fact that the material extracted by CCl_4 was solid indicates that if petroleum had been present, it must have been a heavy oil. The heavy crude oil of 20° Bé. was liquid after it had undergone the same treatment as the extracts from the sediments. Yet it lost only 4 per cent during this process. A higher proportionate loss would not be expected from a petroleum that gave a solid residue.

Sample 119, the mother rock of a future oil shale, contained 68 parts per 100,000 of paraffinaceous substances. As these were solid, it follows from analogy with the experiment with crude oils that less than 4 per cent of the liquid compounds might have evaporated during the process of extraction. This means that less than 3 parts per 100,000 could have been lost from this sample. The three marine sediments have much lower maximum estimates. Sample 418 might have lost 1 part per

100,000; sample 454, 0.3 part; and sample 477, nothing. However, nothing in the analyses indicates the presence of liquid hydrocarbons, and the presumption is that none was present. Consequently, it must be inferred that petroleum probably does not occur in sediments at the time of their accumulation, and if it is present, it is in very small amounts, certainly less than 3 parts per 100,000, a quantity too small to be of importance in the formation of commercial oil pools.

DISCUSSION OF THE CONSTITUENTS OF THE EXTRACT

Solid hydrocarbons.—Paraffines are reported in significant quantities from all the sediments except sample 477 from Pamlico Sound. The paraffine fraction, however, may include wax-like substances. One would expect wax-like materials present in the sediment to be soluble to a certain extent in carbon tetrachloride; consequently, higher alcohols resulting from saponification of the waxes should have been detected by the analyses, but none except sterols was found. However, if any wax or wax-like substance that was extracted by carbon tetrachloride resisted saponification and acetylation, it would be included in the part reported as paraffine. We are inclined to regard the high figure of 68 parts of paraffine per 100,000 from sample 119 as in part probably representing wax or wax-like substances, which are known to be present in algal deposits of the type it represents. It is also possible that the marine sediments contain similar waxes. Consequently the figures for the paraffine part can only be regarded as the maximum value for the content of paraffine. The actual amount perhaps is lower. However, the analyses indicate that some paraffine is present, from which it is to be inferred that some normal components of petroleum may be present in fresh deposits.

Fatty acids.—Fatty acids are normal constituents of the sediments. They occur in a concentration ranging from 2 to 6 parts per 100,000 in the marine deposits, and a concentration of 16 parts per 100,000 in the algal sample. With the exception of caproic acid ($C_6H_{12}O_2$) observed in the limestone-forming sediment from Florida Bay, the acids in all the samples are similar. They consist essentially of three acids, cerotic ($C_{26}H_{52}O_2$), montanic ($C_{28}H_{56}O_2$), and melissic ($C_{30}H_{60}O_2$). In the only sample for which we have data on the manner of occurrence of the fatty acid (sample 477), the evidence indicated that the acid occurred in the sediment in the free state rather than as a soap or fat. In sample 454 from Florida Bay, both a fatty acid and what appears to be a paraffine have the same chemical radical, $C_{30}H_{60}$. Whether this is a coincidence or is of genetic importance is not certain.

The content of fatty acid in the sediments may be greater than the figures given, as compounds such as calcium and magnesium soaps would not be soluble in CCl_4, but would remain in the sediment. In view of the equilibrium constants of such soaps it is well within reason to expect some of the fatty acid to occur in such a manner. It is hoped later to ascertain the amounts of such types of compounds.

The sum of the fatty acids and hydrocarbons in the marine deposits forms about 0.04 per cent of the sediment. This is less than 1 per cent of the organic content and seems inadequate to account for all the petroleum that might be generated from a source bed. Presumably some of the remaining 99 per cent of the organic matter must contribute to the formation of oil.

Organic sulphur compounds.—Organic sulphur compounds occur in about 30 parts per 100,000, except in sample 418, in which pigments and sulphur compounds together amount to only 8 parts per 100,000. Nothing is known of the nature of these sulphur compounds, as it was impracticable to analyze them further. The quantity of organic sulphur compounds present in the sediment is probably greater than that indicated by the analysis, as some probably would not have been dissolved by carbon tetrachloride.

Pigments.—Pigments are present in all samples in about 5 parts per 100,000. Practically no attempt was made to identify the pigments, but in sample 454, from Florida Bay, they seem to be of a carotinoid nature and in the sample from the algal lake they are three-fourths green pigment and one-fourth brown pigment.

Sterols.—Small amounts of sterols were present in all the samples. The chief significance of these compounds is that they give clues as to the animal or vegetable origin of the organic matter. Phytosterol is a normal component of plants, and cholesterol, of animals. Samples 454 from Florida Bay and 119 from the algal lake contain phytosterol, which indicates the presence of organic matter of vegetable origin. Sample 418, from the Channel Islands of California, contains cholesterol, which points to an animal origin of some of the organic matter. This is probably due to *Foraminifera*.

Sulphur.—Free sulphur[1] is a common constituent of all the samples. It occurs in quantities ranging from 22 to 104 parts per 100,000. As it

[1]P. D. Trask and C. C. Wu, "Free Sulphur in Recent Sediments," *Bull. Geol. Soc. Amer.*, Vol. 41 (1930), pp. 89-90.

was recovered directly from the CCl_4 extract, it is an original constituent of the deposits, but its initial concentration in the sediments may be lower than the figures given, as some sulphur may have been formed by oxidation of sulphur compounds during drying. Microscopical examination of some of the samples from Florida Bay demonstrated the presence of minute grains of a yellow, waxy substance having a refractive index much higher than 1.74. These grains probably are sulphur.

As the four sediments analyzed represent widely different types of deposits, it is probable that free sulphur is a common minor constituent of many sediments.

3

Reprinted from *Jour. Chem. Education* **37**:554–561 (1960)

Frederick D. Rossini[1]
Carnegie Institute of Technology
Pittsburgh 13, Pennsylvania

Hydrocarbons in Petroleum

Petroleum, one of the most important natural resources in our world today, is a very interesting and intriguing chemical substance. Although it contains small amounts of compounds of sulfur, of nitrogen, of oxygen, and of metallic constituents, petroleum is largely hydrocarbon material, with molecules ranging in size from 1 to 50 or more carbon atoms and comprising a diversity of molecular types. Because petroleum contains many hydrocarbon compounds which are of themselves important chemicals of commerce, and because chemicals derived from petroleum are increasing greatly in number and importance, petroleum has come to be called nature's storehouse of chemicals.

Going back into history, we find that early man long wondered about the thick dark liquid that oozed from the earth or floated on natural waters. He learned many centuries ago that this strange substance would burn and give out great heat and light. Petroleum, or some portion of it, as bitumen or asphalt, is mentioned in the Bible, and records show that asphalt was used by man as early as 3800 B.C. (*1*, *2*). Petroleum was found as an oily liquid exuding from the surface of the earth in tar springs or on shores of lakes or rising from springs beneath the beds of rivers in many places on earth, as Rumania, Iran, Italy, Trinidad, Cuba, India, and later the United States. Surface wells were dug to increase the quantity available.

But it remained for the first drilled petroleum well to start the great flow of crude oil from the earth—the famous Drake well, located near Titusville, Pennsylvania, completed in 1859 to a depth of $69^1/_2$ ft, with a production of about 25 barrels per day. Since then, the annual production of petroleum from the continental United States has grown to truly astonishing amounts, as follows, in billions of barrels per year: 0.000002 in 1860; 0.026 in 1880: 0.064 in 1900; 0.44 in 1920; 1.3 in 1940; nearly 2.0 in 1950; and nearly 3.5 in 1960. The "proved" reserves of petroleum in the continental United States have also increased greatly, as follows in billions of barrels: 3 in 1900; 7 in 1920; 19 in 1940; and 35 in 1960. The ratio of the proved reserves to the annual production in any given year has decreased somewhat over the years: 21/1 in 1910; 16/1 in 1920; 15/1 in 1930; 15/1 in 1940; 13/1 in 1950; and about 10/1 in 1960.

Early Work

In 1906, Coates (*3*), one of the early investigators of the composition of petroleum, stated: "It is exceedingly difficult to get the hydrocarbons (from petroleum) in a state of purity. Indeed, I do not think it can be done by distillation." In 1924, Anderson and Erskine (*4*) published the results of their work on the composition of "natural" gasoline. In 1926, Brown and Carr (*5*) published the results of the first investigation in which appreciable quantities of "pure" hydrocarbons were separated from petroleum by systematic distillation. From 1932 to 1938, notable contributions on the composition of petroleum were published by Fenske and co-workers (*6–10*). The state of the existing knowledge in 1928 was well summarized by Gruse (*11*).

A summary of other work on the composition of petroleum, including hydrocarbon constituents, sulfur compounds, nitrogen compounds, oxygen compounds, and metallic constituents is given by Rossini, Mair, and Streiff (*13*), Lochte (*14*), Gruse and Stevens (*15*), and Smith, Dunning, Rall, and Ball (*16*).

[1] Present address, University of Notre Dame, Notre Dame, Indiana.

The cover

The whole story of the API distillation laboratory is better revealed by this photograph. The 14 distilling assemblies, each with a separating power of 200 theoretical plates, extend through two floors. The laboratory operates 24 hours per day, seven days per week.

The API Research Project 6

In 1927, the American Petroleum Institute Research Project 6 was established at the National Bureau of Standards under the direction of Edward W. Washburn to investigate the hydrocarbon constituents of petroleum. Washburn continued to direct the work until his untimely death early in 1934. The writer then became Director of the Project and has continued actively in it for 26 years. For the past ten years, Beveridge J. Mair has been Associate Director, becoming Director in July, 1960.

An interesting historical account of the API Research Project 6 has been prepared by Hill (*12*). A complete list of the many professional investigators who have been associated with the Project from 1927 to 1952 is given in reference *13*. The purpose of the present report is to give a self-consistent story of hydrocarbons in petroleum as developed by the approximately 500 man-years of work put on this problem by the API Research Project 6.

To get at the heart of the composition of petroleum, and to lay open the components in a way that accurately discloses their identity, the API Research Project 6 has had one representative petroleum continually under exhaustive fractionation since 1928. The work of the Project has established that substantially all of the hydrocarbons in petroleum are composed of only three kinds of component parts, namely, paraffin groups, cycloparaffin (or naphthalene) groups, and aromatic groups. Examples of these three classes of component parts are shown in Figure 1.

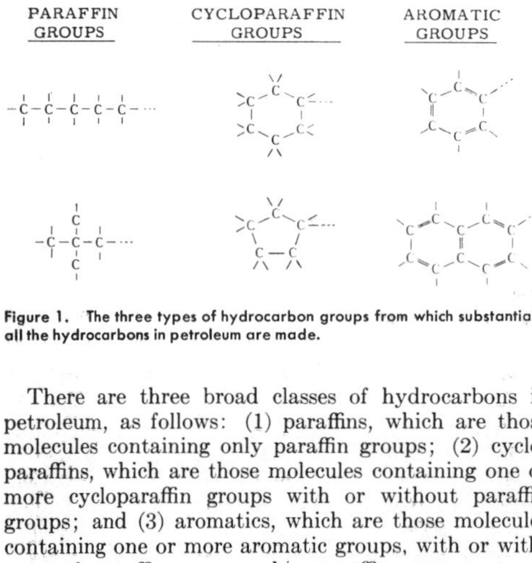

Figure 1. The three types of hydrocarbon groups from which substantially all the hydrocarbons in petroleum are made.

There are three broad classes of hydrocarbons in petroleum, as follows: (1) paraffins, which are those molecules containing only paraffin groups; (2) cycloparaffins, which are those molecules containing one or more cycloparaffin groups with or without paraffin groups; and (3) aromatics, which are those molecules containing one or more aromatic groups, with or without cycloparaffin groups and/or paraffin groups.

The precise manner in which the above groups are combined to form the individual compounds can be originally determined with accuracy only by the actual isolation of the individual compounds. Fortunately, this exhaustive fractionation need be done only once. When this basic information becomes known, short-cut methods can be used to make rapid analyses of the compounds previously established to be components of petroleum.

One of the important aims of the work has been to increase our knowledge of the composition of different petroleums the world over. Some years ago, the Project completed an extended investigation on the composition of the gasoline fraction of petroleum from seven fields in the continental United States, each representing major production. The results of this investigation, supported by related findings on petroleum from other lands, provide a framework which makes it possible to translate to any new petroleum the basic information and data obtained on the one representative petroleum of the API Research Project 6.

In addition to the foregoing information, one of the important products of the work has been the development of fractionating processes of high separating power.

General Plan of Operation

The general plan of operation on the one representative petroleum has been substantially as follows: (1) one representative petroleum, more or less "intermediate" in character, was selected for examination, from the Brett No. 6 well in the Ponca City field in Oklahoma; (2) a sufficient quantity (approximately 500 gal.) of this representative petroleum was taken as starting material, so that after appropriate fractionation the principal components would be expected to be present in an adequate amount to permit identification; (3) the fractionation of the material was (and is) carried out in such a way as to neither destroy constituents already present nor to produce new ones previously absent; (4) the entire petroleum is separated by distillation (at pressures low enough to keep the distilling temperature below 200°C) into convenient broad fractions, as the "gas," "gasoline," "kerosine," "light gas oil," "heavy gas oil and light lubricating distillate," and "lubricant" fractions; (5) the "gas" fraction was separated by distillation alone; (6) each of the remaining broad fractions was (or is being) separated, by the use of the proper combination of fractionating processes, into classes and subclasses of hydrocarbons including, as appropriate, normal paraffins, branched paraffins, mono-cycloparaffins, bicycloparaffins, higher cycloparaffins, mononuclear aromatics, dinuclear aromatics, and higher nuclear aromatics; (7) each subclass of hydrocarbons is separated by distillation with high separating power according to boiling point or molecular size to produce "homogeneous" fractions containing molecules of substantially the same size and class; (8) the homogeneous fractions are further fractionated in so far as possible to yield the principal individual compounds; (9) each of the homogeneous fractions which remains unresolved with respect to individual compounds is examined as to physical and spectrographic properties in order to derive as much information as possible concerning the composition of the unresolved homogeneous fraction.

We have learned that the more thoroughly we examine the lower boiling broad fractions, as the gas, gasoline, and kerosine fractions, the more we can learn from given observations on the higher boiling fractions.

Fractionating Processes Used

The method which has been developed by the Project for separating petroleum into its principal constituents may be described as the systematic interlocking of the basic processes of fractionation and their variations. The order in which the various fractionating processes are used to separate a given fraction of petroleum into its main constituents is determined by the effectiveness of each given process in furthering the separation already attained by the preceding process. The fractionating processes used by the Project have included five forms of distillation, two forms of extraction, four forms of adsorption, two forms of crystallization, and two forms of separation with solid molecular or clathrate compounds:

Distillation (liquid and gas phases): (*a*) regular, at one pressure (30, 55, 214, or 760 mm Hg); (*b*) alternately at different pressures (30, 55, 214, and 760 mm Hg); (*c*) at very low pressures (0.01 to 0.1 mm Hg); (*d*) azeotropically, with regular polar organic compounds; (*e*) azeotropically, with fluoro-organic compounds.

Extraction (two liquid phases): (*a*) with one solvent phase; (*b*) with two solvent phases.

Adsorption (liquid or gas phase and solid-adsorbed phase): (*a*) regular; (*b*) with one or more added components.

Crystallization (liquid and solid phases): (*a*) regular; (*b*) with a selected solvent.

Formation of molecular or clathrate compounds: (*a*) in batch process; (*b*) in column with "reflux."

For details of the fractionating processes used in this work, the reader may consult references *13* and *17–20*.

Summary of the Present Status of the Work

Using the fractionating processes mentioned above, and operating systematically and exhaustively according to plan, the Project has isolated 175 hydrocarbon compounds from its one representative petroleum as of June 30, 1960. A description of the source and character of this representative petroleum is given in reference *13*. Additional details may be found in reference *22*.

Table 1 gives a summary of the amounts of the broad fractions constituted by the hydrocarbons isolated from this representative petroleum. Table 2 gives the distribution, by class and broad fraction, of the hydrocarbons isolated. Table 3 gives the distribution (by class and number of carbon atoms per molecule) of the hydrocarbons isolated, except for the normal paraffins, of which all 32 members (from methane, CH_4, to normal tritriacontane, $C_{33}H_{68}$, inclusive) have been isolated.

Table 4 lists the 175 hydrocarbons isolated, according to class, and gives the normal boiling point and the amount of the given compound in this petroleum.

Table 1. Summary of the Amounts of the Broad Fractions Constituted by the Hydrocarbons Isolated from the Representative Petroleum

Broad fraction	Gas	Gasoline	Kerosine	Light gas oil	Heavy gas oil and light lubricating distillate	Lubricant fraction	Residue	Total
Boiling range at 1 atm. (°C)	<40°	40°–180°	180°–230°	230°–305°	305°–405°	405°–515°	...	...
Range of normal paraffins	C_1–C_5	C_6–C_{10}	C_{11} and C_{12}	C_{13}–C_{17}	C_{18}–C_{25}	C_{26}–C_{38}	...	...
Estimated percentage of the original petroleum constituted by the given fraction	4	33.2	12.7	18.6	14.5	10.0	7	100
Number of compounds isolated	7	101	37	12	10	8	...	175
Estimated percentage of the fraction accounted for by the hydrocarbons isolated	100	82.3	38.4	30.2	20.5	9.9	...	...
Estimated percentage of the original petroleum accounted for by the hydrocarbons isolated	4.0	27.3_3	4.8_4	5.6_1	2.9_7	0.9_9	...	45.7_4

Table 2. Distribution, by Class and Broad Fraction, of the Hydrocarbons Isolated from the Representative Petroleum

Broad fraction	Gas	Gasoline	Kerosine	Light gas oil	Heavy gas oil and light lubricating distillate	Lubricant fraction	Total
Boiling range at 1 atm. (°C)	<40°	40°–180°	180°–230°	230°–305°	305°–405°	405°–515°	...
Range of normal paraffins	C_1–C_5	C_6–C_{10}	C_{11}–C_{12}	C_{13}–C_{17}	C_{18}–C_{25}	C_{26}–C_{38}	...
		Number of Compounds					
Classes of hydrocarbons							
Normal paraffins	5	5	2	5	8	8	33
Branched paraffins	2	35	...	...	...	...	37
Alkyl cyclopentanes	...	22	...	...	...	...	22
Alkyl cyclohexanes	...	13	1	...	...	...	14
Alkyl cycloheptanes	...	1	...	...	...	...	1
Bicycloparaffins	...	4	6	...	...	...	10
Tricycloparaffins	...	...	1	...	...	...	1
Alkyl benzenes	...	20	20	...	...	...	40
Aromatic cycloparaffins	...	1	6	1	...	...	8
Dinuclear aromatics	...	...	1	6	1	...	8
Trinuclear aromatics	...	...	...	...	1	...	1
Total	7	101	37	12	10	8	175

Fractions by Distillation

Gas. The gas fraction, taken as the material normally boiling up to 40°C, has been fully accounted for in terms of all the possible normal paraffins and branched paraffins, C_1 to C_5, except for 2,2-dimethyl propane which is believed to be present only in extremely small or trace amounts. These compounds are listed in Table 4.

Gasoline. The hydrocarbons of this fraction, taken as the material normally boiling from 40° to 180°C consist of normal (C_6 to C_{10}) and branched paraffins, mono- and bicycloparaffins, and mononuclear aromatics. In Table 4 are listed the compounds isolated, together with those compounds normally boiling up to 132°C which are believed to be present in extremely small amounts.

All of the five possible normal paraffins, C_6 to C_{10}, have been accounted for. All of the possible branched paraffins in the range 40° to 132°C have been isolated, except for six compounds which are believed to be present only in extremely small amounts. Also isolated were eight branched paraffins in the range 132° to 180°C. The monocycloparaffins in the gasoline fraction consist of alkyl cyclopentanes, alkyl cyclohexanes, and very small amounts of alkyl cycloheptanes. All of the possible alkyl cyclopentanes normally boiling up to 132°C have been isolated, except for three compounds which are believed to be present only in extremely small amounts. All of the possible alkyl cyclohexanes normally boiling up to 132°C, including one C_6, one C_7, and eight C_8 compounds, have been isolated. In addition, three of the C_9 trimethylcyclohexanes have been isolated. Among the alkyl cycloheptanes, only cycloheptane itself, present in very small amounts, has been isolated to date. So far four bicycloparaffins have been isolated from the gasoline fraction. All of the 21 possible mononuclear aromatics have been isolated from this gasoline fraction. Of these, 20 are alkyl benzenes and one is an aromatic-cycloparaffin indan.

Kerosine. The hydrocarbons of this fraction, taken as the material normally boiling from 180° to 230°C, consist of normal (C_{11} and C_{12}) and branched paraffins, mono-, bi-, and higher cycloparaffins, and mono- and dinuclear aromatics. From this material have been isolated two normal paraffins, one alkyl cyclohexane, six bicycloparaffins, one tricycloparaffin (adamantane), 20 alkyl benzenes, six aromatic-cycloparaffins, and one dinuclear aromatic, as listed in Table 4.

Light Gas Oil. The hydrocarbons of this fraction, taken as the material normally boiling from 230° to 305° C, consist of normal (C_{13} to C_{17}) and branched paraffins, mono-, bi-, and higher cycloparaffins, and mono- and dinuclear aromatics. From this have been isolated five normal paraffins, one aromatic-cycloparaffin, and six dinuclear aromatics, listed in Table 4.

Heavy Gas Oil and Light Lubricating Distillate. The hydrocarbons of this fraction, taken as the material normally boiling from 305° to 405°C, consist of normal (C_{18} to C_{25}) and branched paraffins, mono-, bi-, and higher cycloparaffins, and mono-, di-, and higher nuclear aromatics. From this material, ten hydrocarbons have been isolated, including eight normal paraffins, one dinuclear aromatic, and one trinuclear aromatic, as listed in Table 4.

Lubricant. The hydrocarbons of this fraction, taken as the material normally boiling from 405° to 515°C, consist of normal (C_{26} to C_{38}) and branched paraffins, mono-, bi-, and higher cycloparaffins, and mono-, di-, and higher nuclear aromatics. From this material, only seven normal paraffins have been isolated, as given in Table 4.

Types of Homologs

Normal Paraffins. As indicated earlier, all of the normal paraffins, C_1 to C_{33}, have been isolated. The amount of each in the original petroleum, in per cent by volume, increases from 1.8 for C_6 to 2.3 for C_7 and then decreases gradually to an estimated value of 0.09 for C_{33}.

Branched Paraffins. A total of 37 branched paraffins have been isolated, all from the gas and gasoline fractions. Enough has been learned about this class to indicate that all possible isomers may be present, although many only in trace amounts. The most abundant isomer in any group is usually the 2-methyl isomer, with the 3-methyl and 4-methyl isomers following. In any given group of isomers, the bulk of the material will consist of the singly branched isomers with the amounts of the doubly-branched ones dropping sharply and of the triply-branched ones even more so. For example, in this representative

Table 3. Distribution, by Class and Number of Carbon Atoms per Molecule, of the Hydrocarbons Isolated

Number of carbon atoms per molecule	4	5	6	7	8	9	10	11	12	13	14	Total
Branched paraffins	1	1	4	6	15	7	3	...	...	...	...	37
Alkyl cyclopentanes	...	1	1	5	13	2	...	...	...	...	...	22
Alkyl cyclohexanes	...	...	1	1	8	3	1	...	...	...	...	14
Alkyl cycloheptanes	...	...	...	1	...	...	...	...	...	...	...	1
Bicycloparaffins	...	...	...	...	3	1	5	1	...	...	...	10
Tricycloparaffins	...	...	...	...	...	...	1	...	...	...	...	1
Alkyl benzenes	...	...	1	1	4	8	22	4	...	...	...	40
Aromatic cycloparaffins	...	...	...	...	...	1	4	3	...	...	...	8
Dinuclear aromatics	...	...	...	...	...	...	1	2	3	1	1	8
Trinuclear aromatics	...	...	...	...	...	...	...	...	...	...	1	1
Total	1	2	7	14	43	21	37	10	3	1	2	142
Normal paraffins, C_1 to C_{33}												33
Grand total												175

Table 4. Hydrocarbons in the Representative Petroleum of the API Research Project 6

Formula	Compound	Normal boiling point (°C)	Amount[a] (percentage by volume)
	Normal paraffins		
CH_4	Methane	−161.49	(0.1)
C_2H_6	Ethane	− 88.63	(0.2)
C_3H_8	Propane	− 42.07	(0.4)
C_4H_{10}	*n*-Butane	− 0.50	(0.9)
C_5H_{12}	*n*-Pentane	36.07	(1.6)
C_6H_{14}	*n*-Hexane	68.74	1.8
C_7H_{16}	*n*-Heptane	98.43	2.3
C_8H_{18}	*n*-Octane	125.66	1.9
C_9H_{20}	*n*-Nonane	150.80	1.8
$C_{10}H_{22}$	*n*-Decane	174.12	1.8
$C_{11}H_{24}$	*n*-Undecane	195.89	1.6
$C_{12}H_{26}$	*n*-Dodecane	216.28	1.4
$C_{13}H_{28}$	*n*-Tridecane	235.43	1.2
$C_{14}H_{30}$	*n*-Tetradecane	253.52	1.0
$C_{15}H_{32}$	*n*-Pentadecane	270.61	0.8
$C_{16}H_{34}$	*n*-Hexadecane	286.79	0.7
$C_{17}H_{36}$	*n*-Heptadecane	302.15	0.6
$C_{18}H_{38}$	*n*-Octadecane	316.71	0.50
$C_{19}H_{40}$	*n*-Nonadecane	330.6	0.43
$C_{20}H_{42}$	*n*-Eicosane	343.8	0.37
$C_{21}H_{44}$	*n*-Heneicosane	356.5	0.32
$C_{22}H_{46}$	*n*-Docosane	368.6	0.28
$C_{23}H_{48}$	*n*-Tricosane	380.2	0.25
$C_{24}H_{50}$	*n*-Tetracosane	391.3	0.22
$C_{25}H_{52}$	*n*-Pentacosane	401.9	0.20
$C_{26}H_{54}$	*n*-Hexacosane	412.2	(0.18)
$C_{27}H_{56}$	*n*-Heptacosane	422.1	(0.16)
$C_{28}H_{58}$	*n*-Octacosane	431.6	(0.14)
$C_{29}H_{60}$	*n*-Nonacosane	440.8	(0.13)
$C_{30}H_{62}$	*n*-Triacontane	449.7	(0.12)
$C_{31}H_{64}$	*n*-Hentriacontane	458.0	(0.11)
$C_{32}H_{66}$	*n*-Dotriacontane	467.0	(0.10)
$C_{33}H_{68}$	*n*-Tritriacontane	475.0	(0.09)
	Branched paraffins		
C_4H_{10}	2-Methylpropane (isobutane)	− 11.73	(0.3)
C_5H_{12}	2,2-Dimethylpropane (neopentane)	9.50	*b*
"	2-Methylbutane (isopentane)	27.85	(0.5)
C_6H_{14}	2,2-Dimethylbutane	49.74	0.04
"	2,3-Dimethylbutane	57.99	0.08
"	2-Methylpentane	60.27	0.37
"	3-Methylpentane	63.28	0.35
C_7H_{16}	2,2-Dimethylpentane	79.20	0.02
"	2,4-Dimethylpentane	80.50	0.08
"	2,2,3-Trimethylbutane	80.87	*b*
"	3,3-Dimethylpentane	86.07	*b*
"	2,3-Dimethylpentane	89.78	0.15
"	2-Methylhexane	90.05	0.73
"	3-Methylhexane	91.85	0.51
"	3-Ethylpentane	93.48	0.06
C_8H_{18}	2,2,4-Trimethylpentane	99.24	*b*
"	2,2,3,3-Tetramethylbutane	106.47	*b*
"	2,2-Dimethylhexane	106.84	0.01
"	2,5-Dimethylhexane	109.10	0.06
"	2,4-Dimethylhexane	109.43	0.06
"	2,2,3-Trimethylpentane	109.84	0.004
"	3,3-Dimethylhexane	111.97	0.03
"	2,3,4-Trimethylpentane	113.47	0.005
"	2,3,3-Trimethylpentane	114.76	0.006
"	2,3-Dimethylhexane	115.61	0.07
"	2-Methyl-3-ethylpentane	115.65	0.06
"	2-Methylheptane	117.65	0.90
"	4-Methylheptane	117.71	0.20
"	3,4-Dimethylhexane	117.72	0.13
"	3-Methyl-3-ethylpentane	118.26	0.02
"	3-Ethylhexane	118.53	0.09
"	3-Methylheptane	118.92	0.30
C_9H_{20}	2,2,4,4-Tetramethylpentane	122.28	*b*
"	2,2,5-Trimethylhexane	124.08	0.002
"	2,2,4-Trimethylhexane	126.54	*b*
"	2,3,5-Trimethylhexane	131.34	0.03
"	2,6-Dimethylheptane	135.21	0.05
"	2,3-Dimethylheptane	140.5	0.05
"	4-Methyloctane	142.48	0.1
"	2-Methyloctane	143.26	0.4
"	3-Methyloctane	144.18	0.1
$C_{10}H_{22}$	5-Methylnonane	165.1	*b*
"	4-Methylnonane	165.7	0.1
"	2-Methylnonane	166.8	0.3
"	3-Methylnonane	167.8	0.1
	Alkyl cyclopentanes		
C_5H_{10}	Cyclopentane	49.26	0.05
C_6H_{12}	Methylcyclopentane	71.81	0.87
C_7H_{14}	1,1-Dimethylcyclopentane	87.85	0.16
"	1,cis-3-Dimethylcyclopentane	90.77	0.87
"	1,trans-3-Dimethylcyclopentane	91.72	0.21
"	1,trans-2-Dimethylcyclopentane	91.87	0.48
"	1,cis-2-Dimethylcyclopentane	99.53	*b*
"	Ethylcyclopentane	103.47	0.16
C_8H_{16}	1,1,3-Trimethylcyclopentane	104.89	0.30
"	1,trans-2,cis-4-Trimethylcyclopentane	109.29	0.22
"	1,trans-2,cis-3-Trimethylcyclopentane	110.2	0.26
"	1,1,2-Trimethylcyclopentane	113.73	0.06
"	1,cis-2,trans-4-Trimethylcyclopentane	116.73	0.01
"	1,cis-2,trans-3-Trimethylcyclopentane	117.5	0.07
"	1,cis-2,cis-4-Trimethylcyclopentane	118.	*b*
"	1-Methyl-trans-3-ethylcyclopentane	120.8 }	0.12
"	1-Methyl-cis-3-ethylcyclopentane	121.4 }	
C_8H_{16}	1-Methyl-trans-2-ethylcyclopentane	121.2	0.14
"	1-Methyl-1-ethylcyclopentane	121.52	0.03
"	1-cis-2,cis-3-Trimethylcyclopentane	123.0	*b*
"	Isopropylcyclopentane	126.42	0.01
"	1-Methyl-cis-2-ethylcyclopentane	128.05	0.04
"	*n*-Propylcyclopentane	130.95	0.06
C_9H_{18}	1,1,3,3-Tetramethylcyclopentane	117.96	*b*
"	1,1,cis-3,trans-4-Tetramethylcyclopentane	121.6	0.04
"	1,trans-2,cis-3,trans-4-Tetramethylcyclopentane	127.4	0.11
"	1,1,cis-2,trans-3-Tetramethylcyclopentane	130.	*b*
"	1,1,cis-2,cis-3-Tetramethylcyclopentane	130.	*b*
"	1,1,cis-2,trans-4-Tetramethylcyclopentane	130.	*b*
"	1,cis-2,trans-3,cis-4-Tetramethylcyclopentane	131.	*b*
"	1,trans-2,trans-3,cis-4-Tetramethylcyclopentane	131.	*b*
C_6H_{12}	Cyclohexane	80.74	0.71
C_7H_{14}	Methylcyclohexane	100.93	1.6
C_8H_{16}	1,trans-4-Dimethylcyclohexane	119.35	0.25
"	1,1-Dimethylcyclohexane	119.54	0.06
"	1,cis-3-Dimethylcyclohexane	120.09	0.63
"	1,trans-2-Dimethylcyclohexane	123.42	0.31
"	1,cis-4-Dimethylcyclohexane	124.32	0.09
"	1,trans-3-Dimethylcyclohexane	124.45	0.07
"	1,cis-2-Dimethylcyclohexane	129.73	0.06
"	Ethylcyclohexane	131.78	0.37
C_9H_{18}	1,1,3-Trimethylcyclohexane	136.63	0.2
"	1,trans-2,trans-4-Trimethylcyclohexane	141.22	0.2
"	1,trans-2,cis-4-Trimethylcyclohexane	145.6	(0.1)
$C_{10}H_{20}$	*n*-Butylcyclohexane	180.95	(0.22)
	Cycloheptanes		
C_7H_{14}	Cycloheptane	118.79	0.01
	Bicycloparaffins		
C_8H_{14}	Methylbicyclo-[2.2.1]-heptane	124.5	0.001
"	cis-Bicyclo-[3.3.0]-octane	136.5	0.06
"	Bicyclo-[3.2.1]-octane	(138.)	0.008
C_9H_{16}	Bicycloparaffin[c]	146.7	(0.06)
$C_{10}H_{18}$	Bicycloparaffin[c]	180.	(0.01)
"	Bicycloparaffin[c]	183.4	(0.01)
"	Bicycloparaffin[c]	183.7	(0.02)
"	trans-Bicyclo-[4.4.0]-decane (trans-Decahydronaphthalene)	187.25	(0.2)
"	cis-Bicyclo-[4.4.0]-decane (cis-Decahydronaphthalene)	195.69	(0.01)
$C_{11}H_{20}$	Bicycloparaffin[c]	202.5	(0.12)
	Tricycloparaffins		
$C_{10}H_{16}$	Tricyclo-[3.3.1.1^{3,7}]-decane (Adamantane)	190.	0.0004
	Alkyl benzenes		
C_6H_6	Benzene	80.10	0.15
C_7H_8	Methylbenzene (toluene)	110.62	0.51
C_8H_{10}	Ethylbenzene	136.19	0.19
"	1,4-Dimethylbenzene (*p*-Xylene)	138.35	0.10
"	1,3-Dimethylbenzene (*m*-Xylene)	139.10	0.51
"	1,2-Dimethylbenzene (*o*-Xylene)	144.41	0.27
C_9H_{12}	Isopropylbenzene	152.39	0.07
"	*n*-Propylbenzene	159.22	0.09
"	1-Methyl-3-ethylbenzene	161.30	0.17
"	1-Methyl-4-ethylbenzene	161.99	0.06
"	1,3,5-Trimethylbenzene	164.72	0.12
"	1-Methyl-2-ethylbenzene	165.15	0.09
"	1,2,4-Trimethylbenzene	169.35	0.51
"	1,2,3-Trimethylbenzene	176.08	0.19
$C_{10}H_{14}$	tert-Butylbenzene	169.12	0.01
"	Isobutylbenzene	172.76	0.008
"	sec-Butylbenzene	173.30	0.017
"	1-Methyl-3-isopropylbenzene	175.14	0.08
"	1-Methyl-4-isopropylbenzene	177.10	0.04
"	1-Methyl-2-isopropylbenzene	178.15	0.009
"	1,3-Diethylbenzene	181.10	0.029
"	1-Methyl-3-propylbenzene	181.80	0.059
"	*n*-Butylbenzene	183.27	0.026
"	1-Methyl-4-propylbenzene	183.30	0.027
"	1,2-Diethylbenzene	183.42	0.005
"	1,3-Dimethyl-5-ethylbenzene	183.58	0.063
"	1,4-Diethylbenzene	183.78	0.007
"	1-Methyl-2-propylbenzene	184.80	0.046
"	1,4-Dimethyl-2-ethylbenzene	186.83	0.041
"	1,3-Dimethyl-3-ethylbenzene	188.20	0.050
"	1,2-Dimethyl-4-ethylbenzene	189.48	0.057
"	1,3-Dimethyl-2-ethylbenzene	190.01	0.014
"	1,2-Dimethyl-3-ethylbenzene	193.91	0.03
"	1,2,4,5-Tetramethylbenzene	196.80	0.052
"	1,2,3,5-Tetramethylbenzene	198.00	0.11
"	1,2,3,4-Tetramethylbenzene	205.04	0.2
$C_{11}H_{16}$	1-Methyl-3-*n*-butylbenzene	204.1	0.06
"	1,3-Dimethyl-4-*n*-propylbenzene	206.6	0.03
"	1,2-Dimethyl-4-*n*-propylbenzene	208.5	0.03
"	Trimethylethylbenzene	212.3	0.04

[a] The figures in parentheses are less certain than the others. Where no value is given for a compound, it is implied that the compound listed is believed to be present in very small amount, less than was detectable at the time the given material was investigated.
[b] Believed to be present in extremely small amount.
[c] Not yet identified.

Table 4 (continued)

Formula	Compound	Normal boiling point (°C)	Amount[a] (percentage by volume)
	Aromatic-cycloparaffins		
C_9H_{10}	Indan	177.8	0.003
$C_{10}H_{12}$	1-Methylindan	190.6	0.022
"	2-Methylindan	191.4	
"	4-Methylindan	205.5	(0.03)
"	1,2,3,4-Tetrahydronaphthalene	207.57	0.03
$C_{11}H_{14}$	2-Methyl-1,2,3,4-tetrahydronaphthalene	220.7	0.04
"	6-Methyl-1,2,3,4-tetrahydronaphthalene	229.3	0.09
"	5-Methyl-1,2,3,4-tetrahydronaphthalene	234.35	0.08
	Dinuclear aromatics		
$C_{10}H_8$	Naphthalene	217.96	0.06
$C_{11}H_{10}$	2-Methylnaphthalene	241.05	0.2
"	1-Methylnaphthalene	244.64	0.1
$C_{12}H_{12}$	Biphenyl	255.2	(0.03)
"	2,6-Dimethylnaphthalene	262.	(0.35)
"	Dimethylnaphthalene[c]	268.	(0.35)
$C_{13}H_{14}$	Trimethylnaphthalene[c]	288.	(0.20)
$C_{14}H_{16}$	Alkyl naphthalene[c]	312.	(0.20)
	Trinuclear aromatics		
$C_{14}H_{10}$	Phenanthrene	340.	(0.20)

[c] Not yet identified.

petroleum, the amounts of each of the C_9 branched isomers, are, in per cent, of the following orders: singly-branched, 0.1; doubly-branched, 0.01; triply-branched, 0.001; quadruply-branched, trace. Knowing (*a*) the total amount of the branched paraffins of a given carbon number, through analysis by type of the broad fractions, (*b*) the number of isomers which are singly-branched, doubly-branched, etc., and (*c*) the relative amounts of the compounds of different branching, it is thus possible to estimate the order of magnitude of the amount of any given branched paraffin in petroleum.

Monocycloparaffins. Most of the monocycloparaffins in petroleum are alkyl cyclopentanes and alkyl cyclohexanes, with only a small amount of alkyl cycloheptanes. In general, in a given group of isomers of the alkyl cyclopentanes, the most abundant compounds are those with the largest number of substituents, each on a different carbon atom, with the substituents on adjacent carbon atoms being "trans" to one another. Similarly, in a given group of isomers of the alkyl cyclohexanes, the most abundant compounds are those with the largest number of substituents, each on a different carbon atom, with the substituents being "equatorial" rather than "axial" to minimize steric hindrance in the molecules. These are also the isomers which are thermodynamically more stable. One interesting structure, related to other natural products occurring among the monocycloparaffins, is found in 1,1,3-trimethylcyclopentane and in 1,1,3-trimethylcyclohexane, both relatively abundant in petroleum.

Bicycloparaffins. So far, ten bicycloparaffins have been isolated, four from the gasoline fraction and six from the kerosine fraction. In Figure 2 are shown the structures of the nuclei of bicycloparaffin hydrocarbons believed to be the ones occurring in petroleum in significant amount. As may be noted, these are structures which have five or six carbon atoms in each ring, a carbon common to two rings being counted twice. The numbering system is the approved one, with the numbers representing the number of carbon atoms in each bridge.

Mononuclear Aromatics. As may be seen from Table 4, all of the 32 possible aromatic hydrocarbons normally boiling up to 190°C have been isolated from this petroleum. Sixteen additional mononuclear aromatics have also been isolated. In any given group of isomers, the alkyl benzenes which are most abundant are those which are thermodynamically more stable, as the 1,3-dialkyl or 1,2,4-trialkyl benzenes, with less preference for the 1,2-dialkyl benzenes. Also, in a given group, the compounds with more substituents are more abundant, as 1,2,4-trimethylbenzene is more abundant than 1-methyl-3-ethylbenzene. With increase in the number of carbon atoms, more of the mononuclear aromatics will be of the aromatic-cycloparaffin subclass, in which, generally, the benzene ring is joined through two common carbon atoms to a cyclopentane ring or a cyclohexane ring or multi-ring combinations of these.

Dinuclear aromatics. A total of eight dinuclear aromatics have been isolated, including naphthalene, biphenyl, and six alkyl naphthalenes. The alkyl naphthalenes are much more abundant than the alkyl biphenyls. With increasing number of carbon atoms per molecule, the dinuclear aromatics will include an increasing number of aromatic cycloparaffins, with the naphthalene nucleus attached through two common carbon atoms to a mono- or bi- or higher cycloparaffin group.

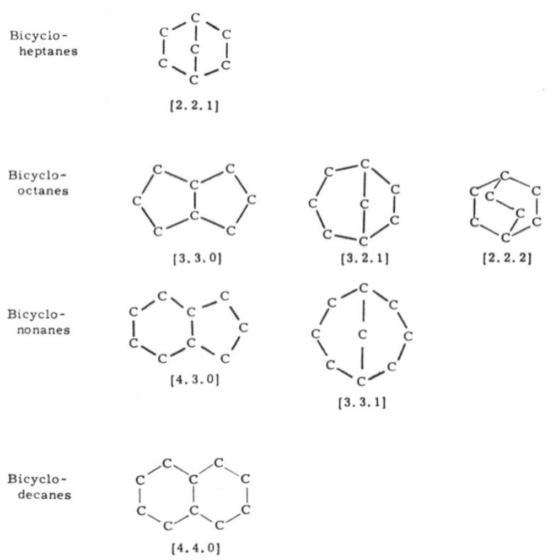

Figure 2. Structures of the nuclei of bicycloparaffin hydrocarbons believed to be the ones occurring in petroleum in significant amounts.

Higher Nuclear Aromatics. Only one compound of this group has so far been isolated, the trinuclear aromatic, phenanthrene. The alkyl phenanthrenes are expected to be more abundant than the alkyl anthracenes. Also, among the higher nuclear aromatics will occur many aromatic-cycloparaffin compounds, with the cycloparaffin portion consisting of several rings, generally condensed together with the aromatic nucleus.

Composition of Petroleums in General

Some years ago, the API Research Project 6 completed a study of the hydrocarbons from different petroleums (*13*, *21*). The Project's one representative petroleum from Oklahoma was intermediate in type. The gasoline fractions from six additional representative petroleums were selected to represent large production in different areas of the United States and to be as different as was known at the time.

The seven representative gasoline fractions were as follows, with the type and source field: A, intermediate, Ponca, Oklahoma; B, intermediate, East Texas; C, high in paraffins, Bradford, Pennsylvania; D, high in normal paraffins, Greendale-Kawkawlin, Michigan; E, high in branched paraffins, Winkler, Texas; F, high in cycloparaffins, Midway, California; G, high in aromatics, Conroe, Texas.

By means of an extended but relatively short procedure involving the fractionating processes of adsorption and distillation, both operated at high separating power, analyses were made of the major paraffin and cycloparaffin components in the 40° to 102°C range of these gasolines and of the alkyl benzenes to 180°C. From this work, the following conclusions were reached:

All petroleums contain substantially the same hydrocarbon compounds.

The principal compounds in the gasoline fraction of each petroleum may be placed in five main classes: normal (straight-chain) paraffins; branched paraffins; alkyl cyclopentanes; alkyl cyclohexanes; alkyl benzenes.

Within each of these five classes, the individual compounds occur in proportions which are usually of the same order of magnitude for different petroleums.

Table 5 gives the relative amounts of the five main classes of hydrocarbons in the 40° to 102°C fraction of the seven petroleums, with the letters, A, B, C, D, E, F, and G representing the same petroleums as above. These figures indicate considerable variation in the relative amounts of the several classes from one petroleum to the other. For example, the ratio of normal paraffins to branched paraffins ranges from 4.8 for D to 0.15 for E, a variation of over 30-fold. Similarly, the ratio of alkyl cyclopentanes to alkyl benzenes ranges from 68 for F to 4.6 for G, a variation of 14-fold.

Table 6 shows how the relative amounts of individual compounds within given classes have nearly the same values in the different gasolines, taking three examples: (1) The relative amounts of benzene, toluene, and the four C_8 alkyl benzenes; (2) the relative amounts of cyclohexane and methylcyclohexane; (3) the relative amounts of the two methylpentanes and the two dimethylbutanes. Additional similar examples may be cited. It is to be noted that the actual amount of the compounds considered varies considerably from one gasoline to another, as indicated by the bottom row of figures for each group of compounds in Table 6.

These results mean that the exhaustive information and data being obtained on the Project's one representative petroleum can be readily translated to other petroleums by making some relatively simple observations on the other petroleums. Although the work on other petroleums has been limited to the gasoline fractions, it is quite clear that the same principles may be applied to compounds of the same class in

Table 5. Relative Amounts of the Main Classes of Hydrocarbons in the 40–102°C Fraction of the Seven Representative Gasolines

Type of gasoline → / Class of hydrocarbons	A	B	C	D	E	F	G
	Relative amount by volume						
Normal paraffins	35.3	24.6	34.1	62.2	9.5	9.9	17.6
Branched paraffins	20.2	27.1	32.0	13.0	61.4	21.4	19.6
Alkyl cyclopentanes	23.1	25.9	13.4	7.8	20.4	40.7	16.6
Alkyl cyclohexanes	20.2	21.9	20.0	15.5	8.2	27.4	42.6
Alkyl benzenes	1.2	0.5	0.5	1.5	0.5	0.6	3.6
Total	100.0	100.0	100.0	100.0	100.0	100.0	100.0

Table 6. Relative Amounts of Individual Compounds within Given Classes in the Seven Different Gasolines

Type of gasoline → / Compounds	A	B	C	D	E	F	G	Range of relative amount	Average relative amount
			Alkyl benzenes, C_6 to C_8						
Relative amounts									
Benzene	0.09	0.04	0.04	0.13	0.10	0.05	0.06	0.04–0.13	0.07
Toluene	0.30	0.29	0.31	0.37	0.20	0.30	0.38	0.20–0.38	0.31
C_8 alkyl benzenes	0.61	0.67	0.65	0.50	0.70	0.65	0.56	0.50–0.70	0.62
Percentage of t e gasoline fractio constituted byn this compound	5.4	5.9	4.9	4.8	1.3	4.3	19.5	...	...
			Alkyl cyclohexanes, C_6 and C_7						
Relative amounts									
Cyclohexane	0.30	0.21	0.24	0.33	0.31	0.29	0.30	0.21–0.33	0.28
Methylcyclohexane	0.70	0.79	0.76	0.67	0.69	0.71	0.70	0.67–0.79	0.72
Percentage of the gasoline fraction constituted by this co npound	6.9	8.4	7.3	6.1	2.1	8.4	13.7	...	...
			Branched paraffins, C_6						
Relative amounts									
Dimethylbutanes	0.14	0.13	0.10	0.14	0.12	0.16	0.16	0.10–0.16	0.14
Methylpentanes	0.86	0.87	0.90	0.86	0.88	0.84	0.84	0.84–0.90	0.86
Percentage of the gasoline fraction constituted by this compound	2.6	4.8	6.0	2.6	6.6	3.4	3.1	...	...

the higher-boiling fractions. For example, the relative amounts found in the Project's representative petroleum for naphthalene, 1-methylnaphthalene, and 2-methylnaphthalene are expected to be of the same magnitude for other petroleums; namely, approximately 1 to 2 to 4. Similarly, the ratio of tert-butylbenzene to 1,2,4-trimethylbenzene (pseudocumene) can be expected to be about 1 to 50 approximately, in all petroleums.

A more complete separation of the petroleum into subclasses would be:

The monocycloparaffins separated into (1) alkyl cyclopentanes and (2) alkyl cyclohexanes.
The mononuclear aromatics separated into (1) alkyl benzenes and (2) aromatic cycloparaffins.
The dinuclear aromatics separated into (1) alkyl naphthalenes and (2) aromatic cycloparaffins.

If this additional separation could be performed, say on the gasoline, kerosine, and light gas oil fractions, one would be led to the following statement: An aliquot sample of all the hydrocarbons of any one subclass from one of these broad fractions, gasoline, kerosine, and light gas oil, will be substantially the same for different petroleums with respect to the magnitude of the individual compounds in the sample. That is to say, for example, a liter of all the branched paraffins from the kerosine fraction of one petroleum will be substantially the same as a liter of all the branched paraffins from the same kerosine fraction of another petroleum. Similarly, a liter of all the alkyl naphthalenes from the light gas oil fraction of one petroleum will be substantially the same as a liter of all the alkyl naphthalenes from the same light gas oil fraction of another petroleum.

Conclusion

In the past 30 years, we have come a long way in our knowledge of what petroleum is chemically. However, we still have facing us several very important scientific questions concerning petroleum, as follows:

How did petroleum get to be *what* it is chemically?

How did petroleum get to be *where* we find it in the earth?

Was petroleum made by nature in one fell swoop, once for all time, or is it being made in the earth continuously?

Because there are numerous investigators working in many different laboratories seeking to uncover these secrets of nature, we are confident that in a few years we will have some answers to these important questions.

Literature Cited

(1) Giddens, P. H., "A Facsimile of a Report on the Rock Oil, or Petroleum, from Venango County, Pennsylvania, by Benjamin Silliman, Jr., in 1855," Meadville, Pennsylvania, **1949**.
(2) Miller, A. E., "The Literature of Basic Research on Hydrocarbons Available from Petroleum," American Chemical Society, Los Angeles, California, March, **1953**.
(3) Coates, C. E., *J. Am. Chem. Soc.*, **28**, 384 (1906).
(4) Anderson, R. P., and Erskine, A. M., *Ind. Eng. Chem.*, **16**, 263 (1924).
(5) Brown, G. G., and Carr, A. R., *Ind. Eng. Chem.*, **18**, 718 (1926).
(6) Fenske, M. R., and Tongberg, C. O., *Ind. Eng. Chem.*, **24**, 814 (1932).
(7) Tongberg, C. O., Quiggly, D., and Fenske, M. R., *Ind. Eng. Chem.*, **28**, 201 (1936).
(8) Tongberg, C. O., Fenske, M. R., and Nickels, J. E., *Ind. Eng. Chem.*, **29**, 70 (1937).
(9) Lawroski, S., Tongberg, C. O., Mazzarola, A. H., and Fenske, M. R., *Ind. Eng. Chem.*, **29**, 674 (1937).
(10) Tongberg, C. O., Fenske, M. R., and Sweeney, W. J., *Ind. Eng. Chem.*, **30**, 166 (1938).
(11) Gruse, W. A., "Petroleum and Its Products," McGraw-Hill Book Co., Inc., New York, **1928**.
(12) Hill, J. B., *Proc. Am. Petroleum Inst.*, *I*, **32**, 183 (1952).
(13) Rossini, F. D., Mair, B. J., and Streiff, A. J., "Hydrocarbons from Petroleum," Reinhold Publishing Corp., New York, **1953**.
(14) Lochte, H. L., *Ind. Eng. Chem.*, **44**, 2597 (1952).
(15) Cruse, W. A., and Stevens, D. R., "The Chemical Technology of Petroleum," McGraw-Hill Book Co., Inc., New York, **1942**.
(16) Smith, H. M., Dunning, H. N., Rall, H. T., and Ball, J. S., *Proc. Am. Petroleum Inst.*, **39**M (1959).
(17) Mair, B. J., Pignocco, A. J., and Rossini, F. D., *Anal. Chem.*, **27**, 190 (1955).
(18) Mair, B. J., *Anal. Chem.*, **28**, 52 (1956).
(19) Mair, B. J., Montjar, M. J., and Rossini, F. D., *Anal. Chem.*, **28**, 56 (1956).
(20) Mair, B. J., and Shamaiengar, M., *Anal. Chem.*, **30**, 276 (1958).
(21) Forziati, A. F., Willingham, C. B., Mair, B. J., and Rossini, F. D., *Proc. Am. Petroleum Inst. (III)*, **24**, 34 (1943); *J. Research Natl. Bur. Standards*, **32**, 11 (1944).
(22) Rossini, F. D., and Mair, B. J., "The Work of the API Research Project 6 on the Composition of Petroleum," Proceedings of the Fifth World Petroleum Congress, New York, N. Y., **1959**.

4

Reprinted from *Science* **116**:437–439 (1952)

The Occurrence of Hydrocarbons in Recent Sediments from the Gulf of Mexico

Paul V. Smith, Jr.
Standard Oil Development Company, Linden, New Jersey

THE ORIGIN OF PETROLEUM has been a widely debated subject for many years. It has been rather generally accepted that organic matter deposited in marine or brackish water sediments served as the source material for eventual petroleum formation. Conceptions of the process from this point have, however, been many and varied. The most prevalent viewpoint, perhaps, has held that burial of the original complex organic matter by several thousands of feet of overburden and a lapse of several millions of years were necessary conditions for its conversion into hydrocarbons (*1*).

More recently a school of thought has developed which believes that oil formation may begin soon after deposition of the organic matter in the sediments (*2*), but the presence of liquid hydrocarbons in Recent marine sediments has never been definitely demonstrated (*3*). Probably the most extensive study ever carried out on Recent sediments was that of Parker D. Trask and his associates under A.P.I. Project 4. They collected samples from almost every kind of environment and with a worldwide distribution. In summarizing his monumental work in this field, Trask has written,

> No liquid hydrocarbons were found, and if any were present they must have occurred in quantities less than . . . 1 part per 100,000 in the marine samples. The sediments investigated represent the richest deposits in the writer's collection of 2,000 samples from all parts of the world; consequently, the absence of petroleum in them indicates that in all probability petroleum does not form in sediments at the time of deposition (*4*).

Several years ago a project dealing with the nature of the organic matter in recent marine sediments was initiated in the Standard Oil Co. (New Jersey). It was aimed at learning something about the types of organic compounds present in the sediments and, if possible, at determining whether a noticeable change in the nature of the organic matter could be observed with increasing age or depth of burial. One of the surprising results of this study has been the discovery of liquid hydrocarbons in Recent sediments from the Gulf of Mexico. Aliphatic and aromatic hydrocarbons have been identified in ten Recent marine sediment samples from four different locations in Texas and Louisiana, in which specimens representing near-shore and off-shore locations (Table 1), and oxidizing and reducing environments, were included. Depths of these sediments ranged from a few inches up to a hundred feet below the water floor, and they were of a mucky or silty nature.

The general procedure involved the following steps: The sediment samples were frozen in dry ice at the collecting site and were maintained in this condition during transit to the laboratories. The material was then dried at 50° C in an oven evacuated to 2 mm Hg and ground in a micropulverizer to an average particle diameter of 2–3 μ. Extraction at refluxing temperatures by a mixture comprising 70 vol % benzene, 15 vol % acetone, and 15 vol % methanol followed. In spite of careful distillation of the solvents prior to use, minute residues were obtained upon evaporation to dryness. In no case did the blank amount to more than 1.6 wt % of the extracted organic matter. After removal of the solvent with a stream of nitrogen, the extracted organic matter was

TABLE 1

RECENT SEDIMENT SAMPLES

Location	Description
7 miles off Grande Isle, La. Long. 90° 01′ W Lat. 29° 08′ N	A core of sediments taken from the floor of the Gulf of Mexico downward to a depth of 106′. Water depth—41.2′. Gray silty clay, occasionally interbedded with gray, fine-grained silty sand.
17 miles off Corpus Christi, Texas Long. 97° 43′ W Lat. 27° 50′ N to 27° 55′ N	A series of samples scooped off the Gulf floor.
Laguna Madre Flats, Texas Sample I Long. 97° 31′ W Lat. 26° 59′ N	Black, organic algal muck. Typical of material that collects in low places in the Laguna Madre flats surface and associated with a strong concentration of hydrogen sulfide. Wind-blown lagoon water remains in these low places for long periods of time after adjacent flats become dry. Sample representative of the uppermost 12 in. of the flats surface.
Sample II Long. 97° 31′ W Lat. 26° 58′ 30″ N	Thinly laminated black and gray algal clay from area which is sometimes dry and sometimes flooded. Uppermost 12 in. of the flats surface.

separated by chromatography over freshly activated, powdered alumina (Harshaw, Al-0101-P). Successive fractions were eluted with *n*-heptane, benzene, pyridine, acetone, and methanol. The heptane cut was designated as the paraffin-naphthene fraction, the benzene cut as the aromatic fraction, and the last three were combined to give an oxygenated or asphaltic fraction.

The concentration of extractable organic matter in many sediments is quite low. In the samples studied, it has ranged from 30 to 450 parts per 100,000 of dried sediments. The quantity of free hydrocarbons has varied from 2 to 19 parts per 100,000.

In Table 2 are found data for several representative samples. In all cases hydrocarbons were isolated, and furthermore, in the case of the Grande Isle core from Louisiana, the amount of hydrocarbons increases and the quantity of complex organic matter decreases with increasing depth of burial below the floor of the Gulf. Since the amount of extractable organic matter is constant for these three sections, the data certainly indicate a change in the nature of the organic matter in the direction of petroleum. It is interesting to note that the Laguna Madre sample (from Texas), in a reducing environment, contains 3.9 times as much free hydrocarbons as the specimen from an oxidizing environment. The former, incidentally, also contains 3.9 times as much sulfur as the latter.

Although the relationship between the hydrocarbon and sulfur contents of the two samples may be fortuitous, it fosters a bit of conjecture. Strains of autotrophic sulfur bacteria are known, of course, which utilize carbonates as a carbon source, oxidize hydrogen sulfide, and deposit free sulfur inside or outside their cells. The sediment sample with low hydrocarbon and sulfur contained 6 times as much carbonate as the hydrocarbon- and sulfur-rich sample. Since the sediments are from the same lagoon and are only a few thousand feet apart, the original organic and carbonate contents were probably quite similar. This combination of circumstances suggests, then, that there may be a direct relationship between sulfur and hydrocarbon production, at least in the case of a particular group of bacteria. Recent work by ZoBell and his co-workers (5), disclosing the production of hydrocarbons by autotrophic, sulfate-reducing bacteria cultivated in an inorganic medium in the laboratory, lends credence to this hypothesis.

Proof that the paraffin-naphthene and aromatic fractions are actually hydrocarbons has been furnished by elemental analysis, infrared absorption spectra, and their behavior on the alumina during the chromatographic separation.

If one were to extrapolate the data obtained on a 106-foot core of sediments taken from the floor of the Gulf of Mexico 7 miles off Grande Isle (186 parts hydrocarbons + asphaltics per million parts of dried sediment), a cubic mile of these sediments would contain 4,500,000 barrels of a paraffin-naphthene, aromatic, and asphaltic mixture resembling crude oil. In smaller units, this would be 7000 barrels per acre for a column one mile deep. Since the concentration of hydrocarbons increased with depth in this core, the estimate is probably conservative.

With the presence of liquid hydrocarbons in Recent sediment samples established, the question resolved itself into, "What is the origin of these hydrocarbons?" They could be present (1) by contamination during either the collection or the analysis of the sediments, (2) by migration or diffusion from greater depths, or (3) by deposition or genesis in the sediments themselves. Great care has been exercised during the whole program to prevent contamination, so it is felt with certainty that this possibility can be discounted. It would be difficult to rule out the possibility of migration or diffusion from greater depths

TABLE 2

EXTRACTION OF ORGANIC MATTER FROM RECENT SEDIMENTS

Sample description	Grams organic matter extracted /100 g dried sediment	Chromatographic analyses of extracted organic matter			
		% Paraffin-naphthene	% Aromatic	% Asphaltic	% Remaining on alumina
The following sections are from a core of Recent sediments taken about 7 miles off Grande Isle, La. The numbers indicate feet below the floor of the Gulf of Mexico					
3–4	0.031	6.0	1.5	14.0	78.5
18–22	.032	17.9	2.5	12.1	67.5
102–103	.031	25.0	5.7	10.6	58.7
Laguna Madre Flats, Texas (0–12" below surface)					
Reducing environment	.185	9.2	1.0	5.1	84.7*
Oxidizing environment	.450	1.1	< 0.5	26.8	72.1*
Sediment free of extractable organic matter	0.0005†				

* Includes free sulfur present in aromatic cut.
† Solvent blank.

were it not now possible to settle the question by means of carbon 14 age determinations.

Ages of 11,800-14,600 ± 1400 years were obtained for the hydrocarbons extracted from several sections of the Grande Isle core of Recent sediments. A composite carbonate sample from the entire core proved to be 12,300 ± 1200 years old, and the nonextractable organic matter, which comprises a major portion of the original organic content, had an average age of 9,200 ± 1000 years. This slightly younger age can be readily explained by the somewhat larger amounts of nonextractable organic matter present in the shallower sections of the core.

For control purposes, a sample of shale was extracted, and the resulting extract was separated by chromatography, using the same technique as was applied to the Recent sediment samples. The hydrocarbons thus obtained were "dead" by carbon 14 analysis, as might be expected, giving a net count, in cpm/g carbon, of 0.03 ± 0.08. In a similar manner, samples of cetane, produced by the hydrogenation of sperm oil, were analyzed before and after going through the same sequence of analytical steps. Net counts of 3.18 ± 0.10 and 3.10 ± 0.06 were recorded, which check the value of 3.14 ± 0.07 obtained for modern wood under the same conditions. All age determinations were carried out by J. Laurence Kulp, of the Lamont Geological Observatory of Columbia University.

The data demonstrate that the analytical procedure used on the Recent sediment samples does nothing to make the hydrocarbons appear either younger or older than they should. Hence it may be concluded that diffusion did not occur to any appreciable extent, and that the isolated hydrocarbons were either deposited with, or generated in, the sediments themselves.

It is interesting to speculate as to the source of these hydrocarbons. Shimkin *et al.* (*6*) have established the presence of polynuclear aromatic compounds in extracts from barnacles. ZoBell and coworkers (*5, 7*) found paraffins and naphthenes in extracts from laboratory cultures of marine bacteria. In our laboratories, hydrocarbons have been detected in concentrations of 45–58 ppm dry weight in oysters and bluefish. Until much more work is done to clarify the picture, it seems plausible to suggest that the petroleumlike mixture being formed in the present era is a composite of the hydrocarbon remains of many forms of marine life. It is a logical extension of this hypothesis to state that crude oils of varying composition might result from changes in the relative contributions of different forms of marine life.

References

1. KNEBEL, G. M. Fundamental Research on Occurrence and Recovery of Petroleum. Report of Progress 1946–47. New York: Am. Petroleum Inst., 89.
2. LINK, W. K. *Bull. Am. Assoc. Petroleum Geol.*, **33**, 1767 (1949).
3. RANKAMA, K., and SAHAMA, T. G. *Geochemistry*. Chicago: Univ. Chicago Press, 355 (1950).
4. TRASK, P. D. In *Problems of Petroleum Geology*. Tulsa, Okla.: Am. Assoc. Petroleum Geol., 32 (1934).
5. Am. Petroleum Inst. Research Project 43A, Third Quart. Rept. for Period ending Mar. 31, 1952.
6. SHIMKIN, M. B., KOE, B. K., and ZECHMEISTER, L. *Science*, **113**, 650 (1951).
7. SISLER, F. D., and ZOBELL, C. E. *J. Bact.*, **62**, 117 (1951).

5

Reprinted from *Carnegie Inst. Washington Year Book* **53**:97–101 (1954)

ORGANIC CONSTITUENTS OF FOSSILS

Philip H. Abelson

Organic material equivalent in quantity to the weight of the earth has been created by living creatures since life originated on this planet. Almost all of this has been metabolized by other creatures, with a few notable exceptions. Under special conditions of deposition and burial, the organic matter represented by coal, oil, and oil shales has been preserved. Some of the chemicals found in these fuels are only moderately altered products of compounds originally part of living organisms.

Examination of a variety of fossil shells and bones has now revealed that small but important quantities of organic materials are often preserved within these fossils. Some of the amino acid building blocks—alanine, glutamic acid, and valine—which are important constituents of today's proteins have been found in fossils as old as 360 million years.

Isolation of the organic compounds from fossils was accomplished through use of a combination of ion-exchange and paper-chromatography techniques. The fossils were first dissolved in acids, and any proteins present hydrolyzed. The solution was freed of mineral constituents by passage through a resin column. After elution from the resin, the mixture of amino acids was resolved on a paper chromatogram. Positive identification of the individual compounds was achieved by addition of radioactive tracers to some of the unknown mixtures.

Recent shells and calcareous materials

of biological origin have been examined as a guide to the understanding of occurrences of organic materials in fossils. The presence of proteins in bones has long been established, but virtually nothing was known about the nature of organic constituents in shells. In table 1 are shown results of examination of some tropical shells. It will be noted that all specimens listed contained amino acids. Twenty other recent, arbitrarily chosen shells were surveyed. All contained amino acids in amounts ranging from 6 to 40 micromoles per gram. The amino acids appearing were those typically found in protein hydrolyzates, with occasional specimens showing additional unidentified components. One may conclude that the occurrence of proteins in recent shells and other hard parts of invertebrates is a general phenomenon.

This survey makes it seem likely that many fossils originally contained proteins. It leads directly to the question of what are the favorable circumstances for preservation of these substances. It seems obvious that the first condition is freedom from bacterial destruction. Organic constituents preserved within a dense inorganic matrix, such as a shell, would be safe from such attack. A further requirement is that the inorganic parts of the fossil should not be dissolved or replaced by other substances. A third requirement is that the organic material should not diffuse from, or be in any way transported away from, the inorganic matrix. Finally, the organic compounds must have sufficient chemical stability to exist unchanged for long periods of time.

The distribution of proteins within typical shells was investigated. By exposure to controlled amounts of dilute acid it is possible to remove layers from shells and thus to determine the location of the protein. Results of such a procedure are shown in table 2. It will be noted that, although the protein was not uniformly distributed throughout the shell, it was present in all portions. Many of the shells are dense, solid structures, and their interiors are therefore quite impervious to bacterial action. Inspection of fossils and X-ray determination of their constituent minerals permit investigation of degree of replacement. The problem of diffusion is more difficult to assess, and no good quantitative approach has yet been devised. It seems likely, however, that diffusion is not too important, and that long-time

TABLE 1

AMINO ACID CONTENT OF VARIOUS RECENT TROPICAL CALCIUM CARBONATE SHELLS

Specimen	Description	Mineral composition	Amino acid content *
Porites lichen	Reef-building coral	Aragonite, 10–20 per cent calcite	6.7
Porolithon onkodes	Coralline alga	Calcite with $MgCO_3$ in solid solution	17.6
Chama lazarus	Reef-dwelling clam	Aragonite	16.4
Codakia punctata	Ribbed clam	Aragonite	8.4
Atactodea glabrata	Intertidal clam	Aragonite	42.7
Cymatium muricinum	Thick-shelled snail	Aragonite	20
Tridacna maxima	Man-eating clam	Aragonite	26.6
Cypraea mauritiana	Tropical snail	Aragonite	6.2

The materials were furnished by Preston Cloud, U. S. Geological Survey. The *Porites lichen* and *Porolithon onkodes* were collected at Saipan. The other specimens were collected at Onotoa Atoll in the Gilbert Islands.

* In micromoles per gram ($\mu M/g$). One $\mu M/g$ is approximately equal to 1 part per 10,000.

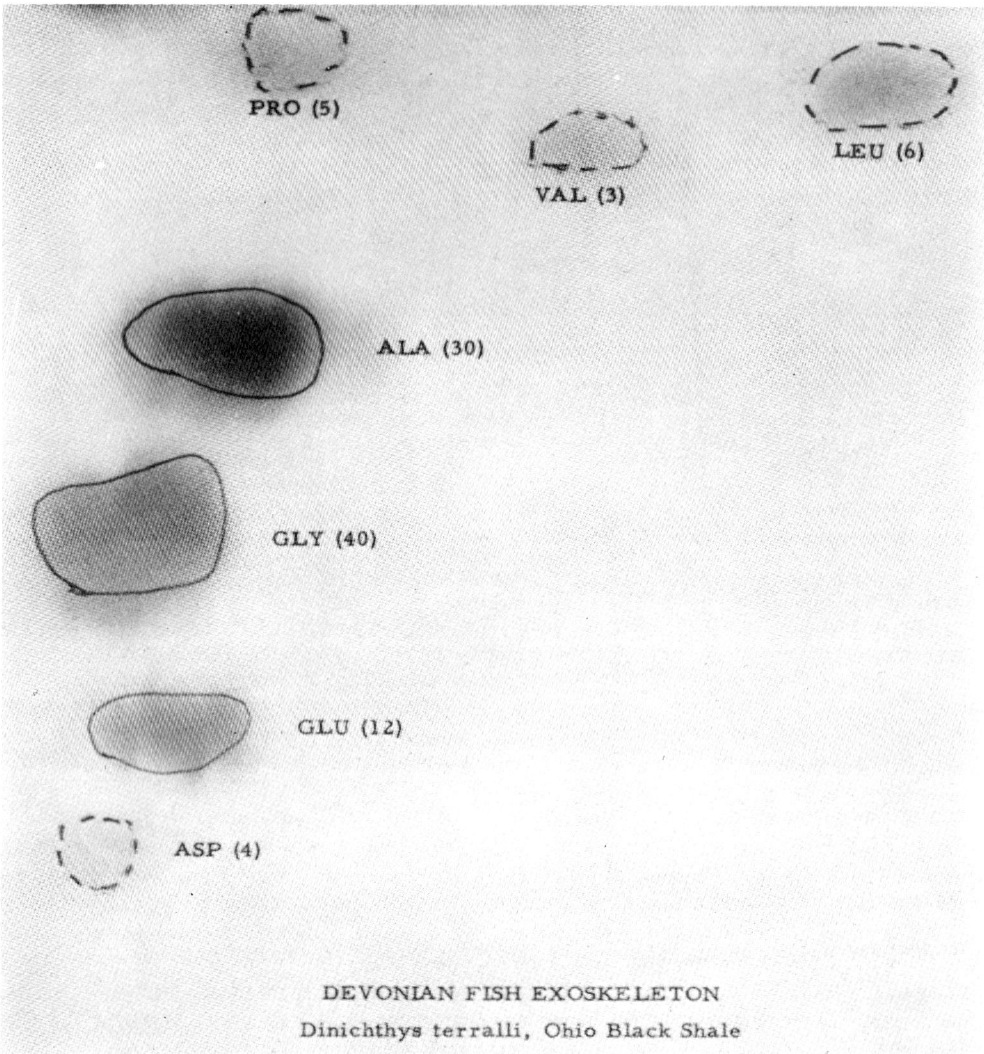

FIG. 1. Two-dimensional paper chromatogram of amino acids separated from a 360-million-year-old fossil. Ala, alanine; asp, aspartic acid; glu, glutamic acid; gly, glycine; leu, leucine; pro, proline; val, valine.

preservation of amino acids is most crucially dependent on the intrinsic thermal stability of the various compounds and on the thermal conditions to which the fossils have been exposed.

With these considerations in mind, a series of fossils showing no silicification or replacement and chosen on the basis of probable mild thermal history were examined. Both shells and bones were studied. In all cases the fossils contained amino acids. Results from some teeth and bones are given in table 3. These results show no trend with time, which is not too surprising, since different burial histories were involved. Actually the oldest fossil in this group was richest in amino acids. It is possible that this was due in part to the preserving black shale matrix. Black shales are formed in anaerobic environments, often under virtually sterile conditions. A chromatogram of amino acids separated from a Devonian fish plate is shown in figure 1. Amino acids from fossils have been shown to be identical with present-day compounds, by use of radioactive tracers. The amino acids found in the fossils listed in table 3 and shown in figure 1 are those which laboratory tests have shown to be particularly thermally stable. They are precisely those amino acids which could survive if a modern protein were to be stored in a fossil for similar periods of time. It therefore seems likely that these fossils originally contained proteins which played the same important role in the lives

TABLE 2

SERIAL SOLUTION OF A RECENT CLAM SHELL

Fraction (%)	Amino acid content (μM/g)	Protein (‰)
10–20	11.7	1.24
20–30	12.3	1.31
30–40	17.8	1.90
40–50	20.3	2.17
50–60	22.0	2.35
60–70	17.8	1.90
70–80	15.0	1.61
80–90	10.2	1.09
Average	15.9	1.70

The specimen used was *Mercenaria mercenaria,* an edible clam grown at Cape May, New Jersey.

TABLE 3

AMINO ACID CONTENT OF VARIOUS FOSSILS

Name	Approximate age (years)	Formation	Amino acid content (μM/g)	Principal constituents
Plesippus (prehistoric horse)	Late Pliocene 5×10^6	Hagerman Lake Beds, Idaho	0.6	Ala, gly
Plesippus (prehistoric horse, tooth)	Late Pliocene 5×10^6	Hagerman Lake Beds, Idaho	1.5	Gly, ala, leu, val, glu
Mesohippus (prehistoric horse, tooth)	Oligocene 40×10^6	White River, Nebraska	0.31	Ala, gly
Mosasaurus (dinosaur)	Cretaceous 100×10^6	Pierre Shale, South Dakota	1.8	Ala, gly, glu, leu, val
Anatosaurus (dinosaur)	Cretaceous 100×10^6	Lance, Lance Creek, Wyoming	2.8	Ala, gly, glu, leu, val, asp
Stegosaurus (dinosaur)	Jurassic 150×10^6	Morrison, Como Bluff, Wyoming	0.26	Ala, gly, glu
Dinichthys (fish)	Devonian 360×10^6	Ohio Black Shale	3.0	Gly, ala, glu, leu, val, asp

Specimens were furnished from the U. S. National Museum by C. Lewis Gazin and David H. Dunkel.

Ala, alanine; asp, aspartic acid; glu, glutamic acid; gly, glycine; leu, leucine; val, valine.

of ancient creatures as they do in today's organisms.

One contingency which must be faced is the possibility that the amino acids found may be adventitious contaminants. Tests in which $CaCO_3$ was precipitated in the presence of amino acids showed that aspartic and glutamic acids are the two amino acids which are readily adsorbed. Alanine and glycine, the principal amino acid constituents of fossils, have almost no degradation, and that alanine, glycine, glutamic acid, leucine, and valine are relatively resistant to heat. Alanine was selected for more detailed study. In dilute solution the rate of degradation of alanine is proportional to concentration. This is the condition for a first-order reaction. Studies at temperatures ranging from 188° to 310° C showed a reaction rate strongly dependent on temperature. Other measurements revealed that the principal reac-

TABLE 4

AMINO ACID CONTENT OF FOSSILS FROM ONE LOCALITY

Specimen	Description	Mineralogy	Amino acid content (μM/g)
Glycymeris parilis	Clam	Aragonite	0.80
Mercenaria mercenaria	Clam	Aragonite	0.75
Melosia staminea	Clam	Aragonite	0.90
Lyropecten madisonius	Scallop	Calcite	1.10
Eucrassatella melina	Bivalve	Aragonite	0.60
Turritella variabilis	Snail	Aragonite	0.40
Ecphora tricostata	Snail	Inner aragonite, outer calcite	1.20
Turritella indenta	Snail	Aragonite	0.50

Fossils collected at Plum Point, Zone 10, Calvert Formation, Maryland, by Wendell Woodring, Kenneth Lohmann, and Philip H. Abelson. Age of shells, about 30 million years.

Amino acids present included alanine, glycine, leucine, valine, glutamic acid, and aspartic acid.

affinity for $CaCO_3$, nor is it likely that they have affinity for bone. It seemed desirable, however, to investigate the amino acid content of a group of fossils collected at one locality. These presumably have been exposed to an identical thermal history and have been similarly exposed to percolating ground waters. The shells showed both quantitative and qualitative differences in amino acid content, a fact which minimizes the possible influence of adsorption from ground water. Results are shown in table 4. This problem should be studied further on material from other localities.

In order to begin to interpret the geological significance of the occurrence of amino acids in fossils, it was necessary to survey the thermal stability of these compounds. Experiments showed that there is considerable variation in rates of thermal

tion occurring at these various temperatures was decarboxylation:

$$CH_3-\underset{\displaystyle H}{\overset{\displaystyle NH_2}{\overset{|}{\underset{|}{C}}}}-COOH \rightarrow CH_3-\overset{\displaystyle NH_2}{\overset{|}{CH_2}}$$

With reasonable confidence, therefore, one can apply the experience relative to first-order reactions which has been accumulated by physical chemists.

It has been shown that $dc/dt = -kc$, $k = \sim 10^{13} e^{-E/RT}$, where c is concentration, E an activation energy, R the gas constant, and T the absolute temperature. The value 10^{13} is a frequency factor deduced from a variety of investigations of first-order reactions performed elsewhere. Data obtained in studies of the degradation of alanine are shown in figure 2. The straight

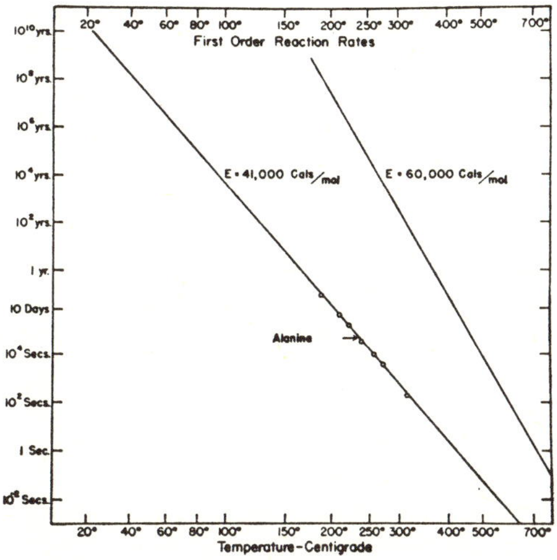

FIG. 2. Thermal degradation of alanine. Time required for decomposition of 63 per cent of original material.

line is based on the equations applicable to a first-order reaction normalized at 258° C. It will be noted that a very large extrapolation is necessary to determine stability at room temperatures. This extrapolation seems justified, however, since the alanine is actually observed in ancient fossils. It indicates that under mild thermal conditions alanine is sufficiently stable to last more than a billion years. Some of the other organic substances are more stable. Molecules degrading by first-order reactions with an activation energy of 60 kilocalories per mole could last a time equivalent to that since the earth first cooled, even if exposed to temperatures of 150° C.

The thermal studies on alanine suggest the possibility of a geological thermometer for sediments. Although the amount of degradation of a substance depends on both time and temperature, rates of reaction are more strikingly influenced by temperature than by time. Ordinarily time can be estimated from the fossil record. Thus, if amino acids are found in a fossil, an upper limit can be set on the temperature to which the fossil has been exposed. This in turn can be related by means of geothermal gradients to an estimate of burial history. Further development of this method may result in a very valuable tool for the geologist.

6

Reprinted from *Science* **149**:722–726 (1965)

Organic Pigments: Their Long-Term Fate

Fossil pigments provide evidence of modification of chemical structure under geological conditions.

Max Blumer

One of the first systematic studies in organic geochemistry (*1*) led to the surprising discovery that organic pigments from plants and animals are preserved in a wide range of geologic materials, often of great antiquity. This marked a decisive step in the development of our present concepts of the long-term preservation and transformation of organic compounds in nature. Much thought and effort has since been devoted to widening the scope of the original study and to interpreting the structural changes of the pigments in terms of the reactivity of the subsurface environment.

With the development of sensitive and selective analytical techniques like thin-layer chromatography, gas chromatography, and mass spectrometry, the study of fossil pigments has entered a new stage. We now have evidence that most sedimentary pigments occur as much more complex mixtures than had been anticipated. By studying the individual chemical structures we can often establish genetic relationships and can reconstruct reaction sequences. Most pigment molecules contain, attached to a central, unsaturated nucleus, a number of different functional groups. By determining how these change in the sediment we arrive at predictions about ways in which other molecules of related structure might react in the same environment.

Thus, the study of fossil pigments—and of other biochemical fossils—provides us with data on the modification of chemical structures under actual geological conditions. It avoids the extrapolation of laboratory results to a different time scale and to poorly known environmental conditions.

The author is senior scientist at the Woods Hole Oceanographic Institution, Woods Hole, Massachusetts.

The Chemistry of a Fossil Crinoid

During fossilization, some biochemicals and their transformation products are imbedded at the source; compounds from other organisms and from the surrounding sediment remain excluded from many fossils. This explains why organic compounds in fossils occur at relatively high concentrations and only in moderately complex mixtures.

A Jurassic crinoid (*Apiocrinus* sp.) from northwestern Switzerland has been studied in detail (*2*). The intensely colored fossil contains an assemblage of numerous crystallized pigments (fringelites) and aromatic hydrocarbons, whose close chemical relationship suggests a common origin. Several considerations lead us to think that the hydrocarbons originated from the pigments, not vice versa. Hydroxyquinone pigments resembling the fringelites commonly occur in plants and animals and are found in many of the recent relatives of the fossil crinoid. The inorganic minerals of the fossil are strongly reduced. In such a reducing environment the pigments could well have been converted into hydrocarbons. The reverse reaction would lead from the stable hydrocarbons to the sterically strained pigments (Fig. 1), and it appears much less probable.

Several closely adjacent hydroxyl groups in fringelites D, E, and F of Fig. 1 considerably strain the aromatic ring, which would be most stable in planar configuration. This strain is most severe in fringelite D; it lowers the stability of this pigment. In the reducing sediments stabilization occurs through stepwise elimination of the sterically hindered groups. This process has been completed in fringelite H, where the only remaining hydroxyl groups are those stabilized by chelation to the quinone oxygens. The increase in stability from fringelite D to fringelite H is reflected in increasing concentrations of the pigments. The rates of these reactions decrease as the strain is lessened; the reactions proceed slowly enough to have preserved some fringelite D and larger amounts of the reduction intermediates E and F in this Jurassic fossil.

The further reduction of the very stable fringelite H proceeds very slowly; pigments still occur at 200 times the concentration of the hydrocarbons. Reaction intermediates which have lost some of their chelated groups are more readily reduced than fringelite H itself; this explains their absence in the fossil.

Reactions like those of the fringelites are geochemically irreversible. The reverse reactions are imperceptibly slow, even on a geological time scale. Given a favorable environment, such irreversible reactions proceed to completion. Reaction intermediates are unstable or "metastable." Typical examples of geochemically irreversible reactions are eliminations of oxygen functions (especially in sterically unfavorable positions), decarboxylations, some hydrogenations and isomerizations, cleavage reactions, and, in general, processes which destroy the ordered building pattern of biochemical products (Table 1).

Acid-base reactions and many reactions involving the exchange of hydrogen, on the other hand, may proceed sufficiently fast in either direction to lead to equilibrium in geological time spans. These we call geochemically reversible reactions. Here the environment determines the ratio of starting material to end product; if conditions in the environment change, this ratio adjusts to the change.

We believe that such reversible redox reactions have affected the composition of the aromatic hydrocarbons in *Apiocrinus*. It was surprising to discover that the major hydrocarbon of the fossil is a partially hydrogenated derivative (Fig. 1, II) of the parent compound of fringelite D. Though easily produced in the laboratory by reduction of the pigments, the parent hydrocarbon (I) itself is not represented in the fossil. In addition to fringelites D, E, F, and H, the fossil contains pigments which we consider precursors of hydrocarbons III and IV. Why, we wondered, are these represented at nearly equal concentrations if hydrocarbon I has been completely reduced to II?

The redox potential of these hydrocarbons cannot be measured in the laboratory. However, correlations between spectra and redox potentials of aromatic hydrocarbons suggest that hydrocarbon I is much more readily reduced than III. Thus, under appropriate reducing conditions, equilibrium would lead to complete reduction of I but to only partial reduction of III.

A similar situation may exist for the polycyclic aromatic and hydroaromatic hydrocarbons of ancient sediments and of crude petroleum. The concentration of oxidized and reduced partners of such redox couples reflects the reduction state of the environment. Because of the slowness of the reactions, the concentration may remain unchanged for a considerable time after changes in the environmental conditions occur—for instance, through uplifting and erosion.

Fossil Porphyrins

Tetrapyrrole pigments derived from chlorophylls and hemins occur in a great variety of sedimentary materials, including crude petroleum. Most abundant are the vanadium and nickel complexes. Frequently these are accompanied by small amounts of uncomplexed pigments. Complexes with gallium and iron as central atoms are thought to occur, though they have not yet been isolated in pure form. Chromatographic separation and structural identification of the porphyrins is simplified by "demetalation" of the complexes. Because of their great stability this requires severe treatment with concentrated acids. Often, artifacts are produced or the pigments are completely destroyed, especially when they are present in small amounts. Newer reagents (lithium or dithiooxamide in ethylene diamine, and butylphenyllithium in tetrahydrofurane) (*3*) promise to bring about "decomplexation" with better yield and without producing artifacts.

We have studied mainly the uncomplexed fossil porphyrins to avoid the uncertainty of demetalation (*4*). As a rule, free porphyrins are scarce in sediments, but the pigment-rich Triassic oil shale of the Grenzbitumen horizon of Serpiano, Switzerland, provides an excellent source. Treibs based much of his early work (*1*) on the analysis of this particular shale. He isolated what appeared to be four major pigments derived from chlorophyll and hemin, together with their metal complexes. When we applied thin-layer chromatography and mass spectrometry to the analysis of these pigments, each of them was resolved into a homologous series of six or more members. In addition, many minor pigment types were found to be equally complex mixtures of homologs.

Chlorophyll and hemin (Fig. 2) both have two acidic substituents; in chlorophyll these are esterified by methanol and the C_{20} isoprenoid alcohol phytol. The majority of the fossil porphyrins, on the other hand, are etioporphyrins—that is, they lack acidic substituents. Treibs thought that the etioporphyrins were derived from the acids by a simple decarboxylation; this would lower by two the number of carbon atoms in the molecule. The molecular-weight distribution obtained by mass spectrometry now suggests two or more additional processes in the transforma-

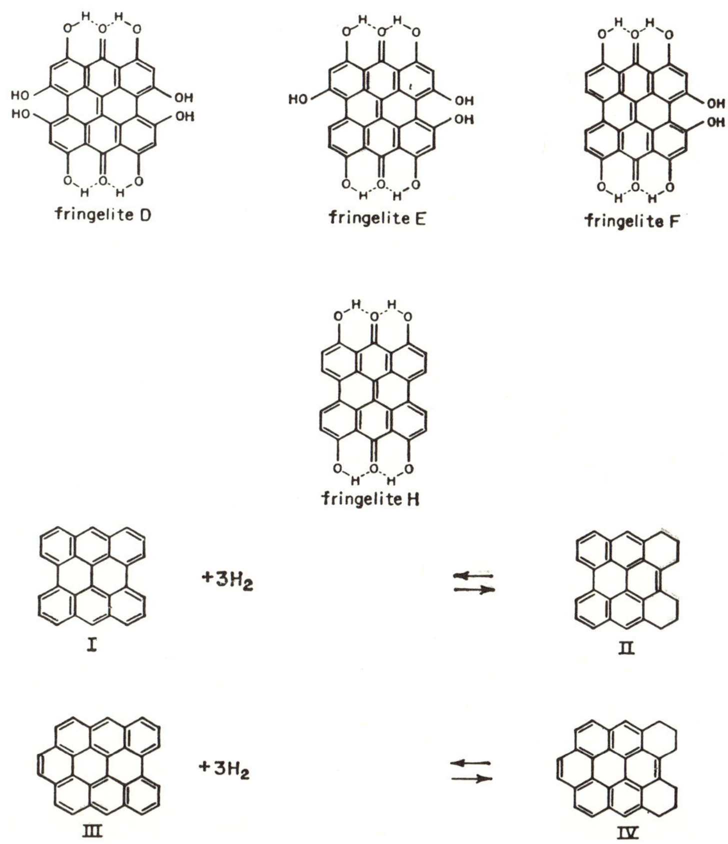

Fig. 1. Pigments and hydrocarbons of a Jurassic crinoid. The closely adjacent —OH groups of fringelite D exert a strain on the normally planar aromatic ring. Reduction in the sediments has lessened this strain in fringelites E and F and eliminated it in fringelite H. Hydrocarbons I–IV are derived from the fossil pigments. Their concentration is influenced by reversible redox equilibria.

tion of the acidic pigments to the etioporphyrins.

A few of the fossil etioporphyrins have the same number of carbon atoms as the original biosynthetic pigments. Others have lost a single carbon atom, but not two, as would be required in a complete decarboxylation. Still, as etioporphyrins these pigments have no carboxyl groups. This implies that one or both carbon atoms of the carboxyl groups have been retained by reduction to methyl groups (Fig. 2, A). This is the first geochemical evidence for the reduction of sedimentary acids. It implies such a low redox potential that it is not surprising that the carbonyl group of chlorophyll and double bonds of hemin and chlorophyll are completely reduced in the fossil porphyrins.

Table 1. Geochemically reversible and irreversible reactions. Only the reactions are listed for which we have evidence from the fringelites, the fossil porphyrins, and the phytol-derived hydrocarbons.

Reaction			Type	Reversibility
R–COOH	⟶	R–H	Decarboxylation	Irreversible
R–COOH	⟶	R–H + shorter-chain acids and hydrocarbons	Cooper-Bray (*5*) degradation	
R–C(C)–R	⟶	R–C–R	Cleavage	Irreversible
R–C–C–R	⟶	R–C	Cleavage	Irreversible
$R_2C = CR_2$	⟶	$R_2HC–CHR_2$	Reduction	Irreversible
R–COH	⟶	RCH_3	Reduction	Irreversible
R–COOH	⟶	RCH_3	Reduction	Irreversible
R_2–C=O	⟶	R_2–CH_2	Reduction	Irreversible
R–OH	⟶	R–H	Reduction	Irreversible
	⇌		Reduction	Reversible
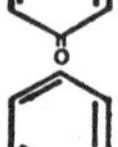	⇌		Reduction	Reversible
	⇌		Reduction	Reversible
R–H	⇌	R–Me	Complexation	Reversible?
R–COOR	⇌	R–COOH	Hydrolysis	Reversible

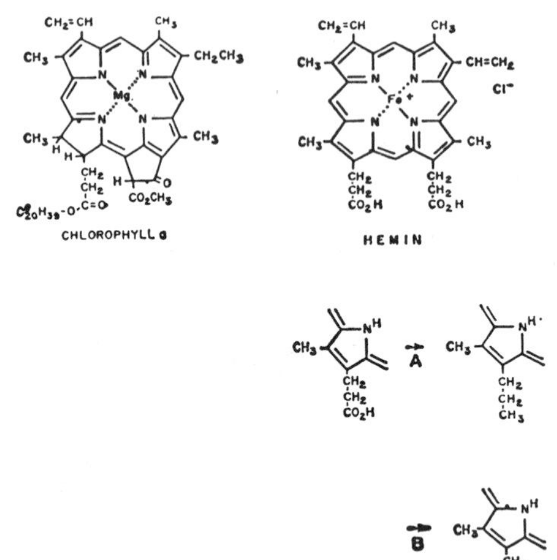

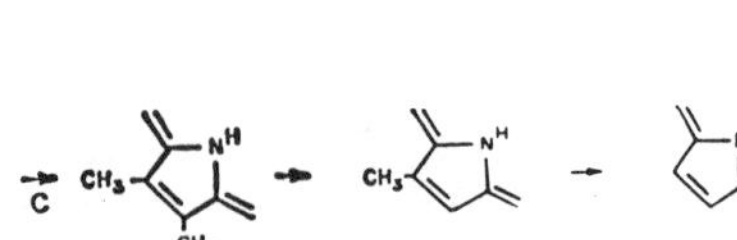

Fig. 2. Chlorophyll *a*, hemin, and some reactions of fossil porphyrins. In the sediments magnesium and iron are replaced with nickel and vanadium; unsaturation in the side chains and the —C=O group disappears, and an equilibrium is established between porphyrins and chlorins (=dihydroporphyrins). The acid side chains are modified by hydrogenation (A), by decarboxylation (B), and by degradative decarboxylation (C). Cleavage of side chains (D) lowers the molecular weight.

Decarboxylation of the acid groups (Fig. 2, B) competes with their reduction; it produces a sizable, but not the principal, fraction of the fossil etiopigments. The majority of the etioporphyrins contain less carbon than the products of reaction B. It has been suggested that fatty acids in sediments are degraded by a low-temperature decarboxylation involving free-radical intermediates and yielding acids and alkanes of lower molecular weight (*5*). Some of the fossil etioporphyrins may well have been produced by this or a similar mechanism (Fig. 2, C).

A few of the fossil etioporphyrins have still lower molecular weights. In these, cleavage of alkyl groups (Fig. 2, D) must have accompanied the complete degradation of the propionic acid side chains.

Sediments and crude petroleum often contain traces of reduced porphyrins. In these chlorins, one of the four pyrrole rings has been saturated by the addition of hydrogen. Chlorophyll itself is a chlorin; thus, we might consider the fossil chlorins to be direct descendants of chlorophyll. On the other hand, they could be reduction products of fossil prophyrins. That we have isolated chlorins structurally related to the fully aromatic pigment hemin suggests the latter. This interpretation is supported by mass spectral analysis of the fossil chlorins derived both from chlorophyll and from hemin. Except for the mass difference caused by the two extra hydrogen atoms, the fossil chlorins and porphyrins have the same molecular-weight distribution. This suggests a reversible redox system at equilibrium. Chlorins probably were formed only after the porphyrins had already undergone their major structural changes.

The fossil chlorins are among the most strongly reduced compounds recovered from sediments. In pure form, they react with the oxygen of the air. In solution, chlorins reduce vanadyl ions to trivalent vanadium. That chlorins survived during uplifting and erosion must be attributed to the very considerable buffering action of many other strongly reduced compounds in this organic-rich shale.

Isoprenoid Hydrocarbons from Phytol

Phytol, a C_{20} isoprenoid alcohol (Fig. 3, I) is a constituent of chlorophyll (Fig. 2). Above, I discussed the geochemical fate of the tetrapyrrole ring fragment of this pigment; now, we shall consider the reactions which phytol undergoes in organisms and in sediments.

Phytol is synthesized by all green photosynthetic plants. An estimated 90 percent of the total production occurs in the sea, mostly in planktonic algae. The small crustaceans, which are among the most important intermediates in the food chain between algae and higher organisms, liberate phytol from chlorophyll by hydrolysis. The phytol is then converted into a wide range of saturated and olefinic hydrocarbons (*6*) (Fig. 3, II–X).

Several isomeric phytadienes are formed (Fig. 3, II–IV), presumably by acid-catalyzed dehydration of phytol in the digestive tract of the animals. Pristane (V) and three isomeric olefins (VI–VIII) with the same carbon skeleton are formed by an alternate degradation mechanism, which causes the loss of one carbon atom. Pristane is stored in amounts up to 5 percent of total body lipids by several species of *Calanus*. It has an unusually low density and melting point and may therefore serve as a buoyancy regulator and may aid in keeping the stored lipids of these species in the liquid state.

In cold water, planktonic crustaceans increase the unsaturation of the fatty acids ingested with their food (*7*); this lowers the melting point of their body fat. A similar process may be responsible for the formation of several C_{19} di- and triolefins (Fig. 3, IX and X) from pristane.

The phytol-derived hydrocarbons are less easily metabolized than other lipids; therefore they are often retained by organisms which feed on zooplankton. Pristane and the monoolefins can be traced throughout the marine food chain; they occur in the fats of large marine fishes and mammals.

Saturated hydrocarbons related to phytol occur in crude oil and in ancient sediments. Their presence in the billion-year-old Nonesuch shale of Michigan has been interpreted as evidence of life in Precambrian times (*8, 9*). No olefins have been reported; their reduction should proceed rapidly in most sediments. Phytane (Fig. 3, XI), the parent hydrocarbon of phytol, is surprisingly abundant; it accounts for 0.2 percent of East Texas crude oil and 2.2 percent of Colorado oil-shale bitumen. It may have been formed by hydrogenation of phytadienes, either those derived directly from organisms or those generated *in situ* by catalytic dehydration of phytol adsorbed on clay minerals. Conversion of phytol to phytane by hydrogenation and reductive elimination of the hydroxyl group also appears possible, though less probable.

Like phytane, pristane (Fig. 3, V) occurs at high concentrations: 0.5 percent in East Texas crude oil and 1.0 percent in Colorado oil-shale bitumen. The derivation of pristane from phytol by oxidation to an acid, followed by decarboxylation and hydrogenation, has been suggested. It could also be formed from norphytene (VII), a C_{19} monoolefin which occurs in organisms and is a by-product of the catalytic dehydration of phytol.

We should not exclude the possibility that some of the ancient pristane was formed by organisms at the time of deposition of the sediments. This is almost certainly the case for the pristane which we have isolated from the topmost sediment layer of a Norwegian fjord (*10*). At the sediment surface, the reductions which could generate pristane from phytol without biochemical mediation proceed too slowly. This is evident from the excellent state of preservation of the highly unsaturated carotenoids in surface sediments.

Several C_{14} to C_{18} isoprenoid hydrocarbons (Fig. 3, XII–XV) have been isolated from petroleum and oil shales. These may be cleavage products of phytol, phytane, or pristane. In addition to these fragments, cleavage reactions also produce low-molecular-weight compounds, with boiling points in the gasoline-boiling-point range. Because of their structural simplicity their biochemical ancestry is not easily recognized.

The fossil tetramethylheptadecane (Fig. 3, XVI) is thought to originate from carotenoids or other isoprenoids with more than 20 carbon atoms. By fragmentation these may also have produced some of the sedimentary phytane, pristane, and lower-carbon-number hydrocarbons.

Isoprenoid acids with the carbon skeleton of phytane, pristane, and farnesane have been isolated from a California petroleum (*11*). Through simple and stepwise decarboxylation with alkyl radical intermediates these acids might produce isoprenoid paraffins of lower molecular weight than pristane.

Phytadienes dimerize to substituted cycloalkenes at moderate temperatures, even without a catalyst. Through cleavage and disproportionation these might give rise to some of the substituted dialkylbenzenes and dialkylcyclohexanes of crude petroleum.

From a structural analysis of fossil pigments we have arrived at tentative reconstructions of the events in their transformations. At present we are re-

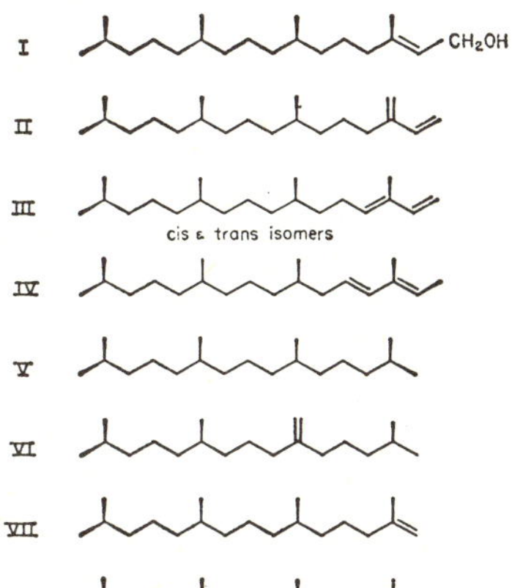

Fig. 3. Phytol (I) and phytol-derived hydrocarbons from organisms, sediments, and petroleum. Compounds II–X have been isolated from marine zooplankton, compounds XI–XVI occur in ancient sediments and petroleum. Pristane (V) is common to the two groups and has been found in a recent marine sediment.

Hydrogenation of the carboxyl group:

$$C_nH_{2n+1}COOH \rightarrow C_nH_{2n+1}CH_3$$

Simple decarboxylation:

$$C_nH_{2n+1}COOH \rightarrow C_nH_{2n+2} + CO_2$$

Stepwise decarboxylation with alkyl radical intermediates (Cooper and Bray):

$$\begin{aligned} C_nH_{2n+1}COOH \rightarrow\ & C_nH_{2n+2} + C_{n-1}H_{2(n-1)+1}COOH + \\ & C_{n-1}H_{2(n-1)+2} + C_{n-2}H_{2(n-2)+1}COOH + \\ & C_{n-2}H_{2(n-2)+2} + C_{n-3}H_{2(n-3)+1}COOH \ \ldots \end{aligned}$$

Cleavage:

$$C_nH_{2n+1}COOH \rightarrow C_{n-x}H_{2(n-x)+2} + C_xH_{2x+2}$$

Fig. 4. Transformation of a saturated straight-chain fatty acid in the sediments. A wide-range mixture of fatty acids and paraffin hydrocarbons is produced. The average molecular weight is lower than that of the starting material.

luctant to interpret these changes in terms of reaction mechanisms, though some convincing steps in this direction have already been taken (*5*). Eventually, we hope, a broader knowledge of sedimentary reactions and their mechanisms will enable us to predict the fate of most organic substances in the sedimentary environment.

Even now we may try to apply the information derived from the pigments to a wider range of compounds. Thus, carboxylic acids in a sediment should undergo transformations resembling those of the propionic acid side chain of the porphyrins. The possible reactions of a saturated, straight-chain fatty acid $C_nH_{2n+1}COOH$ are summarized in Fig. 4. The overall result is the formation of a mixture of fatty acids and paraffins of wide boiling-point range and lower molecular weight than the starting compound, showing a smooth statistical distribution of carbon numbers. This agrees well with the finding that petroleum paraffins have a lower average molecular weight than the biogenic fatty acids of recent source sediments. Also, their carbon-number distribution is smooth, while chains of odd carbon number predominate in the paraffins, and chains of even carbon number predominate in the fatty acids, of recent sediments.

Hydrogenations account for many of the reactions of ancient pigments (Table 1). They also play a dominant role in the processes which are believed to form crude petroleum. What is the source of the hydrogen and what are the driving forces for these reactions?

At the time of deposition the sedimentary organic matter includes a wide range of oxidation states and is far from redox equilibrium. After burial, the most highly oxidized materials are reduced by reaction with compounds of lower reduction state. As geologic time passes, reactions which are too slow to proceed in the laboratory eliminate the least stable compounds and bring the sedimentary organic material closer to equilibrium.

Pigments and petroleum precursors are only a small fraction of the sedimentary organic material. Their hydrogenation is coupled with the dehydrogenation of other compounds. Straight- and branched-chain compounds may lose hydrogen in condensations, cyclizations, and aromatizations. Hydrogen is made available also in the formation of those large aromatic entities which are an important part of the petroleum asphaltenes and of kerogen. These reactions amount to an overall stabilization of the sedimentary organic material.

We have studied sediments of intermediate age. In these, most of the unstable compounds have already disappeared; fringelite D, and to a lesser degree fringelites E and F, are exceptions. Sufficient time has lapsed since the Jurassic period for some reactants to have been brought into equilibrium with the environment. With time, a larger fraction of the sedimentary organic matter will undergo such stabilization and equilibration. The most stable organic compounds survive for extremely long periods, if the temperature of the sediment remains sufficiently low to exclude slow pyrolysis. The discovery of fossil porphyrins in a Precambrian sediment has recently been reported (*9*). It will be most interesting to learn how their structure compares to that of our Triassic pigments.

References and Notes

1. A. Treibs, *Angew. Chem.* **49**, 682 (1936).
2. M. Blumer, *Mikrochemie* **36/37**, 1048 (1951); *Nature* **188**, 1100 (1960); *Geochim. Cosmochim. Acta* **26**, 225 (1962); *ibid.*, p. 228; D. W. Thomas and M. Blumer, *ibid.* **28**, 1467 (1964).
3. U. Eisner, *J. Chem. Soc.* **1964**, 4089 (1964).
4. M. Blumer and G. Omenn, *Nature* **191**, 161 (1961); ———, *Geochim. Cosmochim. Acta* **25**, 81 (1961); D. W. Thomas and M. Blumer, *ibid.* **28**, 1147 (1964).
5. J. E. Cooper and E. E. Bray, *ibid.* **27**, 1113 (1963).
6. M. Blumer, M. M. Mullin, D. W. Thomas, *Science* **140**, 974 (1963); ———, *Helgolaender Wiss. Meeresberichte* **10**, 187 (1964); M. Blumer and D. W. Thomas, *Science* **147**, 1148 (1965); ———, *ibid.* **148**, 370 (1965).
7. T. Farkas and S. Herodek, *J. Lipid Res.* **5**, 369 (1964).
8. R. A. Dean and E. V. Whitehead, *Tetrahedron Letters* **1961**, 768 (1961); J. G. Bendoraitis, B. L. Brown, L. S. Hepner, *Anal. Chem.* **34**, 49 (1962); ———, paper presented at the 6th World Petroleum Congress, Frankfurt, 1963; B. J. Mair, N. C. Krouskop, T. J. Mayer, *J. Chem. Eng. Data* **7**, 420 (1962); J. J. Cummins and W. E. Robinson, *ibid.* **9**, 304 (1964); G. Eglinton, P. M. Scott, T. Belsky, A. L. Burlingame, M. Calvin, *Science* **145**, 263 (1965); J. Oró, D. W. Nooner, A. Zlatkis, S. A. Wikström, E. S. Barghoorn, *ibid.* **148**, 77 (1965).
9. W. G. Meinschein, E. S. Barghoorn, J. W. Schopf, *Science* **145**, 262 (1964).
10. M. Blumer, unpublished results.
11. T. Carson and D. W. Graham, *Tetrahedron* **21**, 471 (1965).

12. The work reported was supported by a research contract with the U.S. Office of Naval Research [contract Nonr-2196 (00)] and a grant from the National Science Foundation (GP-3250). This article is contribution No. 1636 of the Woods Hole Oceanographic Institution.

Part II

MOLECULAR ADVANCES

Editor's Comments on Papers 7 Through 22

7 **BRAY and EVANS**
Distribution of n-*Paraffins as a Clue to Recognition of Source Beds*

8 **DEAN and WHITEHEAD**
The Occurrence of Phytane in Petroleum

9 **COOPER**
Fatty Acids in Recent and Ancient Sediments and Petroleum Reservoir Waters

10 **PARKER and LEO**
Fatty Acids in Blue-Green Algal Mat Communities

11 **RHEAD et al.**
Conversion of Oleic Acid to Saturated Fatty Acids in Severn Estuary Sediments

12 **EGLINTON et al.**
Occurrence of Isoprenoid Fatty Acids in the Green River Shale

13 **SEVER and PARKER**
Fatty Alcohols (Normal and Isoprenoid) in Sediments

14 **ATTAWAY and PARKER**
Sterols in Recent Marine Sediments

15 **ANDERSON et al.**
The Isolation of Steranes from Green River Oil Shale

16 **PALACAS, SWAIN, and SMITH**
Presence of Carbohydrates and Other Organic Compounds in Ancient Sedimentary Rocks

17 **ERDMAN, MARLETT, and HANSON**
Survival of Amino Acids in Marine Sediments

Many different types of organic molecules now have been found to be associated with various kinds of geological materials. Most of these molecules are chemical or molecular fossils that have been derived with or without alteration from their biological precursors. The papers in this section were chosen to illustrate the geochemical occurrence of several different molecules. In general, the papers are early reports and each deals with a given class of compounds. The order of presentation is governed by the structural and compositional complexity of the molecules and begins with the simplest of organic molecules.

n-Alkanes

n-C_{29}

Hydrocarbons are compositionally the simplest organic molecules. Only carbon and hydrogen atoms are present in hydrocarbon structures. Alkanes are one class of hydrocarbons and are characterized by single bonds between carbon atoms. In alkanes, carbon atoms are connected in straight chains (as illustrated above), branched chains (as illustrated by isoprenoid hydrocarbons), and in rings (as illustrated by steranes and triterpanes). Alkanes having straight chains of carbon atoms are called normal alkanes or normal paraffins (*n*-alkanes or *n*-paraffins). These compounds range in chain-length from one to at least seventy-eight carbon atoms in geological materials such as petroleum and petroleum waxes (Speers and Whitehead, 1969). Hundreds of geochemical papers have now been written describing occurrences of *n*-alkanes.

Variations in the relative abundances of odd- and even-carbon-numbered *n*-alkanes in sediments were first noted by Stevens, Bray, and Evans (1956) and Evans, Kenny, Meinschein, and Bray (1957). As a follow-up to these papers, Bray and Evans (Paper 7) detail the distributional properties of *n*-alkanes in recent and ancient sediments and in crude oil and apply these properties to the identification of source beds of petroleum. In order to describe the variations in relative abundances of odd- and even-carbon-numbered *n*-alkanes, the authors define a ratio that was called the carbon preference index (CPI) by Cooper and Bray (1963). Seven years later, another measure of *n*-alkane distribution called odd-even predominance (OEP) was devised by Scalan and Smith (1970). The OEP eliminated certain mathematical limitations found in the CPI. Both measures have been used extensively in geochemical discussions of *n*-alkanes.

ISOPRENOID HYDROCARBONS

PHYTANE (C_{20})

PRISTANE (C_{19})

The discovery of isoprenoid hydrocarbons first in petroleum, later in sediments, marked an important step in the evolution of molecular organic geochemistry. Isoprenoid hydrocarbons have a structural regularity that strongly implicates biological precursors.

Dean and Whitehead (Paper 8) first announced the finding of phytane (2, 6, 10, 14-tetramethylhexadecane) and the tentative identification of pristane (2, 6, 10, 14-tetramethylpentadecane) in petroleum. Immediately following the appearance of this paper, Bendoraitis, Brown, and Hepner (1962) positively identified pristane and tentatively identified phytane in petroleum. On the basis of structural similarities these scientists suggested that these isoprenoid hydrocarbons most likely are derived from the phytol component of chlorophyll, which, as was mentioned earlier, also serves as a possible precursor for certain porphyrins. See Treibs, Paper 1. Pristane and phytane are only two of a series of isoprenoid hydrocarbons commonly found in petroleum and ancient sediments. For example, seven isoprenoid hydrocarbons from C_{14} through C_{21} (C_{17} missing) were found in a crude oil from East Texas (Bendoraitis, Brown, and Hepner, 1963). In Green River Formation oil shale Cummins and Robinson (1964) found C_{15}, C_{16}, C_{18}, C_{19}, and C_{20} isoprenoid hydrocarbons. The rare C_{17} isoprenoid hydrocarbon was first found by McCarthy and Calvin (1967) in a Devonian shale and they suggest squalene as a possible precursor. The first report of an isoprenoid hydrocarbon in recent sediments was by Blumer and Snyder (1965). Since they were first discovered, isoprenoid hydrocarbons, particularly pristane and phytane, have been cited as chemical fossils or biochemical markers in numerous papers. (See Maxwell et al., 1971, for review.) These compounds have been considered in petroleum geochemistry as recommended by Welte (1965) and as important parameters in environmental geochemistry. (See Blumer and Sass, Paper 39.)

n-ALKANOIC AND *n*-ALKENOIC ACIDS

PALMITIC ACID (C_{16})

OLEIC ACID ($C_{18:1}$)

The occurrence of saturated, straight-chain fatty acids (*n*-alkanoic acids) in sediments was first clearly defined by Abelson and Parker (1962) and by Cooper (Paper 9), although Trask and Wu in Paper 2 mentioned observing these kinds of compounds in modern sediments thirty years earlier. The *n*-alkanoic acids differ from *n*-alkanes by having a carboxylic acid group at one end of the carbon chain. Abelson and Parker showed that both recent and ancient sediments contain *n*-alkanoic acids C_{14}, C_{16}, and C_{18}. In the same year Cooper found in sediments not only these molecules but also a complete series of *n*-alkanoic acids that have both even and odd numbers of carbon atoms—from C_{10} to C_{36}. Later work elaborated on these observations (Lawlor and Robinson, 1965; Kvenvolden, 1970) and showed that the distributions of fatty acids varied greatly in sediments, with younger sediments generally showing a greater concentration of acids with even-carbon numbers relative to concentrations of acids with odd-carbon numbers. For an early review of this work see Kvenvolden (1967). Attention was called to the possible relationships that might exist between distributions of *n*-alkanoic acids and *n*-alkanes, and mechanisms were proposed to account for these relationships (Cooper and Bray, 1963).

Straight-chain, unsaturated fatty acids (*n*-alkenoic acids) are also found in modern sediments but usually disappear quickly in the geologic record. The *n*-alkenoic acids are structurally similar to *n*-alkanoic acids except for the double bonds. Parker and Leo (Paper 10) observed both saturated and unsaturated fatty acids in algal mats of recent sediments of the Texas Gulf Coast. The unsaturated fatty acids, oleic acid ($C_{18:1}$) and palmitoleic acid ($C_{16:1}$), were identified. These compounds were most abundant at the surface but their concentrations decreased rapidly with depth. Because their geologic history appears to be short, unsaturated fatty acids became the first candidates for organic geochemical studies utilizing radiolabeled compounds. Rhead et al. (Paper 11) used 9, 10-^{3}H, 1-^{14}C-oleic acid to follow the short-term fate of this molecule in modern sediments. By tracing the labeled products resulting from short-term diagenesis, the pathways of geochemical reactions can be postulated. This work represented a new and innovative approach to molecular organic geochemistry. Early success with the method led to its application in studies of the early diagenesis of sterols (Gaskell and Eglinton, 1975), phytol (Brooks and Maxwell, 1974), and a reevaluation of oleic acid (Gaskell et al., 1976).

BRANCHED-CHAIN ACIDS

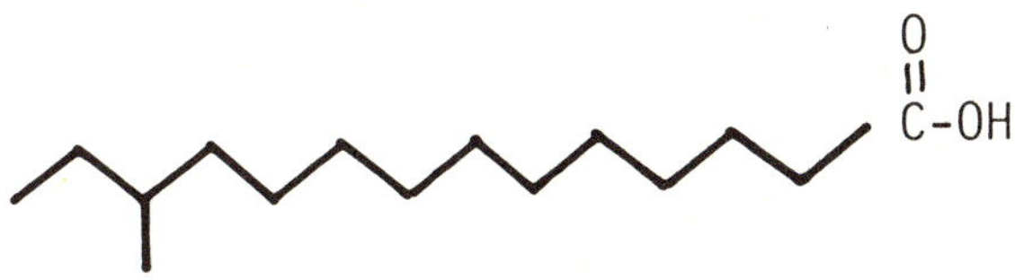

ANTEISO-C_{15} ACID

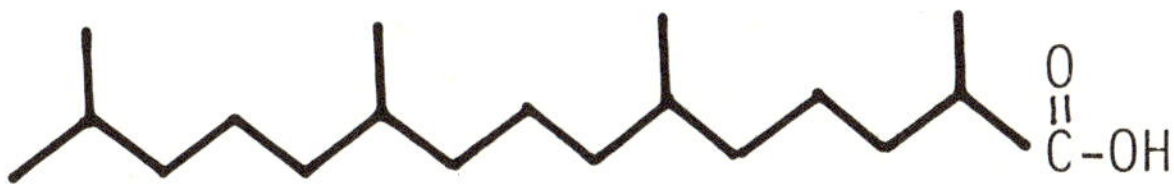

NORPHYTANIC "PRISTANIC" ACID (C_{19})

Straight-chain fatty acids are the simplest of the carboxylic acids. Greater structural complexity is shown by the branched-chain acids such as the iso- and anteiso-acids found by Leo and Parker (1966) in modern marine sediment and in Green River Formation oil shale. These organic acids are characterized by having long chains of carbon atoms with methyl branches at either the second (iso) or third (anteiso) carbon atoms from the end of the chain opposite the carboxyl group. Bacteria are commonly cited as the possible precursors for these compounds in sediments (Cooper and Blumer, 1968).

Another kind of branched-chain fatty acids is the type that has the isoprenoid structure—methyl branches at every fourth carbon atom along the chain. Four isoprenoid acids (C_{14}, C_{15}, C_{19}, and C_{20}) were first discovered by Cason and Graham (1965) in petroleum. The C_{19} ("pristanic") and C_{20} ("phytanic") acids were found the following year in Green River Formation oil shale by Eglinton et al. (Paper 12). Further work has shown that this shale contains a series of isoprenoid acids ranging from C_9 to C_{21} (without C_{13} and C_{18}) (Douglas et al., 1971). Isoprenoid acids were also found in modern marine sediments as first demonstrated by Blumer and Cooper (1967). Like the isoprenoid hydrocarbons, the isoprenoid acids are undoubtedly related to isoprenoid compounds found in organisms living at the time the sediments were deposited.

Many other carboxylic acid structures have been found in various geochemical settings. These structures include dicarboxylic, hydroxy, steroid, terpenoid, and aromatic acids. For a comprehensive review see Seifert (1975).

ALCOHOLS

n-HEXADECANOL (C_{16})

DIHYDROPHYTOL (C_{20})

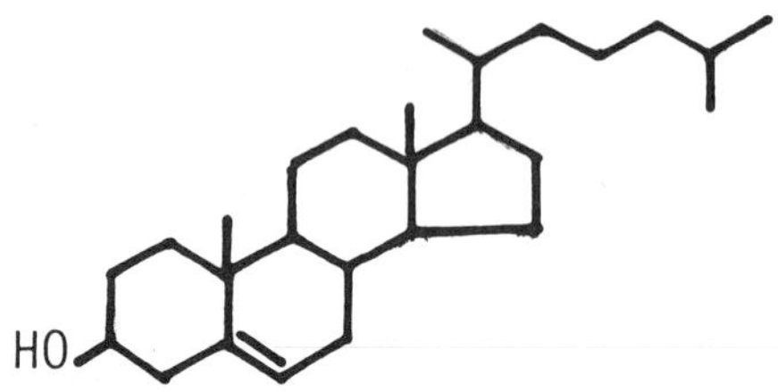

CHOLESTEROL (C_{27})

The straight- and branched-chain alcohols resemble *n*-alkanoic and isoprenoid acids except that their chains terminate with hydroxyl rather than carboxylic acid groups. The discovery of a series of straight-chain alcohols in sediments occurred simultaneously but independently in two different laboratories. Hoering (1968) observed fatty alcohols (C_{16} through C_{28}) in a recent sediment from Tanner Basin, offshore California, and in an ancient sediment, the Green River Formation oil shale. The populations of alcohols were characterized by a dominance of even-carbon-numbered molecules over odd-carbon-numbered molecules, a distributional relationship similar to that of *n*-alkanoic acids. The dominance of even-carbon-numbered alcohols was greater in recent than in ancient sediments. At the same time Sever and Parker (Paper 13) also discovered a similar series of straight-chain alcohols in recent sediments from three different environments and in three different ancient sediments. In addition they observed the isoprenoid alcohol, dihydrophytol, in both recent and ancient sediments, and they suggested that this compound may represent one of the first reaction products in the conversion of phytol to phytane and phytanic acid.

Another class of geochemically important alcohols is the

sterols, also called stenols. These alcohols are cyclic with the basic structural framework of cholesterol. Sterols were mentioned by Trask and Wu in Paper 1, but the first identification by "modern" techniques was by Schwendinger and Erdman (1964) who identified sterols as a class of organic compounds in modern sediments by visible spectroscopy. Six years later Attaway and Parker (Paper 14), using infrared spectroscopy, thin-layer chromatography, gas chromatography, and mass spectrometry, identified four different sterols (cholesterol, stigmasterol, campesterol, and β-sitosterol) in several modern sediments. Later work expanded on the occurrence of sterols in modern marine sediments (Huang and Meinschein, 1976; Lee, Gagosian, and Farrington, 1977) and in lacustrine sediments (Henderson et al., 1971). Besides sterols, triterpene alcohols have also been found in ancient sediments (Mattern, Albrecht, and Ourisson, 1970; Henderson and Steel, 1971).

STERANES, TERPANES, AND AROMATIC HYDROCARBONS

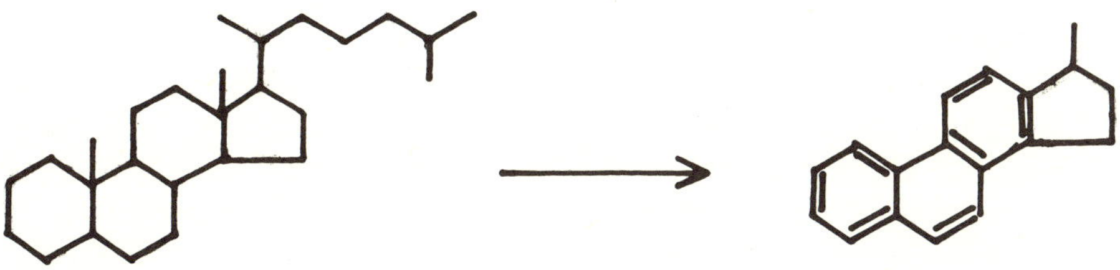

CHOLESTANE (C_{27})

DIELS HYDROCARBON
(AROMATIC HYDROCARBON)

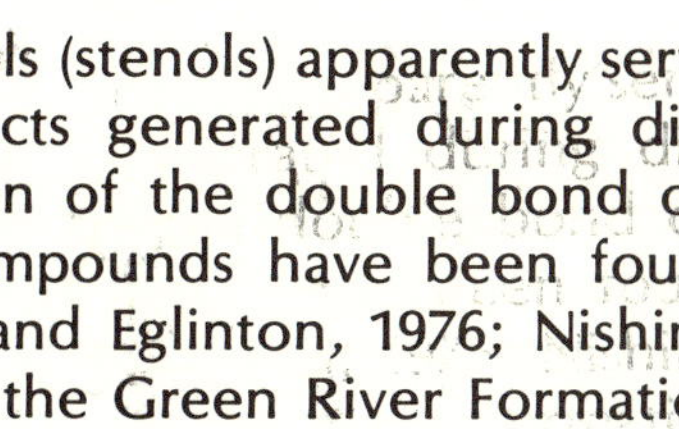

HOPANE (C_{30})

Sterols (stenols) apparently serve as the precursor for a number of products generated during diagenesis. For example, hydrogeneration of the double bond of sterols produces stanols, and these compounds have been found in contemporary sediments (Gaskell and Eglinton, 1976; Nishimura and Koyama, 1976) and in shales of the Green River Formation (Steel and Henderson, 1972).

The short-term diagenesis of stenols to stanols has been followed using ^{14}C-cholesterol as a model (Gaskell and Eglinton, 1974).

The loss of hydroxyl groups from stanols results in steranes (like cholestane). A series of these cyclic hydrocarbons was identified in Green River Formation oil shale by Henderson, Wollrab, and Eglinton (1969) and by Anderson et al. (Paper 15). Cholestane (C_{27}), ergostane (C_{28}), and stigmastane (C_{29}) were positively identified by gas chromatography and mass spectrometry. These same steranes were also identified in petroleum (Hills, Smith, and Whitehead, 1970). The sources of these compounds probably were their corresponding sterols from mainly plant (algal), and, to a lesser extent, animal sources.

Another group of polycyclic alkanes, usually found along with the steranes, are the triterpanes. One example of triterpanes found in ancient sediments and petroleum is the hopane series (Ensminger et al., 1974; Whitehead, 1974). The biological precursors of these compounds are generally thought to have been derived from lower plants and bacteria. For further discussion of these kinds of triterpanes see Hills et al. in Paper 24 and Kimble et al. in Paper 25.

A remarkable similarity in structural framework is evident between certain steranes and aromatic hydrocarbons as well as between certain triterpanes and aromatic hydrocarbons. For example, the similarity between cholestane and Diel's hydrocarbon strongly suggests that the aromatic hydrocarbon may have come from cholestane, or its precursor, cholesterol, by processes involving natural dehydrogenation, and this process may account for many of the aromatic compounds found in sediments (Tissot et al., 1974; Anders, Doolittle, and Robinson 1974) and petroleum (Mair, 1964).

SUGARS

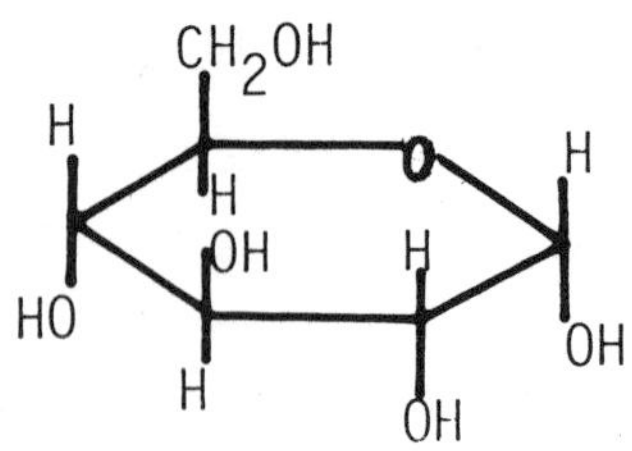

GLUCOSE

Sugars or monosaccharides are polyhydroxyaldehydes or polyhydroxyketones. Sugars must exist in either chain or ring structures in order to account for the many chemical reactions that the sugars undergo. Many sugars joined to each other through glycoside linkages form polysaccharides, and the term carbohydrates refers to a whole family of compounds including sugars and their polymers. Carbohydrates are important constituents of living systems, particularly plant tissue, and therefore have had an opportunity to be incorporated in the geologic record usually as polysaccharides. Analyses of various geological materials have shown that carbohydrates are indeed present. Most of these analyses have involved hydrolysis so that only the individual sugars are recovered. The occurrence and distribution of these sugars have been described in detail. The literature prior to 1963 is well summarized by Vallentyne (1963) who showed that sugars are constituents of soils, fresh water and marine sediments, and ancient woods and lignites. Prashnowsky et al. (1961) describe the sugars in sediments of the Santa Barbara Basin. Ten years later, Modzeleski, Laurie, and Nagy (1971), using gas chromatography–mass spectrometry, confirmed the identity of many of the sugars reported earlier. The occurrence of sugars in ancient sedimentary rocks is discussed by Palacas, Swain, and Smith (Paper 16) whose work led to efforts to use carbohydrates as paleoecologic and stratigraphic markers. (For review, see Swain, 1969.)

AMINO ACIDS

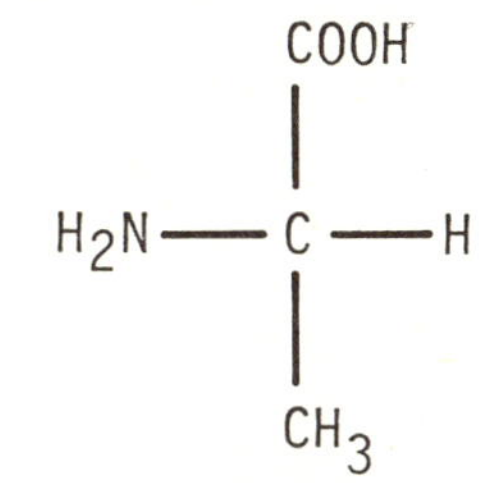

ALANINE

In this book amino acids were introduced in Part I by the paper of Abelson (Paper 5). This class of compounds is reemphasized here in order to make the list of compounds included in Part II as comprehensive as possible. Amino acids are simple molecules having at a minimum one carboxylic acid and one amino functional group. Like the sugars, amino acids are important components of living systems. These acids are the building blocks of proteins that catalyze biochemical reactions and serve as structural components of organisms. Many different amino acids exist, but only about twenty make up proteins, and many of these protein amino acids as well as some nonprotein amino acids are found in geological materials. Abelson in Paper 5 first pointed out the significance of amino acids in fossil shells and bone. Abelson's work was followed up by Erdman, Marlett, and Hanson (Paper 17) who showed that amino acids are also present in sediments and sedimentary rocks. These two papers established the general ubiquity of amino acids in geochemical samples. Since these early observations, the geochemistry of amino acids has received increasing attention. Investigations of amino acids have contributed to a wide range of geochemical topic areas as indicated by the following list:

Degens et al. (1961)—organic material in modern marine sediments
Vallentyne (1964)—thermal reactions
Hare and Mitterer (1967)—nonprotein amino acids in fossil shells
Hare and Abelson (Paper 29)—racemization
Kvenvolden, Ponnamperuma, and Lawless (Paper 45)—amino acids in meteorites
Mitterer (1971)—calcification
King and Hare (1972)—paleotaxonomy
Bada (1972)—geochronology
Mitterer (1972) and Schroeder and Bada (1973)—paleothermometry
Bada, Schroeder, and Carter (1974)—anthropology
Mitterer (1974)—stratigraphy
King (1977)—deep-sea microfossils

The ultimate fate of amino acids in sediments and fossils is not clearly understood. Chemical reactions may lead to low molecular weight hydrocarbons (Erdman, 1961; Thompson and Creath, 1966). Amino acids in sediments may disappear by irreversible uptake by kerogen (Abelson and Hare, 1970) or they may react with other organic materials to yield humic acids and eventually kerogen (Abelson and Hare, 1971; Hoering, 1973).

PORPHYRINS

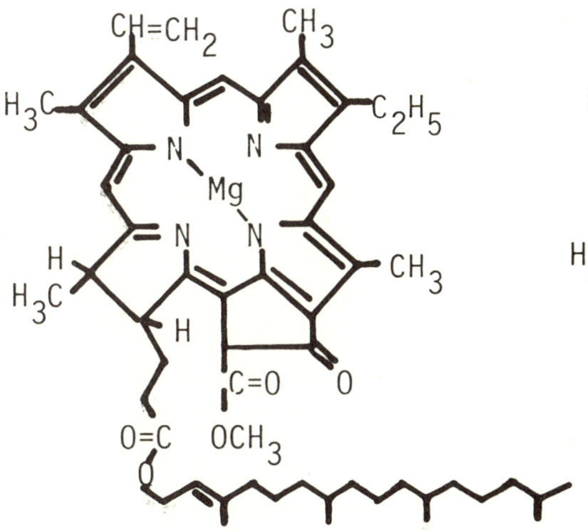

CHLOROPHYLL a

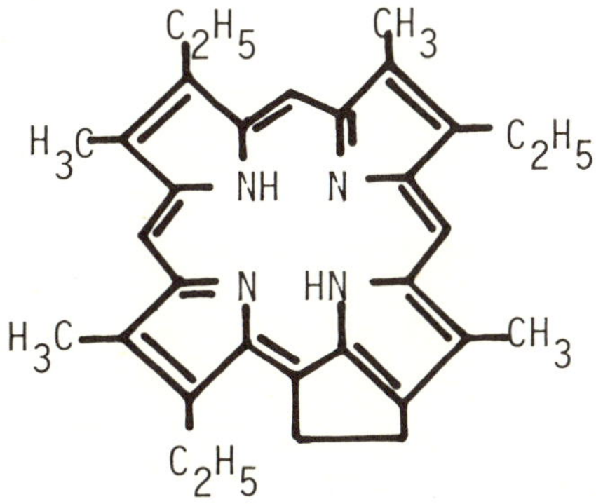

DEOXOPHYLLOERYTHRO-
ETIOPORPHYRIN
(DPEP)

Porphyrins are complex molecules that have a basic structure consisting of four pyrrole rings linked together by methine (–CH=) bridges. The resulting bond system is conjugated, thus making the porphyrin molecule quite stable. Similar tetrapyrrole compounds are the dihydroporphyrins commonly called chlorins (chlorophyll is an example). Chlorins have two additional hydrogen atoms in one of the pyrrole rings. These hydrogen atoms saturate one of the double bonds, disrupting the resonance stability of the porphyrin structure. Because porphyrin compounds are colored, porphyrins are included in the general class of organic compounds called pigments.

The fundamental work of Treibs (see Paper 1) established the presence of porphyrins in petroleum, bituminous shales, and related fossil organic deposits. Metal complexes of deoxophylloerythroetioporphyrins were among the most common porphyrins observed. The major porphyrins were assumed to be derived from chlorophyll through a sequence of intermediate molecules. Much of the research that followed has attempted to define the intermediates and determine the pathways of diagenesis (for example, Orr, Emery, and Grady, 1958). Other work described the occurrence of porphyrins and chlorins in numerous geological materials. (See reviews by Dunning, 1963; Hodgson, Baker, and Peake, 1967).

In early work, the identification of porphyrins was based mainly on their spectroscopic properties. With the application of mass spectrometry, the discovery was made that the porphyrins are commonly present as homologous series and not as individual compounds. Thomas and Blumer (1964) first observed mass spectral evidence for homologous porphyrins in the Serpiano shale. Later, Morandi and Jensen (1966) and Baker (Paper 18) describe in detail the mass spectra of homologous series of porphyrins in oil shale and petroleum. Ten years later Casagrande and Hodgson (1976) showed that homologous series of porphyrins and chlorins occur in near-surface sediments of the Black Sea. The finding of these complex mixtures of porphyrins constitutes one of the major advances in the geochemistry of fossil pigments.

NUCLEIC ACID BASES (PURINES AND PYRIMIDINES)

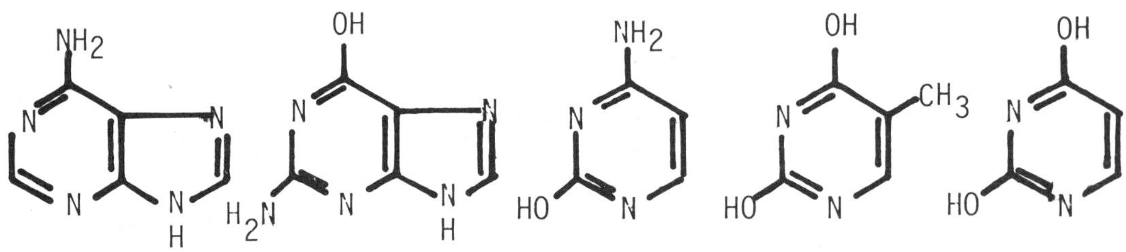

ADENINE GUANINE CYTOSINE THYMINE URACIL

All organisms contain nucleic acid bases that are building blocks for the macromolecules—DNA and RNA—that serve as repositories for genetic information in living systems. In spite of their biological importance, the nucleic acid bases have not been considered extensively in organic geochemistry. Anderson (1961) identified purines (adenine and guanine) and pyrimidines (cytosine and thymine) in humic acids of soils released by hydrolysis, and Rosenburg (1964) reported the presence of these compounds in core samples obtained from the experimental Mohole drilled near Guadalupe Island, Mexico. As with the work on soils, the pyrimidine, uracil, was not detected. Except for a series of papers dealing with the possible presence of purines and pyrimidines in meteorites (Hayatsu, 1964; Folsome et al., 1971, 1973; Hayatsu et al., 1975; and van der Velden and Schwartz, 1977), ten years passed before the geochemistry of nucleic acid bases was again examined. van der Velden and Schwartz (Paper 19) and van der Velden, Chittenden, and Schwartz (1974) reopened this area of organic geochemistry by describing the occurrence of adenine, guanine, cytocine, thymine, uracil, and hypoxanthine in sediments from Lake Erie. Uracil was much less abundant than the other bases, which suggests that uracil is rapidly degraded during early stages of diagenesis. In the few studies that have been done thus far in lake sediments, the total concentrations of purines and pyrimidines decrease rapidly with depth (van der Velden and Schwartz, 1976) and the ultimate fate of these compounds is not yet known.

HUMIC SUBSTANCES AND KEROGEN

Humic substances are the major organic components of soils and modern sediments, while kerogen is the main organic component of ancient sediments. Together these materials constitute the bulk of the organic matter on Earth. These materials are formed from the chemical and biochemical degradation of biological substances that then undergo condensation and polymization reaction to form complex polymeric structures of great stability. Humic substances and kerogen are not well-defined organic compounds. They vary to some extent in composition and properties, depending on the nature of the original organic molecules that formed them. Humic substances are dark-colored materials extracted from soils by alkaline solution and can be partitioned into three fractions: (1) humic acids—base soluble and acid insoluble; (2) fulvic acids—base soluble and acid soluble; and (3) humin—base insoluble and acid insoluble. Kerogen is also dark colored and can be defined as the finely dispersed organic material of sedimentary rocks, which is insoluble in organic and inorganic solvents. Humic substances differ to some extent from kerogen in elemental composition, functional group content, and solubility characteristics. Although these materials differ, a reasonable proposal is that most likely a direct relationship exists between them, with kerogen ultimately being derived from humic substances. (For a review of the geochemistry of humic substances and kerogen, see Welte, 1974.)

Humic substances in soils and terrestrially derived sediments are associations of molecules of polyphenolic, lignin, and condensed lignin origins (Schnitzer, 1976). Oxidative polymerization of phenolic constituents from different sources is favored by many investigators as the appropriate scheme for synthesis (Stevenson and Butler, 1969). But humic substances are also major components of marine sediments where probable source materials lack phenolic constituents. Nissenbaum and Kaplan (Paper 20) recognized that the humic substance in marine sediments could not be accounted for by the conventional explanations used for soils. They proposed that marine humic substances are formed by continuous polymerization of condensed cellular degradation products including mainly amino acids and carbohydrates. Their synthesis scheme proposes kerogen as the end product. Recent work has also shown that kerogenlike material is already present in cell walls of algae and bacteria (Philip and Calvin, 1976).

Despite the fact that kerogen accounts for most of the organic matter dispersed in sedimentary rocks, relatively little attention was given to this substance until Forsman and Hunt (Paper 21) began the preliminary work leading to understanding of possible relationships between kerogen and the formation of petroleum. Out of this work came a recognition of two major types of kerogen. One type has an elemental composition comparable to oil shales and the other is barely distinguishable from coal. These initial observations were followed up by Forsman (1963) who suggested that the organisms that formed noncoaly kerogen may have been the precursors of petroleum, and Abelson (1963) and later, Tissot et al. (see Paper 41), emphasized the importance of kerogen in the origin of petroleum.

CARBON ISOTOPES

The papers considered thus far in this part have been concerned with molecules of differing chemical compositions. Another aspect of organic compounds is their carbon isotopic composition, and this broad subject area is represented by a single paper by Silverman and Epstein (Paper 22). Silverman and Epstein catalogued an extensive list of carbon isotopic analyses of petroleum, other sedimentary organic matter, and plant constituents. From the measurements they were able to use carbon isotopic compositions to distinguish petroleums of marine and non-marine sources, and to define mechanisms for petroleum origins. This work was a natural extension of the classic contributions by Craig (1953), who firmly established studies of carbon isotopes as an important part of the geochemistry of organic molecules. These early papers were followed by many others. Notable advances were made by Park and Epstein (1960) who were able to demonstrate the mechanism for carbon isotopic fractionation during photosynthesis and thus explained many of the relationships among isotopic abundances observed earlier. Sackett et al. (1965) showed that carbon isotopic compositions of marine organisms (plankton) depend on the temperatures of the environment in which they live. Thus the organic materials that are contributed to sediments can have a carbon isotopic composition influenced by both source and temperature. A comprehensive review of the biogeochemistry of carbon isotopes is given by Degens (1969). New directions of research in this area of geochemistry are illustrasted by papers concerned with the inter- and intramolecular isotopic composition of organic molecules (Abelson and Hoering, 1961; Welte, 1969; Meinshein et al., 1974; Galimov, 1974; and DeNiro and Epstein, 1977). Recent reviews by Fuex (1977) and Stahl (1977) show the practical application of the geochemistry of carbon isotopes to the petroleum industry.

REFERENCES CITED

n-Alkanes

Cooper, J. E., and E. E. Bray. 1963. A postulated role of fatty acids in petroleum formation. *Geochim. et Cosmochim. Acta* **27**:1113–1127.

Evans, E. D., G. S. Kenny, W. G. Meinschein, and E. E. Bray. 1957. Distribution of *n*-paraffins and separation of saturated hydrocarbons from recent marine sediments. *Anal. Chem.* **29**:1858–1861.

Scalan, R. S., and J. E. Smith. 1970. An improved measure of the odd-even predominance in the normal alkanes of sediment extracts and petroleum. *Geochim. et Cosmochim. Acta* **34**:611–620.

Speers, G. C., and E. V. Whitehead. 1969. Crude petroleum. In G. Eglinton and M. T. J. Murphy, eds., *Organic Geochemistry—Methods and Results*. Springer-Verlag, Berlin, 638–675.

Stevens, N. P., E. E. Bray, and E. D. Evans. 1956. Hydrocarbons in sediments of the Gulf of Mexico. *Am. Assoc. Petrol. Geol. Bull.* **40**:975–983.

Isoprenoid Hydrocarbons

Bendoraitis, J. G., B. L. Brown, and L. S. Hepner. 1962. Isoprenoid hydrocarbons in petroleum—isolation of 2, 6, 10, 14-tetramethylpentadecane by high temperature gas-liquid chromatography. *Anal. Chem.* **34**:49–53.

Bendoraitis, J. G., B. L. Brown, and L. S. Hepner. 1963. Isolation and identification of isoprenoids in petroleum. *6th World Petrol. Congress Proc.* Sec. V, Paper 15, 1–16.

Blumer, M., and W. D. Snyder. 1965. Isoprenoid hydrocarbons in recent sediments: Presence of pristane and probable absence of phytane. *Science* **150**:1588–1589.

Cummins, J. J., and W. E. Robinson. 1964. Normal and isoprenoid hydrocarbons isolated from oil-shale bitumen. *Jour. Chem. and Eng. Data* **9**:304–307.

McCarthy, E. D., and M. Calvin. 1967. The isolation and identification of the C_{17} saturated isoprenoid hydrocarbon 2, 6, 10-trimethyltetradecane, from a Devonian shale. The role of squalene as a possible precursor. *Tetrahedron* **23**:2609–2619.

Maxwell, J. R., C. T. Pillinger, and G. Eglinton. 1971. Organic geochemistry. *Quart. Rev., The Chem. Soc.* **25**:571–628.

Welte, D. H. 1965. Relation between petroleum and source rock. *Am. Assoc. Petrol. Geol. Bull.* **49**:2246–2268.

n-Alkanoic and *n*-Alkenoic Acids

Abelson, P. H., and P. L. Parker. 1962. Fatty acids in sedimentary rocks. *Carnegie Inst. Washington Year Book* **61**:181–184.

Brooks, P. W., and J. R. Maxwell. 1974. Early stage fate of phytol in a recently deposited lacustrine sediment. In B. Tissot and F. Bienner, eds., *Advances in Organic Geochemistry 1973.* Editions Technip., Paris, 977–991.

Cooper, J. E., and E. E. Bray. 1963. A postulated role of fatty acids in petroleum formation. *Geochim. et Cosmochim. Acta* **27**:1113–1127.

Gaskell, S. J., and G. Eglinton. 1975. Rapid hydrogenation of sterols in a contemporary lacustrine sediment. *Nature* **254**:209–211.

Gaskell, S. J., M. M. Rhead, P. W. Brooks, and G. Eglinton. 1976. Diagenesis of oleic acid in an estuarine sediment. *Chem. Geology* **17**:319–324.

Kvenvolden, K. A. 1967. Normal fatty acids in sediments. *Jour. Am. Oil Chem. Soc.* **44**:628–636.

Kvenvolden, K. A. 1970. Evidence for transformations of normal fatty acids in sediments. In G. D. Hobson and G. C. Speers, eds., *Advances in Organic Geochemistry 1966.* Pergamon Press, Oxford, 335–366.

Lawlor, D. L., and W. E. Robinson. 1965. Fatty acids in Green River Formation oil shale. *Preprint, Dir. Petrol. Chem., Am. Chem. Soc., Detroit meeting,* 5–9.

Branched-chain Acids

Blumer, M., and W. J. Cooper. 1967. Isoprenoid acids in recent sediments. *Science* **158**:1463–1464.

Cason, J., and D. W. Graham. 1965. Isolation of isoprenoid acids from a California petroleum. *Tetrahedron* **21**:471–483.

Cooper, W. J., and M. Blumer. 1968. Linear, *iso* and *anteiso* fatty acids in recent sediments of the North Atlantic. *Deep-Sea Research* **15**: 535–540.

Douglas, A. G., M. Blumer, G. Eglinton, and K. Douraghi-Zadeh. 1971. Gas chromatographic-mass spectrometric characterization of naturally occurring acyclic isoprenoid carboxylic acids. *Tetrahedron* **27**: 1071–1092.

Leo, R. F., and P. L. Parker. 1966. Branched-chain fatty acids in sediments. *Science* **152**:649–650.

Seifert, W. K. 1975. Carboxylic acids in petroleum and sediments. In W. Herz, H. Grisebach, and G. W. Kirby, eds., *Progress in the Chemistry of Organic Natural Products,* **32**. Springer-Verlag, N.Y., 1–49.

Alcohols

Henderson, W., and G. Steel. 1971. Isolation and characterization of a triterpenoid alcohol from the Green River Shale. *Chem. Commun.* **1971**:1331–1332.

Henderson, W., W. E. Reed, G. Steel, and M. Calvin. 1971. Isolation and identification of sterols from a Pleistocene sediment. *Nature* **231**: 308–310.

Hoering, T. C. 1968. Fatty alcohols in sedimentary rocks. *Carnegie Inst. Washington Year Book* **67**:202–203.

Huang, W-Y., and W. G. Meinschein. 1976. Sterols as source indicators of organic materials in sediments. *Geochim. et Cosmochim. Acta* **40**:323–330.

Lee, C. L., R. B. Gagosian, and J. W. Farrington. 1977. Sterol diagenesis in recent sediments from Buzzards Bay, Massachusetts. *Geochim. et Cosmochim. Acta* **41**:985–992.

Mattern, G., P. Albrecht, and G. Ourisson. 1970. 4-Methyl-sterols and sterols in Messel shale (Eocene). *Chem. Commun.* **1970**:1570–1571.

Schwendinger, R. B., and J. G. Erdman. 1964. Sterols in recent aquatic sediments. *Science* **144**:1575–1576.

Steranes, Terpanes, and Aromatic Hydrocarbons

Anders, D. E., F. G. Doolittle, and W. E. Robinson. 1973. Analysis of some aromatic hydrocarbons in a benzene-soluble bitumen from Green River shale. *Geochim. et Cosmochim. Acta* **37**:1213–1228.

Ensminger, A., A. van Dorsselaer, Ch. Spyckerelle, P. Albrecht, and G. Ourisson. 1974. Pentacyclic triterpanes of the hopane type as ubiquitous geochemical markers: Origin and significance. In B. Tissot and F. Bienner, eds., *Advances in Organic Geochemistry 1973*. Editions Technip., Paris, 245–260.

Gaskell, S. J., and G. Eglinton. 1974. Short-term diagenesis of sterols. In B. Tissot and F. Bienner, eds., *Advances in Organic Geochemistry 1973*. Editions Technip., Paris, 963–976.

Gaskell, S. J., and G. Eglinton. 1976. Sterols of a contemporary lacustrine sediment. *Geochim. et Cosmochim. Acta* **40**:1221–1228.

Hills, I. R., G. W. Smith, and E. V. Whitehead. 1970. Hydrocarbons from fossil fuels and their relationship with living organisms. *Jour. Inst. Petrol.* **56**:127–137.

Henderson, W., V. Wollrab, and G. Eglinton. 1969. Identification of steranes and triterpanes from a geological source by capillary gas liquid chromatography and mass spectrometry. In P. A. Schenck and I. Havenaar, eds., *Advances in Organic Geochemistry 1968*. Pergamon Press, Braunschweig, 181–207.

Mair, B. J. 1964. Terpenoids, fatty acids and alcohols as source materials for petroleum hydrocarbons. *Geochim. et Cosmochim. Acta* **28**:1303–1321.

Nishimura, M., and T. Koyama. 1976. Stenols and stanols in lake sediments and diatoms. *Chem. Geology* **17**:229–239.

Steel, G., and W. Henderson. 1972. Isolation and characterization of stanols from the Green River Shale. *Nature* **238**:148–149.

Tissot, B., J. Espitalie, G. Deroo, C. Tempere, and D. Jonathan. 1974. Origine et migration des hydrocarbures dans le Sahara Oriental (Algerié). In B. Tissot and F. Bienner, eds., *Advances in Organic Geochemistry 1973*. Editions Technip., Paris, 315–334.

Whitehead, E. V. 1974. The structure of petroleum pentacyclanes. In B. Tissot and F. Bienner, eds., *Advances in Organic Geochemistry 1973*. Editions Technip., Paris, 225–243.

Sugars

Modzeleski, J. E., W. A. Laurie, and B. Nagy. 1971. Carbohydrates from Santa Barbara Basin sediments: Gas chromatographic-mass spectrometric analyses of trimethylsilyl derivatives. *Geochim. et Cosmochim. Acta* **35**:825–838.

Prashnowsky, A., E. T. Degens, K. O. Emery, and J. Pimenta. 1961. Organic materials in recent and ancient sediments—I. Sugars in marine sediments of Santa Barbara Basin, California. *Neues Jahrb. Geol. Paleontol. Monatsh.* **8**:400–413.

Swain, F. M. 1969. Fossil carbohydrates. In G. Eglinton and M. T. J. Murphy, eds., *Organic Geochemistry—Methods and Results*. Springer-Verlag, Berlin, 374–400.

Vallentyne, J. R. 1963. Geochemistry of carbohydrates. In I. Breger, ed., *Organic Geochemistry*. Macmillan, N. Y., 456–502.

Amino Acids

Abelson, P. H., and P. E. Hare. 1970. Uptake of amino acids by kerogen. *Carnegie Inst. Washington Year Book* **68**:297–303.

Abelson, P. H., and P. E. Hare. 1971. Reactions of amino acids with natural and artificial humus and kerogens. *Carnegie Inst. Washington Year Book* **69**:327–334.

Bada, J. L. 1972. The dating of fossil bones using the racemization of isoleucine. *Earth and Planetary Sci. Letters* **15**:223–231.

Bada, J. L., R. A. Schroeder, and G. F. Carter. 1974. New evidence for the antiquity of man in North America deduced from aspartic acid racemization. *Science* **184**:791–793.

Degens, E. T., A. Prashnowsky, K. O. Emery, and J. Pimenta. 1961. Organic materials in recent and ancient sediments--II. Amino acids in marine sediments of Santa Barbara Basin, California. *Neues Jahrb. Geol. Paleontol. Monatsh.* **8**:413–426.

Erdman, J. G. 1961. Some chemical aspects of petroleum genesis as related to the problem of source bed recognition. *Geochim. et Cosmochim. Acta* **22**:16–36.

Hare, P. E., and R. M. Mitterer. 1967. Nonprotein amino acids in fossil shells. *Carnegie Inst. Washington Year Book* **65**:362–364.

Hoering, T. C. 1973. A comparison of melanoidin and humic acid. *Carnegie Inst. Washington Year Book* **72**:682–690.

King, K., Jr., 1977. Amino acid survey of recent calcareous and siliceous deep-sea microfossils. *Micropaleontology* **23**:180–193.

King, K., Jr., and P. E. Hare. 1972. Amino acid composition of planktonic foraminifera: A paleobiochemical approach to evolution. *Science* **175**:1461–1463.

Mitterer, R. M. 1971. Influence of natural organic matter in $CaCO_3$ precipitation. In O. Bricker, ed., *Carbonate Cements*. Johns Hopkins Univ. Press, Baltimore, 252–258.

Mitterer, R. M. 1972. Calcified protein in the sedimentary environment. In H. R. v. Gaertner and H. Wehner, eds., *Advances in Organic Geochemistry 1971*. Pergamon Press, Braunschweig, 441–452.

Mitterer, R. M. 1974. Pleistocene stratigraphy in southern Florida based on amino acid diagenesis in fossil *Mercenaria*. *Geology* **2**:425–428.

Schroeder, R. A., and J. L. Bada. 1973. Glacial-postglacial temperature differences deduced from aspartic acid racemization in fossil bones. *Science* **182**:479–482.

Thompson, R. R., and W. B. Creath. 1966. Low molecular weight hydrocarbons in recent and fossil shells. *Geochim. et Cosmochim. Acta* **30**:1137–1152.

Vallentyne, J. R. 1964. Biogeochemistry of organic matter—II. Thermal reaction kinetics and transformation products of amino compounds. *Geochim. et Cosmochim. Acta* **28**:157–188.

Porphyrins

Casagrande, D. J., and G. W. Hodgson. 1976. Geochemistry of porphyrins: The observation of homologous tetrapyrroles in a sediment sample from the Black Sea. *Geochim. et Cosmochim. Acta* **40**:479–482.

Dunning, H. N. 1963. Geochemistry of organic pigments. In I. A. Breger, ed., *Organic Geochemistry*. Macmillan, N. Y. , 367–430.

Hodgson, G. W., B. L. Baker, and E. Peake. 1967. Geochemistry of porphyrins. In B. Nagy and U. Colombo, eds., *Fundamental Aspects of Petroleum Geochemistry*. Elsevier Publ. Co., Amsterdam, 177–259.

Morandi, J. R., and H. B. Jensen. 1966. Comparison of porphyrins from shale oil, oil shale, and petroleum by absorption and mass spectroscopy. *Jour. Chem. and Eng. Data* **11**:81–88.

Orr, W. L., K. O. Emery, and J. R. Grady. 1958. Preservation of chlorophyll derivatives in sediments off southern California. *Am. Assoc. Petrol. Geol. Bull.* **42**:925–962.

Thomas, D. W., and M. Blumer. 1964. Porphyrin pigments of a Triassic sediment. *Geochim. et Cosmochim. Acta* **28**:1147–1154.

Nucleic Acid Bases

Anderson, G. 1961. Estimation of purines and pyrimidines in soil humic acid. *Soil Sci.* **19**:156–161.

Folsome, C. E., J. G. Lawless, M. Romiez, and C. Ponnamperuma. 1971. Heterocyclic compounds indigenous to the Murchison meteorite. *Nature* **232**:108–109.

Folsome, C. E., J. G. Lawless, M. Romiez, and C. Ponnamperuma. 1973. Heterocyclic compounds recovered from carbonaceous chondrites. *Geochim. et Cosmochim. Acta* **37**:455–465.

Hayatsu, R. 1964. Orgueil meteorite: Organic nitrogen contents. *Science* **146**:1291–1293.
Hayatsu, R., M. H. Studier, L. P. Moore, and E. Anders. 1975. Purines and triazines in the Murchison meteorite. *Geochim. et Cosmochim. Acta* **39**:471–488.
Rosenberg, E. 1964. Purine and pyrimidines in sediments from the experimental Mohole. *Science* **146**:1680–1681.
Velden, W. van der, G. J. F. Chittenden, and A. W. Schwartz. 1974. Studies on the geochemistry of purines and pyrimidines. In B. Tissot and F. Bienner, eds., *Advances in Organic Geochemistry 1973.* Editions Technip., Paris, 293–304.
Velden, W. van der, and A. W. Schwartz. 1976. Purines, pyrimidines and organic carbon in lake sediments—A comparison of sediments from lakes of varying degrees of eutrophication. *Chem. Geology* **18**:273–284.
Velden, W. van der, and A. W. Schwartz. 1977. Search for purines and pyrimidines in the Murchison meteorite. *Geochim. et Cosmochim. Acta* **41**:961–968.

Humic Substances and Kerogen

Abelson, P. H. 1963. Organic geochemistry and the formation of petroleum. *6th World Petrol. Cong. Proc.*, Sec. 1, 397–407.
Forsman, J. P. 1963. Geochemistry of kerogen. In I. A. Breger, ed., *Organic Geochemistry*. Macmillan, N.Y., 148–182.
Philp, R. P., and M. Calvin. 1976. Possible origin for insoluble organic (kerogen) debris in sediments from insoluble cell-wall materials of algae and bacteria. *Nature* **262**:134–136.
Schnitzer, M. 1976. The chemistry of humic substances. In J. O. Nriagu, ed., *Environmental Biogeochemistry,* vol. 1. Ann Arbor Science Publ., Ann Arbor, Mich., 89–107.
Stevenson, F. J., and J. H. A. Butler. 1969. Chemistry of humic acids and related pigments. In G. Eglinton and M. T. J. Murphy, eds., *Organic Geochemistry—Methods and Results*. Springer-Verlag, Berlin, 534–557.
Welte, D. 1974. Recent advances in organic geochemistry of humic substances and kerogen—A review. In B. Tissot and F. Bienner, eds., *Advances in Organic Geochemistry 1973.* Editions Technip., Paris, 3–13.

Carbon Isotopes

Abelson, P. H., and T. C. Hoering. 1961. Carbon isotope fractionation in formation of amino acids by photosynthetic organisms. *Proc. Natl. Acad. Sci. USA* **47**:623–632.
Craig, H. 1953. The geochemistry of the stable carbon isotopes. *Geochim. et Cosmochim. Acta* **3**:53–92.
Degens, E. T. 1969. Biogeochemistry of stable carbon isotopes. In G. Eglinton and M. T. J. Murphy, eds., *Organic Geochemistry—Methods and Results*. Springer-Verlag, Berlin, 304–329.

DeNiro, M. J., and S. Epstein. 1977. Mechanism of carbon isotopic fractionation associated with lipid synthesis. *Science* **197**:261–263.

Fuex, A. N. 1977. The use of stable isotopes in hydrocarbon exploration. *Jour. Geochem. Exploration* **7**:155–188.

Galimov, E. M. 1974. Organic geochemistry of carbon isotopes. In B. Tissot and F. Bienner, eds., *Advances in Organic Geochemistry 1973*. Editions Technip., Paris, 439–452.

Meinschein, W. G., G. G. L. Rinaldi, J. M. Hayes, and D. A. Schoeller. 1974. Intramolecular isotopic order in biologically produced acetic acid. *Biomedical Mass Spectrometry* **1974**:172–174.

Park, R., and S. Epstein. 1960. Carbon isotopic fractionation during photosynthesis. *Geochim. et Cosmochim. Acta* **21**:110–126.

Sackett, W. M., W. R. Eckelmann, M. L. Bender, and A. W. H. Be. 1965. Temperature dependence of carbon isotope composition in marine plankton and sediments. *Science* **148**:235–237.

Stahl, W. J. 1977. Carbon and nitrogen isotopes in hydrocarbon research and exploration. *Chem. Geology* **20**:121–149.

Welte, D. H. 1969. C^{13}/C^{12} isotope ratios of individual *n*-paraffins and the implication to their origin. *Naturwissenschaften* **56**:133–134.

SUGGESTED ADDITIONAL READINGS

n-Alkanes

Brooks, J. D., and J. W. Smith. 1967. The diagenesis of plant lipids during the formation of coal, petroleum and natural gas—I. Changes in the *n*-paraffin hydrocarbons. *Geochim. et Cosmochim. Acta* **31**:2389–2397.

Clark, R. C., Jr., and M. Blumer. 1967. Distribution of *n*-paraffins in marine organisms and sediments. *Limnology and Oceanography* **12**:79–87.

Dembicki, H., Jr., W. G. Meinschein, and D. E. Hatten. 1976. Possible ecological and environmental significance of the predominance of even-carbon number C_{20}-C_{30} alkanes. *Geochim. et Cosmochim. Acta* **40**:203–208.

Didyk, B. M., and E. D. McCarthy. 1971. Non-marine character of a Chilean petroleum. *Nature Phys. Sci.* **232**:103–105.

Koons, C. B., G. W. Jamieson, and L. S. Ciereszko. 1965. Normal alkane distribution in marine organisms; Possible significance to petroleum origin. *Am. Assoc. Petrol. Geol. Bull.* **49**:301–304.

Kvenvolden, K. A. 1962. Normal paraffin hydrocarbons in sediments from San Francisco Bay, California. *Am. Assoc. Petrol. Geol. Bull.* **46**:1643–1652.

Leythaeuser, D., and D. H. Welte. 1969. Relation between distribution of heavy *n*-paraffins and coalification in Carboniferous coals from the Saar District, Germany. In P. A. Schenck and I. Havenaar, eds., *Advances in Organic Geochemistry 1968*. Pergamon Press, Braunschweig, 429–442.

Martin, R. L., J. C. Winters, and J. A. Williams. 1963. Distributions of *n*-paraffins in crude oils and their implications to the origin of petroleum. *Nature* **199**:110–113.

Isoprenoid Hydrocarbons

Brooks, J. D., K. Gould, and J. W. Smith. 1969. Isoprenoid hydrocarbons in coal and petroleum. *Nature* **222**:257–259.

Gibert, J. M., I. M. R. de Andrade Bruning, D. W. Nooner, and J. Oró. 1975. Predominance of isoprenoids among the alkanes in the Irati oil shale, Permian of Brazil. *Chem. Geology* **15**:209–215.

Han, J., and M. Calvin. 1969. Occurrence of C_{22}-C_{25} isoprenoids in Bell Creek crude oil. *Geochim. et Cosmochim. Acta* **33**:733–742.

Ikan, R., and A. Bortinger. 1971. Normal and isoprenoid alkanes from an Israeli shale. *Geochim. et Cosmochim. Acta* **35**:1059–1065.

Johns, R. B., T. Belsky, E. D. McCarthy, A. L. Burlingame, P. Haug, H. K. Schnoes, W. Richter, and M. Calvin. 1966. The organic geochemistry of ancient sediments—Part II. *Geochim. et Cosmochim. Acta* **30**: 1191–1222.

Powell, T. G., and D. M. McKirdy. 1973. Relationship between ratio of pristane to phytane, crude oil composition and geological environment in Australia. *Nature Phys. Sci.* **243**:37–39.

n-Alkanoic and *n*-Alkenoic Acids

Douglas, A. G., K. Douraghi-Zadeh, G. Eglinton, J. R. Maxwell, and J. N. Ramsay. 1966. Fatty acids in sediments including the Green River Shale (Eocene) and Scottish torbanite (Carboniferous). In G. D. Hobson and G. C. Speers, eds., *Advances in Organic Geochemistry 1966*. Pergamon Press, Oxford, 315–334.

Farrington, J. W., and J. G. Quinn. 1973. Biogeochemistry of fatty acids in recent sediments from Narragansett Bay, Rhode Island. *Geochim. et Cosmochim. Acta* **37**:259–268.

Kvenvolden, K. A. 1966. Molecular distributions of normal fatty acids and paraffins in some Lower Cretaceous sediments. *Nature* **209**: 573–577.

Parker, P. L. 1969. Fatty acids and alcohols. In G. Eglinton and M. T. J. Murphy, eds., *Organic Geochemistry—Methods and Results*. Springer-Verlag, Berlin, 357–373.

Van Hoeven, W., J. R. Maxwell, and M. Calvin. 1969. Fatty acids and hydrocarbons as evidence of life processes in ancient sediments and crude oils. *Geochim. et Cosmochim. Acta* **33**:877–881.

Branched-chain Acids

Eglinton, G., D. H. Hunneman, and K. Douraghi-Zadeh. 1968. Gas chromatographic-mass spectrometric studies of long chain hydroxy acids—II. The hydroxy acids and fatty acids of a 5000-year-old lacustrine sediment. *Tetrahedron* **24**:5929–5941.

Haug, P., H. K. Schnoes, and A. C. Burlingame. 1967. Isoprenoid and dicarboxylic acids isolated from Colorado Green River Shale (Eocene). *Science* **158**:772–773.

Haug, P., H. K. Schnoes, and A. L. Burlingame. 1968. Aromatic carboxylic acids isolated from Colorado Green River Formation (Eocene). *Geochim. et Cosmochim. Acta* **32**:358–361.

Lochte, H. L. 1952. Petroleum acids and bases. *Indus. and Eng. Chem.* **44**:2597–2601.
Seifert, W. K., E. J. Gallegos, and R. M. Teeter. 1971. First identification of a steroid carboxylic acid in petroleum. *Angew. Chem. Internat. Ed.* **10**:747–748.
Simoneit, B. R., and A. L. Burlingame. 1973. Carboxylic acids derived from Tasmanian tasmanite by extractions and kerogen oxidations. *Geochim. et Cosmochim. Acta* **37**:595–610.

Alcohols

Albrecht, P., and G. Ourisson. 1969. Triterpene alcohol isolation from oil shale. *Science* **163**:1192–1193.
Gagosian, R. B. 1975. Sterols in the western North Atlantic Ocean. *Geochim. et Cosmochim. Acta* **39**:1443–1454.
Hatcher, P. G., L. E. Keister, and P. A. McGillivary. 1977. Steroids as sewage specific indicators in New York Bight sediments. *Bull. Environmental Contamination and Toxicology* **17**:491–498.
Schultz, D. M., and J. G. Quinn. 1974. Measurements of phytol in estuarine suspended organic matter. *Marine Biol.* **27**:143–146.
Steel, G., and W. Henderson. 1972. Rapid method for detection and characterization of steroids. *Anal. Chem.* **44**:1302–1304.

Steranes, Terpanes, and Aromatic Hydrocarbons

Aizenshtat, Z. 1973. Perylene and its geological significance. *Geochim. et Cosmochim. Acta* **37**:559–567.
Anders, D. E., and W. E. Robinson. 1971. Cycloalkane constituents of the bitumen from Green River Shale. *Geochim. et Cosmochim. Acta* **35**: 661–678.
Gallegos, E. J. 1971. Identification of new steranes, terpanes, and branched paraffins in Green River Shale by combined capillary gas chromatography and mass spectrometry. *Anal. Chem.* **43**:1151–1160.
Greiner, A. Ch., C. Spyckerelle, and P. Albrecht. 1976. Aromatic hydrocarbons from geological sources—I. New naturally occurring phenanthrene and chrysene derivatives. *Tetrahedron* **32**:257–260.
Mair, B. J., and J. L. Martinez-Pico. 1962. Composition of the trinuclear aromatic portion of the heavy gas oil and light lubricating distillate. *Proc. Am. Petrol. Inst.* **42**:Sec. 3, 173–185.
Meinschein, W. G. 1969. Hydrocarbons—saturated, unsaturated and aromatic. In G. Eglinton and M. T. J. Murphy, eds., *Organic Geochemistry—Methods and Results*. Springer-Verlag, Berlin, 330–356.
Orr, W. L., and J. R. Grady. 1967. Perylene in basin sediments off southern California. *Geochim. et Cosmochim. Acta* **31**:1201–1209.
Reed, W. E. 1977. Biogeochemistry of Mono Lake, California. *Geochim. et Cosmochim. Acta* **41**:1231–1245.
Speers, G. C., and E. V. Whitehead. 1969. Crude petroleum. In G. Eglinton and M. T. J. Murphy, eds., *Organic Geochemistry—Methods and Results*. Springer-Verlag, Berlin, 638–675.

Sugars

Handa, N., and K. Mizuno. 1973. Carbohydrates from lake sediments. *Geochem. Jour.* **7**:215–230.

Rogers, M. A. 1965. Carbohydrates in aquatic plants and associated sediments from two Minnesota lakes. *Geochim. et Cosmochim. Acta* **29**:183–200.

Swain, F. M., and M. A. Rogers. 1966. Stratigraphic distribution of carbohydrate residues in Middle Devonian Onondaga beds of Pennsylvania and western New York. *Geochim. et Cosmochim. Acta* **30**: 497–509.

Amino Acids

Abelson, P. H. 1963. Geochemistry of amino acids. In I. A. Breger, ed., *Organic Geochemistry*. Macmillan, N. Y., 431–455.

Hare, P. E. 1969. Geochemistry of proteins, peptides and amino acids. In G. Eglinton and M. T. J. Murphy, eds., *Organic Geochemistry—Methods and Results*. Springer-Verlag, Berlin, 438–463.

Kvenvolden, K. A. 1975. Advances in the geochemistry of amino acids. *Ann. Rev. Earth and Planetary Sci.* **3**:183–212.

Schroeder, R. A., and J. L. Bada. 1976. A review of the geochemical applications of the amino acid racemization reaction. *Earth Sci. Rev.* **12**: 347–391.

Porphyrins

Baker, E. W. 1969. Porphyrins. In G. Eglinton and M. T. J. Murphy, eds., *Organic Geochemistry—Methods and Results*. Springer-Verlag, Berlin, 464–497.

Blumer, M., and G. S. Omenn. 1961. Fossil porphyrins: Uncomplexed chlorins in a Triassic sediment. *Geochim. et Cosmochim. Acta* **25**: 81–90.

Didyk, B., Y. I. A. Alturki, C. T. Pillinger, and G. Eglinton. 1975. The petroporphyrins of a Cretaceous oil. *Chem. Geology* **15**:193–208.

Hodgson, G. W., B. Hitchon, K. Taguchi, B. L. Baker, and E. Peake. 1968. Geochemistry of porphyrins, chlorins and polycyclic aromatics in soils, sediments, and sedimentary rocks. *Geochim. et Cosmochim. Acta* **32**:737–772.

Nucleic Acid Bases

Dungworth, G., M. Thijssen, J. Zuurveld, W. van der Velden, and A. W. Schwartz. 1977. Distribution of amino acids, amino sugars, purines and pyrimidines in a Lake Ontario sediment core. *Chem. Geology* **19**:295–308.

Humic Substances and Kerogen

Breger, I. A., and A. Brown. 1962. Kerogen in the Chattanooga Shale. *Science* **137**:221–224.

Flaig, W. 1972. Some physical and chemical properties of humic substances as a basis of their characterization. In H. R. v. Gaertner, and H. Wehner, eds., *Advances in Organic Geochemistry 1971*. Pergamon, Braunschweig, 49–67.

Huc, A. Y., and B. Durand. 1974. Étude des acides humiques et de l'humine de sédiments récents considerés comme precurseurs des kerogenes. In B. Tissot and F. Bienner, eds., *Advances in Organic Geochemistry 1973*. Editions Technip., Paris, 53–71.

McIver, R. D. 1967. Composition of kerogen—Clue to its role in the origin of petroleum. *7th World Petrol. Cong. Proc.* **2**:25–36.

Robinson, W. E. 1969. Kerogen of the Green River Formation. In G. Eglinton and M. T. J. Murphy, eds., *Organic Geochemistry—Methods and Results*. Springer-Verlag, Berlin, 619–637.

Schnitzer, M., and S. U. Khan. 1972. *Humic Substances in the Environment*. Marcel Dekker, N.Y., 327 pp.

Stuermer, D. H., and G. R. Harvey. 1974. Humic substances from seawater. *Nature* **250**:480–481.

Swanson, V. E., and J. G. Palacas. 1965. Humate in coastal sands of northwest Florida. *Geological Survey Bull.* 1214–B, B1–B29.

7

Distribution of *n*-paraffins as a clue to recognition of source beds

E. E. BRAY and E. D. EVANS
Magnolia Petroleum Company, Field Research Laboratory, Dallas, Texas

Abstract—A preference for *n*-paraffins with odd numbers of carbon atoms is widespread in Recent sediments. Analyses of numerous Recent sediment samples indicate that molecules with odd numbers of carbon atoms predominate in the heavy *n*-paraffins for at least the first few thousand years after deposition. The depositional environments studied include a continental shelf, some ocean basins, bays, fresh-water lakes, a saline lake and soils. In contrast, the heavy *n*-paraffins in a representative selection of crude oils are evenly distributed between molecules of even- and odd-carbon number.

Analyses of a limited number of postulated source rocks showed no preference in the heavy *n*-paraffins for molecules of either odd- or even-carbon number. Some shales and limestones ranging from Eocene to Mississippian in age have been examined for possible application of the *n*-paraffin distribution to recognition of source beds.

INTRODUCTION

GEOLOGISTS have long recognized the importance of identification of source rocks of petroleum. Although at present there is no completely satisfactory explanation for the origin of petroleum, it has been assumed often that petroleum was generated from source materials during long periods of burial and compaction. There is evidence indicating that large quantities of hydrocarbons have indeed been generated in the sediments during long periods of burial. BUCKLEY *et al.* (1958) made an extensive study of the distribution of dissolved hydrocarbons in subsurface waters of the East Texas and Gulf Coast areas. Methane and heavier hydrocarbons were found at depths below a few thousand feet. The amount of dissolved gas and the percentage of heavier hydrocarbons increased with depth even though the ground water was not saturated in most of the formations studied. ERDMAN *et al.* (1958) found light aromatic hydrocarbons to be abundant in ancient marine shales and limestones and to be uniformly absent in Recent muds from a variety of environments. These hydrocarbons were apparently formed after considerable depth of burial and compaction. FORSMAN *et al.* (1958) studied rock samples representing a wide variety of ages, lithologies and environments of deposition. Both the absolute concentrations of hydrocarbons and the concentrations relative to total organic matter were apparently greater in ancient rocks than in Recent sediments.

On the other hand, reports of the occurrence of hydrocarbons in Recent sediments by SMITH (1954) gave impetus to an idea earlier proposed by WHITEMORE (1944) that crude oils result from selective preservation of these naturally occurring hydrocarbons. BAKER (1959) has proposed that oils may result from selective accumulation of these hydrocarbons. Reports by STEVENS *et al.* (1956), EVANS *et al.* (1957) and MEINSCHEIN (1958) have shown that though there are similarities between hydrocarbons in some Gulf of Mexico Recent sediments and in petroleum, the relative distributions of hydrocarbon types and molecular size vary considerably. For example molecules with odd numbers of carbon atoms were predominant in the *n*-paraffins of the Recent sediments in contrast to corresponding fractions from crude oils and a petroliferous shale which had no predominance with respect to either odd- or even-carbon numbers.

In this paper, we present *n*-paraffin distribution data for a greater variety of Recent sediments and crude oils than have been published previously and include also similar data for some shales and limestones ranging from Eocene to Mississippian in age. These data are examined for possible application in recognition of source beds.

Description of Samples

Recent sediments

Forty-one Recent sediment samples from marine, bay, fresh-water lake, saline lake and soil environments, were examined (Table 1). Samples from marine environments were obtained by coring the sediments to depths of 30 ft, in some cases. Water depths varied from 30 ft on the Texas and Louisiana Gulf Coasts to over 5000 ft in ocean basins off the shore of Southern California where sediments from the Catalina, San Nicolas, Tanner, West Cortez, Santa Barbara, Santa Cruz and Santa Monica Basins were sampled to depths of 10 ft by use of a gravity coring tool. Recent sediments in these California basins have been adequately described by Emery and Rittenberg (1952).

Samples from bay environments were largely soft to firm muds representing sediments up to 65 ft in depth obtained by coring. The core sites were in the shallow waters of Corpus Christi and Galveston bays on the Texas Gulf Coast.

The lake sediments included corings of soft organic mud from fresh-water lakes in Wisconsin and firm dry sediment from a former saline lake in California. The fresh-water lake samples were obtained from Grassy and Trout lakes of Northern Wisconsin. Sediments ranging from 10 ft to 37 ft in thickness were sampled by corings which penetrated the entire Recent sediment deposits. These sediments have been described by Twenhofel *et al.* (1942, 1945). The saline lake samples were from the 61 to 66 ft interval from Searles Lake located in San Bernardino County, California. Flint and Gale (1958) have described these sediments.

The soils examined were from Texas, Wyoming and Sumatra. These have been previously reported by Meinschein (1958).

Crude oils

Thirty-nine crude oils were selected to represent most of the principal oil-producing areas of the Western Hemisphere (Table 2). The selection includes a wide variety of oils with high, medium and low gravities. The reservoir rocks containing these oils ranged from Pliocene to Cambrian in age.

Ancient sediments

Forty-three samples of fine-grained ancient marine sediments, including shales and a few limestones, were examined (Table 3). Outcrop samples were taken from formations of Cretaceous, Jurassic, Pennsylvanian and Mississippian age in the Big Snowy Mountains of Central Montana. These formations are producing oil at depth within distances of a few miles from the outcrops. Other outcrop samples of marine shales were obtained from the Woodford and Chattanooga formations of Mississippian age in Oklahoma and the Chattanooga near Nashville, Tennessee. Additional samples were taken from shallow coreholes which penetrated to maximum depths of a few hundred feet in outcropping formations of North Central

Table 1. Recent sediments

Location	Description	Water depth (ft)	Depth in sediment (ft)	Ratio odd/even carbon-number
Ocean Basins, Off Shore, Southern California				
Catalina Basin	Olive-brown mud, grey interbeds	4300	1–4	4·9
Catalina Basin	Olive-brown mud, grey interbeds	4300	5–9	5·1
San Nicolas Basin	Olive-brown mud	5300	0–4	2·5
San Nicolas Basin	Olive-brown mud	5300	4–6	2·4
Santa Cruz Basin	Olive-brown mud	5700	1·5–5	3·3
Santa Cruz Basin	Olive-brown mud	5700	5–9	3·6
Santa Monica Basin	Dark grey mud	2900	0–3	3·2
Santa Monica Basin	Dark grey mud	2900	3–6	2·9
Tanner Basin	Light greenish-brown mud	5000	0·5–5	4·3
West Cortez Basin	Olive-brown mud	5700	0–4	3·9
West Cortez Basin	Olive-brown mud	5700	0–9	5·1
Santa Barbara Basin	Olive-brown mud, grey interbeds, finely laminated	1900	1–9	3·1
Bays				
Corpus Christi Bay (J)	Soft, grey, shelly mud	12	0–7	2·6
Corpus Christi Bay (J)	Grey plastic shelly mud	12	10–25	2·7
Corpus Christi Bay (J)	Grey firm creamy clay mud	12	25–35	2·6
Corpus Christi Bay (J)	Grey firm creamy clay mud	12	35–45	2·7
Corpus Christi Bay (J)	Grey soft plastic shelly mud	12	45–50	3·1
Corpus Christi Bay (J)	Grey plastic shelly mud	12	50–55	2·8
Corpus Christi Bay (J)	Grey plastic clay mud	12	55–60	3·8
Corpus Christi Bay (J)	Grey plastic clay mud	12	60–65	5·2
Corpus Christi Bay (H,K)	Grey mud	12	0–40	2·4
Galveston Bay	Grey plastic mud	9	20–30	3·5
Galveston Bay	Grey plastic clay mud	9	30–40	3·0
Galveston Bay	Grey plastic clay mud	9	40–50	2·7
Galveston Bay	Grey plastic clay mud	9	50–60	3·4
Gulf of Mexico				
Louisiana, Continental Shelf	Grey mud	40	0–8	5·5
Sabine Pass	Greyish brown mud	39	0–2	3·7

Table 1—(*contd.*)

Location	Description	Water depth (ft)	Depth in sediment (ft)	Ratio odd/even carbon-number
Fresh-water Lakes, Vilas County, Wisconsin				
North Trout Lake	Dark greenish-grey to black organic mud	4	0–2	4·0
South Trout Lake (core 2)	The same as above	58	0–5	4·7
South Trout Lake (core 2)	The same as above	58	5–10	5·3
South Trout Lake (core 3)	The same as above	58	0–3	4·7
South Trout Lake (core 3)	The same as above	58	3–8	4·6
South Trout Lake (core 3)	The same as above	58	8–12	4·6
Grassy Lake	Dark green to yellowish-brown organic mud	4	13–18	2·8
Grassy Lake	Yellowish-brown to black organic mud	4	18–23	3·6
Grassy Lake	Same as above	4	23–28	4·0
Grassy Lake	Same as above	4	28–33	4·6
Soils				
Lusk, Wyoming	Weathered from rocks of Tertiary age, rolling terrain		2–3	2·9
Sago Sumatra	Tropical soil, swampy		2–3	2·7
Saline Lake, San Bernardino Co., California				
Searles Lake	Parting mud between upper and lower salt body		61·8–62·5	3·8
Searles Lake	Same as above		65·8–66·5	3·5

Texas. These included the Mingus and Strawn shales of Pennsylvanian age and the Woodbine formation and Eagle Ford shale of Cretaceous age. Representative shale samples from a 1000 ft interval of fine-grained sediments of Pennsylvanian age were obtained from deep cores near the Scurry Reef in Scurry County, Texas. Cores of Woodbine shale from both above and below the producing zone in the East Texas field were also obtained.

Analytical Procedures and Data

Radiocarbon ages

Radiocarbon dating was used to verify the ages of the Recent sediment samples. The method of dating has been described by Burke and Meinschein (1955).

Extraction of hydrocarbons from sediments

The procedures used for extraction of the sediments have been described previously by Stevens *et al.* (1956) and also by Meinschein and Kenny (1957).

Table 2. Crude oils

Sample designation	Field	Reservoir formation	Age of reservoir formation	Depth (ft)	Ratio odd/even carbon-number
979	Chimire Field, Venezuela	L—2m	Oligocene	6363	1·08
819	West Cat Canyon, California	Sisquoc	Pliocene	3800	1·04
50	Capitan, California	Vaqueros	Miocene	1400	1·09
786	Gaiselberg, Austria	Tortonian	Miocene	7000	1·02
803	Belgian Anticl, California	Oceanic Zone	Oligocene	5530	1·13
1010	Guario, Venezuela	Naranja	Oligocene	5060	1·11
1201	Puchkirchen, Austria	Lattorfian	Oligocene	8100	1·10
832	Roosevelt, Utah	Green River	Eocene	9360	1·10
836	N. Duchesne, Utah	Green River	Eocene	8840	1·02
719	Pruitt, Texas	Olmos	Cretaceous	5000	0·95
781	Rough Range No. 1, Australia	Birdrong	Cretaceous	3610	1·06
697	Sylvestre, Venezuela	Fortuna	Cretaceous	9600	1·09
847	Tow Creek, Colorado	Mancos	Cretaceous	3800	1·03
1505	Walter Fair, Texas	Paluxy	Cretaceous	5400	0·91
1513	Ham Gossett, Texas	Woodbine	Cretaceous	3620	0·95
1517	Ham Gossett, Texas	Rodessa	Cretaceous	6220	0·95
1528	South Currie, Texas	Woodbine	Cretaceous	3000	0·91
1530	Ashby-Ramsey, Texas	Woodbine	Cretaceous	3200	0·92
1554	Grimes-Percilla, Texas	Woodbine	Cretaceous	5900	1·01
1555	Elkhart, Texas	Woodbine	Cretaceous	5800	0·99
1576	West Jacksonville, Texas	Rodessa	Cretaceous	8600	0·99
223	Lost Soldier, Wyoming	Morrison	Jurassic	1900	1·04
729	Omati No. 1, Papua (New Guinea)	—	Jurassic	13,700	1·08
864	North Premier, Saskatchewan	Shaunavon	Jurassic	3500	0·97
826	Rangley, Colorado	Shinarump	Triassic	4500	1·01
923	Virgin, Utah	Moenkopi	Triassic	500	0·99
343	Cross Roads, New Mexico	San Andres	Permian	4840	0·94
743	Dove Creek, Colorado	Hermosa	Pennsylvanian	6150	1·03
1113	Egmanton, England	Millstone Grit	Pennsylvanian	2100	1·05
1296	Kelly Snyder, Texas	Canyon	Pennsylvanian	6900	1·07
285	Garland, Wyoming	Madison	Mississippian	4200	1·02
351	Denton, New Mexico	"Miss. lime"	Mississippian	11,800	1·00
1145	Caddo, Oklahoma	Woodford	Mississippian	4400	1·06
29	Pegasus, Texas	"Dev. lime"	Devonian	11,900	1·00
361	Redwater, Alberta	D—3	Devonian	3200	1·03
25	Pegasus, Texas	Fusselman	Silurian	12,000	1·03
349	Brunson, New Mexico	Montoya	Ordovician	7800	1·03
1192	Willowdale, Kansas	Viola	Ordovician	4400	1·06
26	Pegasus, Texas	Ellenburger	Cambro-Ordovician	12,800	1·05
228	Lost Soldier, Wyoming	Deadwood	Cambrian	5150	0·99

The samples were air dried, and then extracted by a ball milling technique using as solvent a mixture of ten parts benzene and one part methanol. The solvent was subsequently removed from the organic extract by forced evaporation at 40°C. This step had the further effect of stabilizing the extract residues by removing the more volatile constituents in a reproducible manner retaining those *n*-paraffins having eighteen or more carbon atoms per molecule. The crude oils examined were also stabilized in a similar manner.

Isolation of n-*paraffins*

Enriched *n*-paraffin fractions were obtained by use of elution chromatography and in some instances by urea adduction. The methods and reagents used have been described by EVANS *et al.* (1957). The stabilized residues of the crude oils and of the sediment extracts were dissolved in *n*-heptane and placed on activated silica

Table 3. Ancient sediments

Sample designation	Locality	Description	Formation	Age	Hydrocarbon content (wt. %)	Ratio odd/even carbon-number
Core 24, 5–8	Rifle, Colorado	Oil Shale	Green River	Eocene	0·60	2·3
		Marine Shales				
165, 1–9	Rusk Co., Texas	Grey shale, depth 3629–54 ft	Woodbine	Cretaceous	0·0045	1·49
165, 10–15	Rusk Co., Texas	Grey shale, depth 3658–86 ft	Woodbine	Cretaceous	0·0028	1·47
162, 3–2	Denton Co., Texas	Dark grey shale, depth 15–50 ft	L. Woodbine	Cretaceous	0·0021	1·39
162, 4–1	Dallas Co., Texas	Grey marl, depth 25–35 ft	Taylor	Cretaceous	0·0031	1·00
170, 79	Fergus Co., Montana	Black-grey shale, depth 20–30 ft	Colorado	Cretaceous	0·0043	1·79
170, 79	Fergus Co., Montana	Dark grey shale, depth 35–45 ft	Colorado	Cretaceous	0·0053	1·85
173, 32–38	Dallas Co., Texas	Dark grey fissile shale, depth 386–445 ft	Eagle Ford	Cretaceous	0·0025	2·2
173, 55	Dallas Co., Texas	Dark brown lignitic shale, depth 573–575 ft	Eagle Ford	Cretaceous	0·060	2·3
174, 10	Rusk Co., Texas	Dark grey shale, depth 3466 ft	Woodbine	Cretaceous	0·0020	1·32
174, 13–20	Rusk Co., Texas	Grey calcareous shale, depth 3471 ft	Woodbine	Cretaceous	0·0012	1·70
170, 85	Fergus Co., Montana	Greyish black shale, near surface	Morrison	Jurassic	0·0011	1·18
162, 1–1A	Palo Pinto Co., Texas	Grey shale, depth 17–22 ft	Strawn	Pennsylvanian	0·0062	1·04
162, 1–1B	Palo Pinto Co., Texas	Dark grey shale, depth 25–27 ft	Strawn	Pennsylvanian	0·0095	1·01
162, 1–1C	Palo Pinto Co., Texas	Grey brown shale, depth 32–40 ft	Strawn	Pennsylvanian	0·0055	1·03
162, 2–1	Palo Pinto Co., Texas	Grey shale, depth 23–30 ft	Strawn	Pennsylvanian	0·010	1·08
172, 1–2	Scurry Co., Texas	Black shale, depth 6851 ft	Canyon	Pennsylvanian	0·096	1·22
172, 1–2A	Scurry Co., Texas	Black shale, depth 6853 ft	Canyon	Pennsylvanian	0·090	1·22
172, 1–2B	Scurry Co., Texas	Black shale, depth 6856 ft	Canyon	Pennsylvanian	0·060	1·16
172, 3–4	Scurry Co., Texas	Black shale, depth 6859 ft	Canyon	Pennsylvanian	0·025	1·31
172, 3–4A	Scurry Co., Texas	Black shale, depth 6862 ft	Canyon	Pennsylvanian	0·020	1·22
173, 3–4B	Scurry Co., Texas	Dark grey shale, depth 6865 ft	Canyon	Pennsylvanian	0·018	1·26
172, 6–7	Scurry Co., Texas	Black shale, depth 6871 ft	Canyon	Pennsylvanian	0·014	1·56
172, 6–7A	Scurry Co., Texas	Dark grey shale, depth 6874 ft	Canyon	Pennsylvanian	0·013	1·20
172, 8	Scurry Co., Texas	Black fissile shale, depth 6876 ft	Canyon	Pennsylvanian	0·067	1·23
172, 9–10	Scurry Co., Texas	Black shale, depth 6880 ft	Canyon	Pennsylvanian	0·027	1·31
172, 12–13	Scurry Co., Texas	Black fissile shale, depth 6895 ft	Canyon	Pennsylvanian	0·024	1·34
172, 14–16	Scurry Co., Texas	Black fissile shale, depth 6913 ft	Canyon	Pennsylvanian	0·036	1·25
172, 22–24	Scurry Co., Texas	Black fissile shale, depth 6972 ft	Canyon	Pennsylvanian	0·13	1·16
172, 22–24A	Scurry Co., Texas	Black dense shale, depth 6987 ft	Canyon	Pennsylvanian	0·15	1·05
170, 27	Fergus Co., Montana	Grey sandy shale, near surface	Amsden	Pennsylvanian	0·013	1·11
170, 22	Fergus Co., Montana	Grey silty shale, near surface	Otter	Mississippian	0·039	1·44
170, 71	Fergus Co., Montana	Black shale, near surface	Tyler	Mississippian	0·0095	2·2
170, 64	Fergus Co., Montana	Black shale, near surface	Heath	Mississippian	0·0045	1·18
170, 43	Fergus Co., Montana	Black shale, near surface	Heath	Mississippian	0·027	1·03
146, 34	Cherokee Co., Oklahoma	Black shale, near surface	Chattanooga	Mississippian	0·15	0·99
188–B3	De Kalb Co., Tennessee	Dark brown shale, near surface	Chattanooga	Mississippian	0·20	1·00
1191	Kingman Co., Kansas	Dark grey shale, depth 4381 ft	Chattanooga	Mississippian	0·085	1·05
158, core 211	Carter Co., Oklahoma	Dark shale, near surface	Woodford	Mississippian	0·13	1·02
		Limestone				
172, 33	Scurry Co., Texas	Limestone, depth 7917 ft	Canyon	Pennsylvanian	0·30	0·98
170, 3	Fergus Co., Montana	Limestone, near surface	Heath	Mississippian	0·0080	1·15
170, 9, 10	Fergus Co., Montana	Black limestone, near surface	Heath	Mississippian	0·0045	1·00
170, 65	Fergus Co., Montana	Black limestone, near surface	Heath	Mississippian	0·0030	1·15

gel columns pre-wet with *n*-heptane. Additional aliquots of *n*-heptane were then passed through the columns. The resulting saturated hydrocarbon eluates were recovered by blowing away the eluent with filtered air at 40°C. Fractions enriched in *n*-paraffins were obtained by separation of these saturated hydrocarbons on alumina columns using alumina to sample ratios of approximately 10,000 to 1. *n*-Heptane followed by carbon tetrachloride was used as eluent. By use of these

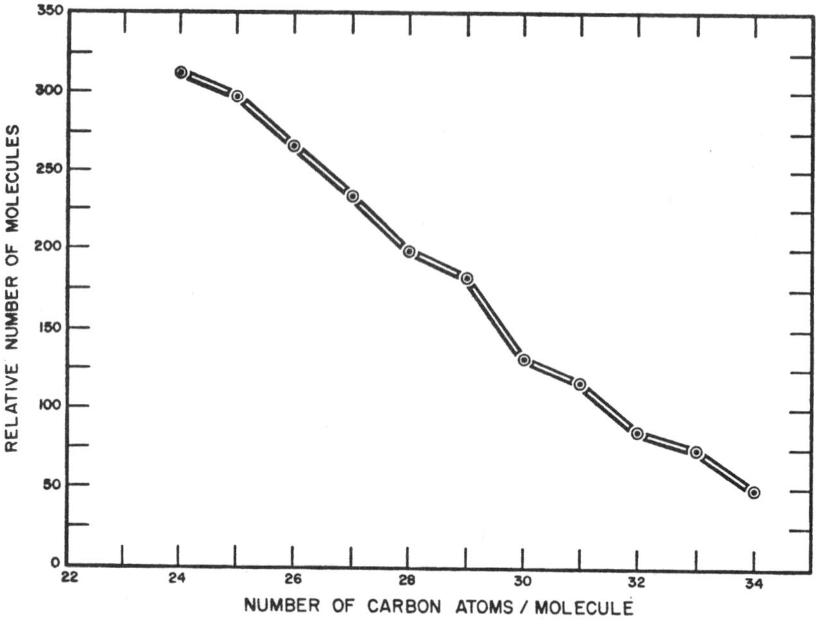

Fig. 1. Distribution of heavy *n*-paraffins recovered from a crude oil (Kelly-Snyder field, Scurry Co., Texas).

procedures, EVANS *et al.* (1957) have separated naphthenes from alkanes. The carbon tetrachloride eluent was shown to recover nearly all the *n*-paraffins ranging in size from twenty-four to thirty carbon atoms per molecule.

Measurement of n-*paraffin distribution*

Distributions of the *n*-paraffins in the fractions were determined from mass spectra. These data were obtained with a Consolidated Electrodynamics Corporation model 21–103 mass spectrometer modified for use with high molecular weight compounds as described by O'NEAL and WIER (1951). The mass spectral data were corrected to a monoisotopic basis for carbon and hydrogen, normalized to the major peak, and arranged in the grid form used by CLERC *et al.* (1955). The heavy *n*-paraffin parent peaks identified in this grid arrangement were calibrated using pure materials. Sensitivity and diffusion corrections obtained from the calibration data were used to adjust *n*-paraffin parent peak intensities to a relative molecular basis. These data give *n*-paraffin distributions such as the one shown for a crude oil in Fig. 1. Numerical values on the ordinate represent relative numbers of molecules. Numbers of carbon atoms per molecule are shown on the abscissa. The graph shows an almost equal distribution between n-paraffin molecules containing odd and even numbers of carbon atoms. A very small prominence may be observed for n-paraffin molecules with 29 carbon atoms. Fig. 2 shows distributions obtained in a similar manner for heavy n-paraffins separated from extracts of a Recent sediment from a marine basin. A striking feature of this graph is the predominance of molecules having odd numbers of carbon atoms,

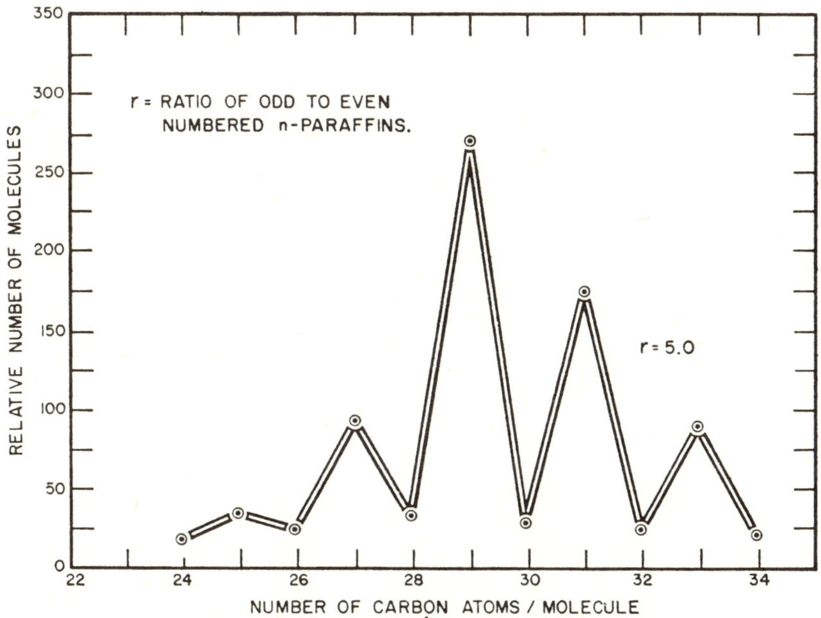

Fig. 2. Distribution of heavy n-paraffins recovered from a Recent sediment extract (Catalina Basin, 5–10 ft).

illustrated by the peaks at 25, 27, 29, 31 and 33. This predominance may be expressed numerically as a ratio obtained by dividing a summation of peak heights for the odd-carbon number n-paraffins by a similar summation of the even-carbon numbers. The value of such a ratio for the sample in Fig. 2 is 5·0. This ratio is the mean of two values computed by using different limits for the summations. One of these values was obtained by dividing the sum of corrected odd-carbon number n-paraffin peak heights at 25, 27, 29, 31 and 33 carbon atoms per molecule by the sum of the even-carbon number n-paraffins at 26, 28, 30, 32 and 34 carbon atoms per molecule. The other value was obtained by changing the denominator to include the even-carbon number n-paraffins at 24, 26, 28, 30 and 32 carbon atoms per molecule. The mean of these two ratios is used to avoid giving a bias to either even- or odd-carbon numbers. This bias could be caused by enhanced peak heights at the ends of the summation range.

Errors

Errors in the odd- to even-carbon-number ratios may arise from either instrumental sources or from impurities in the n-paraffin fractions. Changes in cracking pattern and sensitivity in the mass spectrometer may cause variations in the *n*-paraffin parent peak sensitivities. However, these errors are largely cancelled when the *n*-paraffin distribution is expressed as an odd- to even-carbon-number ratio.

Larger errors may be caused by contamination of the *n*-paraffin fractions with materials such as heptacyclic naphthenes which contribute a background of ions

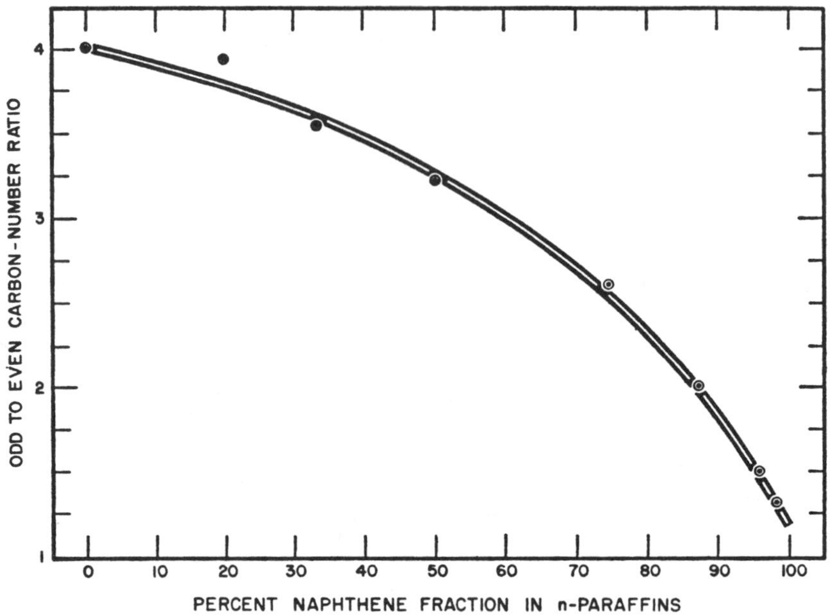

Fig. 3. Blends of naphthene fraction with Recent sediment *n*-paraffins.

to the mass positions of the *n*-paraffin parent peaks. Since these impurities add nearly the same values to both the numerator and denominator of the ratio, their contributions must be sizeable to change the values appreciably. Larger ratio values are subject to greater errors from this cause, since the background is a larger percentage of the denominator. Empirical tests were made to determine how much naphthene contamination could be tolerated in the *n*-paraffin fraction without seriously reducing the value of the odd- to even-carbon-number ratios. The curve of Fig. 3 shows the effects of naphthenic impurities on the odd- to even-carbon-number ratio for *n*-paraffins isolated by urea adduction from a Recent sediment extract. Naphthenes separated from a stabilized crude oil were added in varying amounts to the Recent sediment *n*-paraffins as shown on the abscissa of the curve in Fig. 3. The corresponding ratios obtained from the mixtures are shown on the ordinate. It may be seen from these data that naphthenic impurities up to 30 per cent of the fraction reduced the ratio by only 13 per cent of the difference between the ratio and 1·0. Mass spectrometric hydrocarbon analyses as reported by Clerc *et al.* (1955) were used to determine the amount of contamination in all the *n*-paraffin fractions examined. Although these analyses give only approximate data, they provide a useful empirical guide to the amount of impurities permissible in the n-paraffin fractions. By means of this screening procedure fractions containing naphthenes in concentrations of the order of 30 per cent or more were either further purified or discarded.

Discussion and Results

The identification of source beds may be complicated by prior loss of large portions of oil from the rock, leaving behind the organic debris and hydrocarbon

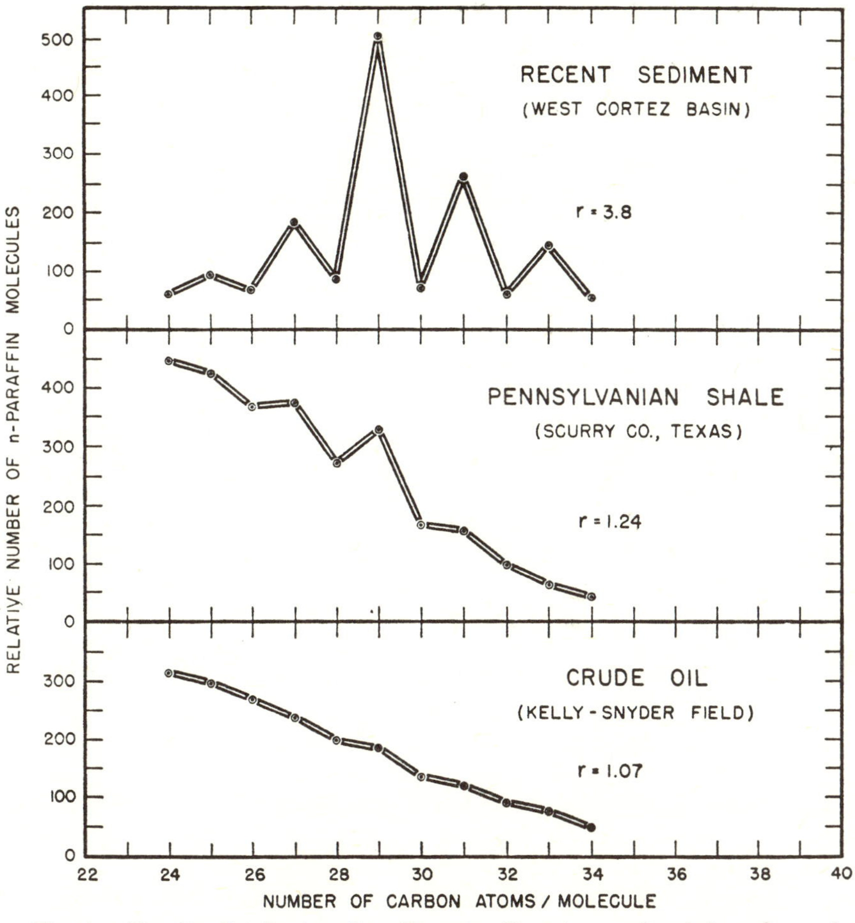

Fig. 4. n-Paraffin distributions for a Recent sediment, a marine shale and a crude oil.

residue possibly weighted with the heavier, less volatile constituents of the oil. These heavy, oily constituents, including n-paraffins, may provide clues to the recognition of source beds since they are a part of both the residue in the rock and the oil.

Examples of heavy n-paraffin distributions, expressed numerically by ratios of odd- to even-carbon-numbered molecules, are shown in Fig. 4. These examples include a Recent sediment from a marine basin, a core of ancient marine shale, and a crude oil with ratios of 3·8, 1·24 and 1·07 respectively. The Recent sediment shows increasing abundance of odd-numbered carbon molecules with size, up to the maximum at 29 carbon atoms. In contrast, the shale extract and the crude oil show a decreasing abundance with size. The example for the crude oil is consistent with data reported previously (STEVENS *et al.* 1956; EVANS *et al.* 1957), showing almost no preference for molecules with odd numbers of carbon atoms.

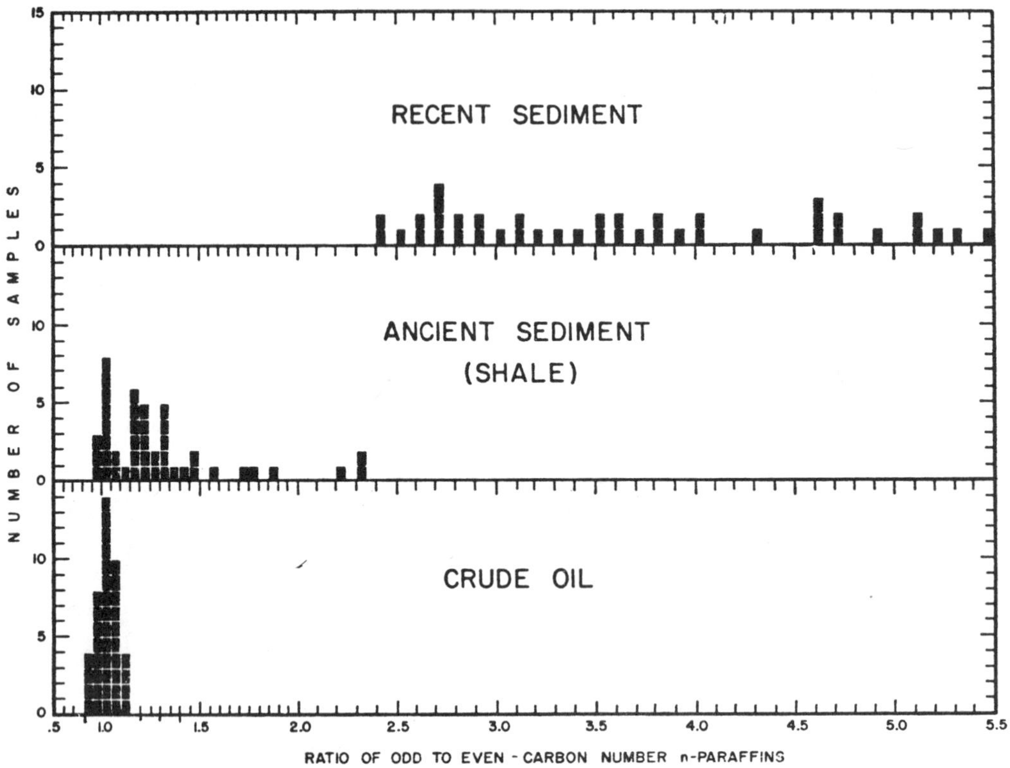

Fig. 5. *n*-Paraffin distributions in crude oils and sediment extracts.

The distribution of odd- to even-carbon-number ratios for *n*-paraffins in a variety of crude oils is shown at the bottom of Fig. 5. The values for 39 representative examples cluster within narrow limits near a value of 1·0, indicating that crude oils in general contain heavy *n*-paraffins which are almost equally distributed between molecules containing even and odd numbers of carbon atoms. The odd- to even-carbon-number ratios for all the Recent sediments examined are also shown in Fig. 5. These ratios are distributed rather uniformly in the range 2·4 to 5·5, indicating that a preference for heavy *n*-paraffins with odd-numbers of carbon atoms is widespread in Recent sediments. The range of values is significantly greater than for crude oils. In the centre of Fig. 5 is shown the distribution of odd- to even-carbon-number ratios for *n*-paraffins extracted from ancient sediments including a variety of shales and some limestones. The ratios, in general, are lower than those of Recent sediments. Some values are close to 1·0 which has been shown to be characteristic of the n-paraffins in crude oils.

The assumption is often made that hydrocarbons originally deposited with sediments of past geologic ages were similar to those in Recent sediments of today. If this is true, the n-paraffin distributions in nearly all of the ancient shales and limestones examined have changed since the time of deposition. The smaller ratios of odd- to even-carbon-number n-paraffins found in most of the rocks appear consistent with the addition of petroleum-like hydrocarbons, either by migration

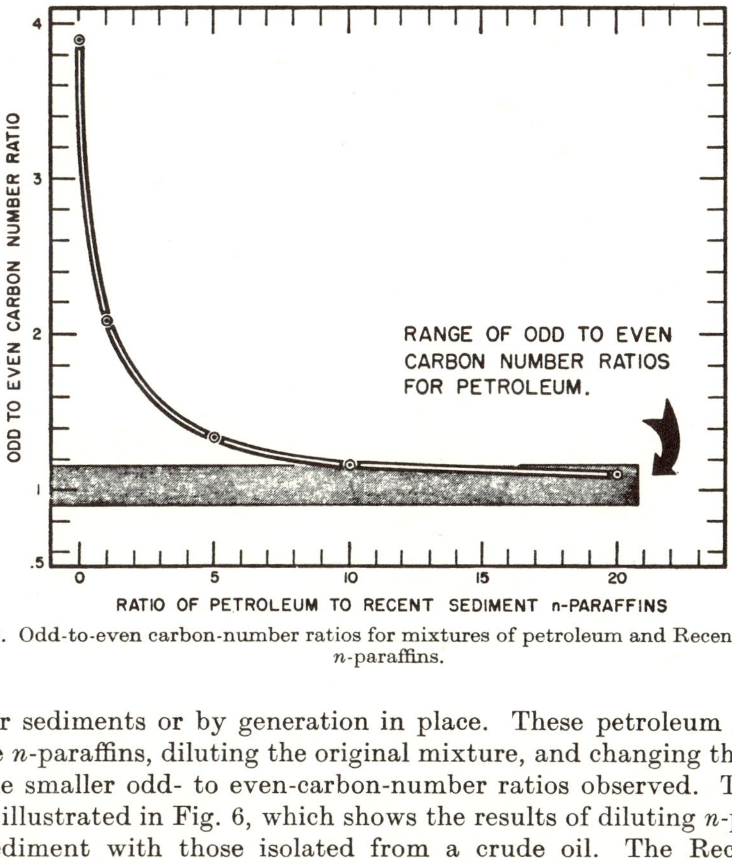

Fig. 6. Odd-to-even carbon-number ratios for mixtures of petroleum and Recent sediment n-paraffins.

from other sediments or by generation in place. These petroleum hydrocarbons contribute n-paraffins, diluting the original mixture, and changing the distribution to give the smaller odd- to even-carbon-number ratios observed. This reduction in ratio is illustrated in Fig. 6, which shows the results of diluting n-paraffins from Recent sediment with those isolated from a crude oil. The Recent sediment n-paraffin had an original odd- to even-carbon-number ratio of 3·9 as shown on the ordinate of Fig. 6. Progressively greater dilutions as shown on the abscissa, caused the ratio to decrease rapidly at first and to finally become diluted to near the values for petroleum after addition of petroleum n-paraffins approximating ten times the amount of sediment n-paraffin. This extinction of the odd-carbon preference by dilution may be useful in the recognition of source beds.

"Source beds", as defined in the A.G.I. Glossary (Howell, 1957), are rocks in which oil or gas has been generated. Although the occurrence of source beds as defined is debatable, we shall assume their existence as a working hypothesis. Beds fitting this definition should contain a mixture of heavy n-paraffins, similar to petroleum which has obscured the predominance of odd-carbon numbered *n*-paraffins, assumed to have been present in the sediment at the time of deposition. Regardless of the level of predominance of odd-carbon sizes in the original sediment, the shales and limestones which now have *n*-paraffins exhibiting ratios significantly greater than those for crude oils have not generated enough hydrocarbons to form a petroleum-like mixture (Fig. 5). This criterion could narrow the search for source beds by a process of elimination. Although the selection of ancient sediment

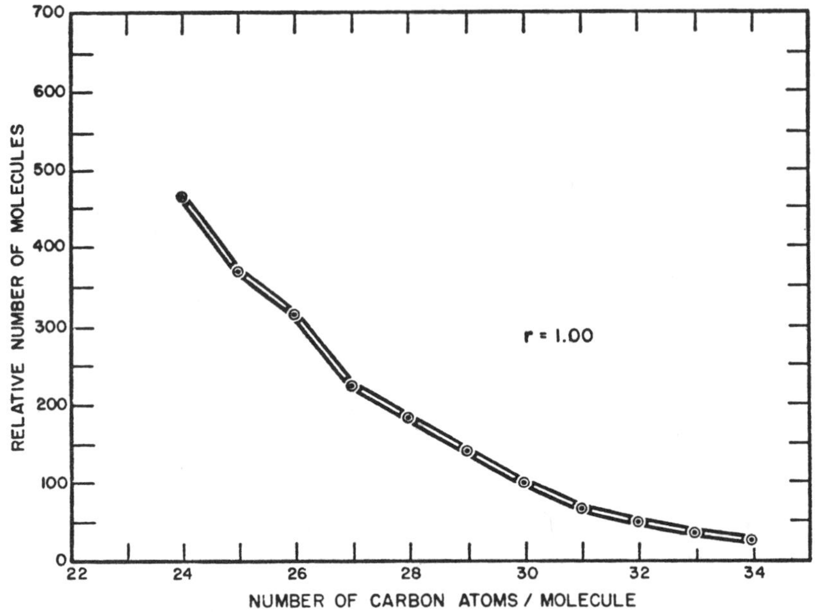

Fig. 7. Distribution of heavy *n*-paraffins recovered from Chattanooga shale (De Kalb County, Tennessee).

samples was biased to include possible source rocks, 65 per cent of those examined had *n*-paraffin odd–even carbon number ratios differing from petroleum (Table 3).

A number of the shales and limestones studied are distinctive in that they have *n*-paraffin odd- to even-carbon-number ratios similar to those for petroleum (Fig. 5). These examples include, among others, postulated "source beds" such as the Chattanooga, Woodford and Heath shales. The distribution of heavy *n*-paraffins obtained from a sample of Chattanooga shale is shown in Fig. 7. The shales bearing petroleum mixtures are not restricted to those containing large quantities of hydrocarbons. There are numerous examples of shales with low hydrocarbon content which exhibit petroleum-like distributions of heavy *n*-paraffins (Table 3). In these instances, the sediments have either yielded their hydrocarbons or were very low in hydrocarbon content in the beginning.

Conclusions

The results of this study may be summarized as follows:

(1) Predominance of heavy *n*-paraffins with odd numbers of carbon atoms per molecule is widespread in Recent sediments.

(2) Crude oils in general contain heavy *n*-paraffins which are almost evenly distributed between molecules of odd and even numbers of carbon atoms.

(3) Postulated source beds analysed in this study are distinctive in that they contain heavy *n*-paraffins which are almost equally distributed between molecules of odd and even numbers of carbon atoms as in petroleum. It should be apparent, of course, that this ratio alone cannot be used to distinguish a source bed from a bed into which oil has migrated.

(4) Most of the shales and limestones examined contain *n*-paraffins with odd- to even-carbon-number ratios significantly greater than that for crude oil. These rocks are regarded as not having generated enough hydrocarbon to become "source beds".

References

Baker E. G., (1959) Origin and migration of oil. *Science* **129,** 871.

Buckley S. E., Hocott C. R. and Taggart M. S., Jr. (1958) Distribution of dissolved hydrocarbons in subsurface waters. *Habitat of Oil* pp. 850–882. American Association of Petroleum Geologists.

Burke W. H., Jr. and Meinschein W. G. (1955) C^{14} dating with a methane proportional counter. *Rev. Sci. Instrum.* **26,** 1137–1140 (December).

Clerc R. J., Hood A. and O'Neal M. J., Jr. (1955) Mass spectrometric analysis of high molecular weight, saturated hydrocarbons. *Analyt. Chem.* **27,** 868–875.

Emery K. O. and Rittenberg S. C., (1952) Early diagenesis of California Basin sediments in relation to origin of oil. *Bull. Amer. Ass. Pet. Geol.* **36,** 735–806.

Erdman J. G., Marlett E. M. and Hanson W. E., (1958) *The Occurrence and Distribution of Low Molecular Weight Aromatic Hydrocarbons in Recent and Ancient Carbonaceous Sediments.* Presented before the Division of Petroleum Chemistry, 134th Meeting, American Chemical Society, Chicago, Illinois, 7–12 September.

Evans E. D., Kenny G. S., Meinschein W. G. and Bray E. E. (1957) Distribution of n-paraffins and separation of saturated hydrocarbons from Recent marine sediments. *Analyt. Chem.* **29,** 1858–1861.

Flint R. F. and Gale W. A. (1958) Stratigraphy and radiocarbon dates at Searles Lake, California. *Amer. J. Sci.* **256,** 689–714.

Forsman J. P. and Hunt J. M. (1958) Insoluble organic matter (kerogen) in sedimentary rocks of marine origin. *Habitat of Oil* pp. 747–778. American Association of Petroleum Geologists.

Howell J. V. (Coordinating Chairman) (1957) *Glossary of Geology and Related Sciences.* American Geological Institute.

Meinschein W. G. (1958) *Sediment and Petroleum Hydrocarbons.* Presented before the Division of Petroleum Chemistry, 134th Meeting, American Chemical Society, Chicago, Illinois, 7–12 September.

Meinschein W. G. and Kenny G. S. (1957) Analyses of a chromatographic fraction of organic extracts of soils. *Analyt. Chem.* **29,** 1153–1161.

O'Neal M. J., Jr. and Wier T. P., Jr. (1951) Mass spectrometry of heavy hydrocarbons. *Analyt. Chem.* **23,** 830–843.

Smith P. V., Jr. (1954) Studies on origin of petroleum: occurrence of hydrocarbons in Recent sediments. *Bull. Amer. Ass. Pet. Geol.* **38,** 377–404.

Stevens N. P., Bray E. E. and Evans E. D. (1956) Hydrocarbons in sediments of Gulf of Mexico. *Bull. Amer. Ass. Pet. Geol.* **40,** 975–983.

Twenhofel W. H., Carter S. L. and McKelvey V. E. (1942) The sediments of Grassy Lake, Villar County, a large bog lake of northern Wisconsin. *Amer. J. Sci.* **240,** 529–546.

Twenhofel W. H., McKelvey V. E., Nelson H. F. and Feray D. E. (1945) Sediments of Trout Lake, Wisconsin. *Bull. Geol. Soc. Amer.* **56,** 1099–1142.

Whitemore F. C. (1944) *Fundamental Research on Occurrence and Recovery of Petroleum*—1943. American Petroleum Institute, New York.

8

THE OCCURRENCE OF PHYTANE IN PETROLEUM

R.A. Dean and E.V. Whitehead

The British Petroleum Company Limited

Research Centre

Petroleum Division

Sunbury-on-Thames, Middlesex

(Received 13 October 1961; in revised form 3 November 1961)

IN the course of recent studies on the composition of petroleum, we had occasion to remove normal paraffins from petroleum distillates boiling in the range 250°-370°C by means of molecular sieve 5A. The gas liquid chromatogram obtained on a non-aromatic fraçtion of an Agha Jari (Iran) distillate from which normal paraffins had been removed, showed very strong peaks having retention volumes on a silicone column very close to those of n-heptadecane and n-octadecane. These peaks were superimposed on the "envelope" which is present in chromatograms of all petroleum distillate fractions boiling above about 200°C, and it was estimated from their areas that the compounds giving rise to them were present, very approximately, to the extent of 0.3 per cent weight of the crude oil. No peaks of comparable magnitude were present at the retention volumes of other n-paraffins (C_{16} and C_{19} to C_{23}). This suggested that these two compounds did not, unlike many components of petroleum, belong to a series in which adjacent members, differing by one carbon atom, are present to approximately the same extent, e.g. the n- and 2-methyl-alkanes.

Our initial postulate was that they were both C_{20} compounds based on isoprene and to substantiate this they were isolated (5 mg of each) using a combination of fractional distillation, liquid thermal diffusion and trapping from gas chromatographic columns using a polyphenyl ether and silicone oil as successive stationary phases. The higher boiling of the two compounds had an infra-red spectrum very similar to that of phytane (2:6:10:14-tetramethylhexadecane, n_D^{20} 1.4418, synthesized from commercial phytol by catalytic hydrogenation with a platinum catalyst at 20°C and 100 atm, conversion of the dihydrophytol to its p-toluene-sulphonate, and reduction of this with lithium aluminium hydride) which in turn was identical within the limits of reproducibility with that recorded by Pliva and Sörensen[1]. Differences in the spectra of the isolated and synthetic compounds were the presence of minor peaks at 8.22, 10.1, 10.98 and 11.26μ and an overall higher absorbance in the former. Mass spectra of the isolated and synthetic materials were also similar, differences being attributable to the presence of other C_{20} branched chain paraffins. These results coupled with the identity of retention volumes (isolated and synthetic respectively: 1.06, 1.06 relative to n-octadecane with silicone oil as stationary phase at 257°C; 0.89, 0.89 relative to n-octadecane with a C_{30} polyphenyl ether as stationary phase at 257°C) make it reasonably certain that the compound almost coincident with n-octadecane on the silicone column is phytane.

The mass spectrum of the material almost coincident with n-heptadecane was, however, that of a C_{19} isoparaffin, not that of a hydrogenated isoprene tetramer. Identity of its retention volumes (0.79 on silicone, 0.65 on polyphenyl ether, conditions as above) with those of 2:6:10:14-tetra-

[1] J. Pliva and N.A. Sörensen, Acta Chem.Scand. 4, 846 (1950).

methylpentadecane[2] (0.79 and 0.64 respectively) suggests that it may have this structure. The quantity of the two hydrocarbons isolated was insufficient to permit the determination of their optical activity, however, a determination on a phytane-containing concentrate showed no measurable activity. Further confirmation of the structures of the two compounds, data on the amount present in a number of different crude oils and a discussion of the significance of their presence in petroleum (cf. page 312 of ref. 3) will form the subject of a future communication.

Acknowledgement - Our thanks are due to the Chairman and Directors of The British Petroleum Company Limited for permission to publish these results.

---

[2] J.S. Sörensen and N.A. Sörensen, Acta Chem.Scand. 3, 939 (1949).

[3] S.F. Birch, T.V. Cullum, R.A. Dean and D.G. Redford, Tetrahedron 7, 311 (1959).

9

Reprinted from *Nature* **193**:744–746 (1962)

FATTY ACIDS IN RECENT AND ANCIENT SEDIMENTS AND PETROLEUM RESERVOIR WATERS

By J. E. COOPER

Socony Mobil Oil Co., Inc., Field Research Laboratory, Dallas, Texas

SATURATED fatty acids have been found in recent sediments, ancient sediments, and waters produced with petroleum. Acids having odd numbers of carbon atoms have been found along with those having even numbers of carbon atoms.

Acids were obtained from sediments by extraction with a refluxing solution of potassium hydroxide in methanol. Water was added to the extract, which then was concentrated by distillation. The sample was washed with carbon tetrachloride to remove non-acidic material. The sample was divided into two equal portions; a known amount of stearic acid was added to one as an internal standard. Both portions were afterwards treated identically. The portions were acidified, and acids were extracted with carbon tetrachloride. The solvent was evaporated cautiously. The acids in the residue were esterified using boron trifluoride in methanol, and straight-chain esters were separated by urea adduction. After decomposition of the adducts with water, the esters were taken up in benzene and determined by gas chromatography. Beeswax was analysed similarly for comparison.

The same general procedure was used for determination of fatty acids in formation water samples. The samples were not divided, however, and no internal standard was added. Instead, the volume of the benzene solution of the methyl esters was determined, and an aliquot was taken for gas chromatographic analysis. This procedure indicates only minimum concentrations of fatty acids.

An F and M model 500 gas chromatograph was used for the gas chromatographic analyses. The instrument was equipped with a 2-ft. column packed with 0·3 per cent silicone grease on 60–80 mesh glass beads. The initial temperature in each run was 75°, and it was increased to 300° at the rate of 5·6°/min. The temperature was held at 300° until all the esters were eluted. The gas chromatograph was calibrated, using a standard solution of esters of the fatty acids having 8, 10, 12, 14, 16, 18 and 20 carbon atoms. Retention times and instrument sensitivities for esters of the other saturated, straight-chain acids were obtained by interpolation and extrapolation. Gas chromatograms of some of the unknown solutions showed small amounts of unidentified esters in addition to esters of saturated, straight-chain acids.

Mass spectra were obtained of esters of fatty acids from beeswax and from Chattanooga Shale. No internal standard was added to either sample. A Consolidated 21–103 mass spectrometer, modified and operated as described by Meinschein and Kenny[1], was used in determining the mass spectra.

Gas chromatographic data in Table 1 show the presence of fatty acids having 8–36 carbon atoms. The mass spectral data (Fig. 1) provide further support for the identification of the acids in the 12–20 carbon atom range. These data show a parallelism between relative abundance of the esters as given by parent peaks, parent-minus-a-methoxyl-group peaks, and gas chromatographic data. Mass spectral peaks corresponding to the odd-numbered acids are much lower in the beeswax sample than in the Chattanooga Shale sample. This supports the assignment of gas chromatographic peaks to esters of odd-numbered acids having 13, 15, 17 and 19 carbon atoms.

Control experiments show that results obtained in this preliminary work are only semi-quantitative.

Table 1. SATURATED, STRAIGHT-CHAIN FATTY ACIDS FROM SEDIMENTS, WATERS, AND BEESWAX

Carbon atoms in acid	Recent sediments: San Nicholas Basin (Pacific Ocean)	Recent sediments: Santa Barbara Basin (Pacific Ocean)	Ancient sediments: Chattanooga shale, outcrop, Oklahoma	Ancient sediments: Eagle Ford shale, core, Texas	Pet. res. waters: Dry Mesa field, Arizona	Pet. res. waters: Panhandle field, Texas	Beeswax
14	1·05*	6·92*	0·44*	0·18*	8·14‡	70·6‡	0·01§
15	0 17	0·74	0·37	0·20	9·60	62·7	—
16	4·43	17·02	0·50	0·70	13·3	65·7	1·00
17	0·16	0·70	0·29	0·32	10·6	56·5	—
18	1·39	3·83	0·39	0·38	12·0	51·8	0·49
19	0·20	0·70	0·14	0·18	10·9	45·9	—
20	0·52	2·36	0·10	0·14	9·89	41·4	0·04
21	0·09	0·65	0·06	0·12	8·90	36·8	—
22	0·29	2·49	0·04	0·10	7·98	34·4	0·01
23	0·10	0·75	0·02	0·08	6·97	30·0	—
24	0·16	4·00	0·01	0·08	7·10	28·6	0·01
25	0·20	0·81	+†	0·08	6·05	23·5	—
26	0·46	3·71	+	0·08	5·54	22·2	+
27	0·24	0·47	+	0·06	3·90	16·3	−
28	0·61	3·11	+	0·06	3·34	12·1	+
29	0·17	0·24		0·04	2·89	10·8	−
30	0·35	1·39		0·08	2·55	8·85	+
Smallest acid found	C_{14}	C_{10}	C_8	C_{11}	C_{10}	C_8	C_{12}
Largest acid found	C_{36}	C_{32}	C_{28}	C_{34}	C_{36}	C_{38}	C_{32}
$\frac{C_{16}+C_{18}}{2C_{17}}$¶	16·7	14·9	1·56	1·71	1·19	1·04	

* μgm./gm.
† Observed peak too small to measure.
‡ μgm./l.
§ Relative to palmitic acid (C_{16}).
¶ Ratio of the mean of concentrations of palmitic and stearic acids to that of heptadecanoic acid.

Relative peak heights are correct to within 20 per cent over the range bounded by pentadecanoic and arachidic acids; however, relative heights of adjacent peaks in this range are correct to within 5 per cent. Absolute concentrations of acids obtained from the sediments are accurate to within 50 per cent. The concentrations determined for acids in the formation waters indicate, however, only an order of magnitude. The consistently low values obtained for those acids having less than 15 carbon atoms are due to losses incurred during the concentration steps and to the increasingly unfavourable equilibrium for urea adduct formation with lower acids. Control experiments have not been performed for acids having more than 20 carbon atoms.

As shown in Table 1, recent sediments, ancient sediments and formation waters, in contrast to most biological systems[2], contain fatty acids having odd numbers of carbon atoms along with those having even numbers. Williams[3] has given evidence for the presence of both odd- and even-numbered fatty acids in samples of Pacific Ocean water, and Hewitt, Kipping and Jeffery[4] have found acids of both types in montan wax. The sediments and the sea water do resemble biological systems, however, in having palmitic and stearic acids as the predominant acids.

Even-numbered acids predominate over odd-numbered ones in every case studied. The ratios of the mean of the concentrations of palmitic and stearic acids to the concentration of heptadecanoic acid (Table 1) provide a rough indication of the degree of predominance of even-numbered acids. In general, the relative abundance of odd-numbered acids increases from recent sediment to ancient sediment to formation water. One exception to this rule, a Palæozoic shale resembling a recent sediment, will be re-examined.

The source of the odd-numbered acids is not known. The analytical procedure does not show whether the acids are free or are parts of larger molecules such as esters. Contamination is an unlikely source for odd-numbered acids, because contamination with naturally occurring fats or oils would result in the introduction of even-numbered acids only. Production of odd-numbered acids by oxidation during collection

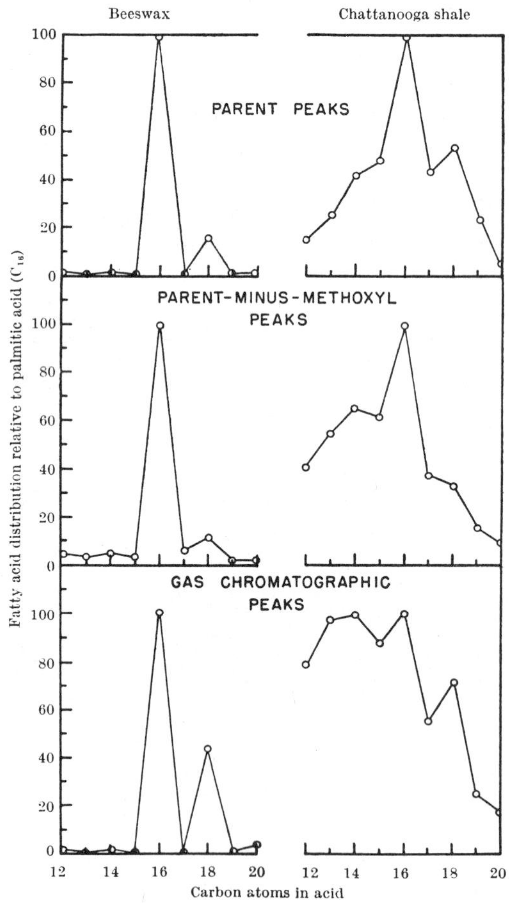

Fig. 1. Comparison of mass spectral and gas chromatographic analyses of fatty acids from beeswax and Chattanooga shale

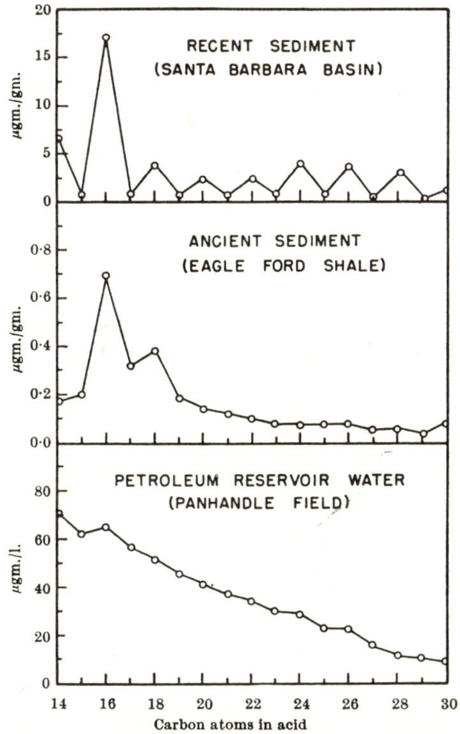

Fig. 2. Comparison of the distributions of fatty acids in a recent sediment, an ancient sediment, and in water from a petroleum reservoir

and storage of samples seems unlikely; but it cannot be disregarded. Since saponification in an atmosphere of nitrogen gives the same results as saponification in air, acid production by oxidation during this step is not important.

The appearance of the odd-numbered acids might arise from a large increase in the relative concentrations of odd-numbered acids initially present in the sediments in very small concentrations. Such an apparent increase could result from the selective removal of the even-numbered acids. Selective solubility is an unlikely cause of such an effect, because enrichment is observed in both waters and sediments. Microbial activity, however, cannot be excluded, because under certain conditions micro-organisms consume even- but not odd-numbered acids[5].

The generally higher relative concentrations of odd-numbered acids in older sediments (Fig. 2) is consistent with the operation of a process which generates odd-numbered acids. The increase in abundance of odd-numbered acids seems to parallel that of even-numbered paraffins as reported by Bray and Evans[6]. This parallelism suggests related processes for the formation of these acids and paraffins. One possible process goes through a reactive intermediate formed by decarboxylation of a fatty acid. Intermediates of this type derived from even-numbered acids could react to give a mixture of odd-numbered paraffins as a product. The odd-numbered acids produced would react similarly to form even-numbered acids and paraffins. The operation of such a process in a sediment containing even-numbered acids would cause the introduction first of odd-numbered acids and paraffins and later even-numbered paraffins. Under conditions favouring acid formation over paraffin formation a secular equilibrium would be attained eventually in which there is no preference between even- or odd-numbered paraffins or acids.

This investigation shows the presence of saturated, straight-chain fatty acids in recent sediments, ancient sediments and petroleum reservoir waters. Acids containing odd numbers of carbon atoms have been found along with those containing even numbers. The distributions of these acids suggest a relationship between the fatty acids and the normal paraffins in sediments.

I wish to express thanks to Mr. E. E. Bray for samples of the recent sediments, to Mr. A. J. Miller for samples of the ancient sediments and formation waters, and to Mr. W. W. Shaw for the performance of much of the experimental work. I am grateful also to the Socony Mobil Oil Co. for sponsoring this work.

[1] Meinschein, W. G., and Kenny, G. S., *Anal. Chem.*, **29**, 1153 (1957).
[2] Shorland, F. B., *Nature*, **174**, 603 (1954).
[3] Williams, P. M., *Nature*, **189**, 219 (1961).
[4] Hewett, D. R., Kipping, P. J., and Jeffery, P. G., *Nature*, **192**, 65 (1961).
[5] Silliker, J. H., and Rittenberg, S. C., *J. Bact.*, **64**, 197 (1952).
[6] Bray, E. E., and Evans, E. D., *Geochim. Cosmochim. Acta*, **22**, 2 (1961).

10

Reprinted from *Science* **148**:363–374 (1965)

FATTY ACIDS IN BLUE-GREEN ALGAL MAT COMMUNITIES

Patrick L. Parker and Richard F. Leo

Abstract. *The concentrations of the major fatty acids were determined for the varied layers of an algal mat community. The living mat contains substantial amounts of the unsaturated acids, while the underlying layers become progressively depleted in unsaturated molecules. A similar progressive increase in the ratio of saturated to unsaturated acids was detected in a sediment core from a hypersaline lagoon.*

The fatty acids are major components of all living organisms. As such they may be an important source material for petroleum and perhaps for the ubiquitous kerogen. Fatty acids have been found in recent and ancient sediments by Cooper and Bray (*1*) and by Abelson *et al.* (*2*). However, the chemical procedure of Cooper and Bray excluded unsaturated acids. The algal mats and their associated sediments provide a natural laboratory in which to test the relative survivability of different fatty acids.

Extensive algal mat communities in the Texas lagoons and mud flats develop in water from 1 to 50 centimeters deep. Many organisms are present in the mat but the biomass is largely made up of the filamentous blue-green algae, *Lyngbya confervoids* (*3*). Well-developed communities have a structure which is of use to the geochemist. The living mat covers the mud bottom like a sheet of leather. Directly under the living mat is a layer of black mud at times a few centimeters thick. The remains of a former mat are under the first mud layer, and they too have a mud layer under them. This pattern of mat and mud repeats itself. In the samples taken for study, the living mat, two former mats, and two mud layers were distinct. All samples were frozen at the time of collection and kept frozen until used.

The analytical procedure was designed to detect fatty acids containing 12 to 20 carbon atoms. Care was taken to minimize decomposition of the unsaturated acids. The frozen sample was treated with dilute HCl to remove carbonate. After filtration the moist sample was treated with methanol for 15 minutes while being stirred and subjected to ultrasonic vibrations. The methanol was recovered by filtration. The sample was treated with chloroform and again stirred and subjected to ultrasonic vibrations. The recovered chloroform was combined with the methanol, and the mixture almost completely dried with a rotating evaporator. This residue was taken up in chloroform and washed with 1*N* HCl to remove inorganic salts. The chloroform was dried under a stream of dry nitrogen. After saponification with potassium hydroxide. impurities were removed by extracting the basic alkaline solution with chloroform. Finally the alkaline solution was acidified, and the fatty acids were extracted into chloroform. The methyl esters were prepared with BF_3 (*4*).

The methyl esters were identified and measured by gas chromatography on a diethelyene glycol succinate column (*5*). Either an Aerograph 202 or an F & M model 400 instrument was used. The chemical yields for the procedure were good, 75 to 95 percent as determined with the use of carbon-14–labeled palmitic acid. The organic carbon content of the sediments was determined by combustion with CuO on a vacuum line, freezing out the CO_2, and allowing it to expand against a calibrated manometer.

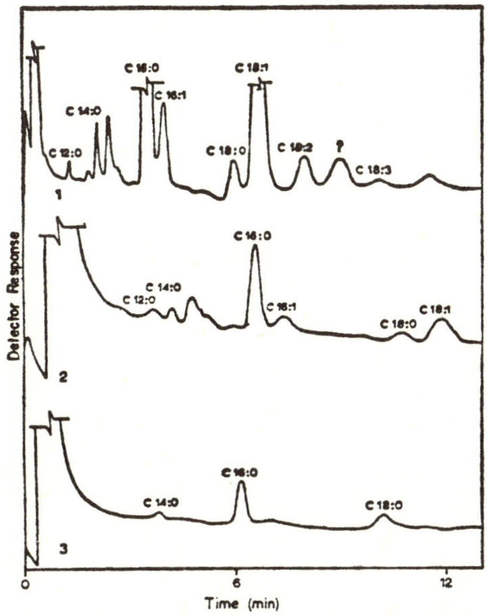

Fig. 1. Gas chromatogram of methyl esters of fatty acids from algal mat: 1, the living mat; 2, first mud layer; 3, second mud layer.

All recent sediments examined thus far are rich in fatty acids (Table 1). All the molecules identified are those which have even numbers of carbon atoms, and which are usually found in organisms. A few unidentified peaks have been encountered, but they do not correspond to any straight-chain fatty acids with even or odd carbon numbers. In the algal mat acids the most striking observation is the systematic change in the ratio of saturated to unsaturated acids (Fig. 1).

The living mat is rich in the unsaturated acids, especially oleic acid. The relative survivability of the different acids is proportional to the degree of unsaturation. This suggests that chemical interactions, such as polymerization, among the organic components are important in the early modification of biochemicals. Nevertheless, biological activity probably determines the overall organic content in the sediments.

Table 1. Concentration of the major fatty acids in recent sediments expressed as parts of fatty acid per million parts of organic carbon of each sediment. n.d., not detected.

Location	Organic carbon (%)	C_{16} (0)*	C_{16} (1)	C_{18} (0)	C_{18} (1)
Harbor Island algal mat					
Living mat †	32	3100	560	130	1200
1st mud layer	1.1	1200	330	180	330
2nd mud layer	0.84	200	23	97	n.d.
Baffin Bay core					
0–10 cm	2.0	149	53	53	89
37–41	0.85	192	35	35	50
60–64	1.1	154	6	58	20

* Numbers in parentheses indicate the number of double bonds. † The living mat contained 296 ppm $C_{18}(2)$ and 31 ppm $C_{18}(3)$.

The same pattern was found in a second algal mat from the middle Laguna Madre. This mat was much larger in area than the first, but the layered structure was not as well developed. Preliminary data from a sediment core show the early disappearance of highly unsaturated acids (Table 1). However, oleic acid has not fallen to the very low level of the algal mat, although the sediment is much older. This may be due to the very different type of environment.

A detailed study of the geochemistry of fatty acids in recent sediments is being carried out in this laboratory. A more complete report will be published later.

References and Notes

1. J. E. Cooper and E. E. Bray, *Geochim. Cosmochim. Acta* **27**, 1113 (1963).
2. P. H. Abelson, T. C. Hoering, P. L. Parker, in *Advances in Organic Geochemistry*, U. Colombo and G. D. Hobson, Eds. (Pergamon, New York, 1963).
3. L. O. Sorensen and J. T. Conover, *Publ. Inst. Mar. Sci. Univ. Tex.* **8**, 61 (1962).
4. L. D. Metcalfe and A. A. Schmitz, *Anal. Chem.* **33**, 363 (1961).
5. S. R. Lipsky, R. A. Landowne, J. E. Lovelock, *ibid.* **31**, 852 (1959).
6. Supported by NSF grant GP 2515.

3 March 1965

11

Reprinted from *Nature* **232**:327–330 (1971)

CONVERSION OF OLEIC ACID TO SATURATED FATTY ACIDS IN SEVERN ESTUARY SEDIMENTS

M. M. Rhead, G. Eglinton, G. H. Draffan, and P. J. England

THE upper, biologically active, layers of Recent aquatic sediments contain considerable quantities of saturated and unsaturated long chain fatty acids[1-3], presumably from the living and dead organisms found there[4,5]. Although fatty acids are major constituents of organisms, in the upper layers of sediments, their concentration is of the same order of magnitude as that of sterols[6,7], usually minor constituents of organisms. The ratio of fatty acids to total organic carbon in marine sediment is at least ten times lower than in plankton[1], again indicating that fatty acids are subject to rapid changes in the sediment. Palmitic acid ($C_{16:0}$) is the most abundant saturated fatty acid in organisms, and oleic acid ($C_{18:1}$) the most widely distributed and plentiful of all fatty acids[4,5]. The lower, biologically less active, layers of sediments, contain little or no unsaturated fatty acid; palmitic acid is the most abundant of the fatty acids[1,2,8]. This difference between the upper and lower layers is unlikely to be the consequence simply of hydrogenation of the carbon–carbon double bands, for in this case the proportion of stearic acid ($C_{18:0}$) in the lower layers and the concentrations of saturated fatty acids would be larger. These reactions could occur by chemical and/or biological means, perhaps through the large number of anaerobes present in upper layers of sediments[9,10].

Our studies were designed to discover the fate of oleic acid in an aquatic sediment. The basis of our experiments was to add 9,10-^{3}H, 1-^{14}C-oleic acid to the upper layers of the Recent sediments of the Severn Estuary and then to isolate the labelled products of short term diagenesis. The ratios of ^{3}H to ^{14}C measured for the various saturated fatty acids obtained are compared with the ^{3}H/^{14}C ratio for the original oleic acid. We hope to extend the methods described here to other substances of geochemical and environmental chemical interest.

If oleic acid present in sediments is converted into saturated acids (of chain length C_{12} and above), two biochemical pathways could be operative: (1) the simple hydrogenation of unsaturated carbon–carbon bonds, followed by loss of C_2 units to give the fatty acids of shorter chain length (the degradative pathway); (2) total breakdown of oleic acid by β-oxidation to give C_2 units (probably as acetyl-CoA) followed by a resynthesis through the fatty acid synthetase complex[11] (the resynthesis pathway).

If the oleic acid introduced is labelled in the carboxylic carbon with ^{14}C and in the 9 or 10 positions with ^{3}H (to give 9,10-^{3}H, 1-^{14}C-oleic acid), the isotopic labels will be incorporated differently into the products of the two pathways outlined here.

In the case of the degradative pathway, which involves the simple loss of C_2 units without total breakdown, the ^{14}C will be lost from the molecule (for all acids except stearic), whereas all the ^{3}H will be retained. Thus, for fatty acids of length C_{16} down to C_{10}, the ratio of ^{3}H to ^{14}C becomes infinitely large, whereas for stearic acid the ratio stays the same as that of the added oleic acid. The ^{14}C removed from the long chain acids will appear in acetyl-CoA (as 1-^{14}C-acetyl-CoA), which will have a ratio of ^{3}H to ^{14}C of 0 to 1.

In the case of the resynthesis pathway, the various enzymatic reactions undergone during β-oxidation and resynthesis from the resulting acetyl-CoA will lead to loss of ^{3}H relative to ^{14}C. The first turn of the β-oxidation spiral will remove all the ^{14}C from the molecule as 1-^{14}C-acetyl-CoA. All the ^{3}H from C-9 will be lost (as NAD^{3}H or FAD^{3}H) as this carbon becomes the terminal carboxylic carbon. The fate of the ^{3}H on C-10 may depend on the stereospecificity of the enzymes involved in the reduction step. Enzymes from various sources[12] have been shown to react stereospecifically with fatty acids and it is likely that the β-oxidation enzymes behave similarly, but their stereospecificity is unknown and either all or none of the ^{3}H could be lost from C-10 of the degraded fatty acid. The acetyl-CoA produced through the β-oxidation spiral could therefore contain anything between zero and 0.5 of the amount of the ^{3}H originally present in the oleic acid (all as 2-^{3}H-acetyl-CoA), and all of the ^{14}C. This acetyl-CoA is most likely to be incorporated into long chain fatty acids by the fatty acid synthetase complex which is present in many organisms, including *Escherichia coli* and yeast[11]. All the acetyl-CoA units to be utilized, except one, are first converted to malonyl-CoA before incorporation into the growing fatty acid. The overall effect is to lose approximately 0.67 of the ^{3}H present as 2-^{3}H-acetyl-CoA but to retain all the ^{14}C. In the case of palmitic acid, the overall result should be a ratio of ^{3}H to ^{14}C of 8.34 : 1, based on acetyl-CoA having an

Table 1 Radioactivity in Saturated Long Chain Fatty Acids from Sediments Incubated with 9,10-^{3}H, 1-^{14}C-Oleic Acid (Corrected for Experimental Losses)

	Yields of radioactivity (Ci 10^{-9})									
	A		*B*		*C*		*D*		*E*	
Acid	^{3}H	^{14}C	^{3}H	^{14}C	^{3}H	^{14}C	^{3}H	^{14}C	^{3}H	^{14}C
Stearic	146.6	5.2	71.2	2.95	12.2	0.41	56.4	2.7	18.9	0.5
Palmitic	167.6	3.8	107.6	4.7	13.7	0.40	70.8	1.07	12.8	0.44
Myristic	80.4	0.7	161.4	3.9	10.3	0.19	46.6	0.27	20.8	0.10
Lauric	74.6	0.2	*	*	4.4	0.03	15.4	0.02	*	*

* Not determined.

100 μCi of 9,10-^{3}H-oleic acid and between 2.5 and 10 μCi of 1-^{14}C-oleic acid were injected into the sediment in each experiment. All the results have been normalized to an initial ratio of ^{3}H to ^{14}C for oleic acid of 20 : 1 (100 μCi of ^{3}H). The conditions of extraction and purification of the acids, and the corrections for losses in these procedures, are described in the text. The conditions of incubation were as follows: *A*, laboratory core, incubated for 6 days at 18–20° C; *B*, laboratory core, incubated for 9 days at 18–20° C; *C*, incubated *in situ* for 7 days, August 1970. *D*, incubated *in situ* for 17 days, September–October 1970; *E*, incubated *in situ* for 38 days, September–October 1970. The statistical errors in the ^{3}H values caused by variation in count rates were in all cases less than ±1%. The ^{14}C values will be subject to larger errors because of the lower count rates, and where the quoted yields are less than 10^{-10} Ci the error may be greater than ±5%.

initial ratio of ^{3}H to ^{14}C of 20 to 1. Experimental verification of this expected ratio comes from the work of Wakil[13] with the synthetase complex from avian liver, and other studies[14] in rat adipose tissue. Overall, the resynthesis pathway, starting from oleic acid with a ratio of ^{3}H to ^{14}C of 20 to 1 should generate palmitic acid with a ratio of ^{3}H to ^{14}C between 0 to 1 and 4.17 to 1, the exact value depending on the ^{3}H loss during β-oxidation. The theoretical ratios for lauric, myristic and palmitic acids formed via the resynthesis pathway are very similar.

Thus the two pathways, degradative and resynthetic, should give end products with very different ratios of ^{3}H to ^{14}C. Except for stearic acid, the degradative pathway causes a total loss of ^{14}C, whereas the resynthesis pathway causes an almost total loss of ^{3}H relative to ^{14}C. We are following only the flow of label and not the total flow of material through each pathway. From the actual ratios of ^{3}H to ^{14}C observed in the saturated fatty acids extracted from the sediments injected with labelled oleic acid, it is possible to calculate what proportion of the label has come through the degradative pathway compared with that arising by resynthesis.

Incubations of oleic acid in aquatic sediments were carried out either *in situ* in the mud of the Severn Estuary (National Grid Reference ST534832) at a depth of 10 cm below the surface, or in cores (15 × 3 cm) which had been removed from areas adjacent to the *in situ* sites, and then transferred to the laboratory. The area of study was chosen near the high water mark on a large expanse of mud flats. A thin layer of grey silt covered a 20 cm layer of black thixotropic mud which contained sulphide and elemental sulphur. The sites were covered by the tide for approximately 4 h in each 24 h, and the temperature varied between 5° and 25° C; the laboratory cores were incubated at 18°–20° C. Labelled oleic acid (from the Radiochemical Centre, Amersham, Buckinghamshire) was prepared by mixing aliquots of 1-^{14}C and 9,10-^{3}H-oleic acid (no additional unlabelled acid was added) to give a ratio of ^{3}H to ^{14}C of between 10 to 1 and 40 to 1. The resulting oleic acid (up to 100 μCi of ^{3}H) was injected (as a 'Tween 80' suspension) so as to cause as little disturbance of the sediment as possible. This represents the addition of a maximum of 0.2 μmol of free oleic acid.

After incubation for up to 38 days, the cores were removed and extracted for free fatty acids using a modification of the Dole extraction procedure[15], or with a 3 to 1 mixture of toluene/methanol. Carrier fatty acids were added (3 μmol for each acid) and the extract methylated with BF_3/methanol[16]. The methyl esters of the saturated acids were then separated from other material by preparative thin-layer chromatography on silver nitrate-impregnated silica gel[17], and finally separated into individual esters by preparative gas chromatography. The peaks corresponding to the even chain length saturated acids were collected directly in 10 ml. aliquots of toluene scintillator (NE 220: Nuclear Enterprises Ltd, Edinburgh) and the samples assayed for radioactivity using a Nuclear Chicago mark I liquid scintillation spectrometer. Quench corrections were applied[18].

Control experiments have shown that the isolation procedures used (Dole extraction, methylation, thin-layer chromatography) result in approximately 70% recovery of U-^{14}C-palmitic acid before preparative gas–liquid chromatographic analysis. Toluene–methanol extraction is less efficient giving 25% recovery of label before gas–liquid chromatography. Similar recoveries have been obtained for 9,10-^{3}H-oleic acid. Table 1 shows the amounts of radioactivity (in nCi) in the saturated fatty acid fraction estimated as being present in sediments after various periods of incubation with 9,10-^{3}H, 1-^{14}C-oleic acid. The figures listed are estimates calculated using these measured efficiencies of recovery of the saturated fatty acids.

The most striking feature of Table 1 is the low conversion of oleic acid into saturated fatty acids observed for all the incubations. The combined long chain saturated fatty acid fraction extracted from the mud was never found to contain more than 3% of the label from the added oleic acid, even though the labelled oleic acid rapidly disappeared (for example in experiment *B* of Table 1, only 5% remained after 9 days incubation). Thus, most of the label of the starting oleic acid does not appear in the free saturated fatty acid component of the mud. The usual fate of oleic acid added to this particular sediment is probably breakdown to acetyl units, which may then be oxidized to CO_2 and H_2O or incorporated into other compounds such as the carbohydrate constituents of cell walls.

Table 2 shows the ratios obtained after various incubation times, with 9,10-^{3}H, 1-^{14}C-oleic acid. There is a definite trend in the ^{3}H to ^{14}C ratio, which increases with decreasing chain length. We have already stated that acids produced by the degradative pathway would have a ratio of ^{3}H to ^{14}C of infinity, whereas those formed by the resynthesis pathway would have ratios of ^{3}H to ^{14}C between 0 to 1 and 4.2 to 1, starting with that for oleic acid—20 to 1. A combination of the pathways would give intermediate ratios. A qualitative

Table 2 Ratios of ^{3}H to ^{14}C measured in Saturated Fatty Acids from Severn Estuary Sediments after Injection of 9,10-^{3}H, 1-^{14}C-Oleic Acid

Acid	*A*	*B*	*C*	*D*	*E*
Stearic	29 : 1	24 : 1	30 : 1	21 : 1	38 : 1
Palmitic	44 : 1	23 : 1	34 : 1	66 : 1	29 : 1
Myristic	112 : 1	41 : 1	54 : 1	170 : 1	215 : 1 *
Lauric	440 : 1 *	—	142 : 1 *	627 : 1 *	

Procedures for incubation and isolation of the acids are given in the text. The ratios of ^{3}H to ^{14}C used in the injected oleic acid varied between 10 to 1 and 40 to 1; all the results have been normalized to a starting ratio for oleic acid of 20 to 1. The conditions of incubation (*A*–*E*) are given beneath Table 1.

* These values are subject to errors of greater than ±5%.

picture can be readily inferred from the data in Table 2. Lauric acid and myristic acid, which have high ratios of ^{3}H to ^{14}C, seem to arise chiefly through the degradative pathway, while palmitic acid, which has a ratio nearer that of oleic acid, seems to be formed through both pathways. The position with stearic is more complex, as a ratio of ^{3}H to ^{14}C greater than that of oleic acid is impossible according to the scheme outlined above. Some of the stearic acid, however, may be formed through a recycling pathway involving palmitic acid:

$$\text{oleic acid} \rightarrow \text{stearic acid} \rightleftharpoons \text{palmitic acid}$$

A ratio of ^{3}H to ^{14}C greater than that of oleic acid would then be obtained. Such recycling of metabolites is well known in biochemistry[13,19], and it is quite likely that recycling is also occurring here.

Table 3 Calculated Fractions of Labelled Saturated Fatty Acids derived by the Degradative Pathway

Acid	*A*	*B*	*C*	*D*	*E*
Palmitic	0.67	0.48	0.60	0.76	0.56
	(0.69)	(0.53)	(0.63)	(0.77)	(0.60)
Myristic	0.84	0.65	0.72	0.90	0.92
	(0.85)	(0.66)	(0.73)		
Lauric	0.96	—	0.88	0.97	

The observed ratios of ^{3}H to ^{14}C used to calculate these results are taken from Table 2, and the conditions of incubation (*A–E*) are given in the footnote to Table 1. The values in this table have been calculated using the formula

$$x = \frac{c_{res}t_{obs} - c_{obs}t_{res}}{c_{obs}(t_{deg} - t_{res}) + t_{obs}(c_{res} - c_{deg})}$$

where x is the fraction of acid formed through the degradative pathway. t_{res} to c_{res} is the theoretical ratio of ^{3}H to ^{14}C obtained for an acid formed through the resynthesis pathway. t_{deg} to c_{deg} is the theoretical ratio ^{3}H to ^{14}C through the degradative pathway. t_{obs} to c_{obs} is the observed ratio of ^{3}H to ^{14}C in the purified methyl esters of the long chain fatty acids. The first figure of each value is that assuming minimal loss of ^{3}H through the resynthesis pathway; that in parentheses is assuming total loss of ^{3}H. Where only one figure is given, the two results differ by less than 0.01. Stearic acid has been omitted from this table because recycling between stearic and palmitic acids cannot be estimated.

The results in Table 3 provide an approximate order for the proportion of individual labelled saturated fatty acids formed from labelled oleic acid by the degradative pathway:

$$\text{lauric} > \text{myristic} \gg \text{palmitic}$$

A significant portion of the labelled palmitic acid is formed on the resynthesis pathway, which is probably a reflexion of the fact that the fatty acid synthetases of many organisms produce predominantly palmitic acid[12]. Thus shorter chain acids (lauric and myristic acids) formed through the resynthesis pathway would have been produced by partial degradation of palmitic acid, with concomitant dilution of label by the cellular pool of palmitic acid.

Several assumptions have been made in this treatment of the results, and these will be discussed briefly. It has been assumed that the major pathways for the transition of label from oleic acid to the saturated acids involve either removal of a small number of C_2 units by means of β-oxidation (degradative pathway), or total breakdown to acetyl-CoA followed by resynthesis (resynthesis pathway). These are the principal pathways of fatty acid metabolism in most organisms[20], and we have assumed that other pathways, such as α-oxidation[21,22], are of little importance. Our calculations of the proportions of flow of label through the two pathways also require the assumption that any acid which has been formed by the resynthesis pathway is not then subsequently totally broken down again. If this occurs, our estimates of the flow through the resynthesis pathway will be too low, for the label will be lost from the system. During the whole incubation period the ratio of ^{3}H to ^{14}C for oleic acid should remain constant, and any variation will cause errors in the calculations of the fractions of labelled acids derived by the two pathways. The oleic acid fractions recovered after the incubations had ratios of ^{3}H to ^{14}C within 10% of the starting ratios.

The composition and abundance of biolipids and geolipids present in Recent sediments vary markedly with depth. Measurement of these differences provides one approach to the problem of studying the path of carbon during diagenesis. Farrington and Quinn[23] measured fatty acids at various depths in Recent sediment from Narragansett Bay, Rhode Island. They concluded that $C_{16:0}$ is the predominant acid at the depths studied (surface to 52 cm) and that the concentrations of the unsaturated fatty acids $C_{16:1}$ and $C_{18:1}$ decrease more rapidly with depth than do the saturated fatty acids. Brogden[24] has studied the anaerobic decomposition of zooplankton in a mixture of sediment and seawater which had been sterilized and inoculated with bacteria from marine sediments. The quantities of $C_{16:1}$ and $C_{18:1}$ acids fell rapidly over the period of the experiment, whereas the concentration of saturated fatty acids increased. Brogden suggested hydrogenation of the unsaturated fatty acids.

The novel approach described here involves direct introduction of radiolabelled biolipids into a Recent sediment. It provides specific and detailed information on the fate of a labelled compound introduced into the sediment. The method involves only minor disturbance of the ecosystem and allows estimates to be made of the fluxes through various biochemical pathways. This work substantiates the current view[1,23] that oleic acid rapidly disappears after burial in a Recent aquatic sediment. We have shown that some of the atoms of oleic acid reappear in the saturated normal fatty acid fraction, although the conversion is small. It is probable that most of the oleic acid is metabolized by degradation of the carbon skeleton through β-oxidation to acetyl-CoA, followed by conversion to CO_2 and H_2O or incorporation into non-lipid cellular components. We have calculated the proportions of the various saturated acids which have been produced from labelled oleic acid by the pathways we have considered, namely the degradative and resynthesis pathways (Table 3). The majority of the label was incorporated through the degradative pathway with retention of part or all of the original carbon skeleton of the oleic acid. The proportion is highest for lauric and myristic acids. But there are difficulties in deriving detailed inferences from information of this sort. The two pathways must differ markedly in their capability for dilution of label. Thus, the degradative pathway is unlikely to be subject to a large dilution of label and the radioactivity incorporated can be used to estimate the rate of carbon flow in this pathway. In contrast, in the resynthesis pathway the label passes through the acetyl-CoA pool of the cell. Acetyl-CoA is a key metabolite in most cells, and several pathways lead into and away from this pool. Calculations using the flow of label through the resynthesis pathway probably underestimate the total carbon flow from acetyl-CoA to long-chain fatty acids.

Our experiments certainly show that oleic acid present in a Recent sediment can be converted into saturated fatty acids with retention of carbon skeleton over a period of a few days. The process is almost certainly effected microbiologically. Whether this represents a process which contributes significantly to the content of the saturated fatty acid fraction found in Ancient sediments is uncertain. The rapid loss of oleic acid during the short periods of our experiments, coupled with the slight incorporation into saturated fatty acids through both pathways, suggests, however, that the saturated fatty acids found in compacted Recent and Ancient sediments are not derived directly from the fatty acids originally contributed to surface layers, but are derived from a number of precursors through many metabolic pathways.

On occasions the steady process of sedimentation may be interrupted by the sudden deposition of large quantities of sediment as a result of major floods or earthquakes, and in this case the sediment might be buried complete with a greater abundance of biolipids and their immediate products of alteration. Another mechanism whereby fatty acids might escape direct biological alteration could be encapsulation within a seed or nut[25]. Our prime concern in this study has been to investigate the processes of hydrogenation which are believed to occur early in the formation of a sediment. In deeper layers, as lithification proceeds, free water is gradually removed and bacterial activity ceases. As the temperature and pressure rise, slow maturation processes begin. Hydrogen transfer is a part of these slow maturation processes. Thus radiolabelled cholesterol, when intimately mixed with a sample of pulverized Green River shale (Eocene) and heated in a sealed tube at 200° for 1,000 h, afforded various radiolabelled products which included 5α and 5β-cholestane (unpublished results). The relationships between biolipids incorporated into a freshly forming sediment and those geolipids remaining after the initial period of biological and chemical diagenesis probably now encompass the following: (1) complete retention of structure as in the odd-numbered alkanes originating in the surface waxes of plant debris; (2) partial retention of structure, as in the saturated fatty acids formed by the degradative pathway in the present study, and partial or complete retention of stereochemistry as in isoprenoid acids derived, presumably, from phytol[26]; (3) complete breakdown and resynthesis as in the saturated fatty acids formed by the resynthesis pathway.

This study indicates that hydrogenation processes play a part in the early diagenesis of fatty acids. We are investigating the fate of several unsaturated biolipids for which saturated geolipid analogues are encountered in Ancient sediments. It seems likely that saturated geolipids may have a dual origin—biologically effected hydrogenation during early diagenesis, and thermally and catalytically induced hydrogen transfer processes during maturation. These hydrogenation processes are the only ones tenable unless biological sources can be found for fully saturated geolipids such as steranes, triterpanes and tetraterpanes and related compounds[27,28].

We thank Drs G. Ware, R. Denton, V. Collins, L. Morris and B. Nicholls and Mr F. J. H. Mackereth for advice, equipment and facilities. We thank Mrs Pat Muir for collecting samples and for the microbiological studies and Mr D. Morgan and Miss J. Hay for technical assistance. We thank the SRC, NERC and NASA for financial support. P. J. E. is a Beit Memorial research fellow.

1 Parker, P. L., and Leo, R. F., *Science*, **148**, 373 (1965).
2 Peterson, D. H., thesis, Univ. Washington (1967).
3 Parker, P. L., in *Organic Geochemistry* (edit. by Eglinton, G., and Murphy, M. T. J.) (Springer, Berlin and Heidelberg, 1969).
4 Shorland, F. B., in *Chemical Plant Taxonomy* (edit. by Swain, T.) (Academic Press, London and New York, 1963).
5 Gunstone, F. D., *An Introduction to the Chemistry and Biochemistry of Fatty Acids and their Glycerides* (Chapman and Hall, London, 1967).
6 Parker, P. L., *Contributions in Marine Science*, **4** (1967).
7 Schwendinger, R. B., and Erdman, J. G., *Science*, **144**, 1575 (1964).
8 Rosenfeld, W. D., *Arch. Biochem. Biophys.*, **16**, 263 (1948).
9 Oppenheimer, C. H., *Geochim. Cosmochim. Acta*, **19**, 244 (1960).
10 Zobell, C. E., and Upham, H. C., *Bull. Scripps Inst. Oceanog.*, **5**, 239 (1944).
11 Lynen, F., *Biochem. J.*, **102**, 381 (1967).
12 Morris, L. J., *Biochem. J.*, **118**, 681 (1970).
13 Wakil, S. J., *J. Lipid Res.*, **2**, 1 (1960).
14 Rognstad, R., and Katz, J., *Arch. Biochem. Biophys.*, **127**, 437 (1968).
15 Dole, V. P., and Meinertz, H., *J. Biol. Chem.*, **235**, 2595 (1960).
16 Metcalfe, L. D., Schmitz, A. A., and Belka, J. R., *Anal. Chem.*, **38**, 514 (1966).
17 Morris, L. J., *Chem. Ind.*, 1238 (1962).
18 Williams, M. A., Cope, G. H., Jackson, J. L., and Hill, P., *Biochem. J.*, **118**, 379 (1970).
19 Denton, R. M., and Randle, P. J., *Biochem. J.*, **104**, 423 (1967).
20 Mahler, H. R., and Cordes, E. H., *Biological Chemistry* (Harper and Row, New York, 1966).
21 Mead, J. F., and Levis, G. M., *J. Biol. Chem.*, **238**, 1634 (1963).
22 Hajra, A. K., and Radin, N. S., *J. Lipid Res.*, **4**, 270 (1963).
23 Farrington, J. W., and Quinn, J. G., *Nature*, **230**, 67 (1971).
24 Brogden, W. B., thesis, Univ. Florida (1968).
25 Godwin, H., *The History of the British Flora* (Cambridge University Press, Cambridge, 1956).
26 Cox, R. E., Maxwell, J. R., Eglinton, G., Pillinger, C. T., Ackman, R. G., and Hooper, S. N., *Chem. Commun.*, 1639 (1970).
27 Henderson, W., Wollrab, V., and Eglinton, G., in *Advances in Organic Geochemistry* (edit. by Schenck, P. A., and Havenear, I.) (Pergamon, London, 1968).
28 Streibl, M., and Herout, V., in *Organic Geochemistry* (edit. by Eglinton, G., and Murphy, M. T. J.) (Springer, Berlin and Heidelberg, 1969).

12

Reprinted from *Science* **153**:1133–1135 (1966)

Occurrence of Isoprenoid Fatty Acids in the Green River Shale

Abstract. *The most abundant free fatty acids present in a sample of shale from the Green River Formation (Eocene, about 60 × 10⁶ years) from Sulfur Creek are the acylic C_{19} and C_{20} isoprenoid acids. The dominance of these acids parallels the abundance of the corresponding hydrocarbons pristane and phytane.*

Four isoprenoid acids, including norphytanic (2,6,10,14-tetramethylpentadecanoic acid, $C_{19}H_{38}O_2$ and phytanic (3,7,11,15-tetramethylhexadecanoic acid, $C_{20}H_{40}O_2$), have recently been isolated from a California petroleum (*1*). The naphthenic acid fractions of crude petroleums are highly complex mixtures (*2*) and like all petroleum constituents are difficult to relate to the source rocks and hence to their biological origins. Further progress in organic geochemistry ideally necessitates study of geological situations where the extractable organic compounds are indigenous to the facies. Under such circumstances a chemical correlation with biological matter originally incorporated in the sediment is feasible. The Green River Formation (*3*) is of this type. This oil shale has been extensively studied both geologically (*4*) and chemically (*5*) and is the accumulated organic-rich sediment derived from large inland lakes of Eocene age.

A sample (200 g) of Green River Shale from the 334-m depth of a core taken from Sulfur Creek (*6*) was pulverized (to pass 200 mesh), digested with 40-percent hydrofluoric acid/hydrochloric acid (1 : 1) at room temperature for 24 hours. The residue was filtered off, washed with water, and dried. The repulverized residue (90 g) was extracted ultrasonically three times with benzene/methanol (1 : 1). The total free acids were isolated from the extract (5 g) according to the method of McCarthy and Duthie (*7*) and converted to their methyl esters (methanol/hydrochloric acid). Column and preparative thin layer chromatography afforded a pure methyl ester fraction (12 mg). Figure 1 represents a gas chromatogram of these esters on SE-30; the peaks labeled phytanic and norphytanic acid esters were enhanced when this mixture was coinjected with the authentic isoprenoid acid esters (*8*). All the labeled peaks have been identified by combined gas chromatography–mass spectrometry.

Fractions trapped from 3-percent Versamid on Gas Chrom Q were reinjected into the gas chromatography-mass spectrometry apparatus, equipped with a 6-percent SE-30 column (2 m by 4 mm; temperature, 190°C; scan time 2 seconds). Peaks which coincided on the Versamid column, that is, phytanic and n-C_{17} and norphytanic and n-C_{16} acid methyl esters, were well separated on the SE-30 column. The n-C_{14}→n-C_{18}, phytanic (Fig. 2) and nor-

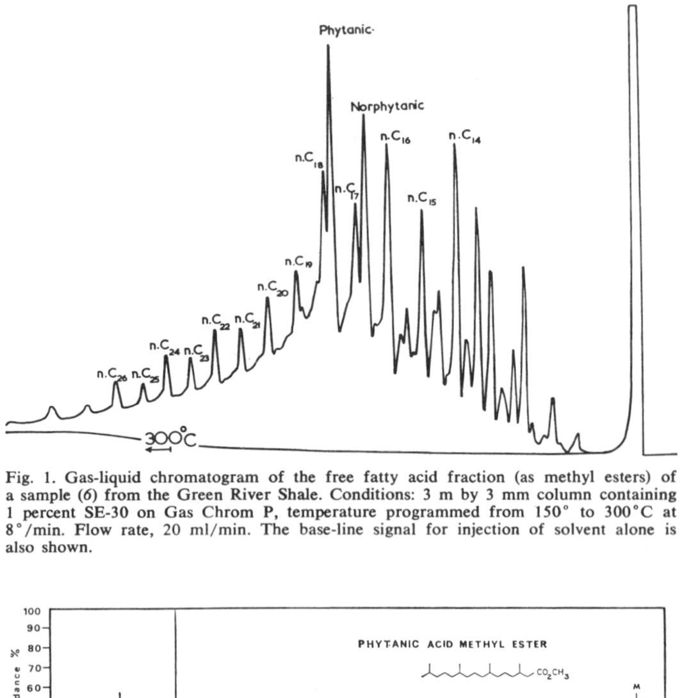

Fig. 1. Gas-liquid chromatogram of the free fatty acid fraction (as methyl esters) of a sample (*6*) from the Green River Shale. Conditions: 3 m by 3 mm column containing 1 percent SE-30 on Gas Chrom P, temperature programmed from 150° to 300°C at 8°/min. Flow rate, 20 ml/min. The base-line signal for injection of solvent alone is also shown.

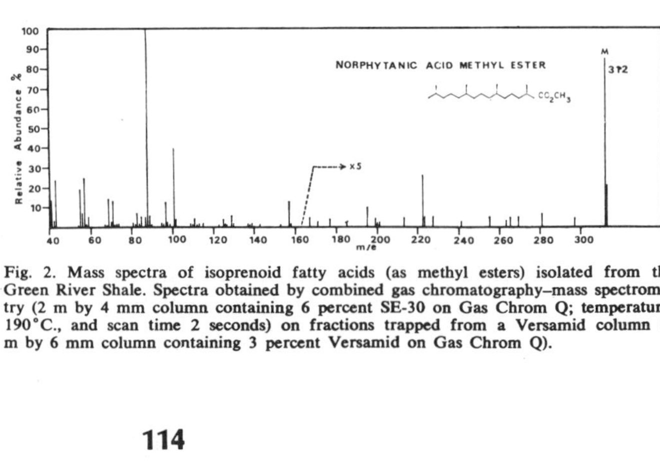

Fig. 2. Mass spectra of isoprenoid fatty acids (as methyl esters) isolated from the Green River Shale. Spectra obtained by combined gas chromatography–mass spectrometry (2 m by 4 mm column containing 6 percent SE-30 on Gas Chrom Q; temperature, 190°C., and scan time 2 seconds) on fractions trapped from a Versamid column (3 m by 6 mm column containing 3 percent Versamid on Gas Chrom Q).

phytanic (Fig. 2) acid esters were identified by comparing the mass spectra so obtained with those of authentic compounds (*8*). In addition, the infrared spectra (film and in CCl_4 solution) of the isolated and the authentic isoprenoid acid esters were identical. The n-$C_{19} \rightarrow n$-C_{26} acid esters were identified without prefractionation by recording the mass spectra of peaks which appeared in a programmed gas chromatogram using the combined gas chromatography–mass spectrometry instrument. Of course, the mass spectrometric identification, though adequate for the recognition of the carbon skeletons of these isoprenoid acids, does not provide any information about the configuration at carbon atoms 3, 7, and 11. The stereochemistry of phytanic acid derived from natural phytol has been defined as 7(R) and 11(R), the geometry at carbon 3 being determined by catalytic reduction (*9*).

There is a need to correlate the stereochemistry of the isoprenoid phytanic and norphytanic acids from geological (*1*) and biological sources (*10*), and laboratory synthesis (*11*). The same situation applies to the corresponding isoprenoid hydrocarbons (*12*). Stereochemical definition will afford a more rigorous correlation with presumed source materials. For some time it has been held that geological isoprenoids probably derive from the phytyl side chain of chlorophyll (see, *inter alia*, *5*, *12*), but Kates (*13*) has recently suggested that they may have arisen from the lipids of halophilic bacteria.

Although in the present study the *n*-alkanoic acids have not been separated from the branched and cylic acids, there is an obvious predominance of the even-numbered normal acids, as has been reported by Lawlor and Robinson (*14*) for the total fatty acids extracted from a sample from the Mahogany Zone of the same formation. These authors have pointed out the close similarity in the distribution of the *n*-alkanes and *n*-fatty acids and argue that this indicated a geogenetic relationship between the two groups of compounds, the acids giving rise to the hydrocarbons. Our results indicate a futher parallelism between the distribution of the isoprenoid acids and alkanes, though here the dominant C_{20} isoprenoid acid is paralleled by the dominance of the isoprenoid alkane with the same carbon number.

GEOFFREY EGLINTON, A. G. DOUGLAS
JAMES R. MAXWELL, J. N. RAMSAY
Department of Chemistry, University of Glasgow, Glasgow, W.2., Scotland
S. STÄLLBERG-STENHAGEN
Department of Medical Biochemistry, University of Gothenburg, Gothenburg, Sweden

References and Notes

1. J. Cason and D. W. Graham, *Tetrahedron* **21**, 471 (1965).
2. H. L. Lochte and E. R. Littman, *The Petroleum Acids and Bases* (Constable, London, 1955).
3. N. de Nevers, *Sci. Am.* **214**, 21 (1966); B. Schaeffer and M. Mangus, *Nat. Hist.* **74**, 10–21 (1965).
4. W. H. Bradley, *Amer. J. Sci.* **262**, 413 (1964), and references therein.
5. W. E. Robinson, J. J. Cummins, G. V. Dineen, *Geochim. Cosmochim. Acta* **29**, 249 (1965); G. Eglinton, P. M. Scott, T. Belsky, A. L. Burlingame, M. Calvin *Science* **145**, 263 (1964); W. E. Robinson, D. L. Lawlor, J. J. Cummins, J. I. Fester, *U.S. Bur. Mines Rept. Invest. No. 6166* (1963); J. J. Cummins and W. E. Robinson, *Chem. Eng. Data* **9**, 304 (1964).
6. Core section KR-63-133 kindly provided by Dr. W. E. Robinson of the Laramie Petroleum Research Center.
7. R. D. McCarthy and A. H. Duthie, *J. Lipid Res.* **3**, 117 (1962).
8. Phytanic and norphytanic acid methyl esters kindly provided by Dr. R. P. Hansen, Fats Research Laboratory, D.S.I.R., Wellington, New Zealand.
9. J. W. K. Burrell, L. M. Jackman, B. C. L. Weedon, *Proc. Chem. Soc.* **1959**, 263 (1959).
10. Butterfat: R. P. Hansen, *Nature* **201**, 192 (1964); ——— and J. D. Morrison, *Biochem. J.* **93**, 225 (1964); R. P. Hansen and F. B. Shorland, *ibid.* **55**, 662 (1953); W. Sonneveld, P. H. Begemann, G. J. van Beere, R. Kenning, J. C. M. Schogt, *J. Lipid Res.* **3**, 351 (1962); R. P. Hansen, F. B. Shorland, J. D. Morrison, *J. Dairy Res.* **32**, 21 (1965); R. P. Hansen and F. B. Shorland, *Biochem. J.* **50**, 358 (1951); N. Bjurstam, B. Hallgren. R. Rhyage, S. Ställberg-Stenhagen, referred to by E. Stenhagen, *Z. Anal. Chem.* **181**, 642 (1961). Ox fat: R. P. Hansen, F. B. Shorland, N. J. Cooke, *J. Sci. Food Agr.* **9**, 391 (1958); R. P. Hansen, *Chem. Ind. London* **1965**, 303 (1965). Ox serum: W. R. H. Duncan and G. A. Garton, *Biochem. J.* **89**, 414 (1963); A. K. Lough, *ibid.* **86**, 14P (1963); **91**, 584 (1964). Sheep fat: R. P. Hansen, *New Zealand J. Sci.* **8**, 158 (1965); *Chem. Ind. London* **1965**, 1258 (1965). Refsum's syndrome: D. Steinberg, S. Avigan, C. Mize, L. Eldjarn, K. Try, S. Refsum, *Biochem. Biophys. Res. Comm.* **19**, 783 (1965) and references therein; R. P. Hansen, *Biochim. Biophys. Acta* **106**, 304 (1965). Rumen bacteria: R. P. Hansen, personal communication.
11. R. P. Hansen, F. B. Shorland, I. A. M. Prior, *Biochim. Biophys. Acta* **116**, 178 (1966).
12. The isoprenoid hydrocarbons pristane ($C_{19}H_{40}$) and the phytadienes ($C_{20}H_{38}$) are established constituents of certain living organisms. Their presence in marine organisms may in part explain the occurrence of pristane and phytane in recent marine sediments. (For leading references see A. G. Douglas and G. Eglinton, "The distribution of alkanes," chap. 4 in *Comparative Phytochemistry*, T. Swain, Ed. (Academic Press, New York, 1966).
13. M. Kates, personal communication to Prof. M. Calvin; M. Kates, L. S. Yengoyan, P. S. Sastry, *Biochim. Biophys. Acta* **98**, 252 (1965).
14. D. L. Lawlor and W. E. Robinson, information presented before the Division of Petroleum Chemistry, American Chemical Society, Detroit meeting, April 1965.
15. We have recently learned that Dr. M. Blumer, Woods Hole Oceanographic Institution, has isolated phytanic acid, norphytanic acid, and some higher and lower homologs from a sample of Serpiano Oil Shale (Triassic, about 200×10^6 years).
16. Supported in part by grants from the Natural Environmental Research Council (B/SR/655), Shell International, and NASA (NsG101-61).

13

Reprinted from *Science* **164**:1052–1054 (1969)

Fatty Alcohols (Normal and Isoprenoid) in Sediments

Abstract. *Normal long-chain alcohols were isolated from Recent marine sediments from several environments. The isoprenoid alcohol dihydrophytol, which is thought to originate from phytol, the side chain of chlorophyll, by hydrogenation in the reducing environment of the sediment, was also present in most of the samples. Both the normal and isoprenoid alcohols were found in the Green River shale (Eocene). Geochemical implications are considered.*

Biochemicals and recognizable degradation products of biochemicals, although present only in trace amounts in sediments, are in most cases the only source of data on the chemical composition of ancient organisms. It is useful to study the organic compounds contained in Recent sediments and to note biological sources for various compounds and some of the chemical transformations that specific biochemicals undergo in geologically short periods of time (*1*). We now report the finding of normal and isoprenoid fatty alcohols in sediments from several environments. Although the fatty acids (*2, 3*) and corresponding long-chain hydrocarbons (*4*) have been found in marine sediments, the finding of alcohols has not been reported.

Results of analyses on fatty alcohols in samples from three different Recent environments and three ancient sediments are presented in Table 1. The sample of Miocene age is from an undescribed outcrop in the Philippine Islands. The sample of shale from the Green River (Eocene) is from the U.S. Bureau of Mines oil-shale mine near Rifle, Colorado. The Eagleford shale (Upper Cretaceous) is from an outcrop near Austin, Texas (*5*). Baffin Bay is a hypersaline arm of the Laguna Madre located about 30 miles (48 km) south of Corpus Christi, Texas. The core from the Gulf of Mexico was taken in 25 m of water off Port Aransas, Texas. The sample from the San Nicholas Basin is a grab sample taken off the coast of southern California (*5*). The fatty-acid composition of some of these samples has been reported (*1, 2, 6*). All of the Recent samples of sediment were frozen soon after the time of collection and were kept frozen until used.

The following analytical procedure yields a sample suitable for gas-chromatographic analysis, but it does not exclude any type of long-chain alcohol. The sample was treated with dilute hydrochloric acid to remove carbonate and was washed with water to remove inorganic salts. After filtration, the moist mud was treated with methanol for 20 minutes while being stirred and subjected to ultrasonic vibrations. The methanol extract was recovered by filtration. The mud was again extracted with chloroform while being stirred and subjected to ultrasonic vibrations. The methanol and chloroform filtrates were combined and dried on a rotary evaporator. The organic residue was taken up in chloroform and washed with 1*N* hydrochloric acid to remove any last traces of inorganic salts. The chloroform was then dried, and the residue was weighed. The residue was saponified with 0.5*N* potassium hydroxide in methanol for 1 hour on a steam bath. The nonsaponifiable fraction containing the hydrocarbons, alcohols, and sterols was then separated from the saponifiable fatty-acid fraction by extraction of the alkaline solution with benzene. The nonsaponifiable benzene fraction was dried under a stream of dry nitrogen and then was taken up in 1 ml of *n*-hexane. This was carefully placed on a chromatography column (2.5 by 20 cm) packed with prewashed silica gel (Woelm, activity I). Straight-chain hydrocarbons were eluted from the column with 75 ml of hexane. Elution of the column with 75 ml of benzene removed most of the aromatics of high molecular weight. The alcohols were eluted with 75 ml of methanol. The methanol fraction was then dried under a stream of dry nitrogen and was taken up in a known volume of benzene. This benzene solution was used for preliminary identification of the alcohols by means of gas-liquid chromatography (GLC).

A gas chromatograph (F & M Biomedical model 400) with a hydrogen flame-ionization detector, operated isothermally, and a copper column, 1.5 m long, packed with 10-percent FFAP coating (Varian) on 60- to 80-mesh Chromosorb G support were used. The column oven was operated at 215°C for chromatographing alcohols of lower molecular weight, and at 250°C for alcohols of high molecular weight. Fatty alcohol standards consisting of alcohols containing an even number of carbons from C_{14} to C_{28} (no double bonds) and dihydrophytol were used for calibration. The alcohols containing 14 to 22 carbon atoms and dihydrophytol were easily identified and measured by this procedure.

Since retention times in gas-liquid chromatography are not unique, the identifications were verified by several methods. A second set of retention times was obtained by use of a nonpolar SE-30 column. The infrared spectra of the isolated alcohol fractions (methanol fraction off silica gel) were taken in CCl_4 for several of the more alcohol-rich samples. Absorption bands due to the OH functional group were present. Finally, the fatty alcohols were converted to their acetate esters by refluxing with acetic anhydride; the retention times of the alcohol peaks shifted to those expected for the corresponding esters (Fig. 1). The alcohol fraction was further purified by column chromatography on aluminum oxide, but this was proved unnecessary for most samples.

The Recent sediments contain normal saturated alcohols with 12 to 26 carbon atoms (Table 1). Alcohols with both even and odd numbers of carbon atoms are present. In addition, normal,

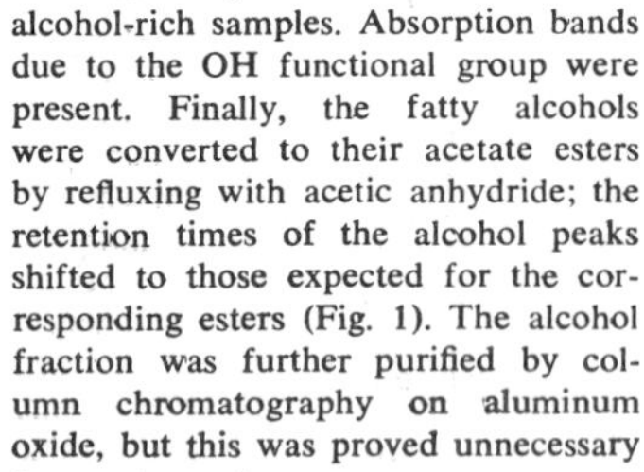

Fig. 1. Gas-liquid chromatogram of the fatty alcohols of lower molecular weight from the Eagleford shale on a FFAP substrate. The upper curve shows peaks for the free alcohols; the lower curve is for the acetate esters. In the lower curve peak *C* is enlarged due to the addition of pure dihydrophytol acetate spike. *A*, C_{14} (no double bonds); *B*, C_{16} (no double bonds); *C*, dihydrophytol; *D*, C_{18} (no double bonds).

Table 1. Concentration (parts of alcohol per million parts of dry sediment) of the normal alcohols and dihydrophytol in sediments.

Alcohol*	Samples from					
	Baffin Bay	Gulf of Mexico	San Nicholas Basin	Miocene age	Green River shale	Eagle-ford shale
12:0	1.00					
14:0	1.40	0.36	3.00	1.75	3.20	1.30
15:0	0.94	.52	1.50	0.31	2.10	0.88
16:0	.99	1.08	2.20	.14	0.50	1.10
16:1	.08		0.54			
17:0	.55	0.18	.80	.07	.35	0.92
Dihydrophytol	.39	.32	1.57	.08	1.10	.09
18:0	.64	.82	1.75	.17	0.96	.90
18:1	1.06		3.25			
20:0	0.17	.60	3.90	.20	2.00	.65
21:0			1.35	.14		.42
22:0	.85	1.35	7.05	.15	9.50	.32
24:0	.61	0.48				
26:0	1.05					
Total	9.7	5.7	26.9	3.0	19.7	6.6
	Percentage of total organic carbon					
	1.8	1.0	6.7	4.2	18.	15.

* The numbers before the colon indicate the numbers of carbon atoms; the numbers after the colon indicate the numbers of double bonds.

unsaturated alcohols are present in the Recent muds but are absent in the ancient sediments. The isoprenoid, dihydrophytol (3,7,11,15-tetramethylhexadecanol), is present in substantial concentrations in both Recent and ancient sediments. Several unidentified peaks present in the GLC traces may be isoprenoid alcohols of lower molecular weight.

Normal, monomethyl branched-chain and isoprenoid long-chain hydrocarbons, alcohols (Table 1), and fatty acids are present in sediments (*1, 2, 4, 7, 8*). The relations among these three different types of carbon skeletons involving three different functional groups are beginning to emerge. We confirmed (*9*) the presence of the C_{20} isoprenoid acid, phytanic, and the C_{19} isoprenoid acid, pristanic, in Recent sediments (*7*). Eglinton *et al.* (*8*) found phytanic acid and pristanic acids in the Green River shale. Blumer (*7*) found that, whereas the C_{19} isoprenoid hydrocarbon, pristane, is present in Recent sediment, the C_{20} isoprenoid hydrocarbon, phytane, is absent. Both these hydrocarbons are present in ancient sediment and crude oil (*10*). The C_{20} isoprenoid alcohol, dihydrophytol, present in both Recent and ancient sediments (Table 1), may be the first reaction product in the conversion of phytol into phytane and phytanic acid. We did not determine whether the C_{19} isoprenoid alcohol was present in our samples.

Because, in the biosphere, fatty alcohols occur less frequently and generally in much lower concentration than the corresponding fatty acids, it was expected that the concentration of fatty alcohols would be much lower than that of fatty acids in Recent sediments. However, the amounts of alcohols found in Recent sediment were from one order of magnitude less to the same order of magnitude as the concentrations of fatty acids. Baffin Bay mud contained 9.7 μg of total alcohol per gram of dry sediment (Table 1). Parker (*6*) reported 40 μg of fatty acid per gram of sediment over the same range of molecular weight. These high concentrations are surprising but difficult to explain. In order to be sure that the sonification used in the extraction procedure did not generate fatty alcohols from other organic material present in the sediment, several control experiments were run. In one experiment, 50 mg of palmitic acid was added to a sample of Baffin Bay sediment which had been extracted with organic solvents. The mixture was extracted with methanol by the use of sonification. Fatty alcohols and hydrocarbons were analyzed according to the standard procedure. No alcohols or hydrocarbons were detected. In other control experiments split samples were extracted by Soxhlet extraction and sonic energy, and the alcohol patterns were compared. In general, the total yield of alcohols was slightly higher with sonification. In qualitative composition the recovered fatty alcohols were not significantly different. The extraction procedure based on sonic energy does not create any artifacts.

Fatty alcohols in sediments probably have their origin in the marine life of the environments studied. Because Baffin Bay normally receives very little fresh water as runoff, it is often twice as saline as seawater. Such restricted runoff probably could not transport enough terrestrial organic matter to account for the uniform concentrations of alcohols observed in the bay sediments. Inasmuch as the same types and concentrations of alcohols are present in all the Recent samples, we assume Baffin Bay to be typical. We cannot specify which marine organisms are producing the normal fatty alcohols, although the bacteria may be involved. The isoprenoid alcohol, dihydrophytol, probably is derived from phytol, the side chain of the chlorophyll molecule. The reducing environment of the sediment must be sufficient to hydrogenate the double bond of phytol. Likewise, reduction of the corresponding fatty acid would yield the fatty alcohols. However, our data do not allow us to evaluate this mechanism.

Hoering (*11*) has isolated and identified fatty alcohols in Recent sediment while we were conducting our investigations. His results, based in part on mass spectra, are qualitatively similar to ours. The fatty alcohols promise to be a rewarding subject for organic geochemical studies. Dihydrophytol, in particular, should serve as a useful biological marker.

JUDY SEVER
P. L. PARKER

University of Texas,
Marine Science Institute,
Port Aransas 78373

References and Notes

1. R. F. Leo and P. L. Parker, *Science* **152**, 649 (1966).
2. J. E. Cooper and E. E. Bray, *Geochim. Cosmochim. Acta* **27**, 1113 (1963).
3. P. H. Abelson, T. C. Hoering, P. L. Parker, in *Advances in Organic Geochemistry*, U. Colombo and G. D. Hobson, Eds. (Pergamon, New York, 1963).
4. E. E. Bray and E. D. Evans, *Bull. Amer. Ass. Petrol. Geol.* **49**, 248 (1965).
5. We thank the Esso Production Research Company for our sample of Miocene age, and W. E. Robinson for the Green River shale. Dr. T. C. Hoering collected and donated the sample from San Nicholas Basin.
6. P. L. Parker, *Contrib. Mar. Sci. Univ. Tex.* **12**, 135 (1967).
7. M. Blumer and W. D. Snyder, *Science* **150**, 1588 (1965); M. Blumer and W. J. Cooper, *ibid.* **158**, 1463 (1967); M. Blumer, *ibid.* **149**, 722 (1965).
8. G. Eglinton, A. G. Douglas, J. R. Maxwell, J. N. Ramsay, S. Ställberg-Stenhagen, *ibid.* **153**, 1133 (1966).
9. T. Whelan, thesis, University of Texas (1968).
10. W. E. Robinson, J. J. Cummins, G. V. Dineen, *Geochim. Cosmochim. Acta* **29**, 249 (1965); J. G. Bendoraitis, B. L. Brown, L. S. Hepner, *Anal. Chem.* **34**, 49 (1962).
11. T. C. Hoering, *Carnegie Inst. Wash. Year B.* **67**, 202 (1968).
12. Supported by NSF grant GA-299.

24 October 1968; revised 11 March 1969

14

Reprinted from *Science* **169**:674–676 (1970)

STEROLS IN RECENT MARINE SEDIMENTS

David Attaway and P. L. Parker

Abstract. *Sterolic fractions have been isolated from Recent marine sediments representing two different environments. The fractions were characterized by infrared spectroscopy and thin-layer chromatography. The major sterols in the fractions were identified by gas chromatography and mass spectrometry. The demonstrated survival of these common plant and animal sterols for several thousand years suggests that these molecules will be useful geochemical indicators.*

Most organic matter produced by a marine community its quickly re-oxidized by the organisms or by inorganic processes. A small fraction escapes oxidation and is deposited in sediments. This small fraction is the object of study for organic geochemistry. The part of this preserved organic matter not bound into kerogen contains most of the preserved biochemicals (*1*).

The high degree of structural diversity of sterols suggests that the sterols might carry significant information with regard to the type of biological community associated with various sediments and perhaps even to specific organisms of the community (*2*). Using a colorimetric method, Schwendinger and Erdman (*3*) made a survey for sterols and reported positive tests for them in a series of Recent sediments, varying from freshwater to deep marine. They stated that high sterolic concentrations of 60 to 300 parts per million in their samples make sterols possible precursors for some of the compounds found in petroleum. Others (*4*) think that sterols or their degradation products are major contributors to the optical activity of petroleum because much of this activity is confined to that fraction of compounds having 27 to 30 carbon atoms. We now report the finding of several specific sterols from marine sediments from two different marine environments.

The bottom meter of a three-meter

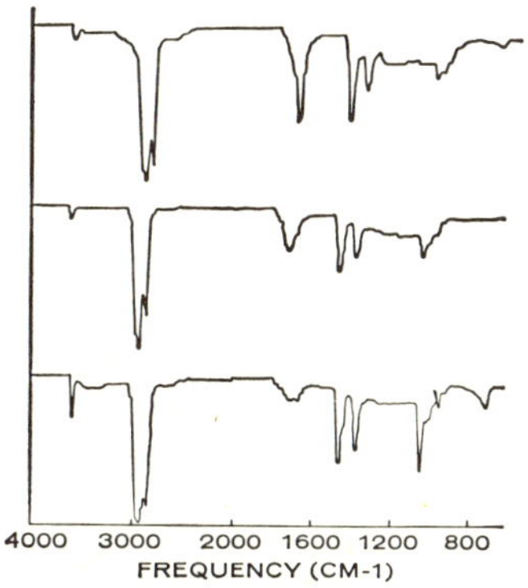

Fig. 1. Infrared spectra of sterol fraction from Baffin Bay sediment (upper), sterol fraction of sediment from San Pedro Basin (middle), and mixture of campesterol and β-sitosterol (lower). Measurements were made on solutions in carbon tetrachloride.

piston core from Baffin Bay, Texas, and a grab sample from San Pedro Basin, offshore from southern California, were analyzed. The section of core was estimated to be from 2000 to 3000 years old. The hydrocarbons, long chain alcohols, and fatty acids in these samples have been well studied (see *5*).

The samples were extracted with mixtures of distilled chloroform and distilled methanol by stirring at room temperature. The resulting extracts were saponified in methanolic KOH. The unsaponifiable fractions were extracted from the reaction mixtures in ethyl ether and chromatographed in a mixture of ethyl ether and heptane (1:9) on columns of silica gel (Woelm). Fractions were collected, and those in which sterolic materials were indicated by thin-layer chromatography, were combined and chromatographed again on fresh columns of silica gel. The results of thin-layer chromatography indicated that one fraction from each of the second columns contained primarily sterols. These fractions gave the characteristic lavender color of sterols when on thin layers of silica gel G. They were sprayed with 5 percent sulfuric acid in ethanol and heated at 95°C for 15 minutes and they exhibited R_F values corresponding to that of a mixture of authentic sterols.

The infrared spectra of the two sterolic fractions were measured in carbon tetrachloride (7 mg/ml). Figure 1 shows the infrared spectrum of the Baffin Bay and San Pedro sterolic fractions and the infrared spectrum of a mixture of campesterol and β-sitosterol. The band at 3580 cm^{-1} (oxygen-hydrogen stretching), the 1050 cm^{-1} (carbon-hydroxyl stretching), and the general similarity of the spectra suggested that the fractions were sterolic.

The lipid yield from the Baffin Bay sample was 400 ppm on a dry-weight basis, and the yield of sterolic material was 6 ppm. The San Pedro sample gave 10,000 ppm of lipids and 40 ppm of sterols which is near the range of concentrations reported by Schwendinger and Erdman (*3*).

Some of the components of these fractions were tentatively identified by comparative gas chromatography on columns (1.8 m by 0.3 cm) of 3 percent JXR on Chromosorb Q (80/100 mesh). The temperature was programed from 175° to 250°C at 1 deg/min. Helium, flowing at 35 ml/min, was the carrier gas. Dual columns with flame detectors were used. Four components of each sterolic fraction exhibited retention times the same as those of cholesterol, campesterol, stigmasterol, or β-sitosterol. Coinjection of the sterolic fractions with authentic samples of those four sterols indicated that retention times corresponded exactly.

The sterolic fractions were treated with refluxing acetic anhydride for 3 hours, and resulting steryl acetates were isolated by the method of Vishniac (*6*). Comparative gas chromatography of the acetyl derivatives and of authentic steryl acetates also indicated that four major components of each of the fractions were cholesterol, campesterol, stigmasterol, and β-sitosterol. Figure 2 shows the gas chromatogram of the sterolic fraction of the Baffin Bay sediment. Suggested identities of four of the components are shown. Figure 3 shows the corresponding gas-chromatogram of the sterolic fraction from the San Pedro sample. It has a pattern similar to that of the Baffin Bay sample; although it contains a major concentration of component *X* which, if present in the Baffin Bay sample, is a minor component.

Mass spectra data of the Baffin Bay sample indicate molecular ions at mass 386, 388, 400, 412, 414, 416, and 428. Ions of mass 386, 400, 412, and 414 correspond to cholesterol, campesterol, stigmasterol, and β-sitosterol, respectively.

When all components of these sterolic fractions are identified, it should be possible to assess the relative contributions of phytogenic and zoogenous sterols and perhaps to suggest types of source organisms for them.

Since a triterpene alcohol, similar in structure to sterols, has been isolated (*7*) from the Messel oil shale (50 million years old) it is likely that intact

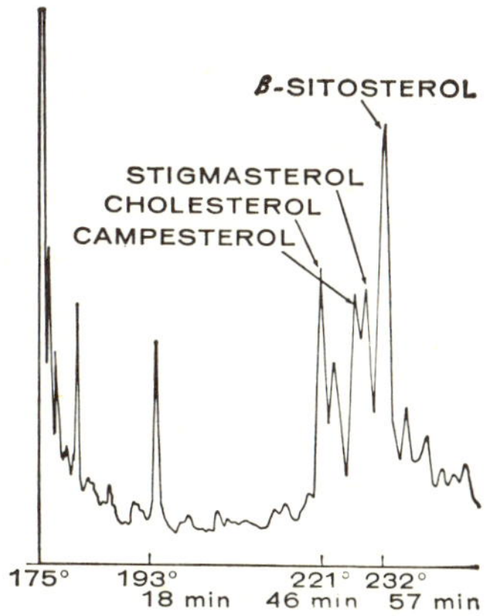

Fig. 2. Gas chromatogram of sterol fraction of Baffin Bay sediment.

sterols are also present in similar ancient sedimentary rocks. Further study of sterol geochemistry may yield clues to the nature of contributing organisms which in turn partly identifies the environments at times of deposition. For

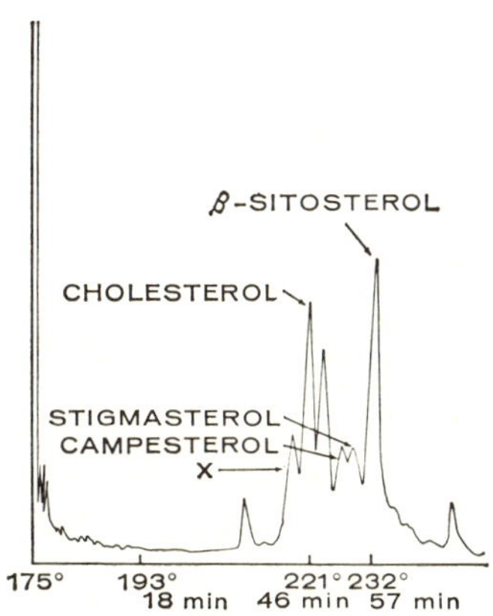

Fig. 3. Gas chromatogram of sterol fraction from San Pedro sample.

example, blue-green algae and bacteria lack sterols so that an environment dominated by these forms might have an unusual sterol pattern.

References and Notes

1. G. Eglington and M. T. J. Murphy, Eds., *Organic Geochemistry: Methods and Results* (Springer-Verlag, New York, 1969).
2. W. Bergmann, in *Comparative Biochemistry*, M. Florkin and H. S. Mason, Eds. (Academic Press, New York, 1962), vol. 3.
3. R. B. Schwendinger and J. G. Erdman, *Science* **144**, 1575 (1964); W. Bergmann, in *Organic Geochemistry*, T. A. Breger, Ed. (Pergamon Press, New York, 1963), p. 515.
4. M. A. Bestougeff, in *Fundamental Aspects of Petroleum Geochemistry*, B. Nagy and U. Colombo, Eds. (Elsevier, New York, 1967).
5. R. F. Leo and P. L. Parker, *Science* **152**, 649 (1966); J. Sever and P. L. Parker, *ibid.* **164**, 1052 (1969); P. L. Parker, *Contrib. Mar. Sci.* **12**, 135 (1967); T. C. Hoering, *Carnegie Inst. Wash. Year B.* **67**, 202 (1968).
6. H. S. Vishniac, *Biochim. Biophys. Acta* **26**, 430 (1957).
7. P. Albrecht and G. Ourisson, *Science* **163**, 1192 (1969).

8. Supported by NSF grant GA-299. One of us (D.A.) was a University of Texas postdoctoral fellow during the research period. We thank Dr. T. C. Hoering for the sample of San Pedro Basin sediment.

* Present address: University of Kansas, State Geological Survey, Lawrence 66044.

27 March 1970; revised 17 June 1970

15

Reprinted from *Geochim. et Cosmochim. Acta* **33**:1304–1307 (1969), with the permission of Microforms International Marketing Corporation as exclusive copyright licensee of Pergamon Press journal back files

The isolation of steranes from Green River oil shale

P. C. Anderson, P. M. Gardner and E. V. Whitehead
The British Petroleum Co., Ltd., Sunbury-on-Thames, Middlesex

and

D. E. Anders and W. E. Robinson
Laramie Petroleum Research Center, Bureau of Mines
U.S. Department of the Interior, Laramie, Wyoming 82070

(*Received* 14 *April* 1969; *accepted in revised form* 6 *June* 1969)

Abstract—The C_{28} sterane was identified and the presence of the C_{27} and C_{29} steranes in Green River oil shale was confirmed. This was accomplished by comparison of the spectra of the isolated components with authentic standards.

The Green River oil-shale deposits of Colorado (U.S.A.) have been extensively investigated. It has been suggested that the n-alkanes (Cummins and Robinson, 1964; Johns *et al.*, 1966; Belsky *et al.*, 1965; Han *et al.*, 1968) and isoprenoid alkanes (Cummins and Robinson, 1964) present in the bitumen are derived from primitive blue-green algae and wind-borne pollen from terrestrial plants. The phytane and pristane (Cummins and Robinson, 1964; Douglas and Eglinton, 1966; Eglinton *et al.*, 1966a) present in the shale could have been derived from chlorophyll, while the perhydro-β-carotene (Robinson and Anders, 1966; Eglinton *et al.*, 1966b; Murphy *et al.*, 1967) could have been connected with the pigments β-carotene and xanthophyll. Gammacerane (Hills *et al.*, 1966) has also been identified in the shale.

Burlingame *et al.* (1965) have tentatively identified the C_{27} and C_{28} steranes in Green River oil shale and have compared the mass spectra of the C_{29} sterane from Green River oil shale with an authentic C_{29} sterane (stigmastane). Henderson *et al.* (1968) compared the mass spectra of the C_{27} sterane from Green River oil shale with an authentic C_{27} sterane. In this investigation the C_{28} sterane was identified and the presence of the C_{27} and the C_{29} steranes was confirmed in Green River oil shale. This was accomplished by comparing the mass spectra and the gas chromatographic emergence times of the three steranes with authentic standards. The significance of these findings is that these steranes can be related to the principal phytosterols.

Robinson and Anders obtained a sterane concentrate (KR 66-146) by molecular distillation of a branched- and cyclo-alkane fraction of Green River oil-shale bitumen (Cummins and Robinson, 1964). Preliminary mass spectrographic analysis of this concentrate at 11 and 70 eV indicated the presence of the C_nH_{2n-6} and C_nH_{2n-8} hydrocarbon series with molecular ions at 372, 386, 400 and 412. The molecular ion at m/e 412 and the prominent fragment at 191 could be attributed to triterpane "G" (Hills and Whitehead, 1966).

Further examination by gas–liquid chromatography (GLC) at 250°C using a 148-ft glass-capillary column (SE 52 Silicone gum and a hydrogen carrier gas) showed that KR 66-146 was comprised of four major components, labeled A, B, C and D (Fig. 1). Coinjection of KR 66-146 with cholestane, ergostane, stigmastane and hydrocarbon "G" (Hills and Whitehead, 1966) resulted in the heights of A, B, C and D being increased relative to the rest of the mixture. Hydrocarbon "G" appears to be the same as peak 20 reported by Henderson *et al.* (1968). Similarly, the major peaks were reinforced by these compounds on a 108-ft SE 30 capillary column. A small quantity of gammacerane was also detected.

The mixture was separated into its four main components (A, B, C and D) by preparative scale GLC using a Varian "Autoprep 705" containing a 6-ft by $\frac{3}{8}$-in.-i.d. SE 30 column at 220° to 230° C with a nitrogen carrier gas. By this method A (4 mg), B (15 mg) and C (15 mg) were isolated in about 60% purity.

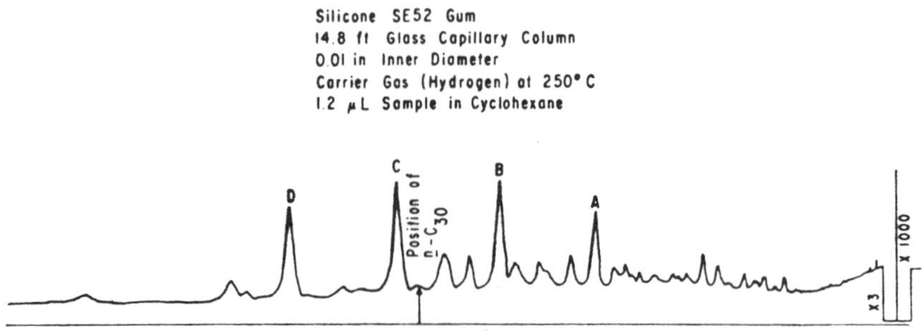

Fig. 1. Gas–liquid chromatogram of a sterane–triterpane concentrate (Sample No. KR66-146).

Components A, B and C were adducted with thiourea using a chloroform-methanol solvent (MURPHY *et al.*, 1967). The adducted portion was washed with chloroform, dissolved in water and the hydrocarbon extracted with chloroform. Further purification was effected by elution with purified n-pentane through a 3-ft by 0·1-in. alumina column (Peter Spence and Sons, Ltd., Type A alumina, 70–100 mesh). The purities of the fractions, as determined by capillary chromatography, are indicated in Table 1:

Table 1

Component	Weight	Purity	Principal impurities	
A	0·7 mg	80%	2·1% B	
B	4·2 mg	80%	7·5% A	1·4% C
C	0·8 mg	46%	33% B	3% "G"

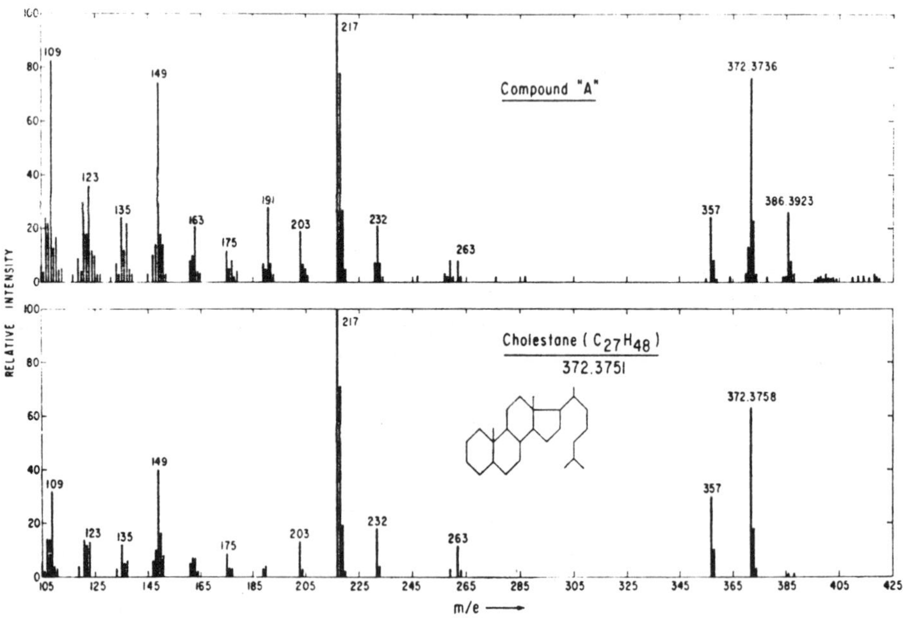

Fig. 2. Mass spectra of compound "A" and authentic cholestane.

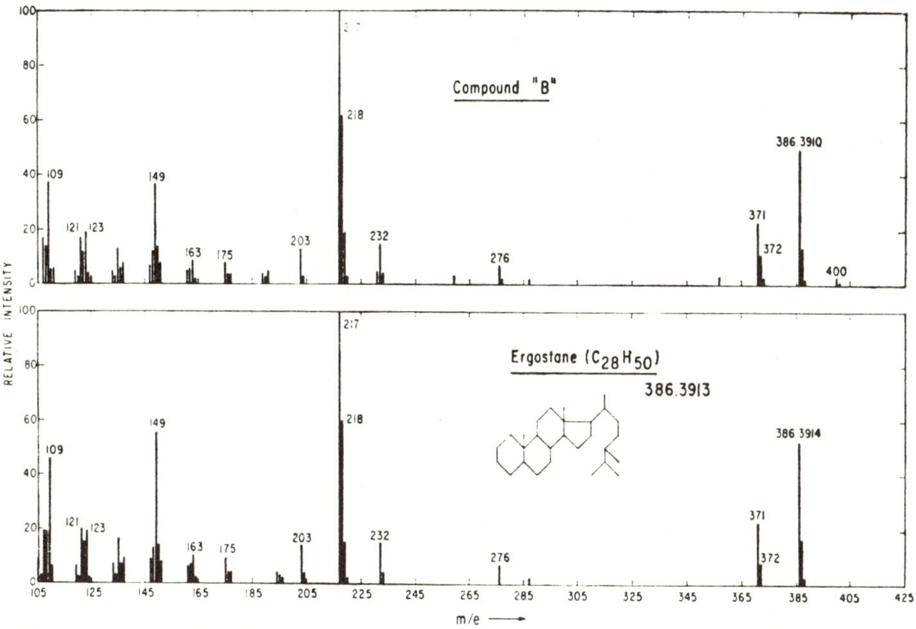

Fig. 3. Mass spectra of compound "B" and authentic Ergostane.

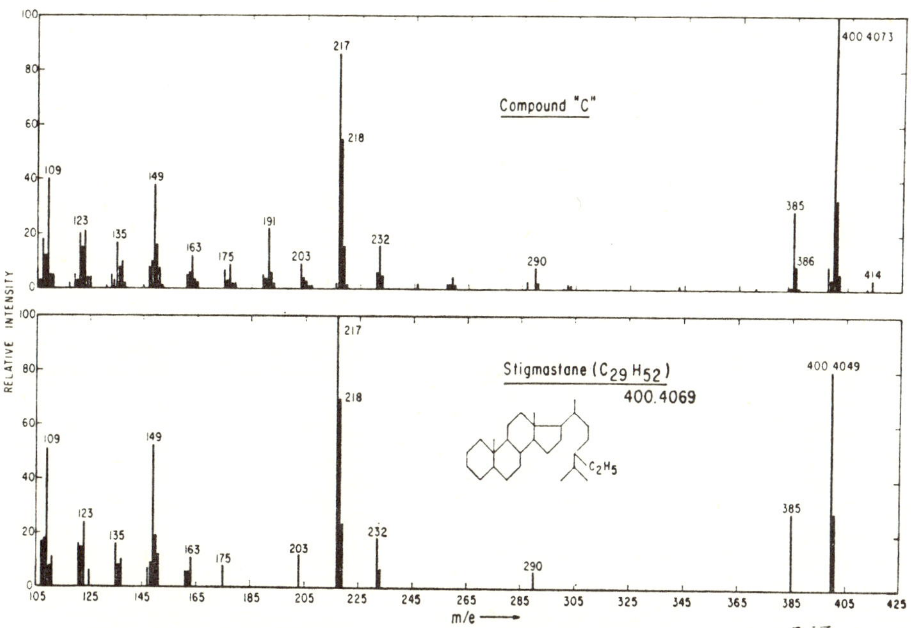

Fig. 4. Mass spectra of compound "C" and authentic stigmastane.

To increase the purity of the fraction containing component C, an additional quantity of KR 66-146 was placed on a column of alumina (100 parts of alumina to 1 part of sample) and eluted with n-pentane. The material percolating from the column in the 250–275 ml volume range contained mainly component C and pentane. After removal of the pentane, the fraction was further purified by GLC. The purified fraction contained about 93% of component C.

Each of three fractions of maximum purity was subjected to GLC and, once again, A, B and C were reinforced by cholestane, ergostane and stigmastane on SE 30 and SE 52 stationary phases.

The mass spectra of A and cholestane were recorded consecutively on an AEI MS 9 high resolution mass spectrometer and similar spectra were obtained ; B and C were treated in the same manner and again the spectra were very similar to those of ergostane and stigmastane (Figs. 2–4). The prominent peak at m/e 191 in KR 66-146 was reduced in these purified sterane fractions.

The evidence leads to the conclusion that the C_{27}, C_{28} and C_{29} steranes are present in the bitumen from the Green River oil shale and that they were derived from cholesterol, ergosterol, stigmasterol or related sterols.

Acknowledgment—The authors are grateful to Professor D. H. R. Barton, Imperial College, London, who kindly donated samples of authentic ergostane (m.p. 82°C) and stigmastane (m.p. 84°C). The cholestane (Batch No. 28517) was obtained from Koch-Light Laboratories, Limited, Colnbrook, Buckinghamshire, London.

References

Belsky T., Johns R. B., McCarthy E. D., Burlingame A. L., Richter W. and Calvin M. (1965) Evidence of life processes in a sediment two and a half billion years old. *Nature* **206**, 446–447.

Burlingame A. L., Haug P., Belsky T. and Calvin M. (1965) Occurrence of biogenic steranes and pentacyclic triterpanes in an Eocene shale (52 million years) and in an early Precambrian shale (2.7 billion years): A preliminary report. *Proc. Nat. Acad. Sci. U.S.* **54**, 1406–1412.

Cummins J. J. and Robinson W. E. (1964) Normal and isoprenoid hydrocarbons isolated from oil-shale bitumen. *J. Chem. Engng. Data* **9**, 304–307.

Douglas A. G. and Eglinton G. (1966) *Comparative Phytochemistry*, pp. 57–73. Academic Press.

Eglinton G., Douglas A. G., Maxwell J. R., Ramsey J. N., and Stallberg-Stenhagen S. (1966a) Occurrence of isoprenoid fatty acids in Green River shale. *Science* **153**, 1133–1135.

Eglinton G., Maxwell J. R., Murphy Sister M. T. J., Henderson W. and Douraghi-Zadeh K. (1966b) Hydrocarbons and fatty acids in algal shales and related materials. Annual Meeting Geochemical Society, San Francisco, California.

Han J., McCarthy E. D., Van Hoeven W., Calvin M. and Bradley W. H. (1968) Organic geochemical studies II. A preliminary report on the distribution of aliphatic hydrocarbons in algae, in bacteria, and in a recent lake sediment. *Proc. Nat. Acad. Sci. U.S.* **59**, 29–33.

Henderson W., Wollrab V. and Eglinton G. (1968) Identification of steroids and triterpanes from a geological source by capillary gas–liquid chromatography and mass spectrometry. *Chem. Commun.* 710–712.

Hills I. R. and Whitehead E. V. (1966) Triterpanes in optically active petroleum distillates. *Nature* **209**, 977–979.

Hills I. R., Whitehead E. V., Anders D. E., Cummins J. J. and Robinson W. E. (1966) An optically active triterpane, gammacerane, in Green River Colorado oil-shale bitumen. *Chem. Commun.* 752–754.

Johns R. B., Belsky T., McCarthy E. D., Burlingame A. L., Haug P., Schones H. K., Richter W. and Calvin M. (1966) The organic geochemistry of ancient sediments—Part II. *Geochim. Cosmochim. Acta* **30**, 1191–1222.

Murphy Sister M. T. J., McCormick A. and Eglinton G. (1967) Perhydro-β-carotene in the Green River shale. *Science* **157**, 1040–1042.

Robinson W. E. and Anders D. E. (1966) Green River oil-shale kerogen. Annual meeting Geochemical Society, San Francisco, California.

16

Reprinted from *Nature* **185**:234 (1960)

Presence of Carbohydrates and Other Organic Compounds in Ancient Sedimentary Rocks

Most investigators have been of the opinion that proteinaceous and carbohydrate materials that are deposited in marine and lake sediments cannot survive for long periods of time but are rapidly decomposed. However, in recent years several investigators have shown the presence of amino-acids in Tertiary shales[1] and in fossil shells and bones as old as the Devonian[2].

Preliminary studies have now demonstrated the presence of minute amounts of carbohydrate as well as proteinaceous substances in bituminous sedimentary rocks ranging in age from Ordovician to Tertiary. The reducing environment under which these sediments accumulated, coupled with the association between the organic compounds and finely divided inorganic matter, probably account for the preservation of these compounds.

Outcrop and well-core samples were thoroughly cleaned by immersion in chromic acid and subsequent washing with water. After being powdered, each sample (about 500 gm.) was treated with 0·5 *N* sulphuric acid for 8–10 hr. on a boiling water-bath. The hydrolysis mixture was neutralized with barium carbonate and freed from inorganic matter by ethanol precipitation and by ion-exchange resins. The hydrolysate from each sample was concentrated *in vacuo* and examined by paper chromatography. The following sugars were tentatively identified: glucose, xylose and arabinose, and possibly galactose and rhamnose (see Table 1). The concentration of glucose, which was the dominant component, expressed in mgm./kgm. of rock sample, varied from 0·14 for the Simpson to 1·3 for the Marcellus sample. In certain of the rock samples there was some indication of the presence of glycerol.

The glucose (180×10^6–300×10^6 years old), isolated from the Leonard and the Marcellus formations by the above treatment, was separated by paper chromatography (solvent: pyridine–ethyl acetate–water (2 : 5 : 7)) and characterized as the D-isomer by transforming it into N-*p*-nitrophenyl-β-D-glucosylamine, m.p. and mixed m.p. 184° C., $[\alpha]_D^{20}$ – 200° (approx.) in pyridine (*c*. 0·1)[3].

The mode of occurrence of the carbohydrate compounds in the rock formations is not yet known. It has been shown, however, that in the case of a sample from the Marcellus formation, extraction with water yielded about 10 per cent of the total carbohydrate material as the free sugars, arabinose, glucose and xylose; the remaining undissolved carbohydrate material may be present in the polymeric form.

The amino-acids present in the acid hydrolysates of several rock samples were also investigated by paper chromatography. In comparing two core samples of shale of late Palæozoic age, the highly fossiliferous one of Pennsylvanian age appeared to contain alanine, glycine, valine, glutamic acid, aspartic acid, proline, leucine, cystine and tyrosine, whereas the non-fossiliferous shale of Permian age yielded only one amino-acid tentatively identified as aspartic acid. The amino-acids in the Pennsylvanian sample may be derived from fossil shell fragments.

Table 1. Carbohydrate Components of Sedimentary Rocks

Formation	Age	Component sugars				
		Glucose	Galactose	Arabinose	Xylose	Rhamnose
Stonehenge	Ordovician	+		(? +)	(? +)	
Simpson	Ordovician	+	(? +)	+	+	
Upper Chambersburg	Ordovician	+		(? +)		
Marcellus	Devonian	+	(? +)	+	+	+
Woodford	Upper-Devonian	+		+	+	
Des Moines	Pennsylvanian	+		+	+	
Missouri	Pennsylvanian	+				(? +)
Leonard	Permian	+	(? +)		+	
Schuler	Jurassic	+		+	+	
Green River	Eocene	+	(? +)		(? +)	
Elko	Miocene	+				

The rock was treated with hot 0·5 *N* sulphuric acid and the sugars identified chromatographically.

Control experiments showed that neither the sugars nor the amino-acids arose from the materials used in the isolation techniques. These organic substances are believed to be original constituents of the rocks, although in the outcrop samples, despite careful collection and cleaning treatment, the possibility of surface contamination cannot be completely ruled out at present.

J. G. Palacas
F. M. Swain

Department of Geology,
University of Minnesota,
Minneapolis, 14.

F. Smith

Department of Agricultural Biochemistry,
University of Minnesota,
St. Paul, 1.

[1] Erdman, J. G., Marlett, E. M., and Hanson, W. E., *Science*, **124**, 1026 (1956).

[2] Abelson, P. H., "Organic Constituents of Fossils", Carnegie Inst. Washington Year Book, **53**, 97 (1954).

[3] Weygand, F., Perkow, W., and Kuhner, P., *Ber.*, **84**, 594 (1951).

17

Reprinted from *Science* **124**:1026 (1956)

Survival of Amino Acids in Marine Sediments

Proteinaceous material is carried into marine sediments upon the death and burial of aquatic organisms. The fate of this material has been a subject of conjecture, most investigators believing that it is rapidly decomposed (*1*). Studies of the thermal stability of an amino acid have recently shown that these compounds may survive intact for a long time at ordinary temperatures (*2*). In addition, protective mechanisms may play a part in preserving protein from early microbiological decomposition in terrestrial and marine muds. Recent work (*3*) has demonstrated the formation of complexes of protein with lignins and with clays, which are resistant to attack by microorganisms. Traces of proteins or peptides have been detected in fossil clams from the Pleistocene (*4*), and amino acids have been found in invertebrate and vertebrate fossils as old as the Ordovician (*2*).

Table 1. Comparison of a Recent and an Oligocene marine mud.

Content	Recent	Oligocene
Carbonate carbon (%)	0.71	1.49
Organic carbon (%)	0.53	0.27
Organic nitrogen (%)	0.044	0.032
Amino acids (μ mole/g)*	3.0	0.51
Principal amino acids in order of decreasing abundance†		
Identification certain	valine +++	alanine +++
	leucines +++	glutamic acid ++
	alanine ++	glycine +
	glutamic acid ++	proline +
	aspartic acid ++	leucines +
	glycine ++	aspartic acid < +
	proline ++	
	tyrosine ++	
	phenylalanine +	
Identification tentative	arginine ++	valine +

* For an amino acid mixture, 1 μmole/g is approximately 0.01 percent.
† Order of abundance: +++, high; ++, medium; +, low.

In the present work, the amino acid content was determined for a typical recent shallow-water, marine deposit and for a similar sediment laid down during the Oligocene, approximately 30 million years ago. The recent sample was collected by coring on the inner continental shelf of the Gulf of Mexico and represents a narrow sedimentary layer at a depth of approximately 120 cm; this is below the zone of major bacterial activity (*5*). The age of the sediment is probably not more than a few thousand years. The ancient, or Oligocene, sample comprised a section of a marine shale core that was cut from the lower Anahuac formation in Fort Bend County, Tex., at a depth of 5000 ft.

Immediately upon collection, both sediment samples were frozen in dry ice. In the case of the shale sample, the outer portion of the core was removed by turning on a lathe while the core was still frozen; the center was then crushed, suspended in distilled water, and milled to yield a thin slurry. On freeze-drying, both the recent and ancient samples were reduced to a powder with a flourlike texture.

The two samples were hydrolyzed, and the solutions were freed of salts electrolytically and by passage through Dowex 50 cation-exchange resin. Amino acids in the salt-free concentrates were resolved by one- and two-dimensional paper chromatography, using t-butanol-water-formic acid and phenol-water-ammonia as irrigation liquids. Positive identifications of individual amino acids were obtained by comparison of R_f values, using pure amino acids for reference, and by means of specific color reagents—for example, ninhydrin, isatin, 1,2-naphthoquinone-4-sulfonic acid and 1-nitroso-2-naphthol.

Data in Table 1 show that the modern and ancient marine samples selected for study are quite similar with respect to concentrations of carbonate carbon, organic carbon, and nitrogen. In contrast, the concentrations of free or combined amino acids are less in the Oligocene sample by a factor of 6.

Despite the difference in concentration, the arrays of amino acids in the two samples, modern and ancient, are strikingly similar, with the exception of the absence of tyrosine, phenylalanine, and arginine in the latter. Tyrosine, containing a chemically reactive phenolic group, would not be expected to be as stable as the other amino acids under earth conditions. Phenylalanine and arginine, likewise, are among the less stable of these acids.

Hitherto the detection of amino acids in ancient material has been limited to fossil shells or bones. Little or nothing is known about finely grained marine sediments. Although shell fragments were not eliminated from the samples, amino acids were also obtained on alkaline hydrolysis, where the shell structure was not disintegrated. Using Abelson's values for ancient shells (*4*), it would appear that more than 80 percent of the amino acids are associated with the noncalcareous portion of the Oligocene sample. Thus the amino acids of living organisms, in addition to those in large fragments of bone and shell, also may survive for long periods of time.

J. Gordon Erdman*
Everett M. Marlett*
William E. Hanson*

Mellon Institute,
Pittsburgh, Pennsylvania

References

1. A. I. Bogomolov, *Zhur. Priklad. Khim.* 27, 1012 (1954); C. E. ZoBell *Marine Microbiology* (Chronica Botanica, Waltham, Mass., 1946), pp. 142–143.
2. P. H. Abelson, *Science* 119, 576 (1954); also *Carnegie Inst. Wash. Year Book No. 53*, 97–101 (1954).
3. S. A. Waksman and K. R. N. Iyer, *Soil Sci.* 36, 69 (1933); L. A. Pinck, R. S. Dyal, F. E. Allison, *ibid.* 78, 109 (1954).
4. P. H. Abelson, *Carnegie Inst. Wash. Year Book No. 54*, 107–109 (1955).
5. C. E. ZoBell, *J. Sediment. Petrol.* 12, 127 (1942).

* Multiple fellowship of Gulf Research and Development Company.

28 September 1956

18

Reprinted from *Jour. Amer. Chem. Soc.* **88**:2311–2315 (1966)

Mass Spectrometric Characterization of Petroporphyrins[1]

Earl W. Baker

Contribution from Mellon Institute, Pittsburgh, Pennsylvania 15213.
R ceived January 13, 1966

Abstract: Mass spectra of the porphyrins obtained by acid extraction from a number of asphaltenes showed that the petroporphyrins are not homogeneous. Up to a dozen homologs of both the deoxophyllo and etio series are present in a Gaussian-like distribution with a band width of three to five methylene groups. High-resolution mass spectrometry on selected major peaks showed only the presence of alkyl porphyrins with no oxygen functional groups. Chromatographic separation of the petroporphyrins confirmed the presence of etio and phyllo series porphyrins and indicated the presence of a small amount of material of the rhodo series. The possible significance of these findings on the geochemistry of formation of the petroporphyrins is discussed.

The structure of petroporphyrins[2] has been a subject of investigation since Treibs' discovery of this class of compounds 30 years ago. In a series of papers Treibs described the separation of porphyrins from a variety of natural bituminous materials including oil, shale, and coal.[3] He identified the materials as etioporphyrin III (Etio) and deoxophylloerythroetioporphyrin (DPEP). On the basis of this identification, Treibs then put forward a degradative scheme by which chlorophyll a or b could be converted to DPEP and hemin to Etio. Treibs described his scheme in terms of six types of reactions: (1) demetalation, (2) saponification, (3) reduction, (4) aromatization, (5) decarboxylation, and (6) chelation. Application of these six steps to chlorophyll a (Figure 1a) would yield vanadyl DPEP (Figure 1b). The same types of reactions would convert chlorophyll b and bacteriochlorophyll to the same end product. The important point to note is that this scheme yields only one product, DPEP. If the starting material were hemin, Etio would be the final product.

Following Treibs' discoveries, many workers have devised methods of separation and identification of the porphyrins from bituminous materials. These studies in the main have broadened and tended to confirm Treibs' ideas as to the source and method of formation of the petroporphyrins, and his thesis has stood the test of time without any major revisions or extensions. There have, however, been scattered indications that the porphyrins in petroleum were not a single species but rather a series of compounds, some with molecular weights considerably higher than vanadyl Etio or vanadyl DPEP.[4–8] However, no definite proof has been offered, and, therefore, a mass spectrographic investigation was undertaken.

(1) Presented at the 151st National Meeting of the American Chemical Society, Pittsburgh, Pa., March 1966.

(2) The use of the term "petroporphyrins" to describe the plurality of porphyrins found in petroleum was originated by Professor A. H. Corwin: see, for example, Paper V-10, A. H. Corwin, 5th World Petroleum Congress, New York, N. Y., 1959; also, A. H. Corwin, W. S. Caughey, A. M. Leone, J. E. Danieley, and J. F. Bagli, 132nd National Meeting of the American Chemical Society, New York, N. Y., Sept 1957, Preprints, 2, No. 4, p A-35; A. H. Corwin, a report to the Petroleum Research Fund Advisory Board of the American Chemical Society, March 1957. Professor Corwin suggested that a special name for these materials was justified because of indications that at least some of the components were not identical with known porphyrins. The work reported here confirms that these materials do constitute a new class of naturally occurring compounds. The designation of this class of porphyrins as "petroporphyrins" seems appropriate and conforms to the accepted practice of allowing the person who first recognizes the unique nature of a material the right to name it.

(3) A. Treibs, *Angew. Chem.*, **49**, 682 (1936), and references therein.

(4) L. K. Beach and J. E. Shewmaker, *Ind. Eng. Chem.*, **49**, 1157 (1957).

Results

The porphyrins used in this study were obtained by extraction from asphaltenes with methanesulfonic acid (MSA).[9] Mass spectra were obtained on an AEI MS-9 double-focusing mass spectrometer. The conditions in the spectrometer source were selected so that no thermal rearrangements occurred, and essentially only molecular ions were recorded. Reference to Table I shows that the knowledge of the mass of an unknown porphyrin makes it possible to assign it to certain classes. For example, porphyrins of the etio and phyllo series must have molecular weights of $310 + 14n$, where n is an integer. Likewise, porphyrins with an isocyclic ring (deoxophylloerythro series) must have molecular weights of $308 + 14m$, where m is an integer 2 or greater.

Table I. Molecular Weights of Alkyl-Substituted Porphyrins

n[a]	Etio or phyllo series	Deoxophylloerythro series[b]
8	422	420
9	436	434
10	450	448
11	464	462
12	478	476[c]
13	492	490
14	506	504
15	520	518
16	534	532
17	548	546
18	562	560

[a] Number of methylene groups attached to the porphyrin nucleus. [b] The presence of an isocyclic ring leads to molecular formulas with two less protons than the etio type. [c] Deoxophylloerythroetioporphyrin.

In Figure 2 the mass spectrum of a sample of Boscan petroporphyrin is given. The major peaks in the

(5) M. F. Millson, D. S. Montgomery, and S. R. Brown, *Geochim. Cosmochim. Acta*, **30** (2) 207 (1966).

(6) G. Costantinides and G. Arich, Paper V-11, 6th World Petroleum Congress, Frankfurt, Germany, 1963.

(7) R. A. Dean and E. V. Whitehead, Paper V-9, 6th World Petroleum Congress, Frankfurt, Germany, 1963.

(8) A. H. Corwin and E. W. Baker, 147th National Meeting of the American Chemical Society, Philadelphia, Pa., March 1964, Preprints 9, No. 1, p 19, and references therein.

(9) J. G. Erdman, U. S. Patent 3,190,829 (June 22, 1965).

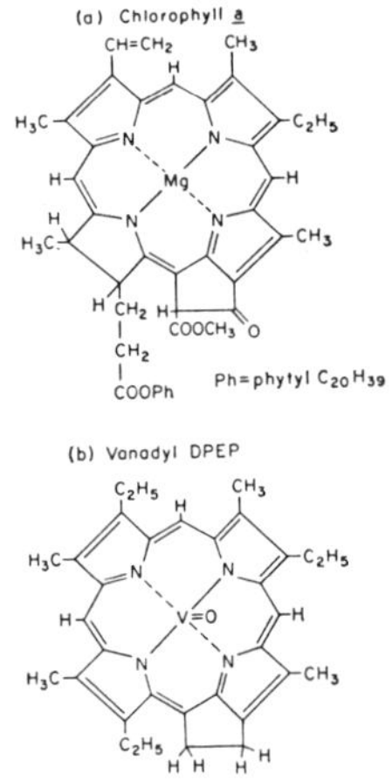

Figure 1. Structure of chlorophyll a and vanadyl DPEP.

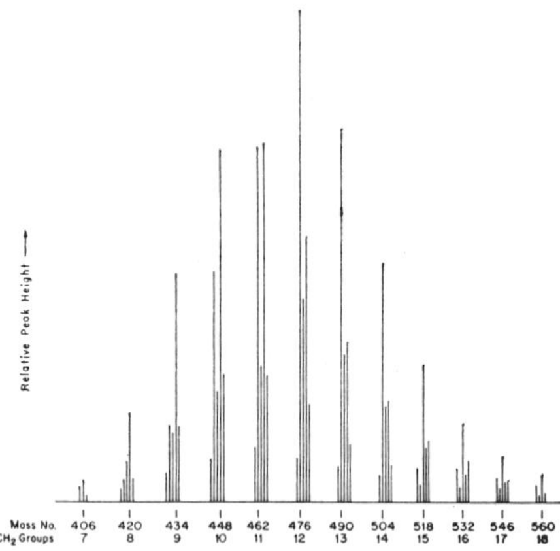

Figure 2. Mass spectrum of Boscan petroporphyrin.

spectrum are those of porphyrins of the deoxophylloerythro series. Approximately a dozen different porphyrins of this series are present encompassing those with as few as eight methylene groups to as great as 18 or even 19. DPEP itself (12 CH_2 groups) is the largest component comprising *ca.* 25% of the mixture. The quantities of the other porphyrins of this series fall off in a symmetrical way with increasing or decreasing molecular weights. Accurate mass measurements of a number of selected peaks (Table II) showed that these parent peaks represent alkyl porphyrins, and specifically that they contain no oxygen. The closest molecular formula of an oxygen containing porphyrin is some 30 millimass (mM) units from the observed mass, whereas the accuracy of the mass measurement appears to be about 2.5 mM units. Thus, we conclude that the higher molecular weight components of the mixture are not the result of incomplete loss of oxygen, but are actually

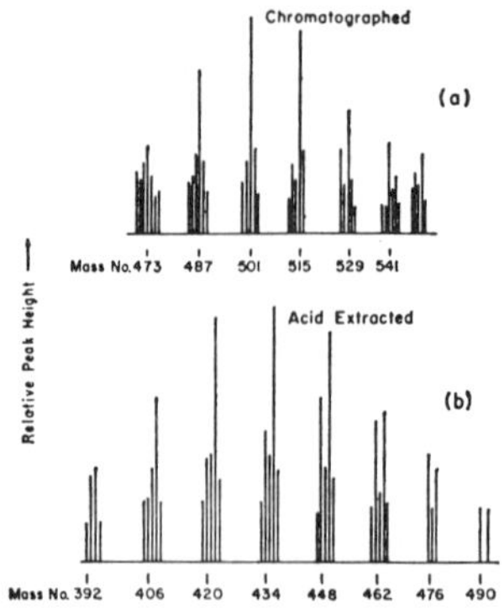

Figure 3. Low-voltage mass spectrum of porphyrins from Wilmington crude.

alkyl porphyrins, some of which must possess side chains longer than two carbons, since 18 or 19 methylene groups must be distributed over the eight β positions on the porphyrin.

Table II. Accurate Mass Determination of Selected Peaks in Boscan Petroporphyrins

Nominal peak mass			Δ, mM[b]
434	Obsd mass[a]	434.2446	
	Calcd for $C_{29}H_{30}N_4$	434.2473	2.7
	Calcd for $C_{28}H_{26}N_4O$	434.2108	33.8
476	Obsd mass	476.2928	
	Calcd for $C_{32}H_{36}N_4$	476.2906	2.2
	Calcd for $C_{31}H_{32}N_4O$	476.2576	35.2
504	Obsd mass	504.3233	
	Calcd for $C_{34}H_{40}N_4$	504.3253	2.0
	Calcd for $C_{33}H_{36}N_4O$	504.2889	34.4
506	Obsd mass	506.3392	
	Calcd for $C_{34}H_{42}N_4$	506.3409	1.7
	Calcd for $C_{33}H_{38}N_4O$	506.3045	34.7

[a] Resolution *ca.* 1 in 20,000, mass measured by peak matching against perfluoro-*t*-butylamine. [b] Difference between observed mass and calculated mass in millimass units.

The question of whether acid extraction of the porphyrins leads to the production of artifacts or loss of peripheral groups has been raised.[10] Therefore, experiments were undertaken to obtain samples of native porphyrin which could be compared spectrometrically with that obtained by acid extraction.

A sample of petroporphyrin from Wilmington crude was obtained by chromatographic methods. To avoid the possibility that degradation might occur during the separation, the chromatographic substrates were selected so that no highly active (acidic) sites would be present. Thus, only deactivated alumina and silica gel were used with relatively nonpolar solvents. Five chromatographic separations with intermediate partitioning between dimethylformamide and hexane produced a material of a purity of about 20%. It is possible to obtain a mass spectrum on such an impure

(10) M. Blumer, *Anal. Chem.*, 33, 1288 (1961).

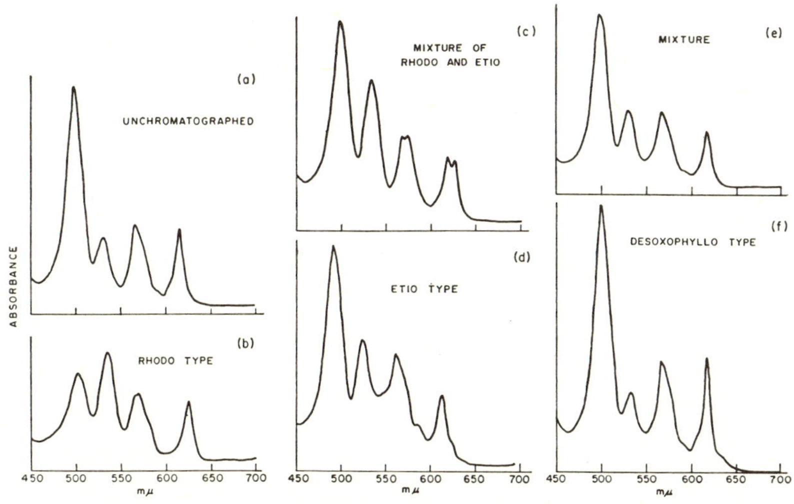

Figure 4. Visible spectra of petroporphyrins from Boscan asphaltene (see text for details).

material since the porphyrin peaks are readily recognizable above the background. A portion of the mass spectrum is shown in Figure 3a. During the separation every effort was made to prevent fractionation of the vanadyl porphyrins so that the sample would be representative of the natural porphyrin mixture.

A second sample of Wilmington petroporphyrin was prepared by the MSA extraction method. The significant portion of its mass spectrum is shown in Figure 3b. These spectra are compared in Table III. In the first column are shown the masses of the major peaks. In the next column, 65 mass units (the mass of VO minus two protons) are subtracted to give the mass of the porphyrin from which the chelate was derived. In the final column are the masses of the major peaks of the free base porphyrin recovered by MSA extraction. It will be noted that the correspondence between these peaks is one to one.

Table III. Comparison of Molecular Weights of Petroporphyrins Separated by Acid Extraction and by Chromatography

Method of separation	Chromatographic		Acid extraction
Form of porphyrin	Vanadyl chelate		Free base
	m/e	-65^a	*m/e*
	543	478	478
	541	476	476
	529	464	464
	515	450	450
	501	436	436
	487	422	422
	473	408	408

[a] Subtraction of 65 mass units from the mass of the vanadyl chelate gives the mass of the free base porphyrin from which this chelate was formed.

The conclusion may be drawn that at least in Wilmington crude and probably in all others no significant alteration of the porphyrins occurs during the acid extraction and subsequent work-up.

Chromatographic Separation of Boscan Petroporphyrins. A partial chromatographic separation of the Boscan petroporphyrins into spectral types has been accomplished. The visible spectra of the cuts from silica gel chromatography are shown in Figure 4. In Figure 4a, the spectrum of the unchromatographed petroporphyrin is shown. This spectrum is similar to that reported by a number of workers as being typical of the petroporphyrin.[11] The spectrum in Figure 4b is that of a rhodoporphyrin. This represents a relatively small amount of material and might be dismissed as an artifact of the separation scheme, but other evidence suggests that it is real. This fraction with a rhodo-type spectrum was present in all of the petroleum asphaltene samples examined. A fraction with a similar type spectrum taken from the Athabasca tar has been the subject of an extensive investigation by Millson, Montgomery, and Brown.[5] Cut 2 (Figure 4c) is clearly a mixture and is the result of incomplete resolution of the rhodo type (cut 1) and etio type (cut 3). Cut 4 (Figure 4e) is probably also an unresolved mixture. Cut 5 (Figure 3) is a typical deoxophylloerythroetio type. This is one of the largest fractions obtained and has been crystallized.

In Figure 5 is shown a mass spectrum of cut 3 which was referred to as an etio type by reference to its visible spectrum. The wide spread in molecular weight from 394 to 562 representing *ca.* 6 to 18 methylene groups will be noted. The mass spectrum confirms that the material which has major peaks at 464, 478, 492, etc., is mainly of the etio series, while the peaks at 2 mass units less, *i.e.*, 462, 476, 490, show the presence of a considerable amount of the deoxophylloerythro series remaining. The fact that the maximum is at mass 450

(11) H. N. Dunning, J. W. Moore, and A. T. Myers, *Ind. Eng. Chem.*, **46**, 2000 (1954).

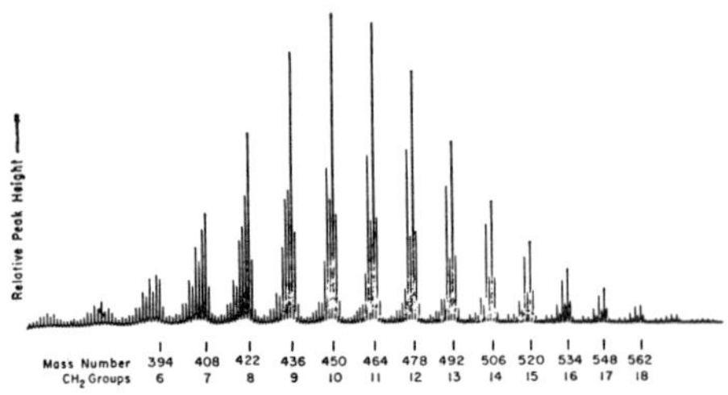

Figure 5. Low-voltage mass spectrum of Boscan petroporphrin enriched Etio fraction (see text for details).

rather than at 478 also is noteworthy. Such a drift toward lower average molecular weight is typical of the results as the samples were further purified by chromatography. Apparently some fractionation did occur with the higher molecular weight materials being concentrated in the tails of the peaks and, thus, were lost as the chromatography was terminated before the last trace was recovered from the columns.

This figure illustrates the degree of separation that has been achieved on single columns, the width of the envelope of molecular weights present, and some of the pitfalls in the purification.

It was felt that a comparison of the porphyrins from different sources would be interesting. The materials selected and their origins are given in Table IV. Samples of porphyrins suitable for spectrometric examination were obtained from a number of different asphaltenes by the MSA extraction and chromatographic purification method.

Table IV. Mass Distribution of Some Petroporphyrins

Source	*m/e* of most abundant porphyrin	Approximate half-width of envelope[a]: Mass units	CH_2 units
Boscan asphaltene[b]	476	70	5
Mara asphaltene[b]	450	70	5
Mara gas oil[c]	436	56	4
Gilsonite[d]	476	No envelope	
Wilmington asphaltene[e]	436	42	3
Boscan asphaltene (phyllo fraction)	462	42	3
Boscan Asphaltene (etio fraction)	450	70	5

[a] Taken as the number of mass units up and down from the most abundant peak to the grouping approximately $^1/_{10}$th the concentration of the most abundant peak. [b] Boscan and Mara crudes are of Venezuelan origin; the asphaltene fraction was obtained by precipitation with pentane. [c] Vacuum distillate of equivalent atmospheric boiling point 700–1050°F. [d] Asphaltene fraction; obtained from Cowboy Vein, Uinta Basin, Utah. [e] California origin.

A condensation of these data is presented in Table IV. Interestingly, the porphyrin obtained from gilsonite did not exhibit the Gaussian-like mass distribution characteristic of the petroleum porphyrins. In fact, one predominant peak at 476, the mass of deoxophylloerythroetioporphyrin, with small amounts of materials at lower masses was observed. It has been reported that only nickel chelates are present in gilsonite.[12]

All of the other samples are similar in that both etio-type and deoxophyllo-type materials are present. There are, however, some interesting differences in the average mass of the porphyrins from different sources. For example, the *average* porphyrin from Boscan contains 12 methylene groups while that from Wilmington contains only 9. Also, the range of the masses of porphyrins is somewhat different, the range in Boscan being the largest. This difference is indicated in Table IV as the approximate number of methylene groups seen on the mass spectrum above and below the most abundant peak.

Discussion

These studies show that the porphyrin fractions of crude petroleum are extremely complex mixtures. Not only do the relative amounts of rhodo, etio, and phyllo porphyrins vary with source, but the molecular weights vary both in average and distribution. In addition, each molecular peak is probably due to many isomers rather than a single compound.

As pointed out earlier, the degradation scheme proposed by Treibs leads only to one final product from chlorophyll, that is DPEP. This scheme does not accommodate the presence of a number of homologs which we and others have shown to be present. A number of other investigators have also noted the multiplicity of porphyrins present. Morandi and Jensen[13] state that the porphyrins in oil shale are compounds of at least three homologous series. Dean and Whitehead[7] have noted that vanadyl porphyrins from a crude petroleum showed parent mass peaks in the range of C_{28}–C_{33}. Thomas and Blumer[14] showed that porphyrin extracted from an oil shale of Swiss origin contained pigments with a range of C_{27}–C_{31}. These authors also noted that such a discovery compelled an explanation beyond that offered in Treibs' scheme. They suggested that the compounds of masses 490–492 and 504–506 (those containing one or two more carbon atoms than DPEP and Etio) might be accounted for by a reduction of the carboxyl groups to an alkyl group with the same number of carbons. However, the presence of porphyrins with masses of 518 to 560 cannot be accounted for by such a mechanism.

We suggest that the observed Gaussian-like distribution of molecular weights requires an additional process beyond those listed by Treibs. After consideration of the possibilities, we feel that the most likely explanation is a transalkylation reaction.

The nearest analogy, although admittedly somewhat tenuous, is the Jacobsen rearrangement.[15] The Jacobsen rearrangement, generally considered as the acid-catalyzed migration of methyl groups from one position to another in aromatic systems, has an additional very interesting feature; *i.e.*, some products are the result of a cross migration of alkyl groups. Thus, tetramethyl-

(12) J. M. Sugihara and L. R. McGee, *J. Org. Chem.*, **22**, 795 (1957).
(13) J. R. Morandi, and H. B. Jensen, *J. Chem. Eng. Data*, **11**, 81 (1966).
(14) D. W. Thomas and M. Blumer, *Geochim. Cosmochim. Acta*, **28**, 1147 (1964).
(15) P. deMayo "Molecular Rearrangements," Part I, Interscience Publishers, Inc., New York, N. Y., 1963, p 301.

benzene reacts to produce some trimethyl- and pentamethylbenzenesulfonates. These rearrangements proceed rapidly under conditions where the aromatic system is protonated. In the case of the aromatic systems strong acids are required; the porphyrins, on the other hand, are much stronger bases, requiring only organic acids for protonation. It would appear that sufficiently high temperatures and geologic time could produce the observed migrations. A further example of the relative ease of the loss of peripheral groups from the porphyrins is the production of deuteroporphyrin from protoporphyrin by the loss of two vinyl groups. This reaction is carried out in about 0.5 hr in fused resorcinol at 160–180°. The mechanism has not been established but is generally referred to as a transvinylation reaction, either ionic or free radical. While no direct evidence in favor of any mechanism can be offered, transalkylation *via* an ionic or free-radical mechanism could help to account for the multitude of reaction products observed.

Experimental Section

Asphaltenes were prepared in the standard way by precipitation from the crude petroleum with 10 parts of pentane at room temperature. The precipitated asphaltenes were rewashed with pentane and recovered by filtration or centrifugation. Some samples were further purified by exhaustive Soxhlet extraction. The petroporphyrins were prepared by extraction from the asphaltenes with methanesulfonic acid.[9] The porphyrins were recovered from the methanesulfonic acid by extraction with methylene chloride and purified by chromatography over silica gel with cyclohexane–benzene–ether. Visible absorption spectra were taken on a Beckman DK-1 recording spectrophotometer. Mass spectra were taken on an AEI MS-9 double-focusing mass spectrometer, fitted with a direct sample insertion lock. Source temperatures and ionizing potential were selected to give only parent molecular ions. Typical conditions were 230–250° and 7–9 ev.

Acknowledgments. This work was sponsored by Gulf Research Development Co., as part of the research program of the Multiple Fellowship on Petroleum. The author thanks Mr. R. E. Rhodes of Research Services, Mellon Institute, for obtaining the mass spectral data.

19

Reprinted from *Science* **185**:691–693 (1974)

PURINES AND PYRIMIDINES IN SEDIMENTS FROM LAKE ERIE

W. van der Velden and Alan W. Schwartz

Abstract. *Quantitative analyses of purines and pyrimidines in sequential sections of cores from the central and eastern basins of Lake Erie show steeply increasing concentrations in the youngest sediments. This may be related to increased loading of nutrients and recent cultural eutrophication of the lake. The purine and pyrimidine distributions suggest the operation of a specific degradative process for uracil at an extremely early stage in, or prior to, sediment formation.*

In spite of the ubiquity of the purines and pyrimidines in biological material, there have been very few publications on the quantitative distribution of these compounds in sediments. As the repository of genetic information, the purines and pyrimidines are, potentially, of enormous importance in organic geochemistry. Yet we are aware of only one report of purines and pyrimidines in core samples [from the experimental Mohole drilling off Guadalupe Island, Mexico (*1*)] and a number of isolated reports of their occurrence in soils (*2*). Part of the explanation for this scarcity of data may be the lack of an appropriate extraction method, since the usual procedures for extraction of nucleic acid bases from biological material are not applicable to sediments. In addition, reliable and highly sensitive methods for quantitative analyses have not generally been available until fairly recently. To deal with these problems we have made use of an extraction method involving destruction of the mineral matrix, combined with minimum handling of the sample and a highly sensitive and selective method of analysis.

All operations on sediment samples were carried out in Teflon receptacles which could be attached to reflux condensers for solvent extraction or centrifuged. A sediment sample of 5.0 g was weighed out into a receptacle and stirred overnight with 1*M* HCl (35 ml), at room temperature, with a Teflon-coated, magnetic stirring bar. After centrifugation this extract was discarded and the sediment was washed with a second portion of 1*M* HCl (25 ml), which was also discarded. The purpose of this preliminary extraction was to decompose any carbonates which might be present and to eliminate any loosely bound contaminants. The proportion of purines and pyrimidines extracted in this manner was less than 1 percent of that found in the final extraction. Hydrogen fluoride (60 ml, 40 percent) was then added to the residue and the receptacle was heated on a steam bath (under N_2) overnight. The dry residue was extracted with 1*M* HCl (35 ml) under reflux (1 hour), with stirring. After cooling and centrifugation, the supernatant was removed and percolated through a column of activated charcoal (0.6 by 5 cm). The residue was washed with 1*M* HCl (25 ml) and this washing was added to the column. The column was washed with 25-ml portions of 1*M* HCl and H_2O, and purines and pyrimidines were eluted with 40 ml of HCOOH (100 percent). The eluate was evaporated to dryness under vacuum, and the residue was taken up in H_2O (25 ml) and transferred to a column of Dowex 1 X 8 (Cl^-) (0.6 by 2 cm). The column was washed with additional portions of H_2O and the combined eluate and washings (40 ml) were evaporated under vacuum.

Ion-exclusion chromatography was performed on both anion and cation exclusion columns (*3*). The columns used were Aminex A-6 (0.6 by 50 cm) in 0.02*M* NH_4CO_3 plus 0.1*M* NH_3 (*p*H 10.0) and Aminex A-25 (0.6 by 25 cm) in 0.1*M* sodium acetate (*p*H 4.0), both columns operating at 60°C and a flow rate of 24 ml/hour. Detection was carried out by measurement of the optical density of the column effluents at 254 and 280 nm with a Chromatronix model 230 dual channel absorbance detector. Determination of retention times and integration of peaks were done with a Hewlett-Packard 3370B integrator. A more complete description of the methods and some preliminary results have already been reported (*4*). Blank samples carried through the entire procedure demonstrated the absence of significant contamination. The detection limit for an individual purine or pyrimidine was approximately 1 part per billion (10^{-9} g), and the level of contamination was less than 10 parts per billion. An estimate of the recovery efficiency of each base was made by adding standard samples labeled with ^{14}C to sediment samples before the HF digestion.

The core samples studied were collected in late 1972 in Lake Erie by the Canada Centre for Inland Waters. Core LE19 (eastern basin) was collected at 42°33′00″N, 79°53′36″W in a water depth of 63 m. Core LE43 (central basin) was collected at 41°56′06″N 81°28′42″W in a water depth of 24 m. Both cores were divided into sections of 5 or 10 cm each and freeze-dried. Table 1 shows the concentrations of the purines and pyrimidines that were identified in the surface sections of the cores. The totals found in the surface sediments at these two locations represent 1 to 2 percent of the total nitrogen reported in samples from the same areas (*5*). The very low concentration of uracil is particularly noteworthy. In the sediments of the experimental Mohole no uracil was detected, even in surface samples (*1*). The sedimentation time in oceans is, of course, very much greater than that in lakes, and the possibility exists that organic matter is already devoid of uracil before reaching the bottom. It has been shown, for example, that chlorophyl[ls] are almost completely decomposed in Lake Erie and Lake Ontario befo[re]

Table 1. Concentrations of purines and pyrimidines, in parts per million (ppm), in surface sediments of Lake Erie. Identification of the indicated purines and pyrimidines was based on retention times on standardized anion and cation exclusion columns and on ratios of the absorbance at 254 and 280 nm (4). The orders of elution as well as the optical ratios were different on the two columns, and the quantitative results reported are based on analyses in both systems. Average recoveries (percentages) of ^{14}C-labeled purines and pyrimidines added to sediment samples were as follows: cytosine (C), 92; uracil (U), 86; thymine (T), 92; adenine (A), 75; guanine (G), 66; and hypoxanthine (HX), 53. The reported concentrations are not corrected for recovery.

Core	Section (cm)	Concentration (ppm)						
		C	U	T	A	G	HX	Total
LE19	0–10	24	5	16	29	40	9	123
LE43	0– 5	27	5	16	25	29	6	108

reaching the lake bed (6). On the other hand, degradation of uracil at the sediment-water interface, where microbial activity is greatest, cannot be ruled out (7). Preliminary results on cores sectioned at 1-cm intervals show that uracil is already markedly underabundant in the first centimeter of sediment. Rapid degradation, biological or nonbiological, must be occurring in the water column or at the sediment-water interface.

Figure 1A shows the variation of the total nucleic acid base content with depth in the two cores. The concentration of each base follows, in general, the same profile as the total concentration. The sedimentation rates in the eastern and central basins of Lake Erie, and consequently the ages corresponding to given depths, are quite different. This is indicated by the locations of the *Castanea* and *Ambrosia* horizons, which have been determined in cores taken close to those in the present study (5). These horizons represent the depths in the cores in which pollen counts of chestnut (*Castanea*) and ragweed (*Ambrosia*) are observed to decrease and increase, respectively. The decrease of *Castanea* pollen has been correlated with an epidemic fungus disease, which is dated at approximately 1935 for the vicinity of Lake Erie. The increase of *Ambrosia* pollen represents the effect of forest clearance by early colonists and is dated at 1850 (5). An approximate relation between depth in the core and age is based on average sedimentation rates for the periods delineated by these horizons, that is, 1850 to 1935 and 1935 to the present. The data in Fig. 1A have been replotted against estimated dates of deposition in this manner in Fig. 1B. The use of average sedimentation rates is, of course, only an approximation made for this analysis, and no corrections have been applied for increasing compaction of the core with depth. In the case of core LE43, dates prior to 1850 (precolonial) have been assigned by assuming the same average sedimentation rate as in the period 1850 to 1935. Consequently, the actual ages of the lower sections of this core are probably considerably greater than indicated.

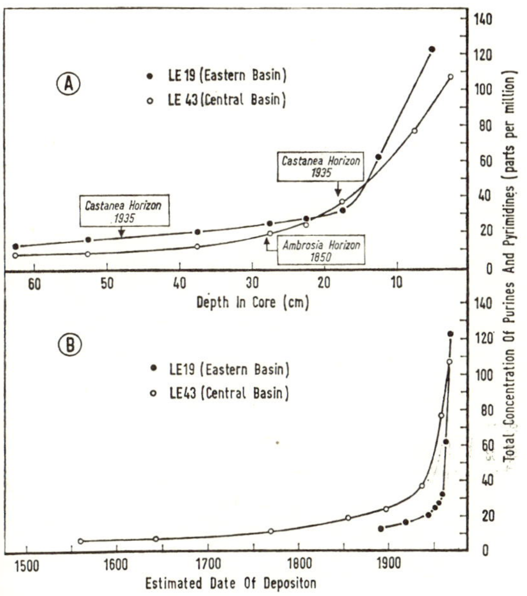

Fig. 1. Total concentration of identified purines and pyrimidines in sediments of the central and eastern basins of Lake Erie (not corrected for losses). (A) Concentration versus depth in the cores; (B) concentration versus estimated date of deposition.

When the total concentrations of purines and pyrimidines are replotted against estimated date of deposition instead of depth in the core, a more revealing picture emerges (Fig. 1B). An extremely rapid increase in purine and pyrimidine contents of sediments of both the central and eastern basins of Lake Erie is seen to have taken place since 1900. The results from the central basin, in particular, are in close agreement with the observations of Kemp *et al.* (8). These authors interpreted enrichment of organic C, N, and P as being due mainly to increased loading of nutrients since about 1900. More recent studies suggest that the most important increases in input to the sediments of both basins have occurred since 1950 (5). Our data are also in agreement on this point. Although the sedimentation rate is higher at the location of core LE19, the total concentration of extracted purines and pyrimidines is, in general, somewhat lower, except for the upper 15 cm of sediment. It has been suggested that the eastern basin has "aged" less rapidly than other regions of Lake Erie because of its depth, but is (as of 1968) on the verge of becoming eutrophic (9). Eutrophication is typified by (among other factors) greatly increased productivity, depletion of dissolved oxygen, an increase in dissolved solids, and an increasing rate of sedimentation of organic material. The data of Fig. 1B, showing a dramatic increase in purine and pyrimidine concentrations beginning around 1950 in the eastern basin, certainly seem to be in agreement with this hypothesis, although rapid alteration of these compounds within the sediment cannot yet be eliminated as an alternative explanation.

References and Notes

1. E. Rosenberg, *Science* **146**, 1680 (1964).
2. See, for example, G. Anderson, in *Soil Biochemistry*, A. D. McLaren and G. H. Peterson, Eds. (Dekker, New York, 1967), p. 67; J. M. Bremner, in *ibid.*, p. 19; W. Flaig, *Soil. Sci.* **111**, 19 (1971).
3. R. P. Singhal, *Arch. Biochem. Biophys.* **152**, 800 (1972); ——— and W. Cohn, *Biochemistry* **12**, 1532 (1973).
4. W. van der Velden, G. J. F. Chittenden, A. W. Schwartz, paper presented at the Sixth International Congress of Organic Geochemistry, Paris, 18 to 21 September 1973.
5. A. L. W. Kemp, T. W. Anderson, R. L. Thomas, A. Mudrochova, *J. Sediment. Petrol.* **44**, 207 (1974).
6. A. L. W. Kemp and C. F. M. Lewis, *Proc. 11th Conf. Great Lakes Res.* (1968), p. 206.
7. A. S. Menon, C. V. Marion, A. N. Miller, *Proc. 14th Conf. Great Lakes Res.* (1971), p. 768.
8. A. L. W. Kemp, C. B. J. Gray, A. Mudrochova, in *Nutrients in Natural Waters*, H. E. Allen and J. R. Kramer, Eds. (Wiley, New York, 1972), p. 251.
9. G. L. Harlow, *Proc. 23rd. Ind. Waste Conf. Purdue Univ.* (1968), p. 856.
10. We thank W. M. J. Strachan and the Canada Centre for Inland Waters, Burlington, Ontario, for the core samples and for supplying valuable background data. We also thank A. L. W. Kemp for advice and critical discussion of the results.

3 April 1974

20

Reprinted from *Limnology and Oceanography* **17**:570–582 (1970)

CHEMICAL AND ISOTOPIC EVIDENCE FOR THE *IN SITU* ORIGIN OF MARINE HUMIC SUBSTANCES[1,2]

Arie Nissenbaum[3]
Isotope Department, Weizmann Institute, Rehovot, Israel

and

I. R. Kaplan[4]
Department of Geology and Institute of Geophysics and Planetary Physics, University of California, Los Angeles 90024

ABSTRACT

Humic and fulvic acids were extracted from marine and nonmarine Recent sediments and from soils. These acids are shown to be major components of the organic matter from marine and nonmarine sediments—some marine sediments may contain 70% of their organic carbon in the humic and fulvic acid fraction. Marine and terrestrial humic acids have similar carbon and hydrogen content, but the former generally contain more sulfur and nitrogen. $\delta^{34}S$ values of marine humic acid indicate that the sulfur is introduced into the organic matter as hydrogen sulfide produced by sulfate reduction. Marine humates have a rather constant $\delta^{13}C$ value of −20 to −22‰, whereas the $\delta^{13}C$ of soil humic acid is related to its plant source material and usually ranges around −25 to −26‰. The evidence shows that marine humic acids can be formed *in situ* from degradation products of plankton and are not necessarily transported from the continent. The suggested pathway of marine humic acid formation and transformation in the sediment is (1) degraded cellular material→(2) water-soluble complex containing amino acids and carbohydrates→(3) fulvic acids→(4) humic acids→(5) kerogen.

INTRODUCTION

The terms humic acid or humic substances were originally applied to the dark brown material extracted from soils by alkaline solutions. The most commonly used classification (and the one used here) divides the alkaline extract into two fractions, an acid-soluble, base-soluble fraction called fulvic acid and an acid-insoluble, base-soluble fraction called humic acid.

The chemistry of soil humic and fulvic acids has been under investigation for several decades (*see* Kononova 1966; Manskaya and Drozdova 1968; Hurst and Burgess 1967). It is generally assumed that soil humic acid is formed from lignin produced by higher plants (Kononova 1966), by a transformation at least partly catalyzed by microorganisms (Flaig 1964); white-rot fungi are particularly active (Oglesby et al. 1967). As an alternative source for humic acid formation, the Maillard or "browning" reaction has been suggested (Enders and Theis 1938). This hypothesis is based on the formation of a base-soluble and acid-insoluble brown polymeric solid (melanoidin) when amino acids and carbohydrates are cohydrolyzed in acid solution.

Several workers have reported the existence of humic acids in oceanic sediments: Degens et al. (1964) claimed humic acid represents 30–60% of the organic matter in sediments from the basins off southern California; Bordovskiy (1965) found humic acid composes 40% of the organic matter in Bering Sea sediments; Kasatochkin et al. (1968) reported humic acids from sediments in the Indian Ocean.

Degens et al. (1964) suggested that the

---

[1] Publication No. 973 of the Institute of Geophysics and Planetary Physics, University of California, Los Angeles.

[2] This study was carried out under U.S. Atomic Energy Commission Contract AT(04-3)-34 P.A. 134.

[3] Financially supported by the Oceanographic and Limnological Research Company, Israel.

[4] Partial support was provided by a Guggenheim Foundation fellowship.

humic acid in the basins off southern California is of continental origin. Skopintsev (*cited in* Bordovskiy 1965) suggested a dual origin for the marine humic acids, an autochthonous part formed from marine plankton and an allochthonous part brought in from the continent. Bordovskiy (1965) suggested that the existence of humic acid in the central Atlantic, in areas far removed from the continent, indicates that humic substances are mostly, if not wholly, autochthonous.

In this study we have tried to trace the origin of humic acids by using chemical and isotopic criteria. On the average, marine plankton is depleted in ^{13}C compared with terrestrial plants from temperate zones (Craig 1953), so it may be assumed that humic acids from marine sources would also be depleted in ^{13}C. Furthermore, if marine humic acid is related to plankton, which do not contain lignin, marine humates may be chemically different from soil humic acid.

We thank C. Petrowski and E. Ruth for their help in various stages of this study; H. King performed most of the chemical analyses; A. Reuveni helped with the ESR work. The following persons provided samples: M. A. Rashid, A. G. Carey, P. Fan, A. Otsuki, V. E. Swanson, R. J. Gibbs, and P. Zubovic.

EXPERIMENTAL

The sediment and soil samples were shaken with 0.1 N NaOH on a mechanical rotary shaker for 4–5 hr at 37C under a nitrogen atmosphere. The samples were then rapidly centrifuged and the solids again treated as described; this was repeated until the supernatant was virtually colorless. Samples rich in $CaCO_3$ were decalcified first with dilute HCl. The combined supernatants were acidified with 3 N HCl and the fulvic acid separated by centrifugation. The humic acid was purified by recycling between acid and basic solutions and was finally precipitated from an acid solution and separated by high-speed centrifugation. The humic acid pellet was dialyzed against double-distilled water, and the purified acid was then dried by lyophilization. The acid solutions, containing fulvic acids, were demineralized by dialysis against double-distilled water and lyophilized to recover the solid acid. This procedure caused the loss of part of the fulvic acid, which passed through the membrane.

Elemental analyses of the humic acids were made in the Microanalytical Laboratories, Department of Chemistry, UCLA. Samples for carbon isotope analysis were prepared by combustion at 1,050C in an oxygen atmosphere (Kaplan et al. 1970).

In preparation for sulfur isotope measurements, the humic acids were oxidized by bromine and aqua regia, the excess bromine and nitric acid were removed, and the sulfur precipitated as $BaSO_4$. The $BaSO_4$ was reduced by graphite and converted to Ag_2S, which was combusted with CuO at 800C to produce SO_2, which was used for $^{34}S:^{32}S$ mass spectrometric measurement.

Two separate Nuclide Corp. 6-inch (15 cm) dual-collecting mass spectrometers were used to analyze CO_2 and SO_2. The $\delta^{13}C$ is reported vs. the Chicago PDB standard and the $\delta^{34}S$ vs. Canyon Diablo meteorite standard.

Electron spin resonance (ESR) spectra were measured on a Varian E-12 spectrometer. The E value (the ratio of optical density at 645 nm over that of 665 nm) was measured on 1% humic and fulvic acid solution in 0.05 N $NaHCO_3$ on a Zeiss PNQ-2 spectrophotometer.

Sampling

We attempted to study fresh sediment samples from a number of characteristic environments. Most marine samples were taken from the upper 10 cm of the sediment column to minimize the effect of diagenesis on the composition of humic acids. Of course, there are still differences of several orders of magnitude between sediment accumulation rates in deep-sea samples and those from the littoral or continental environment.

TABLE 1. *Chemical and isotopic composition of humic acids from marine sediments* (all results on ash-free basis; oxygen by difference)*

Sampling locations	%C	%H	%N	%S	%O	C:H	C:N	$\delta^{13}C$ ‰
Off southern California								
San Pedro Basin	53.72	6.42	3.88	1.24	34.74	8.4	13.9	−22.7
Santa Cruz Basin	51.84	6.56	5.17	1.01	35.42	7.9	10.0	−21.8
Santa Barbara Basin	54.11	6.47	4.54	2.12	32.76	8.4	13.1	−22.5
Tanner Basin	52.28	6.14	4.72	1.20	35.66	8.5	11.1	−22.0
Long Basin	51.69	6.11	4.64	0.87	36.69	8.5	11.1	−22.3
Santa Monica Basin	58.88	5.93	6.24	1.79	27.16	9.9	9.4	−27.4
Gulf of California								
Mazatlan Basin	48.89	5.12	—	—	—	9.6	—	−20.2
Pacific Ocean, N of Hawaii								
4,500-m depth	—	—	—	—	—	—	—	−23.1
Pacific Ocean, off Oregon								
44 km offshore	—	—	—	—	—	—	—	−22.0
72.5 km offshore	—	—	—	—	—	—	—	−23.0
100 km offshore	—	—	—	—	—	—	—	−22.1
120 km offshore	55.56	4.60	—	—	—	12.1	—	−22.1
160 km offshore	56.29	5.43	—	—	—	10.4	—	−22.5
Atlantic Ocean, N of Paramaribo								
100 km offshore	—	—	—	—	—	—	—	−23.6
Saanich Inlet								
Surface of core 4	56.7	5.9	2.2	5.9	29.3	9.6	25.8	−22.0†
Core 3B, 34-m depth	57.8	5.4	1.5	3.0	32.3	10.8	38.6	−23.4†
Atlantic Ocean, Cariaco Trench								
Oxidizing	55.7	5.8	4.8	—	—	9.6	11.6	−17.2‡
Reducing	59.8	5.5	4.2	—	—	10.9	14.2	−17.2‡
Atlantic Ocean, Emerald Basin, Nova Scotia shelf								
Sample 1 (68-2-2-1)	52.8	6.6	5.6	—	—	8.0	9.4	−22.2§
Sample 2 (68-2-1-1)	—	—	—	—	—	—	—	−19.7§

* Source of analytical data is this work unless otherwise noted.
† Brown et al. 1972.
‡ Rashid and King 1970.
§ Rashid and King 1969.

Marine samples

In Table 1 we have grouped samples from locations where the water depth was more than 200 m. In every case, the water column above the sediment was of the oceanic type. The only exception is that of Saanich Inlet (British Columbia) where the salinity is a few permille lower than that of seawater (Nissenbaum et al. 1972).

We analyzed samples from the following areas: 1) Basins off southern California (Fig. 1), including both nearshore and distant basins; these environments ranged from highly reducing to oxidizing areas. 2) Pacific Ocean off Newport, Oregon—this set of samples from an east–west traverse was taken by A. Carey (Fig. 2). 3) Nova Scotian shelf—these samples were supplied by M. A. Rashid; they were taken about 60–70 km offshore (at water depth of 200–300 m). The environment is oxidizing (Rashid and King 1969, 1970). 4) Gulf of California—a single sample was analyzed from a reducing basin in the southernmost part of the gulf. 5) Pacific Ocean—a sample from the Pacific Ocean north of Hawaii was supplied by P. Fan. 6) Saanich Inlet, Vancouver Island—this fjord is separated from the Georgia Strait by a narrow sill, which causes an intermittent formation of anoxic water. The sediments are highly reducing and rich in organic carbon (up to 5%: Brown et al. 1972). 7) Atlantic Ocean, Cariaco Trench

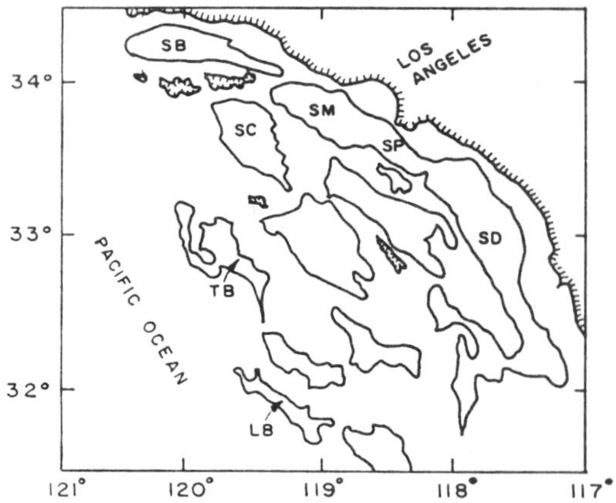

Fig. 1. Sampling location of humic acids collected in the California borderland. SB—Santa Barbara Basin; SM—Santa Monica Basin; SD—San Diego Trough; SP—San Pedro Channel; SC—Santa Cruz Basin; LB—Long Basin; TB—Tanner Basin.

—two samples were obtained from M. A. Rashid. One is from the sill of the basin area at 800-m depth and the other is from 1,390-m depth in the deepest part of the basin. Both sites are presently reducing and are about 35 km offshore (Rashid and King 1970).

Coastal and littoral samples

The samples in Table 2 represent humic acids from an aquatic environment adjacent to the continent. All samples are from depths not exceeding a few meters. 1) Kaneohe Bay—two samples (collected by P. Fan) from 3-m depth, about 0.5 km apart. 2) Newport Marsh. 3) Humate-cemented sandstone from the northeastern coast of the Gulf of Mexico (Swanson and Palacas 1965). 4) A tidal marsh area in the Florida panhandle (Palacas et al. 1968). 5) Choctawhatchee Bay estuary—two samples were collected from the east and west end of the bay. The humic acids were extracted from the top 15 cm of sediments (V. E. Swanson, personal communication). 6) Amazon River estuary—sample collected by R. J. Gibbs. 7) Surface sediment from New Dragon's Mouth, Gulf of Paria, Venezuela, collected by R. J. Gibbs. 8) A reducing lagoon (5–30-cm depth) in Musquodoboit Harbor (Rashid and King 1970).

Continental samples

The continental samples listed in Table 3 include humic acids extracted from soil and lake sediments. 1) Humic acid from Rendzina and Terra Rosa soil from central Israel (Nissenbaum and Kaplan 1966). 2) Humic acid from recent sediments of the Dead Sea, Israel; the samples were collected from the deep north basin of the lake (Nissenbaum 1969). 3) Soil humic acid from the Ein Feshcha Oasis on the

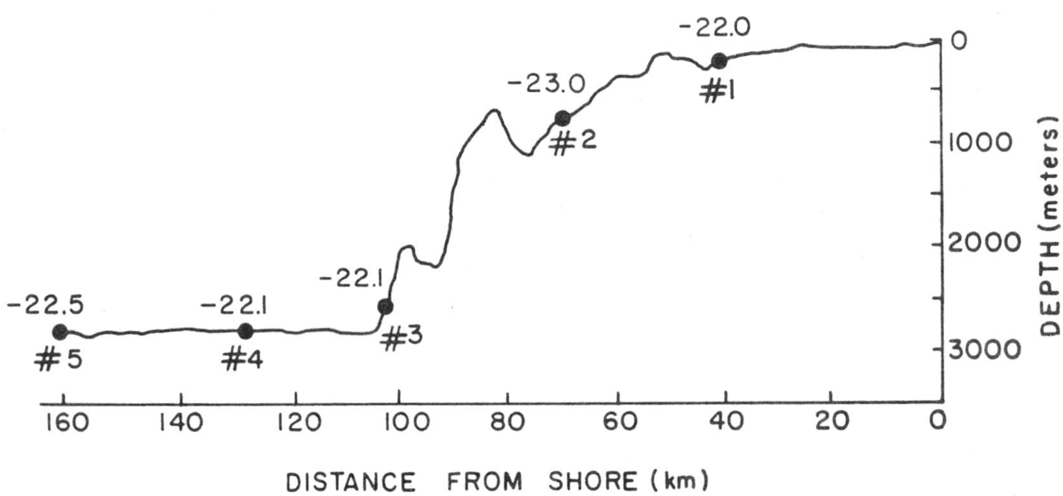

Fig. 2. Location and $\delta^{13}C$ of humic acids collected in traverse off the Oregon coast.

TABLE 2. *Chemical and isotopic composition of humic acids from coastal and littoral environments* (results on ash-free basis; oxygen by difference)*

Sample location	%C	%H	%N	%S	%O	C:H	C:N	$\delta^{13}C$ ‰
Kaneohe Bay, Hawaii, No. 1	—	—	—	—	—	—	—	–24.2
No. 2	—	—	—	—	—	—	—	–24.9
Humate-cemented sands, northwest Florida, No. 1	50.15	5.57	1.04	0.52	42.72	9.0	48.2	–25.7†
No. 2	52.49	4.27	0.83	0.88	41.53	12.3	63.5	–25.7†
Newport Marsh, so. Calif.	51.11	6.61	7.06	2.27	32.95	7.8	7.3	–19.1
Choctawhatchee Bay, Gulf of Mexico, sample 1	47.94	—	0.46	0.28	—	—	100.4	–25.3
Sample 2	—	—	—	—	—	—	—	–25.5
Tidal marsh, NE Gulf of Mexico, sample 1	52.40	5.75	—	—	—	9.1	—	–19.3‡
Sample 2	47.94	4.59	—	—	—	10.4	—	–21.2‡
Gulf of Paria, New Dragon's Mouth	52.64	5.84	5.02	70.2	—	9.0	10.4	–25.5
Estuary of the Amazon River	54.37	6.04	—	—	—	9.4	—	–27.3
Musquodoboit Harbor lagoon, Nova Scotia	53.8	6.5	3.8	—	—	8.3	14.2	–23.7§

* Source of analytical data is this work unless otherwise noted.
† Swanson and Palacas 1965.
‡ Palacas et al. 1968.
§ Rashid and King 1970.

northwestern side of the Dead Sea. 4) Forest soil from Vancouver Island (Brown et al. 1972). 5) Humic acid from Lake Haruna, Japan, and from soil around the lake (Otsuki and Hanya 1967). 6) Humic acid from Hawaiian canefield soil. 7) Humic acids from Hula peat, Israel (Nissenbaum and Kaplan 1966) and Minnesota peat (collected by P. Zubovic, U.S.G.S.). 8) Humic acid from podzolic forest soil, Musquodoboit Harbor, Nova Scotia (Rashid and King 1970). 9) Sediment from the Amazon River, Brazil (collected by R. J. Gibbs). 10) Melanoidin, a brown polymerized solid, base-soluble and acid-insoluble, obtained by an acid hydrolysis of equimolar solution of glycine and glucose. 11) Tertiary lignite from North Dakota.

RESULTS

Elemental analyses

Results of elemental and $\delta^{13}C$ analyses are given in Tables 1, 2, and 3. Carbon and nitrogen showed no systematic distribution in the humic acids extracted from various environments, although samples from the same locality generally yielded similar results. The majority of samples fell in the range C $\simeq$ 52–56%, H $\simeq$ 5.0–6.1%, and C:H $\simeq$ 8.4–10. It is therefore apparent that these two elements are not useful for establishing criteria to differentiate marine from nonmarine humic acids, contrary to the previous report by Bordovskiy (1965). The nitrogen content, however, showed a much wider range of concentrations, and the overall spread in C:N values is 7.3 to 100. The greatest variation is in soil and in coastal environments, where biological activity by microorganisms and burrowing animals may cause mineralization and loss of ammonia. With the exception of the two Saanich Inlet samples, the C:N ratios of all other marine humic acids analyzed here are in the range 9.4–14.2.

Surprisingly high amounts of sulfur (up to 6%) were found in marine samples, even from areas that are oxidizing in their surface sediments such as the basins off southern California (Santa Barbara and Tanner Basins, however, are reducing even at the top: Emery 1960). Some of the sulfur found may be an artifact introduced

TABLE 3. *Chemical and isotopic composition of humic acid from terrestrial environments* (results on ash-free basis; oxygen by difference)*

Sample locations	%C	%H	%N	%S	%O	C:H	C:N	$\delta^{13}C$ ‰
Terra Rosa soil, Israel	46.7	5.2	—	<0.2	—	9.0	—	−25.6†
Rendzina soil, Israel	48.1	4.8	—	<0.2	—	10.0	—	−26.1†
Minnesota peat	56.1	5.7	1.7	0.5	36.0	9.0	34	−24.9
Hula peat, Israel	51.5	5.0	2.7	—	—	10.3	19	−19.2†
Canefield soil, Hawaii	—	—	—	—	—	—	—	−14.8
Forest soil, Saanich Inlet	55.13	6.25	3.56	0.47	34.6	8.8	15.5	−29.1‡
Haruna Lake sediment, Japan	53.64	6.14	5.70	—	—	8.7	9.4	−21.0§
Amazon River sediment	52.64	6.04	5.00	<0.2	36.3	8.7	10.5	−26.5
Soil around Lake Haruna	56.64	6.04	5.0	<0.2	32.3	8.7	10.5	−28.2§
Dead Sea, Israel, sediment (beneath 165-m water depth)	56.32	5.25	1.6	2.0	34.5	10.7	35.2	−25.1‖
Dead Sea, sediment (beneath 330-m water depth)	58.94	6.20	1.9	3.4	29.6	9.5	31.0	−25.0‖
Soil from around the Dead Sea	52.57	5.64	2.93	1.89	37.0	9.0	18.0	−25.2
North Dakota lignite	57.73	4.34	0.98	2.44	34.7	13.5	59	−23.8
Forest soil, Nova Scotia	—	—	—	—	—	—	—	−26.2
Melanoidin	57.50	7.39	4.60	—	—	7.8	12.7	—

* Source of analytical data is this work unless otherwise noted.
† Nissenbaum and Kaplan 1966.
‡ Brown et al. 1972.
§ Otsuki and Hanya 1967.
‖ Nissenbaum 1969.

into the humic acid by the NaOH treatment. However, the samples from Saanich Inlet and the Dead Sea had all elemental sulfur extracted by solvents and the unstable sulfides removed by acid treatment before the humic acid extraction, and the humic acid was still rich in sulfur.

The samples rich in sulfur are from areas where the biological sulfur cycle is very active and therefore the sulfur may not be related to the organic matter from the water but may have been introduced from the sediment.

The amount of sulfur that we found is lower than that reported by Swanson (unpublished) for three samples from the basins off southern California. This is particularly true of the Santa Barbara samples: theirs had 6.43% sulfur and ours 2.12%. The reason for this is not clear, but may be related to differences in sample preparation. The mode of occurrence of this sulfur is completely unknown. Some may exist as sulfur-containing amino acids which can be incorporated into the humic acid "molecule" (Scharpenseel 1962), but the amounts found are too high to be accounted for by this mechanism alone.

Visible spectroscopy

The spectra of the humic acids reported here had essentially the same shape, regardless of origin, showing a featureless increase in absorption with decreasing wavelength. The ratio of the extinction at 465 nm to that at 665 nm (E value, or E_4:E_6 ratio as it is sometimes termed) has been used as an indicator of the degree of condensation of humic substances (Schnitzer and Skinner 1969).

Generally, the E value is lower with increased condensation. Thus, higher aromaticity exhibits smaller E values. Table 4 gives the results for several humic acid samples. The marine samples (except for Long Basin) exhibit fairly constant values between 5.20 and 6.01. Coastal and littoral samples had slightly higher E values whereas continental samples had a wide range of values. The highest ratio was that of melanoidin.

The uniformity of marine samples indicates their origin from a common source by similar chemical reactions. No statistically significant correlation was found between the C:H ratios and E values, so the

TABLE 4. *Values of* $E_{465}:E_{665}$ *of humic and fulvic acids (FA—fulvic acid; the rest are humic acids)*

Sample	E value
Marine	
Santa Cruz Basin	5.65
Santa Barbara Basin	5.17
San Pedro Channel	5.20
San Pedro Channel (FA)	8.48
Long Basin	7.43
Long Basin (FA)	8.94
Tanner Basin	6.01
Gulf of California	5.48
Pacific Ocean, off Oregon	5.66
Saanich Inlet, surface, core 4	5.35
Saanich Inlet, core 3B, 26-m depth (FA)	5.90
Saanich Inlet, core 3B, 17-m depth	5.72
Coastal and littoral	
Newport Marsh	6.92
Kaneohe Bay	9.29
Florida humate, No. 1	6.30
Florida humate, No. 1 (FA)	11.69
Continental	
North Dakota lignite	4.55
Minnesota peat	6.48
Hula peat	6.46
Dead Sea, 330-m depth	4.05
Melanoidin	10.15

E value is not a function of the degree of condensation alone.

Fulvic acids have higher *E* values than the corresponding humic acids, possibly related to their lower molecular weight. A similar effect was noted by Ladd (1969) who found an inverse relationship between molecular weight and absorbance at 450 nm.

ESR spectra

The ESR spectrum of humic and fulvic acids has been previously investigated (Steelink and Tollin 1967; Schnitzer and Skinner 1969). Most humic acids extracted from soils had spectra of 1.8 to 1.9 G linebreadth (Atherton et al. 1967). Humic acids from acid soils gave four line spectra whereas humic acids from basic soils gave fairly ill-defined spectra.

The spectra of some powdered humic acids and related materials that we measured (at 9.4 GHz operating frequency) are shown in Fig. 3. The humic acids all have similar spectra. The *g* values are close to that of the diphenyldipicrylhydrazyl (DPPH) standard and fall in the range of 2.0035–2.0042. Those values are markedly different from that of plankton which was very low, or that obtained for humic acid produced by fungi ($g = 2.0021$: Schnitzer and Skinner 1969). The melanoidin preparation had a value of 2.0041.

The *g* value of plankton is different enough from that of the humic acid to indicate that, although plankton may be a source for humic substances, considerable structural and chemical rearrangement of the organic compounds involved must take place during their formation.

The linewidth of the humic acids analyzed is usually within the range of 3.5 to 6.5 G, similar to that of fulvic acids (Schnitzer and Skinner 1969) or low rank coals (Retcofsky et al. 1968). The spectra typically consists of a single fairly ill-defined peak and no hyperfine structure. Soil humic acid generally has a narrower and better-defined lineshape.

Carbon isotope distribution

The results are tabulated in Tables 1, 2, and 3. The values fall between –14.8 and –29.1‰. Marine samples have a much more limited range than samples from other environments. Most are –22 to –23‰; only two sampling locations (Santa Monica Basin and Cariaco Trench) are markedly different. The $\delta^{13}C$ values are close to the isotopic value for total organic carbon in Recent marine sediments (around –19 to –23‰: Eckelmann et al. 1962; Sackett 1964; Degens 1969). Sackett (1964), Degens (1969), Rogers and Koons (1969), and Scalan and Morgan (1970) all concluded that the isotopic composition of the organic carbon in the sediment is determined by the isotopic composition of the plankton in the water column, for which values of around –19‰ have been found (Degens et al. 1968*a*). Plankton in Saanich Inlet gave $\delta^{13}C = -19.2$‰ and the organic carbon in the sediment gave values of –20 to –21‰ (Brown et al. 1972). The reason

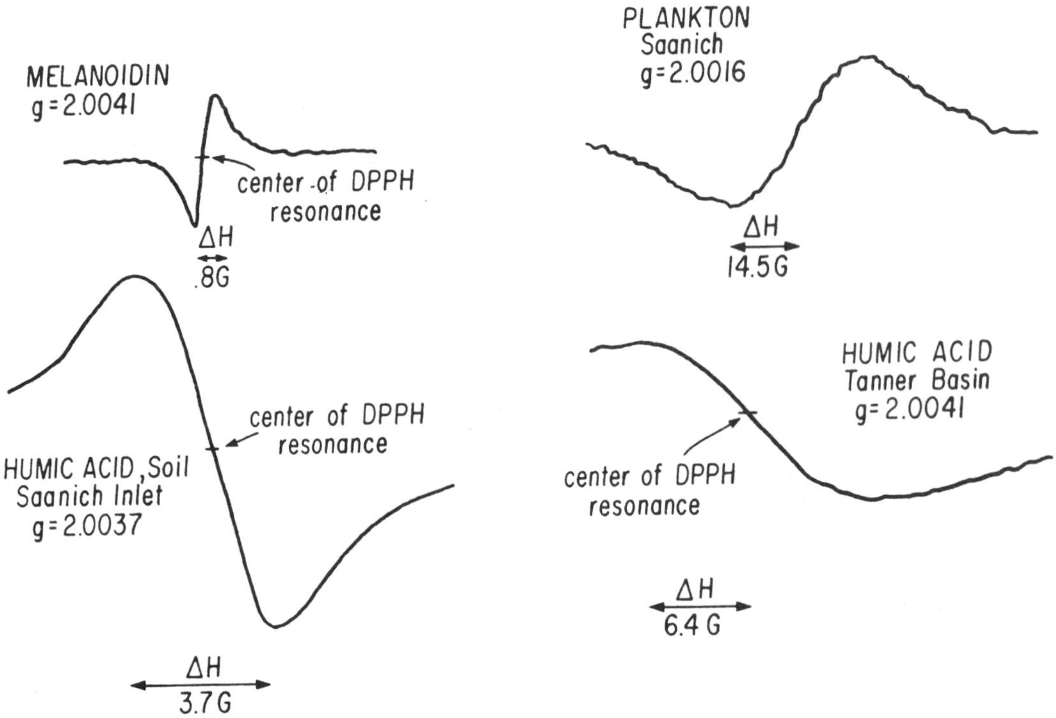

FIG. 3. ESR spectra of melanoidin, soil humic acid, marine plankton, and humic acid.

for the higher values of $\delta^{13}C$ in the Cariaco Trench samples is unknown. Eckelmann et al. (1962) measured $\delta^{13}C = -20.1‰$ in total organic matter in the sediment at 100-cm depth from a core taken in the trench at a water depth of 460 m; this value decreased to $\approx -23‰$ at 700-cm depth in the same core. The depletion of $\delta^{13}C$ in the Santa Monica Basin sample is readily explained by a large influx of terrestrial material. The east (landward) wall of this basin is cut by valleys and small canyons which enables transport of material from the continent into the basin by slumping or by flow (D. Gorsline, personal communication).

Coastal and littoral samples show a $\delta^{13}C$ range of –19 to –27‰; most fall into the –25 to –27‰ range of continental plants (Degens 1969). Two samples which came from a marsh environment measured –19‰—a value similar to that of –19.9‰ for salt marsh plants (Smith and Epstein 1970). The samples with values of –23 to –24‰ probably represent a mixed environment with contributions from both marine and continental sources.

The continental samples were mostly depleted in ^{13}C as compared to the coastal and marine samples, and most of the values were between –25 and –28‰—within the range of carbon from most trees and temperate-zone plants. The value for the Hula peat (–19‰) may be related to ^{12}C depletion in some of the source plant material. We would like to call particular attention to the value of –14.8‰ obtained for humic acid from the soil of a canefield in Hawaii, where the sugar cane yielded a value of –14‰. Sugar cane is one of the continental plants that follows the Hatch-Slack pathway (Smith and Epstein 1971) and is depleted in ^{12}C. In fact, $\delta^{13}C$ values clearly differentiate between cane sugar and beet sugar. This low value for the humic acid shows that humic acid isotopic values are almost the same as their source plant material.

TABLE 5. *The $\delta^{34}S$ values of sulfur from humic acids*

Sample	$\delta^{34}S$ (‰)
Marine	
San Pedro Basin	−6.4
Tanner Basin	−8.0
Santa Cruz Basin	−3.0
Saanich Inlet, core 5, 2.0–2.1 m	−2.1
Saanich Inlet, core 1, 1.7–1.8 m	+3.6
Continental	
Hula peat	+2.3
Florida humate	+4.2

Sulfur isotopes

Several samples were analyzed for their ^{34}S:^{32}S ratios (Table 5). The values ranged from −8 to +4‰. The marine samples were all on the low side. The exceptional sample from Saanich Inlet core 1 actually represents detrital material from the continent, as is also shown by numerous other evidence (Brown et al. 1972). The values for humic acids from continental sources are similar to the +4.6‰ value obtained by R. Krouse (unpublished results) for humic acid from peat moss.

The $\delta^{34}S$ of the marine samples is close to the values obtained by Kaplan et al. (1963) for the free and organic sulfur in sediments from the basins off southern California—quite different from that of marine algae. We may therefore assume that the sulfur in marine humic acids has been largely introduced during diagenesis in the sediment rather than through assimilation by the parent organic material, as may be the case for soil humic acids. This introduction of sulfur from external sources (possibly through reaction of H_2S or elemental sulfur with organic compounds) may explain the high sulfur content of marine humic acids (Table 1).

DISCUSSION

Humic and fulvic acids are major components of recent marine sediments; unfortunately quantitative data concerning them are scarce. A summary of our data and some from the literature is given in Table 6. Values reported are not always precise because the published results may not always have considered the ash content of humic acid. Furthermore, reported concentrations are generally given relative to the total organic matter, which has to be estimated from the organic carbon content of the sediment. The values for the percentage of humic acid-bound carbon would depend on several assumptions as to the carbon content of the total organic matter and that bound in humic substances. However, the data, imprecise as they are, definitely indicate humic and fulvic substances to be a major reservoir for organic carbon in Recent marine sediments.

It has been frequently assumed that humic substances are of terrestrial origin and

TABLE 6. *Humic substances expressed as percent carbon of the total organic carbon in Recent sediments*

Location	Total org C (%)	Source
San Diego Basin	35*	Degens et al. (1964)
Santa Barbara Basin	10.6	Swanson et al. (unpublished)
Santa Monica Basin	13.8	Swanson et al. (unpublished)
San Pedro Channel	13.2	Swanson et al. (unpublished)
Saanich Inlet, surface, core 1	44	Brown et al. (1972)
Saanich Inlet, surface, core 4	68	Brown et al. (1972)
Black Sea	8–20	Bordovskiy (1965)
Bering Sea	17–40	Bordovskiy (1965)
Northwest Pacific Ocean clay	34	Bordovskiy (1965)
Dead Sea, 165-m depth	40	Nissenbaum (1969)
Dead Sea, 330-m depth	49	Nissenbaum (1969)
Choctawhatchee Bay	4–31	Nissenbaum (1969)

* Humic acid only. The rest of the data include both humic and fulvic acid fractions.

that humic acid in marine sediments must be allochthonous (e.g. Degens et al. 1964). However, according to Sieburth and Jensen (1968), "Gelbstoffe" (humic materials) from continental sources rapidly precipitate on reaching the sea. The occurrence of humic substances in marine sediments distant from the continental masses had led several Russian scientists to suggest that humic acids may be authigenic in the marine environment (Bordovskiy 1965; Kasatochkin et al. 1968). Usually, no evidence other than geographical was brought forward to support this idea, although Bordovskiy (1965) showed differences in the C:H ratios between continental and marine humates. Recently evidence has accumulated that even in landlocked lakes, humic acids are not necessarily brought in from the surrounding soils. Otsuki and Hanya (1967) showed, on the basis of infrared data, that humic acid from Lake Haruna (Japan) is different from soil humic acid and is produced from the lacustrine biota. Similar conclusions were reached by Stevenson and Goh (1971), who studied humic acid from Mud Lake (Florida).

Our data also strongly suggest that humic substances in marine sediments may be largely authigenic. The most conclusive evidence comes from the ^{13}C:^{12}C ratios. The uniformity of this ratio in marine humic acids suggests a large and isotopically constant source material. The largest constant primary reservoir available in the –19 to –23‰ range is marine plankton.

The occurrence of humic acids with $\delta^{13}C$ values in the marine range of –20 to –23‰ in areas where sampling was close to continental masses supports the hypothesis that terrigenous humic acids are probably not transported far into the oceans. The traverse off Oregon (Fig. 2), collected opposite the mouth of the Columbia River, shows no obvious continental influence in organic matter distribution on the basis of the carbon isotope index, even in the samples closest to shore. The same is true in the California borderland (Fig. 1) where marine values were measured in almost all basins.

TABLE 7. *Isotopic composition of fulvic acid from various environments ($\Delta\, \delta^{13}C$ is $\delta^{13}C$ fulvic acid minus $\delta^{13}C$ of humic acid from the same sample)*

Sampling location	$\delta^{13}C$ (‰)	$\Delta\, \delta^{13}C$ (‰)
Canefield soil, Hawaii	–18.2	–3.4
Kaneohe Bay, No. 1	–23.3	+0.9
Kaneohe Bay, No. 2	–24.4	+0.5
Forest soil, Nova Scotia	–26.3	–0.1
Santa Cruz Basin	–20.7	+1.1
San Pedro Basin	–21.7	+1.0
Santa Barbara Basin	–19.1	+3.4
Tanner Basin	–20.3	+1.7
Long Basin	–20.3	+2.0
Newport Marsh	–17.4	+1.7
Florida humate sands, No. 1	–24.1	+1.6
Florida humate sands, No. 2	–24.1	+1.6
Forest soil, Saanich Inlet	–27.0	+2.1
Hula peat	–20.8	–1.6

A particularly convincing case is that of the Amazon River area. Humic acid in sediments from the river and its estuary had a $\delta^{13}C$ value of around –26.5‰ (Tables 2 and 3), but samples collected in the Atlantic, 100 km north of Paramaribo (Surinam), within the zone of influence of the river, were isotopically heavier and approached marine values ($\delta^{13}C = -23.6$‰).

Degens et al. (1968*b*) demonstrated a considerable isotopic fractionation between various chemical components of plankton. Marine plankton is, on the average, 1–2‰ enriched in ^{13}C relative to the humic acid fraction, indicating humic substances are probably a compilation of diagenetically transformed cellular material. This hypothesis is supported by the finding of nonprotein amino acids in humic acid from Santa Barbara Basin (Degens et al. 1964).

The position of fulvic acid in the transformation sequence is not clear. Fulvic acid is probably not a separate chemical entity but is closely related to humic acid, having a lower molecular weight and slightly different chemical composition (Kononova 1966). The ^{13}C:^{12}C ratios we measured do not show a clearcut pattern (Table 7). However, most of the samples have a $\delta^{13}C$ value 1–2‰ heavier than those of the corresponding humic acid, so that in the marine samples, at least, values for fulvic acids correspond more closely to their unaltered plankton precursor. We suggest, therefore, that fulvic acids are an intermediate product in humic acid formation. Further evidence for this was found in sediment from Saanich Inlet, where fulvic acid decreases markedly with depth, presumably through transformation to humic acid (Brown et al. 1972).

In addition to base-extractable complexes of high molecular weight of the humic and fulvic acid types, other complexes are present either in solution or as colloidal micelles in the interstitial water of marine sediments (Nissenbaum et al. 1972). These soluble complexes apparently represent condensation of cellular degradation products, including amino acids, carbohydrates, and possibly polycarboxylic acids and lipids. We suggest that continuous polymerization of this material, accompanied by some chemical modification (such as loss of –COOH functional groups) leads to the formation of insoluble fulvic acid, which is removed from solution possibly by adsorption on clay particles. Continuing diagenesis increases the molecular weight of the fulvic acid, and this, in addition to chemical modification, produces molecules of the humic acid type. If the isotopic data can be used as a criterion for chemical changes, then the fulvic acid→humic acid transformation involves loss of a fraction enriched in ^{13}C, presumably the carboxyl moiety of amino acids (Abelson and Hoering 1961). In addition, this transformation presumably involves dehydrogenation, ring closure, and increase in aromaticity. The process probably continues until all available water-soluble complexes are exhausted. Subsequent diagenesis causes further loss of oxygenated groups, perhaps as CO_2, and the formation of the carbon-rich kerogen. In addition to the chemical composition data, evidence for this can be cited again from the ^{13}C:^{12}C ratios which are always lower in marine kerogen than in marine humic acids.

To summarize our discussion, terrigenous and marine humic acids appear to be similar from both chemical and spectroscopic evidence. However, they can be differentiated: Marine humic acids contain more nitrogen and sulfur, and the ^{13}C:^{12}C ratios in normal marine humic acids are enriched in ^{13}C over normal continental humic acids.

The similarity of marine and terrigenous humic substances is tentatively explained by the following mechanism: Terrigenous humic acids are formed by degradation of lignins into quinonoid and phenolic compounds. These compounds react with amino acids and other molecules and eventually polymerize into humic acids. In contrast, marine humates are formed by condensation, probably of the Maillard type, of carbohydrates, amino acids, and possibly other simple molecules. The condensation is accompanied by cyclization of sugars to hydroaromatic and hydroxyaromatic acids, compounds probably intermediate in both degradative and synthetic pathways. Little is presently known about the existence and abundance of the hydro- and hydroxy-aromatic compounds in sediments (Degens et al. 1964).

REFERENCES

ABELSON, P. H., AND T. C. HOERING. 1961. Carbon isotope fractionation in formation of amino acids by photosynthetic organisms. Proc. Nat. Acad. Sci. **47**: 623–732.

ATHERTON, N. M., P. A. CRANWELL, A. J. FLOYD, AND R. D. HAWORTH. 1967. Humic acid—I.E.S.R. spectra of humic acids. Tetrahedron **23**: 1653–1667.

BORDOVSKIY, O. K. 1965. Accumulation of organic matter in bottom sediments. Mar. Geol. **3**: 33–82.

BROWN, F. S., M. J. BAEDECKER, A. NISSENBAUM, AND I. R. KAPLAN. 1972. Early diagenesis in Saanich Inlet, a reducing fjord. 3. Geochim. Cosmochim. Acta **36**: In press.

CRAIG, H. 1953. The geochemistry of the stable carbon isotopes. Geochim. Cosmochim. Acta **3**: 53–92.

DEGENS, E. T. 1969. Biogeochemistry of stable carbon isotopes, p. 304–329. *In* G. Eglinton and M. T. J. Murphy [eds.], Organic geochemistry. Springer.

———, M. BENRENDT, B. GOTTHARDT, AND E. REPPMAN. 1968*a*. Metabolic fractionation of carbon isotopes in marine plankton. 2. Deep-Sea Res. **15**: 11–20.

———, R. R. L. GUILLARD, W. M. SACKETT, AND J. A. HELLEBUST. 1968*b*. Metabolic fractionation of carbon isotopes in marine plankton. 1. Deep-Sea Res. **15**: 1–9.

———, J. H. REUTER, AND N. F. SHAW. 1964. Biochemical compounds in offshore California sediments and sea water. Geochim. Cosmochim. Acta **28**: 45–66.

ECKELMANN, W. R., W. S. BROECKER, D. W. WHITLOCK, AND J. R. ALLSUP. 1962. Implications of carbon isotope composition of total organic carbon of some Recent sediments and ancient soils. Bull. Amer. Ass. Petrol. Geol. **46**: 699–704.

EMERY, K. O. 1960. The sea off southern California. Wiley. 357 p.

ENDERS, C., AND K. THEIS. 1938. Die melanoidine une ihre beziehung zu den huminsauren. Brennst. Chem. **19**: 360–439.

FLAIG, W. 1964. Effects of microorganisms in the transformation of lignin to humic substances. Geochim. Cosmochim. Acta **28**: 1523–1535.

HURST, H. M., AND N. A. BURGESS. 1967. Lignin and humic acids, p. 260–286. *In* A. D. McLaren and G. H. Patterson [eds.], Soil biochemistry. Dekker.

KAPLAN, I. R., K. O. EMERY, AND S. C. RITTENBERG. 1963. The distribution and isotopic abundance of sulfur in Recent sediments off southern California. Geochim. Cosmochim. Acta **27**: 297–331.

———, J. W. SMITH, AND E. RUTH. 1970. Carbon and sulfur concentration and isotopic composition in Apollo 11 lunar samples. Proc. Apollo 11 Lunar Sci. Conf., v. 2. Geochim. Cosmochim. Acta, Suppl. 1, p. 1317–1329.

KASATOCHKIN, V. I., I. K. BORDOVSKIY, N. K. LARINA, AND K. CHERKINSKAYA. 1968. Chemical nature of humic acids in Indian Ocean floor sediments. Dokl. Akad. Nauk SSSR **179**: 690–693.

KONONOVA, M. M. 1966. Soil organic matter, 2nd English ed. Pergamon. 544 p.

LADD, J. N. 1969. The extinction coefficients of soil humic acids fractionated by Sephadex gel filtration. Soil Sci. **107**: 303–306.

MANSKAYA, S. N., AND T. V. DROZDOVA. 1968. Geochemistry of organic substances. Pergamon. 354 p.

NISSENBAUM, A. 1969. Studies on the geochemistry of the Jordan River-Dead Sea system. Ph.D. thesis, Univ. Calif. Los Angeles. 288 p.

———, AND I. R. KAPLAN. 1966. Origin of the Beeri (Israel) sulfur deposit. Chem. Geol. **1**: 295–316.

———, B. J. PRESLEY, AND I. R. KAPLAN. 1972. Early diagenesis in Saanich Inlet, a reducing fjord. 1. Geochim. Cosmochim. Acta **36**: In press.

OGLESBY, R. T., R. F. CHRISTMAN, AND C. H. DRIVER. 1967. The biotransformation of lignin to humus—facts and postulates. Advan. Appl. Microbiol. **9**: 171–184.

OTSUKI, A., AND T. HANYA. 1967. Some precursors of humic acid in Recent lake sediments suggested by infra-red spectra. Geochim. Cosmochim. Acta **31**: 1505–1516.

PALACAS, J. G., V. E. SWANSON, AND A. H. LOVE. 1968. Organic geochemistry of Recent sediment in the Choctawhatchee Bay, Florida—a preliminary report. U.S. Geol. Surv. Prof. Pap. 600-C, p. C97–C106.

RASHID, M. A., AND L. H. KING. 1969. Molecular weight distribution measurements on humic and fulvic acid fractions from marine clays on the Scotian shelf. Geochim. Cosmochim. Acta **33**: 147–151.

———, AND ———. 1970. Major oxygen-containing functional groups present in humic and fulvic acid fractions isolated from contrasting marine environments. Geochim. Cosmochim. Acta **34**: 193–202.

RETCOFSKY, H. L., J. M. STARK, AND R. A. FRIEDEL. 1968. Electron spin resonance in American coal. Anal. Chem. **40**: 1699–1704.

ROGERS, M. A., AND C. B. KOONS. 1969. Organic carbon δC^{13} values from Quaternary marine sequences in the Gulf of Mexico: A reflection of paleotemperature changes. Trans. Gulf Coast Ass. Geol. Soc. **19**: 529–534.

SACKETT, W. M. 1964. The depositional history and isotopic organic carbon composition of marine sediments. Mar. Geol. **2**: 173–185.

SCALAN, R. S., AND T. D. MORGAN. 1970. Isotope ratio mass spectrometer instrumentation and application to organic matter contained in Recent sediments. Int. Mass Spectrom. Ion Physics **4**: 267–281.

SCHARPENSEEL, H. W. 1962. Studies with humic acids labeled with carbon-14 and hydrogen-3 as to the mode of linkage of their amino acid fractions and with sulfate, methionine, and cystine labeled with sulfur-35 regarding the sulfur turnover in humic acid and soil. Radioisotopes Soil-Plant Nutr. Stud., Proc. Symp., Bombay, 1962, p. 115–133.

SCHNITZER, M., AND S. I. M. SKINNER. 1969. Free radicals in soil humic compounds. Soil Sci. **108**: 383–390.

SIEBURTH, J. McN., AND A. JENSEN. 1968. Studies on algal substances in the sea. 1. J. Exp. Mar. Biol. Ecol. **2**: 174–189.

SMITH, B. N., AND S. EPSTEIN. 1970. Biogeochemistry of the stable isotopes of hydrogen and carbon in salt marsh biota. Plant Physiol. **46**: 738–742.

———, AND ———. 1971. Two categories of C13:C12 ratios for higher plants. Plant Physiol. 47: 380–384.

STEELINK, C., AND G. TOLLIN. 1967. Free radicals in soils, p. 147–169. *In* A. D. McLaren and G. H. Patterson [eds.], Soil biochemistry. Dekker.

STEVENSON, F. J., AND K. M. GOH. 1971. Infrared spectra of humic acids and related materials. Geochim. Cosmochim. Acta 35: 471–484.

SWANSON, V. E., AND J. G. PALACAS. 1965. Humates in coastal sands of Northwest Florida. U.S. Geol. Surv. Bull. 1214-B. 29 p.

21

Reprinted from *Geochim. et Cosmochim. Acta* **15**:170–182 (1958), with the permission of Microforms International Marketing Corporation as exclusive copyright licensee of Pergamon Press journal back files

INSOLUBLE ORGANIC MATTER (KEROGEN) IN SEDIMENTARY ROCKS

J. P. Forsman and J. M. Hunt

Abstract—Kerogen was isolated, by techniques developed in this study, from twenty-one rock samples representing a wide variety of geologic ages, rock types, and environments of deposition. These techniques make it possible to separate relatively mineral-free kerogen from rocks containing trace amounts of organic matter and to obtain reliable measurements of physical and chemical properties of kerogen. The two types of kerogens recognized in this study appear to be controlled by the nature of the original living matter deposited with the sediments, although metamorphic forces also appear to have affected the properties of kerogen. Hydrogenolysis of different kerogens results in soluble products varying appreciably in amount and type.

PREVIOUS studies (HUNT, STEWART and DICKEY, 1954; HUNT and JAMIESON, 1956) have shown that a large majority of sedimentary non-reservoir rocks contain from a few hundredths to 5 per cent of organic matter. Two general types of organic matter are recognized in these rocks. *Soluble organic matter*, usually 5 to 15 per cent of the total, is isolated by extraction of the rock with organic solvents. This material, which resembles petroleum in constitution as well as in appearance, can be separated into paraffin–naphthene hydrocarbons, organic compounds containing nitrogen, sulphur or oxygen, and "asphaltenes." These types of compounds all have their counterparts in petroleum (HUNT and JAMIESON, 1956).

Relatively little attention has been given to the *insoluble organic matter* or *kerogen* of a sedimentary rock, despite the fact that it accounts for most of the total organic content. Several investigators have carried out chemical studies of kerogen in oil shales and torbanites (reviewed by HIMUS, 1949), but only a few papers have appeared on kerogen in less organic-rich marine rocks (BREGER and DEUL, 1953–1956; TAKAHASHI, 1935).

The present study of kerogen in sedimentary rocks of marine origin was undertaken to determine certain of its chemical and physical properties that could lead to a better understanding of the processes involved in the formation of petroleum. However, before any of the various properties of kerogen could be measured, it was first necessary to isolate the kerogen from the inorganic constituents of the rock.

PROCEDURES

The isolation procedures used in this study (Fig. 1) were based on the methods first used by the Bureau of Mines (GUTHRIE, 1938) and later developed by DANCY and GIEDROYC (1950). Briefly, the method consisted of treating with an excess of acid a quantity of pulverized and solvent-extracted rock (HUNT and JAMIESON, 1956) sufficient to yield at least 1 g of kerogen. The initial acid treatment varied according to the lithology of the samples. Carbonates and calcareous shales were

* Paper read before the Division of Analytical Chemistry, 131st Meeting, ACS, Miami, Florida, April 1957.

† Present address: Esso Research and Engineering Co., Linden, N.J., U.S.A.

best digested with hydrochloric acid to remove soluble calcium and magnesium which otherwise would form insoluble precipitates in subsequent treatments. Following the hydrochloric acid step, the insoluble residue was successively digested with cold and with hot concentrated hydrofluoric acid to remove alumina-silicates and quartz.

Early in the investigation, it was recognized that, in contrast to kerogen-rich oil shales, the isolation of organic matter from the more common types of marine

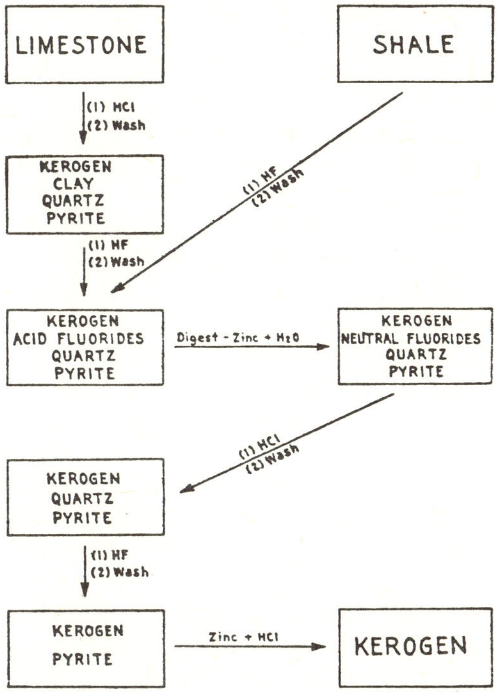

Fig. 1. Flowsheet for isolation of kerogen.

shales presented special problems. These were caused by the fact that the more typical marine shales contain much less organic matter than oil shales, and thus the hydrofluoric acid-insoluble minerals generally exceed the organic content. Accordingly, a considerable amount of effort was expended in improving techniques of isolation and in devising new methods for reducing the ash content.

The more difficultly soluble inorganic material encountered in this investigation was of two general types. These were: (1) insoluble minerals such as pyrite and zircon and (2) complex fluorides formed during the reaction of hydrofluoric acid with clay minerals and feldspars.

Of the first type, pyrite proved to be the most troublesome because it was abundant in nearly all of the samples. Pure pyrite is readily soluble in concentrated nitric acid, but the drastic effects of the reagent on organic matter precluded its use in this investigation. For a few samples, the size of the pyrite particles was large enough that physical methods such as heavy liquid flotation could be used in their separation. In all other cases, an improvement of the zinc–hydrochloric acid method devised by HIMUS and BASAK (1949) was used. This technique was found to be very effective in dissolving pure pyrite, but the presence of organic material appeared to inhibit the reaction. Consequently, several treatments with this reagent were generally necessary.

The second class of insoluble material consisted of a number of complex fluorides of various types. In a few cases, the identity of these products was established by the comparison of X-ray diffraction spectra of digestion residues with synthetic material. These compounds that were identified included potassium fluosilicate, aluminum oxyfluoride, and potassium fluoaluminate. X-ray diffraction lines were found in some samples that could not be ascribed to definite compounds. However, indirect evidence was obtained that some of the spectra might be attributed to mixtures of complex fluorides composed of ferrous iron, ammonium ion, and hydrogen with such metals as aluminium, calcium, magnesium, and potassium.

Several techniques were used for removing insoluble fluorides. Each of these methods, which included digestion with hot hydrochloric acid, hot water extraction, and treatment with zinc and water or with aqueous alkalis, were found to be specific for certain types of compounds. Thus, the treatments used and their sequences depended on the nature of inorganic material in the residues.

DANCY and GIEDROYC (1950) recommend the use of sodium carbonate for the decomposition of insoluble fluorides. However, the present work indicated that the addition of potassium or sodium salts to the reaction mixtures causes the formation of the corresponding insoluble fluosilicates (M_2SiF_6). Ammonium carbonate is preferred since it does not form an insoluble fluosilicate.

Table 1 lists the descriptions and analyses of the rock samples used in this investigation; Table 2 summarizes the elemental analytical data on the kerogen isolates.

Three examples of the combined chemical and analytical attack on the problem of removing the more insoluble inorganic components are given in the following sections. The first example, a black shale from the Eocene Wilcox of Louisiana, illustrates the use of X-ray diffraction in the identification of inorganic impurities and in following their removal at various stages of concentration. The second sample, an argillaceous limestone from the Duvernay shale of Alberta, gives some of the refinements in technique developed during the course of the work. The isolation of kerogen from a metamorphosed rock of Texas is included as the third example because it shows a modification in technique which is useful for the more graphitic samples.

Black shale

Portions of pulverized shale were placed in 2 l. Monel beakers, and 50 per cent hydrofluoric acid was added slowly in order to prevent loss of material by frothing. After digestion on a steam bath, the contents of the two beakers were diluted with water; the solid residue was collected in a basket-type centrifugal filter and washed with water until only traces of fluoride were detected in the filtrate (lead chloride test). The moist filter cake was transferred to Monel beakers, and, after a second treatment with hydrofluoric acid, the gelatinous, dark-brown insoluble residue was centrifuged and washed with dilute hydrochloric acid in polyethylene bottles.

Table 1. Description and analyses of rocks used in kerogen isolation

Lab. serial no.	Formation	Sample locality: Well	Sample locality: Location	Sample locality: State	Depth (ft)	Lithology	Age	Analyses of extracted rock: Organic C %	Analyses of extracted rock: Kjeldahl N %	Analyses of extracted rock: Total S %
6529	Cherokee	TCO # 1 Winchell	6–36S–5E	Kansas	3136–47	blk. sh.	Pennsylvanian	1·73	0·11	1·01
6412	Hoxbar	TCO # 1 Garrett	32–11N–25W	Oklahoma	10687–8	dk. gry. sh.	Pennsylvanian	0·65	0·15	0·03
593	Woodford	TCO # 2 Moore-Daniel	36–20N–5W	Oklahoma	5785–92	blk. sh.	Mississippian	4·19	0·20	3·30
13210	Frontier	TCO # 1 State Burke	36–37N–79W	Wyoming	5800*	dk. gry. sh.	Upper Cretaceous	1·79	0·13	1·61
609	Caney	TCO # 1 Moss Wilson	8–1N–1W	Oklahoma	7495–500	blk. sh.	Mississippian	3·51, 3·45	0·27, 0·28	1·9, 2·0
6875	Wilcox	TCO # 2 Clyde Graham	48–6N–9E	Louisiana	6120–24	blk. sh.	Eocene	0·86	0·08	0·13
13211	Monterey	Ohio Oil Co.	12–25S–12E	California	4497–517	gry. sh.	Miocene	2·03	0·14	1·57
12967	Lower Duvernay	Calif. Std.	4–53N–25W4	Alberta	5530	calc. sh. to	Devonian	7·63	0·2	1·73
		Winterburn Province				arg. ls.		0·74	0·064	0·52
7678–80	Morrison	TCO Govt. Steele Creek	28–26N–82W	Wyoming	3429–47	red to grn. sh.	Jurassic	0·11	0·12	0·16
12976	Phyllite	Humble # 1 Carroll	Falls Co.	Texas	3677	phyllite		0·22	0·10	0·07
12974	Green River	NH Canyon (outcrop)		Utah	—	mahogany "shale" (marl)	Eocene	12·40	0·48	1·72
12975	Green River	(Outcrop)	27–11–10E	Utah	—	papery "shale" (marl)	Eocene	18·33	0·39	1·50
982	Traverse	TCO # 2 Louis Stevens	15–13N–18W	Michigan	1710–20	ls.	Devonian	0·53	0·04	0·02
7762	Ste. Genevieve	TCO # 2 Sellers	19–0–23	Kentucky	2510	ls.	Mississippian	0·036	0·017	0·18
7695	Madison	TCO Govt. Steele Creek	28–26N–82W	Wyoming	5794–5	pink dolo.	Mississippian	0·04	0·06	0·01
12973	Collier	(quarry spl.)	Near Mt. Ida	Arkansas	—	blk. ls.	Cambrian	0·16	0·012	0·14
9561	La Luna	Totumal 141–3	—	Colombia	8177	blk. ls., arg.	Cretaceous	3·73	0·26	1·12

*Approximate

To remove insoluble acid fluorides, the dried residue was taken up with water, an excess of powdered zinc was added, and the mixture was digested over-night on the steam bath. After washing by centrifugation and decantation, the sample was acidified with hydrochloric acid and washed again.

Sieving of the moist residue through a 200-mesh screen separated the material into two fractions: a very fine, dark-brown solid, and coarse, very black particles. Subsequent treatments of the fine fraction included digestion with zinc, a digestion

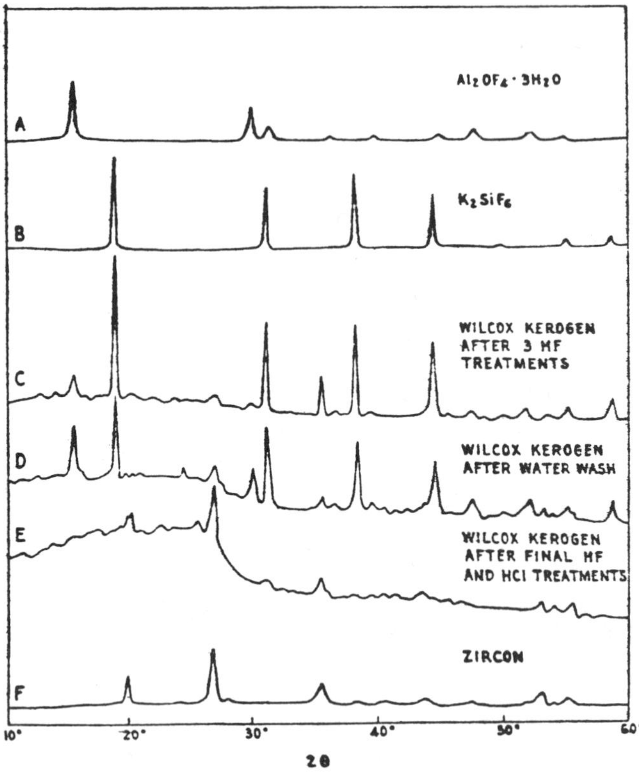

Fig. 2. X-ray diffraction spectra of Wilcox kerogen and inorganic impurities. Cu*K* radiation.

with hydrofluoric acid, and digestions with hydrochloric acid. A qualitative X-ray diffraction analysis showed the presence of about 30 per cent of potassium fluosilicate (K_2SiF_6) and a small amount of aluminium oxyfluoride ($Al_2OF_4 \cdot 3H_2O$). (Compare Figs. 2A, B, and C.)

To remove the potassium fluosilicate, the kerogen concentrate was washed with hot water until no precipitate was obtained on addition of *iso*propyl alcohol to the filtrate. Spectrum (Fig. 2D) of the kerogen showed the presence of zircon, aluminium oxyfluoride, and some potassium fluosilicate.

After additional digestions with hydrofluoric and dilute hydrochloric acid, an X-ray diffraction analysis of the residue showed the absence of fluorides but revealed the presence of a considerable amount of zircon (see Fig. 2F). Separate experiments with zircon indicated that chemical dissolution of the mineral could be effected only under drastic conditions, and no attempt was made to remove this mineral from the kerogen sample.

The coarse black material retained by the sieve was digested successively with hydrofluoric acid, powdered zinc, and hydrochloric acid. The X-ray diffraction spectrum of the material showed the presence of a considerable amount of unidentified inorganic material, here assumed to be insoluble fluorides.

Elemental analyses obtained on the fractions isolated from the Wilcox shale are given in Table 2.

Table 2. Elemental analyses of kerogen

Lab. serial no.	Name	Organic carbon %	Organic factor* %	Hydrogen %	Kjeldahl nitrogen %	Oxygen %	Total sulphur %	Ash %	Total† %
6529	Cherokee sh.	71·93	1·16	4·17	2·01	5·43	2·43	21·44	107·41
6412	Hoxbar sh.	30·04	—	2·40	0·84	—	1·71	—	—
593	Woodford sh.	65·60	1·17	4·96	2·51	3·68	10·73	16·29	103·77
13,210	Frontier sh.	68·88	1·22	5·14	0·14	10·16	5·36	8·53	98·21
609	Caney sh.	57·70	1·29	5·08	2·24	9·33	11·46	13·28	99·09
6875	Wilcox sh.								
	Kerogen fines	71·17	1·25	4·82	0·16	12·77	0·35	4·04	93·31
	Coarse kerogen	54·63	1·31	4·38	2·22	10·62	1·92	20·91	94·68
	Lignites	59·86	—	5·37	0·35	—	—	23·79	—
13,211	Monterey sh.	40·72	1·24	4·12	1·35	4·24	20·52	33·33	104·28
12,967	Duvernay								
	Calc. sh.	64·19	1·16	6·19	1·08	3·29	16·22	13·19	104·16
	Arg. ls.	63·03	1·18	6·98	1·41	3·06	8·68	19·32	102·48
7678–80	Morrison sh.	23·05	1·39	2·24	1·37	5·40	15·15	58·18	105·39
12,976	Phyllite	87·30	1·08	2·82	0·18	3·58	0·29	5·16	99·33
12,974	Green River (mahogany) sh.	73·46	1·25	9·09	2·51	6·97	2·33	8·11	102·47
12,975	Green River (papery) sh.	76·65	1·25	9·53	1·46	8·22	1·91	10·17	107·94
982	Traverse ls.	68·63	1·22	8·03	1·32	5·77	8·98	16·02	108·75
		74·36	1·15	5·46	1·96	3·52	12·72	17·35	115·37
7762	Ste. Genevieve ls.								
	Kerogen in ls.	53·48	1·26	4·21	1·88	7·56	7·87	22·90	97·98
	Kerogen in stylolite	40·97	1·29	3·65	2·03	6·28	7·38	41·95	102·26
7695	Madison dolo.	65·06	1·16	3·68	1·60	5·12	2·16	22·25	98·87
12,973	Collier ls.	76·58	1·05	2·17	0·79	1·22	0·14	13·49	94·39
9561	La Luna ls.	64·35	1·14	5·18	1·44	2·26	7·26	9·29	89·78

* Organic factor equals percentage of total organic matter divided by percentage of organic carbon.
† Totals in excess of 100 are probably caused by the contribution of inorganic sulphur (as pyrite) in both the ash and the sulphur analyses. Values less than 100 are probably due to inorganic matter (fluorides) that volatilize during ashing.

Argillaceous limestone, Duvernay shale, Alberta

Portions of the pulverized, organic solvent-extracted rock were placed in 1 l. polyethylene beakers; and, after wetting the samples with *iso*propyl alcohol, concentrated hydrochloric acid was added in portions to the stirred alcohol-rock slurry. The moist acid residues obtained by centrifugation were combined in a polyethylene beaker, and 50 per cent hydrofluoric acid was added in portions. Distilled water was added, and the liquid phase was separated from the residue by centrifuging and decanting.

To remove insoluble fluorides formed during the hydrofluoric acid treatment, excess sodium carbonate solution was added to the residue, and the mixture was allowed to stand until evolution of carbon dioxide had ceased (1 day). The centrifuged residue was digested for 2 days with dilute hydrochloric acid and washed with distilled water. Inorganic substances detected by X-ray diffraction analyses were pyrite (large amount) and calcium fluoride.

To remove the pyrite, *iso*propyl alcohol and concentrated hydrochloric acid were added to the washed kerogen residue. To this mixture an excess of powdered zinc was added in approximately 5 g portions. The pyrite was dissolved as evidenced by the evolution of hydrogen sulphide and appearance of ferrous iron in the water–alcohol mixture. After evolution of gas had ceased, the liquid phase was removed by centrifugation and the zinc–hydrochloric acid reaction was repeated three times. The formation of hydrogen sulphide took place during the fourth treatment. The kerogen residue from the fourth zinc–hydrochloric acid reaction was washed with water, centrifuged, and then dried at 100°C. Pyrite was the only impurity detected by X-ray diffraction analysis.

Table 3. Elemental material balance
(Argillaceous Limestone, Duvernay shale, Alberta)

	Organic carbon	Total nitrogen	Total sulphur
Wt. (%) of element in extracted rock	0·74	0·062	0·52
Wt. (g) of element in 237·5 g rock	1·8	0·15	1·2
Wt. (%) of element in kerogen concentrate	64·19	1·08	16·22
Wt. (g) of element in 3·0 g concentrate	1·9	0·032	0·49
Amount of element recovered (%)	105	21	41

The data in Table 3 show that reasonable agreement was obtained in the carbon material balance, and the decrease in the weight of sulphur was consistent with loss of pyrite and other sulphides during various stages of concentration of the kerogen. Most of the nitrogen lost during isolation of the kerogen was probably ammonia, and possibly some extraction of basic nitrogen organic compounds occurred.

Metamorphosed shale (phyllite)

Portions of the pulverized and organic solvent-extracted rock were placed in 1 l. polyethylene beakers; 35 per cent hydrofluoric acid was added, and, after heating for several days on the steam bath, the residue was centrifuged. Zinc and water were added and the mixtures were allowed to stand. After the slow evolution of gas had ceased, the liquid was removed by centrifugation and the excess zinc was dissolved by treatment with concentrated hydrochloric acid. This treatment dissolved some of the pyrite as evidenced by the evolution of hydrogen sulphide and the appearance of ferrous iron (thiocyanate test after oxidation with hydrogen peroxide) in the reaction mixture.

After additional treatments with hydrofluoric acid and with zinc, a mechanical separation of the kerogen from remaining inorganic matter was accomplished in the following manner. The sample was agitated in a Waring Blendor with a mixture of *n*-heptane and water and poured into a separatory funnel. On standing, most of the inorganic material settled to the bottom, whereas the black organic matter remained at the interface of the water–heptane layers. The solid inorganic matter was drawn off, and the kerogen was filtered and dried. An X-ray diffraction spectrum of the sample showed the presence of about 10 per cent of quartz plus some unidentified material.

The remaining quartz was removed by digestion with 50 per cent hydrofluoric acid in the steam bath. The X-ray diffraction pattern of the washed residue (Fig. 3) indicated that some graphite was present.

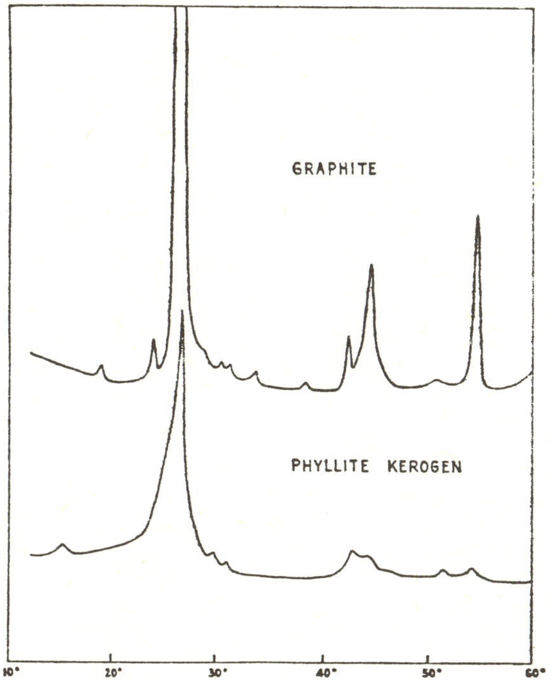

Fig. 3. X-ray diffraction spectra of phyllite kerogen and pure graphite. Cu*K* radiation.

Results and Discussion

Physical characteristics

Kerogens isolated from typical marine rocks were obtained as very fine, amorphous, soft powders that varied from dark brown to jet black in colour, many showing a marked resemblance to coal dust. When the samples were heated in open test tubes, usually little or no oily distillate was formed. This behaviour is distinctly different from that exhibited by oil shales, which under the same conditions yield substantial amounts of oily distillates.

Refractive indices of the powdered kerogen samples were determined by the oil immersion method at 450× magnification. Table 4, which lists these measurements, also includes indices of a few coals, asphaltites, and mineral waxes.

Typical coals and asphaltites show a fairly uniform increase in refractive index with decreasing hydrogen–carbon atomic ratio. While a few kerogen samples appeared to follow this relationship, the refractive indices of most of the kerogens

Table 4. Refractive indices of kerogens, coals and asphaltites

Kerogen	Refractive index	Asphalt and coal	Refractive index†
Frontier	1·73	Anthracite coal	1·80
Traverse*	1·73	Bituminous coal	1·75
Caney	1·64	Subbituminous coal	1·70
Woodford	1·61	Albertite	1·66
Green River (mahogany)	1·60	Lignite	1·63
Traverse*	1·57	Gilsonite	1·61
Green River (papery)	1·50	Wurtzilite	1·57
		Ozokerite	1·53

* Separate fractions obtained during acid treatment, in which part of the kerogen floated on the surface of the acid mixture

† Hunt *et al.* (1954): Lowry (1945).

were lower than predicted. This effect may have been caused by fluoride contaminants, whose refractive indices are somewhat lower than those of kerogen.

Specific gravity measurements were made by the pycnometer method in *iso*propyl alcohol at 25°C. Table 5 lists the specific gravities, corrected for mineral content,* of several kerogen samples, together with corresponding data for a few coals, asphaltites, and mineral waxes.

Table 5. Corrected specific gravities of kerogens, coals and asphaltites

Sample	Specific gravity†	Sample	Specific gravity†
Traverse*	1·51	Anthracite coal	1·4 –1·8
Cherokee	1·39	Bituminous coal	1·2 –1·5
Green River (papery)	1·38	Albertite	1·07–1·10
Traverse*	1·32	Wurtzilite	1·05–1·07
Green River (mahogany)	1·22	Gilsonite	1·03–1·10
Frontier	1·22	Ozokerite	0·85–1·0
		Peat, dried	0·84

* See footnote, Table 4.

† Kerogen data corrected for mineral content. Data for asphalts and coals taken from Abraham (1945) and Lowry (1945).

For recording the infra-red spectra of kerogens, scattering was minimized by grinding samples under *iso*propyl alcohol in a mortar. The resultant fine powders were then deposited in the form on thin films on salt block cells, using a technique previously developed in this laboratory (Hunt and Turner, 1953). Absorption peaks were noted at several wavelengths; however, because of the weak absorption,

* Corrections were made with the assumption that the major impurities were pyrite (sp. gr., 5·0) and inorganic fluorides (mean sp. gr., 2·8). Pyrite in the samples were estimated from iron and sulphur determinations; the identity and quantity of fluorides were determined from X-ray diffraction and emission spectrographic analysis.

no marked differences were observed between any of the spectra. The two non-marine (Green River) kerogens had slightly more intense peaks at wavelengths of 3·5 and 6·9 μ, indicating that these samples have less condensed and more paraffinic structures than typical marine kerogens.

Chemical composition

The results of the elemental analyses show that, in all of the kerogen samples, carbon, hydrogen, and oxygen are the major constituents. On an atomic basis, nitrogen is less abundant, but nevertheless it constitutes an integral part of the

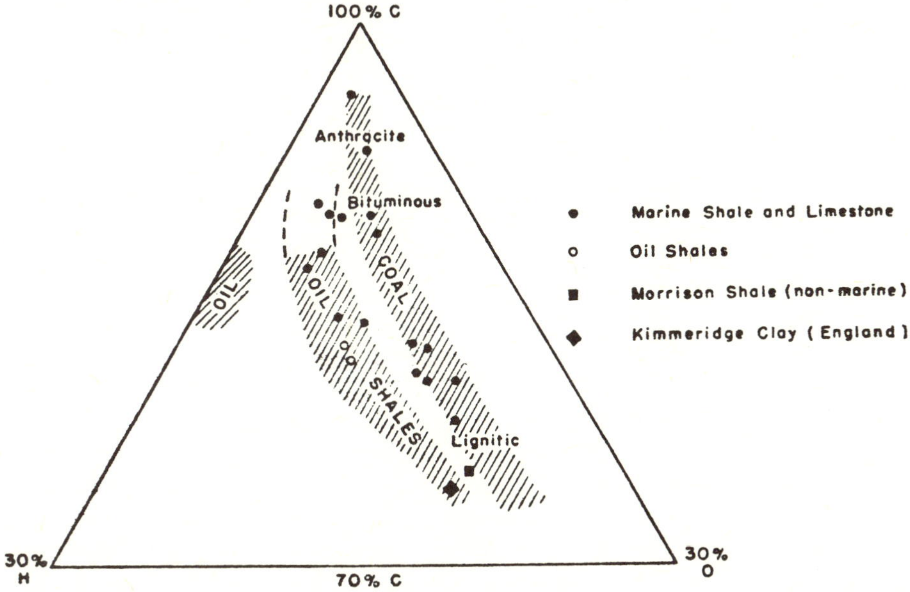

Fig. 4. Distribution of carbon, hydrogen and oxygen in kerogen.

kerogen molecule. Organic sulphur was presumed to be present in all of the kerogens but, because of residual pyrite, the organic sulphur content could not be determined accurately. In a few instances, it was established that the total sulphur was considerably greater than could be attributed to inorganic sulphur alone.

All the samples analysed by emission spectrography contained a number of metallic constituents some of which may have been combined chemically with the kerogen. No attempt was made in the present study to demonstrate the presence of metallo–organic substances.

In the triangular diagram (Fig. 4), the solid circles express the composition of the kerogens on an ash-, nitrogen-, and sulphur-free basis. Also shown in the diagram is a coal band, which represents the composition of coals in various stages of maturity or metamorphism (Bangham, 1950). The shaded area entitled "oil shales," which lies parallel to and left of the coal band, represents literature data for kerogens isolated from approximately thirty samples of oil shales and torbanites of various ages and localities (Himus, 1951).

Two points of geologic interest are provided by the data in Fig. 4. First, it is apparent that there are two broad types of kerogen in marine rocks. Members of one type have elemental compositions that are comparable to oil shales and torbanites, whereas the other type is scarcely distinguishable from coal. Both limestone and shale kerogens are included in each of these types.

The second point of importance is the suggestion that kerogen changes in elemental composition in a manner similar to that which has taken place in coals. The evidence for the change in rank is: (1) the distribution of members of each of the two types along the lengths of the oil shale and coal bands, respectively; (2) the fact that anthracite-like kerogens were found in two highly metamorphosed rocks but not in the relatively unmetamorphosed rocks; and (3) the progressive change in hydrogen–carbon atomic ratios and fixed carbon values of kerogen (FORSMAN and HUNT, in press).

A few of the kerogen samples did not fall into either of the two categories. The first of these, the Morrison shale kerogen, appears to have a composition intermediate to that of a very low-rank coal or lignite and that of the Kimmeridge clay of England (DANCY and GIEDROYC, 1950). The three other unclassified samples lie well to the left of the coal band, but almost directly above the oil shale region. Although oil shale analyses that fall in this region are lacking, it is probable that the oil shale band should be extended upward toward 100 per cent carbon, the hypothetical end-product of all sedimentary organic matter. The proposed extension, which is shown as dashed lines in Fig. 4, includes two of the three remaining samples.

During the concentration of kerogen, large losses of the original nitrogen content of the rock occurred, although the recovery of organic carbon generally exceeded 90 per cent. Because of the high ratio of nitrogen lost to carbon lost, most of the nitrogen must have been removed during acid extraction as ammonium ion rather than as organic nitrogen. For shales, the average nitrogen loss based on rock weight was 0·09 per cent; for calcareous shales, 0·05 per cent; and for limestones, 0·02 per cent. These values agree with the prediction that the amount of ammonia in a rock increases with increasing clay content.

Carbon/nitrogen ratios of the kerogen samples generally fell in the range of 25–60. A few kerogen samples, notably those from metamorphosed rock, contained only traces of nitrogen.

Although kerogens differ appreciably in elemental composition, it is apparent that elemental analyses reflect only gross differences between kerogens. Because of the possibility that two kerogens of identical elemental composition might have entirely different structures, it was desirable to find a method that would yield some information concerning structural differences between kerogens. One of the more promising methods appeared to be hydrogenolysis at high temperatures, a process that is employed in the commercial liquefaction of coal.

In the present work, seven representative samples, including four kerogens, one coal, and two shales, were subjected to uncatalysed hydrogenolysis in a 1000 ml stainless steel Aminco super-pressure vessel. Preliminary experiments indicated that hydrogenolysis in the presence of *n*-heptane would yield up to 200 per cent more soluble material than in the absence of solvent. With pure heptane, extensive cracking and some polymerization took place at 450°C; whereas, at

380°C, no significant change in boiling point occurred. Consequently, hydrogenolysis of kerogen was carried out at the latter temperature. Other reaction conditions were: maximum pressure, 2700–3100 lb/in^2; time, 12–20 hr. The soluble products, which were obtained in yields varying between 18 and 87 per cent, were further separated into types of organic compounds by the standard solubility-alumina chromatography method used at this laboratory (Hunt and Jamieson, 1956).

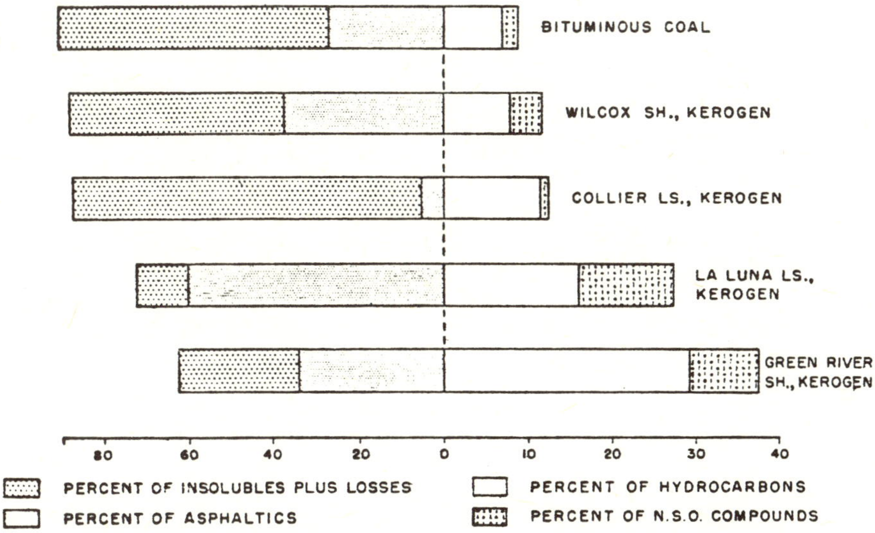

Fig. 5. Distribution of hydrogenolysis products of kerogen and coal.

The results of the hydrogenolysis of four kerogens and one coal, presented graphically in Fig. 5, show that there are very marked differences to be found between kerogens. For example, in the total amount of kerogen converted to soluble products, both the La Luna and Green River (mahogany shale) samples are clearly different from the Wilcox kerogen. The hydrogenolysis products of the La Luna and Green River kerogens are, in turn, dissimilar in their hydrocarbon and asphaltic contents. Unlike the other kerogens, the soluble products from the Cambrian Collier limestone consisted predominantly of hydrocarbons, and over four-fifths of this kerogen resisted attack by hydrogen.

Table 6. Distribution of hydrocarbon types in kerogen hydrogenolysis products

Sample	Paraffin–naphthenes/aromatics
Bituminous coal	0·13
Wilcox kerogen	0·10
Collier kerogen	1·60
La Luna kerogen	0·65
Green River kerogen	1·20

In view of the wide variations in the results obtained with different kerogen samples, the marked similarity between the hydrogenolysis products of the Wilcox kerogen and of the coal appears especially significant.

In Table 6, differences between the kerogens may also be seen in the ratio of paraffins plus naphthenes to aromatics in the soluble products. The resemblance between the bituminous coal and the Wilcox kerogen again appear in this data.

Acknowledgements—Special thanks are due: G. E. C. Wear, of the Products Research Division, Esso Research and Engineering Company, for the Unterzaucher oxygen analyses of the kerogen samples; to P. V. Smith, Jr. and M. C. Brenneman, Chemicals Research Division, Esso Research and Engineering Company, for preliminary hydrogenolysis work; and to C. G. Claypool and F. E. Dollarhide, of the Jersey Production Research Company for the X-ray and hydrogenolysis data, respectively. The authors wish to express their gratitude to P. A. Dickey for his many constructive suggestions.

References

Abraham H. (1945) *Asphalts and Allied Substances* (5th Ed.). Van Nostrand, New York.

Bangham D. H. (Editor) (1950) *Progress in Coal Science*. Interscience, New York.

Breger I. A. and Deul M. (1953) *U. S. Geol. Surv. T. E. I. Prog. Repts.* **390,** 161–2; (1953–4); **440,** 141–2; (1954) **490,** 178–82; (1954–5) **540,** 187–90; (1955) **590,** 264–7; (1955–6) **620,** **264–6.**

Deul M. (1956) *Bull. Amer. Assoc. Petrol. Geol.* **40,** 909–17.

Dancy T. E. and Giedroyc V. (1950) *J. Inst. Petrol.* **36,** 593–603; 607–23.

Forsman J. P. and Hunt J. M. *Insoluble Organic Matter (Kerogen) in Sedimentary Rocks of Marine Origin*. In press.

Guthrie B. (1938) *U.S. Bur. Mines, Bull.* **415,** 105–17.

Himus G. W. (1951) Observations on the composition of kerogen rocks and the chemical constitution of kerogen. *Oil Shale and Cannel Coal* Vol. 2. Institute of Petroleum, London.

Himus G. W. and Basak G. C. (1949) *Fuel* **28,** 57–65.

Hunt J. M. and Turner D. S. (1953) *Analyt. Chem.* **25,** 1169–74.

Hunt J. M., Stewart, F. and Dickey P. A. (1954) *Bull. Amer. Assoc. Petrol. Geol.* **38,** 1671–98.

Hunt J. M. and Jamieson G. W. (1956) *Bull. Amer. Assoc. Petrol. Geol.* **40,** 477–88.

Lowry H. H. (Editor) (1945) *Chemistry of Coal Utilization* Vol. 1, pp. 310–27. John Wiley, New York.

Takahashi Jun-ichi (1935) *Sci. Repts. Tohoku Univ.* Ser. 3. **1,** 63–156.

22

Reprinted from *Am. Assoc. Petrol. Geol. Bull.* **42**:998–1012 (1958)

CARBON ISOTOPIC COMPOSITIONS OF PETROLEUMS AND OTHER SEDIMENTARY ORGANIC MATERIALS[1]

SOL R. SILVERMAN[2] AND SAMUEL EPSTEIN[3]
La Habra and Pasadena, California

ABSTRACT

Carbon isotope analyses of petroleums, modern plant constituents, and other sedimentary organic materials indicate that the range of C-13/C-12 ratios in organic carbon is about 4.5 per cent. The range of variation in petroleums alone is about one per cent. Ratios of marine organisms average about 10 per mil higher than non-marine organisms; petroleums of marine origin are correspondingly higher in C-13 content than non-marine petroleums. Petroleums have lower C-13/C-12 ratios than their biologic sources have. The lipid fractions of plants, however, are isotopically lighter than the whole plants, so that isotope fractionation is not necessary if petroleum is derived from lipid materials. Analyses of soluble organic matter in Recent sediments imply that isotope fractionation may be involved and transformation of non-lipid materials, therefore, may occur during petroleum formation.

The C-13/C-12 ratio of the aromatic-enriched fraction of a petroleum does not differ by more than one per mil from the ratio of the paraffinic-naphthenic fraction. This indicates that changes in chemical composition alone are not responsible for the 10-per-mil range of isotopic composition noted for petroleums. Natural gases from petroliferous areas have considerably lower C-13/C-12 ratios than associated petroleums. In view of the narrow range of isotopic composition in petroleums, gas formation does not appear to be an important natural mechanism for depleting oil fields. The isotope ratios of organic carbon in ancient sediments fall within the isotope ratio range of petroleums.

INTRODUCTION

Previous investigations of relative abundances of the stable carbon isotopes in natural carbonaceous materials show that C-13/C-12 ratios differ by as much as 5 per cent. In their studies of carbon isotope abundances in nature, Nier and Gulbransen (1939), Murphy and Nier (1941), Craig (1953), and Landergren (1954) found this ratio to be high in carbonates and low in organic carbon. Urey (1947), Craig (1953), and others discussed the thermodynamic criteria underlying some of these variations. For example, in the carbonate-carbon dioxide system in equilibrium at room temperature and atmospheric pressure, the C-13/C-12 ratio of the carbonate phase is greater than that of the carbon dioxide. Methane, in equilibrium with carbon dioxide, tends to become enriched in C-12 while the carbon dioxide is correspondingly depleted in C-12. In general, knowledge of the isotope fractionation factors involved in reactions among carbon compounds can throw considerable light on the physical and chemical processes responsible for the origin of natural carbonaceous materials.

[1] Most of the information in this paper was presented on November 3, 1954, at the Los Angeles meeting of the Geological Society of America. Manuscript received, September 3, 1957.

[2] California Research Corporation.

[3] Division of the Geological Sciences, California Institute of Technology.

The writers are indebted to K. O. Emery and W. L. Orr of the Geology Department, University of Southern California, for supplying the Santa Barbara basin cores, to R. B. Park of the Division of Biological Sciences, California Institute of Technology, for making available the modern plant fractions analyzed in this survey, and to W. H. Bradley, chief geologist of the U. S. Geological Survey, for his helpful comments on the paleolimnology of Lake Uinta. Thanks are also due to the various petroleum companies who provided the petroleum samples. The writers are grateful for Atomic Energy Commission support under Contract No. AT(11-1)-208 to the California Institute of Technology.

The present study of carbon isotopic compositions of petroleums and related organic deposits was undertaken to determine whether isotopic fractionation occurs during petroleum formation or petroleum maturation. The carbon isotopic composition of petroleums is a relatively unexplored phase of the carbon cycle. West (1945) analyzed the normal propane and normal butane fractions of petroleums from oil fields in Kansas, Oklahoma, Texas, and Wyoming. His results implied that C-13/C-12 ratios of these samples are essentially constant within his experimental error (± 1 per cent). West found, however, that the C-13/C-12 ratios of hydrocarbons from well cuttings are slightly higher than those of the petroleum fractions. He attributed these differences to selective adsorption and evaporation.

Craig's (1953) summary includes analyses of two petroleum samples, one natural asphalt, and the organic carbon of several shales. The C-13/C-12 ratios of the petroleum samples differed by slightly more than 6 per mil (parts per thousand). More recently, Wickman (1956) measured C-13/C-12 ratios of ten petroleums from widely separated localities and also found a 6 per mil range in C-13/C-12 ratios.

EXPERIMENTAL PROCEDURE

In this study, samples were converted to carbon dioxide according to the procedure described by Craig (1953). This involves combustion over copper oxide in an oxygen atmosphere at 800°–900°C. and collection of the carbon dioxide at liquid nitrogen temperature. Petroleum samples containing volatile constituents were combusted in small Vycor tubes open at one end. The tube containing the sample was frozen in liquid nitrogen and inserted into the cold part of the combustion apparatus. The system was sealed and evacuated rapidly (to 10^{-3} mm Hg pressure) while the sample was frozen. Oxygen was admitted into the system and the sample tube then was moved into the hot part of the combustion tube by means of an iron plunger magnetically controlled from outside the tube. An automatic Toepler pump was used to cycle the gases through the copper oxide-packed combustion tube. Combustion products were condensed in a trap cooled with liquid nitrogen, and the carbon dioxide was purified by condensing contaminants in a trap cooled with a dry ice-chloroform-carbon tetrachloride mixture.

The carbon dioxide was analyzed in a Nier, 60°, sector-type mass spectrometer, which incorporates the modifications described by McKinney *et al.* (1950) for measuring small variations in C-13/C-12 and O-18/O-16 ratios to a precision of ± 0.1 per mil. Mass spectrometric analyses are reported as per mil deviations (δ-values) of the C-13/C-12 ratio of a sample from that of the carbon dioxide prepared from a Cretaceous belemnite, *Belemnitella americana*, from the Peedee formation of South Carolina. This is the same carbonate standard used by the University of Chicago group in their study of C-13/C-12 and O-18/O-16 ratio variations in nature (Urey *et al.*, 1951). C-13/C-12 ratio differences are reported as δ-values, which are defined as:

$$\delta \text{ in per mil } (^\circ/_{\circ\circ}) = \left(\frac{\text{C-13/C-12 sample} - \text{C-13/C-12 standard}}{\text{C-13/C-12 standard}}\right) \times 1000.$$

Positive and negative δ-values indicate higher and lower C-13/C-12 ratios, respectively, than that of the standard. The reported δ-values have been corrected for O-17 contribution to the mass 45-beam intensity and for the error that results from mixing of sample and standard gases at the analyzer tube inlet (see Craig, 1953, p. 56).

The present survey includes a study of C-13/C-12 variations in chromatographically separated petroleum fractions. These fractions were prepared by permitting a crude oil to penetrate a silica-gel-packed column for 11 months. Samples

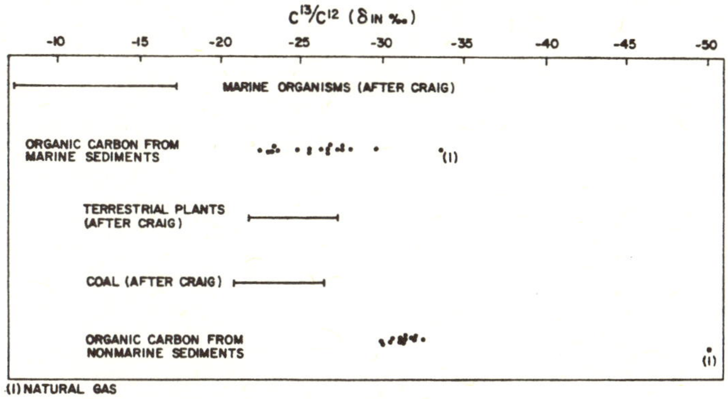

FIG. 1.—C-13/C-12 δ-values of various organic materials.

were prepared by extruding the packing and cutting it into seven sections. The adsorbed hydrocarbons in each section were extracted with carbon disulfide, which was later removed by evaporation. After removal of solvent, the extracts were combusted in the manner described previously for preparation for C-13/C-12 analysis. In addition, the infrared absorption spectrum of each extract was obtained with a Perkin-Elmer Model 21 spectrometer to provide information on the chemical compositions of the fractions and to verify complete removal of solvent.

DISCUSSION OF RESULTS

Approximately fifty crude oils and other sedimentary organic materials were analyzed in the present study. They exhibit a 30-per-mil range in isotopic composition. Results of these analyses are summarized and compared with Craig's data for organisms and coals in Figure 1. Petroleums alone, from various geographic localities and geologic ages, occupy a more restricted (10 per mil) range of isotopic composition. From the limited number of petroleum samples analyzed, it appears that C-13/C-12 ratios are related to environmental conditions

TABLE I. CARBON ISOTOPIC COMPOSITION OF ORGANIC MATTER OF MARINE ORIGIN

Number	*Sample*	*Age*	*C-13/C-12* δ(°/₀₀)
—	Marine plants	Modern	−12.9[1]
—	Marine invertebrates	Modern	−13.3[2]
9654	Extract—organic shale, Calif.	Pliocene	−22.2
9652	Total organic matter—organic shale, Calif.	Miocene	−22.8
9493	Southern Segregated crude oil, Calif.[3]	Miocene and Pliocene	−23.0
11580	Crude oil—Oxnard field, Calif.	Oligocene	−23.1
2622-C	Extract—recent offshore sediment, Santa Barbara basin, Calif.	Recent	−23.3
11705	Crude oil—Bay Marchand field, La.	Miocene	−24.6
11061	Crude oil—Minas field, Sumatra	Miocene	−25.3
12929	Crude oil—Carraño Valley field, Venez.	Eocene	−25.3
9488	Seep oil—Whiskey Creek, Utah	Eocene	−26.0
9960-41	Extract—recent offshore sediment, Gulf Coast, La.	Recent	−26.4
10012	Crude oil—Lagunillas field, Venez.	Miocene	−26.4
11730	Crude oil—Barataria field, La.	Miocene	−26.5
13327	Crude oil—Bents Fort field, Colo.	L. Pennsylvanian	−27.0
9368	Crude oil—Rangely field, Colo.	U. Cretaceous	−27.3
13125	Crude oil—Gas City field, Mont.	Silurian	−27.3
9351	Crude oil—Ashley Valley field, Utah	Pennsylvanian	−27.8
11325	Crude oil—Cabin Creek field, W.Va.	Mississippian	−29.2
12915	Crude oil—Poplar field, Mont.	Mississippian	−29.4
200-SP	Gas—South Coles Levee field, Calif.	Miocene	−33.4

[1] Average value; range is −7.6 to −16.5 °/₀₀ (after Craig, 1953).
[2] Average value; range is −10.2 to −17.4 °/₀₀ (after Craig, 1953).
[3] Southern Segregated crude oil is a mixture containing major amounts of West Coyote, Seal Beach, Signal Hill, and Santa Fe Springs productions and minor amounts of Rosecrans and Alondra productions. All of these fields are in Los Angeles County, California.

under which the source materials accumulated. Petroleums and organic extracts derived from marine sediments have δ-values ranging from −22.2 to −29.4 per mil (Table I). Oils and bituminous materials derived from non-marine sediments (Green River shale) range from −29.9 to −32.5 per mil (Table II). The locations and producing depths of the oils and bitumens are given in Table III.

A similar difference exists in the isotopic compositions of present-day non-marine and marine organism. Plants growing in marine environments have higher C-13/C-12 ratios than do terrestrial plants. Wickman (1952) discusses these differences and their relations to the carbon dioxide cycle. He concludes

TABLE II. CARBON ISOTOPIC COMPOSITION OF ORGANIC MATTER OF NON-MARINE ORIGIN

Number	*Sample*	*Age*	*C-13/C-12* δ(°/₀₀)
—	Fresh-water plant (*Rhizoclonium* sp.)	Modern	−22.2[1]
—	Coal	Various	−23.7[2]
—	Land plants (wood)	Modern	−24.9[3]
—	Land plants (leaves and plants)	Modern	−25.8[4]
8878	Gilsonite, Utah	Eocene	−29.9
10886	Crude oil—Duchesne field, Utah	Eocene	−30.0
5429	Extract—Green River shale, Utah	Eocene	−30.5
11521	Crude oil—Brennan Bottom field, Utah	Eocene	−30.6
—	Retorted oil—Green River shale, Utah	Eocene	−31.0
11066	Crude oil—Red Wash field, Utah	Eocene	−31.0
10885	Crude oil—Duchesne field, Utah	Eocene	−31.2
10424	Crude oil—Red Wash field, Utah	Eocene	−31.3
10884	Crude oil—Duchesne field, Utah	Paleocene	−31.3
10169	Crude oil—Flat Mesa field, Utah	Eocene	−31.7
—	Total organic matter—Green River shale	Eocene	−31.9
8764	Ozocerite, Utah	Paleocene	−32.0
6337	Ozocerite, Utah	Paleocene	−32.5
11342	Gas—Red Wash field, Utah	Eocene	−49.3

[1] After Craig, 1953.
[2] Average value, range is −21.1 to −26.7 (after Craig, 1953).
[3] Average value, range is −22.0 to −27.4 (after Craig, 1953).
[4] Average value, range is −23.1 to −28.6 (after Craig, 1953).

TABLE III. LOCATIONS AND PRODUCING DEPTHS OF SAMPLES ANALYZED

Sample No.	*Well*	*Field*	*Location*	*Producing Depth (Feet)*	*Producing Formation*
9654	General Petroleum, McNally 1	—	Los Angeles Co., Calif.	11,102–11,110*	Pliocene shale
9652	Standard of California, Rowland Estate 1	Puente Hills	Los Angeles Co., Calif.	4,196–4,211*	Lower Puente shale
9493	Standard of California, Southern Segregated Crude Oil**	—**	Los Angeles Co., Calif.	—	—
11580	Standard of California, Lucas R. D. 1	Oxnard	Ventura Co., Calif.	8,727–8,835	Sespe
11705	The California Company, Louisiana State 1367 E-9	Bay Marchand	Lafourche Parish, La.	6,213–6,228	Miocene sand
11061	Minas 1	Minas	Pakanbaroe region, Sumatra	2,107–2,121	Minas
12929	Surface oil seep	Carraño Valley	State of Trujillo, Distr. of Escuque, Venez.	Surface	Misoa
9488	Whiskey Creek seep oil	—	Garfield Co., Colo.	Surface	Wasatch
10012	Composite—number of unspecified wells	Lagunillas	State of Zulia, Distr. of Bolivar, Venez.	Unknown	Lagunillas Tar sand
11730	The California Company, Fleming Plantation A-6	Barataria	Jefferson Parish, La.	8,182–8,198	Miocene "A" Sand
13327	M. J. Stransbury Cordes, Oberlander 1	Bents Fort	Bents Co., Colo.	4,619–4,632	Cherokee
9368	R. and H. California Co. Lease	Rangely	Rio Blanco Co., Colo.	Unknown	Mancos
13125	Shell Gas City 33-21	Gas City	Dawson Co., Mont.	8,619–8,676	Interlake
9351	Equity Ashley Valley 1	Ashley Valley	Uintah Co., Utah	4,122	Weber
11325	Pure Oil William Coal 202	Cabin Creek	Boone and Kanawka Cos., W.Va.	3,000	Berea
12915	C. H. Murphy Unit 40	Poplar	Roosevelt Co., Mont.	5,699–5,719	Charles
200-SP	Standard of California, SCL 76-5	South Cole's Levee	Kern Co., Calif.	8,940–9,500	Stevens
8878	Gilsonite	—	Uintah Co., Utah	Surface	Green River
10886	Carter Eyner Neilson 1	(Duchesne)	Duchesne Co., Utah	4,520–4,600	Green River
11521	Gulf Brennan-Federal 1	Brennan Bottom	Uintah Co., Utah	6,865–7,495	Green River
11066	The California Company, Red Wash 16	Red Wash	Uintah Co., Utah	5,458–5,529	Green River
10885	Carter, Ute Tribal 2	Duchesne	Duchesne Co., Utah	7,552–7,661	Green River
10424	The California Company, Red Wash 4	Red Wash	Uintah Co., Utah	5,606–5,710	Green River
10884	Carter, Ute Tribal 1	Duchesne	Duchesne Co., Utah	7,498–7,549	Wasatch
10169	Carter, Joseph Smith 1	Flat Mesa	Duchesne Co., Utah	8,820–9,103	Green River
8764	Ozocerite	—	Carbon Co., Utah	Surface	Wasatch
6337	Ozocerite	—	Carbon Co., Utah	Surface	Wasatch
11342	The California Company, Red Wash 2 (gas)	Red Wash	Uintah Co., Utah	4,150–4,200	Green River

* Core interval.
** Southern Segregated crude oil is a mixture containing major amounts of West Coyote, Seal Beach, Signal Hill, and Santa Fe Springs productions and minor amounts of Rosecrans and Alondra productions.

that the preferential assimilation of C-12O_2 over C-13O_2 during photosynthesis is influenced by the plant's local environment. His analyses show the average C-13/C-12 ratio of marine hydrophytes to be about 9 per mil higher than the average value for lacustrine plants. Craig (1953) reports a range of -7.6 to -16.5 per mil for marine plants and a range of -22.0 to -27.4 per mil for terrestrial plants. Craig also finds that the organic carbon of fresh-water species containing no carbonate is isotopically similar to terrestrial plant carbon. The organic carbon of fresh-water hydrophytes containing carbonate is higher in C-13 content corresponding with the proportion of associated carbonate. Craig regards this as evidence that the organic carbon of marine plants is partly derived from bicarbonate ions. Wickman (1952) also mentions this possibility, although both are aware that plant physiologists are not convinced of bicarbonate utilization in photosynthesis.

These considerations are pertinent to the present study because dolomite and calcite are the predominant mineral constituents of the Green River shale (Bradley, 1931, pp. 29–30). Thus, Craig's data relating high C-13/C-12 ratios of fresh-water plants with associated carbonates suggest that the organic material in Green River shale was derived largely from terrestrial plant debris carried into ancient Lake Uinta. This possibility is dismissed by Bradley (1931) who states (pp. 52–53):

The evidence appears conclusive that the structureless material which makes up by far the greater part of the total organic matter of the Green River oil shale was derived from the partial putrefaction of aquatic organisms that grew in the lakes. The organic matter is most abundant in the center of the lake basin and diminishes in all directions toward the shore phases, from which it is practically absent. Aquatic microorganisms dominate the microflora and microfauna of the oil shale. Fragments of the tissues of higher land plants are exceedingly rare, though their spores and pollens are not uncommon. And finally, . . . there is a remarkably close parallelism between the oil shale and recent lacustrine organic oozes or sapropels.

Bradley concludes, after comparisons with organic oozes from modern lakes, that blue-green algae were the dominant inhabitants of Lake Uinta. Modern blue-green algae inhabit the surfaces of stagnant waters. The abundance of fungi (Bradley, 1931, p. 51) identified in Green River shale also indicates that the organic ooze of Lake Uinta was exposed to the air during some periods, at least, of its existence. Such accumulations in the surface layer (epilimnion) of a stagnant lake are characteristic of thermally stratified lakes. Bradley (1948) gives a detailed account of the factors responsible for temperature stratification of lakes during the summer season. Under these conditions, the epilimnion is isolated from the deeper, colder layer (the hypolimnion). The higher temperatures of the epilimnion tend to decrease the solubility of calcium and magnesium carbonates. Photosynthesis by the algae depletes the carbon dioxide content of the epilimnion, thereby raising the pH of its waters and accelerating the precipitation of carbonates. The carbonate particles settle out of the epilimnion into the hypolimnion. These deeper waters, which are acidic because of their high carbon dioxide and hydrogen sulphide concentrations, can dissolve the carbonate and further raise the carbon dioxide and bicarbonate ion concentrations in the hypolimnion. Such changes may increase the stability of thermal stratification and create a condition in which the algae live in relatively bicarbonate-free waters during most of the biologically productive season. In this manner, Bradley's conception of Eocene paleolimnology in the Rocky Mountain region allows for a plausible explanation for the co-existence of low C-13/C-12 ratio (−31.9 per mil) organic matter and high ratio (+2.0 per mil) carbonate in the Green River shale. Wickman's (1952) data for lacustrine plants support this interpretation insofar as the C-13/C-12 ratios of plants living in stagnant water average 8.8 per mil lower than those of plants from non-stagnant water. It appears, therefore, that the observed differences in C-13/C-12 ratios of marine and non-marine oils can be inherited from initial differences in the source materials. The detailed consideration of Lake Uinta paleolimnology implies that these differences may be more directly related to the degree of water circulation than to the salinity of the water. Unfortunately, the only non-marine petroleums or gases available in the present study were those produced from the Green River shale. Analyses of other petroleums whose non-marine origin is assured by adequate geologic control are needed to confirm this possible explanation.

Natural gases from petroliferous areas are isotopically lighter (lower C-13 content) than petroleums. A gas sample produced from marine sediments (Stevens sand) in the South Cole's Levee field, California, has a δ-value of −33.4 per mil. Gas derived from nonmarine shale (Green River shale) in the Red Wash field, Uinta Basin, Utah, analyzed −49.3 per mil. Apparently, the C-13/C-12 ratio of natural gas is related to its source in the same manner that petroleum ratios are; that is, gases derived from non-marine sources are isotopically lighter than those of marine origin. It is also significant that the isotope ratios of gases are very much lower than those of petroleums. On the assumption that gases from petroliferous areas are derived from petroleum, this implies that gas formation involves a relatively large isotope fractionation. Inasmuch as oils of wide latitudes of age, composition, and location show a relatively small range of isotopic composition (10 per mil), it may be concluded that the oil pools represented by the samples analyzed in the present survey, as well as those analyzed by Craig (1953) and Wickman (1956), were not markedly depleted by gas formation since their accumulation. Relatively minor losses of gas, however, may be partly responsible for the existing isotopic differences in petroleums.

Another important relation indicated by the present study is that petroleum is isotopically lighter, by about 10 per mil, than the organisms from which it is supposedly derived. If the carbon isotope distribution in ancient organisms was equivalent to that in modern fauna, then their transformation into petroleum must be accompanied by loss of materials relatively rich in C-13. Loss of heavy carbon can occur during the transformation by elimination of C-13-rich materials and preservation of C-12-rich products. On the other hand, petroleum may be formed by selective accumulation of specific C-12-rich constituents of organisms. For example, if fatty acids and hydrocarbons, which are present in minute concentrations in plants and animals, have lower C-13/C-12 ratios than the relatively less stable carbohydrates, and if petroleum is derived exclusively from fatty acids or hydrocarbons, then isotope fractionation is not necessarily involved in petroleum formation. This possibility is in accord with current theories claiming that petroleum is derived from fatty acids (Brooks, 1952) or hydrocarbons (Smith, 1954) originally present in the source material.

Some evidence along these lines exists in the work of R. O. Belkengren (cited by Craig, 1953), who reported that the C-13 content of plant fats is about 10 per mil lower than that of plant cellulose. Because of the importance of this information to the understanding of petroleum origin, a brief study was made of the carbon isotopic compositions of modern plant constituents. The ether-soluble fractions of three varieties of marine algae and one terrestrial plant were separated and analyzed. This fraction is known as the "lipid" fraction. Lipids are an important family of biochemical constituents of organisms; they are distinctly different from the carbohydrates and proteins. In a strict sense, lipids are esters of higher aliphatic acids, but the terms "lipid fraction" or "fat-soluble fraction" generally refer to all ether-soluble organic matter in plants. This includes, in addition to the aliphatic esters, a wide variety of water insoluble compounds such as hydrocarbons, fatty (olefinic) acids, aliphatic acids, aliphatic and olefinic alcohols, aliphatic and olefinic glycols, hydroxy-acids and their esters, terpenes, carotene pigments, and the "essential oils" (derivatives of cyclic hydrocarbons). As a group, the lipid fraction is the most stable fraction of plants and animals and therefore is the portion of organic debris most likely to survive geologic conditions. It includes the fatty acids, which Brooks (1952) believes to be the source of petroleum, and the hydrocarbons, which Smith (1954) claims to be the source of petroleum.

Preliminary results of this study indicate that the δ-values of the lipid fractions of green algae (*Ulva* sp.), red algae (*Gelidium* sp.), brown algae (*Macrocystis* sp), and tomato plant (*Lycopercicum* sp.) are 7.1, 3.5, 7.9, and 4.9 per mil lower, respectively, than those of the entire plants. Thus the C-13/C-12 ratios of plant lipids may be sufficiently similar to that of petroleum so that extensive isotope fractionation is not necessary if lipids are the sole source of petroleum. This observation is examined further in the discussion of organic matter in Recent sediments.

Although no consistent relation between C-13/C-12 ratio and age of petroleum can be recognized, geologically older (Paleozoic) oils of marine origin generally have lower ratios than do the Tertiary oils. The present survey offers no adequate explanation for these differences. It may be that the carbon isotope ratios of marine organisms were lower during Paleozoic time or that maturation processes reduce the C-13 content of petroleums. Analyses of many more Paleozoic and younger oils are needed before a reliable interpretation of the influence of age or maturity on isotopic composition of petroleums will be forthcoming. An extended study may reveal also whether ratios below 30 per mil are restricted to non-marine oils or whether age and petroleum composition can also be responsible for low δ-values.

INFLUENCE OF CHEMICAL COMPOSITION

It appears, therefore, that many factors may be responsible for the range of isotope ratios noted for petroleums. For example, Barton (1937) and McNab, Smith, and Betts (1952) observed that geologically older oils tend to be more paraffinic than younger oils. This generalization, plus the fact that Green River formation oils are highly paraffinic, suggests that chemical composition may have a strong influence on the carbon isotope ratios in oils. The analyses listed in Table I, however, do not substantiate this possibility. For example, Minas crude, which is as paraffinic as any, and more paraffinic than most, of the Green River formation oils, has a relatively high δ-value (−25.3 per mil). This matter was investigated further by an experimental study to establish the relation between classes of hydrocarbon constituents and isotopic composition.

Although petroleums are complex chemical mixtures, they are composed mainly of three classes of hydrocarbon compounds, namely, paraffins, aromatics, and naphthenes. Aromatics can be separated readily from paraffins and naphthenes by established procedures. One method, chromatographic separation on silica gel, was used to separate these constituents from a California crude oil (Southern Segregated crude). The fractions prepared by this method, and the measured C-13/C-12 δ-values of the fractions are shown in Figure 2. In addition, each fraction was analyzed by infrared absorption spectrometry to ascertain the extent of separation. As described in the section on experimental procedure, the silica gel column was divided into seven sections. These sections are designated B, C, D, E, F, G, and H, consecutively, from top to bottom of the column (Fig. 2). Fraction A refers to the unadsorbed liquid remaining above the silica packing.

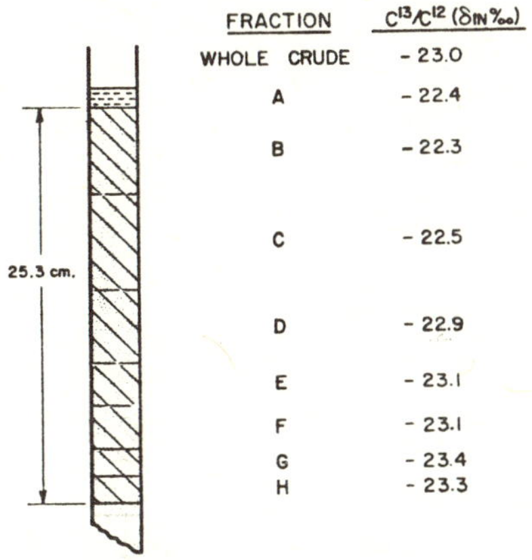

FIG. 2.—C-13/C-12 δ-values of chromatographic fractions prepared from Southern Segregated crude

The infrared spectra of the fractions (Fig. 3) reveal that aromatic compounds (indicated by absorption maxima at 6.25, 12.35, 13.45, and 14.35 μ) are most abundant (more than 30 per cent) in fraction B (top of column) and essentially absent in fraction H (bottom of column). Conversely, the intensity of the straight-chain paraffin peak at approximately 13.85 μ indicates lowest paraffin content in fraction B and highest paraffin content in fraction H.

The C-13/C-12 δ-values of the chromatographic fractions (Fig. 2) show a progressive decrease with increasing paraffinic (decreasing aromatic) character. Fraction B, highest in aromatic content, is 0.7 per mil heavier than the whole crude. The C-13/C-12 δ-value of the aromatic-free fraction (H) is 0.3 per mil lower than that of the whole crude. Thus, the over-all difference between aromatic-enriched and aromatic-free fractions is only one per mil.

The tendency for paraffins to be slightly enriched in C-12 content was substantiated by a study of waxes separated from crude petroleums. C-13/C-12 δ-values of waxes separated under natural conditions and also by standard laboratory procedures are shown in Table IV. Natural separation results from crystallization of high-melting-point constituents because of reduced temperatures, relative to reservoir temperatures, during storage or production. Infrared absorption

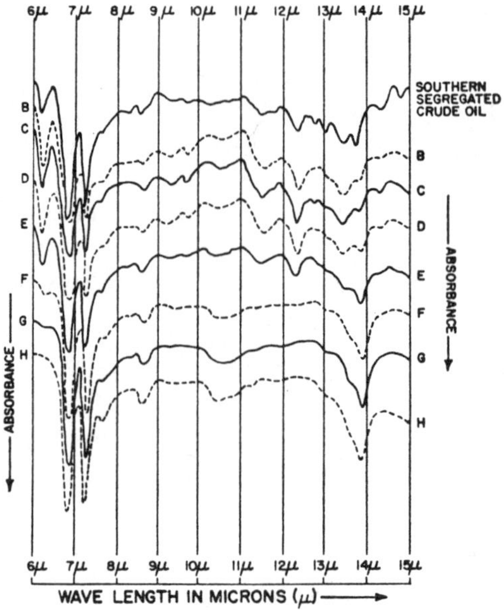

FIG. 3.—Infrared spectra of Southern Segregated crude and chromatographic fractions B to H.

studies of the separated waxes in Table IV reveal them to be composed almost entirely of paraffinic hydrocarbons. The wax segregates are consistently lower (by 0.2 to 0.9 per mil) in C-13 content than the fluid oil.

The chromatographic and wax segregate data reveal that C-13/C-12 ratio differences between aromatic and paraffinic constituents of oils are not greater than one per mil. Thus, chemical compositional differences of petroleums can not be entirely responsible for the 10-per-mil range of variations noted for all crude oils analyzed. This information also leads to the conclusion that the chemical processes responsible for the transformation of source material to petroleum, if this transformation occurs, differ from the processes responsible for the specific chemical compositions (proportions of different hydrocarbon groups present) of petroleums.

TABLE IV. C-13/C-12 δ-VALUES OF FLUID OIL AND WAX SEGREGATES OF VARIOUS PETROLEUMS

Sample	*Fluid Oil C13/C12* δ(°/₀₀)	*Solid Wax Segregate C13/C12* δ(°/₀₀)
Royal Coal Mine oil, Carbon Co., Utah	−26.3	−27.2
Joseph Smith No. 1 crude oil, Duchesne Co., Utah	−31.7	−31.9
Ute Tribal No. 1 crude oil (Duchesne Unit), Duchesne Co., Utah	−31.3	−31.9*
Eyner Neilson No. 1 crude oil, Duchesne Co., Utah	−30.0	−30.4*

* Chemically separated wax.

ORGANIC MATTER IN RECENT SEDIMENTS

In view of the isotopic similarity between petroleum and the lipid fractions of plants, it was desirable to determine C-13/C-12 ratios of organic matter extracted from Recent offshore sediments. Such analyses may provide further information regarding the role of lipid materials in petroleum formation. If fatty acids or plant hydrocarbons accumulate after burial to form petroleum, then the organic matter extracted from Recent sediments should have a C-13/C-12 ratio that does not differ markedly from that of the final petroleum accumulation. On the other hand, if transformation of non-lipid plant debris is involved in petroleum formation, the C-13/C-12 ratios of Recent sediment extracts should vary according to the extent by which such transformation has progressed.

A calcareous mud core from the south side of Cross Bank, under 3 feet of water, in Florida Bay (Florida Keys) was the first Recent sediment core selected

TABLE V. C-13/C-12 δ-VALUES AND CONCENTRATIONS OF EXTRACTABLE ORGANIC MATTER, FLORIDA OFFSHORE CORE*

Core Interval (*Inches from Surface*)	*Extractable Organic Matter* (*Grams per 100 Grams of Dried Sediment*)	*C-13/C-12* $\delta(°/_{oo})$
60–66	0.07	−20.3
66–69	0.17	−21.6
69–73	0.47	−26.2
73–80	0.25	−26.0

* Location—80°33′25″ E. Long.
24°59′45″ N. Lat.

for carbon isotope studies. A 20-inch interval (from 60 to 80 inches below the top of the core) was divided into four sections. Each section was extracted with a 4:1 chloroform-methanol mixture which was removed by evaporation before the extract was analyzed. The results shown in Table V indicate a progressive increase in extractable organic matter and a progressive decrease in C-13/C-12 ratio with relatively small increase in depth. The 69–73-inch interval contains peat layers which are composed almost entirely of marine plant (*Thallasia testodenarius*) remains. Craig's analysis of a present-day specimen of this plant is −12.0 per mil. The remaining three sections are composed of light gray calcareous mud. This difference probably accounts for the exceptionally high extractable organic content of the 69–73-inch section.

Craig (1953) reports that the total organic matter in the surface mud off Lignum Vitae Island in the Florida Keys averages −13.5 per mil. Thus, all available C-13/C-12 data for the Florida samples indicate that the isotopic composition of extracted organic matter undergoes a change with small increased depth of burial which is comparable in direction and magnitude with the previously noted difference between organisms and petroleum.

Extracts of the two other offshore cores, one from the Santa Barbara Basin, California, and the other from the Gulf of Mexico near the Louisiana coast, analyzed −23.3 and −26.4 per mil, respectively. The Santa Barbara basin samples are isotopically similar to California petroleums (Table I). The C-13/C-12 ratio of the Louisiana sample also is similar to the δ-values reported for Southern Louisiana crude oils (sample Nos. 11730 and 11705, Table I). This sample represents the total extractable content of the top 1.5 feet of a core obtained from the Eugene Island area (28° 15′ 53° N. Lat., 91° 45′ 12° E. Long.).

The quantity and isotopic composition of extractable organic matter in the Santa Barbara basin are not related to depth of burial in the manner noted for the Florida Bay samples. In the Santa Barbara basin the extractable matter decreases and C-13/C-12 ratios remain constant with increasing depth (Table VI).

Thus, whereas the Florida core data suggest that organic matter at shallow depths is undergoing compositional changes which may be construed as transformation of source material to petroleum, this change does not occur in the California and Louisiana cores. An evaluation of these differences is hampered by the known occurrence of petroleum in Louisiana offshore areas and ocean-bottom tar seeps in some California offshore localities (Emery and Rittenberg, 1952, p. 784). It is not possible, at this time, to exclude the possibility that core extracts in these areas may be subject to contamination from deep-seated petroleum accumulations. Also, the possible influence of inorganic mineralogical differences among the sediments should be considered in connection with the observed isotopic differences. For example, Brooks (1950) proposes that petroleum formation by means of carbonium ion reactions is catalyzed by acid silicate minerals. Limestone reservoirs do not provide such catalytic action; therefore, transformation of organic matter in limestone muds would be retarded relative to transformation in a clay or silicate mineral matrix.

TABLE VI. C-13/C-12 δ-VALUES AND CONCENTRATIONS OF EXTRACTABLE ORGANIC MATTER, SANTA BARBARA BASIN CORE*

Core Interval (Inches from Surface)	*Extractable Organic Matter (Grams per 100 Grams of Dried Sediment)*	C-13/C-12 δ(°/₀₀)
6–8	0.38	−23.3
36–38	0.34	−23.3
76–78	0.24	−23.3

* Location—120°01′55″ W. Long.
34°13′28″ N. Lat.
Water depth is 328 fathoms.

Landergren (1954) analyzed the total organic matter in a number of deep-sea cores from the Pacific Ocean and obtained δ-values ranging from −24.6 to −19.5 per mil. These values are below the known range for marine organisms and approach the carbon isotopic compositions of petroleums and organic matter from ancient marine sediments. Landergren's deep-sea core data (Landergren, 1954, Table VI, p. 109) include analyses of organic matter from two depths in a single core (Core 70); C-13/C-12 δ-values of organic matter from the 250.0 to 251.5 cm and 270.0 to 271.5 cm. intervals are −20.8 and −22.8 per mil, respectively. Landergren's values for two cores (Nos. 71 and 72) from the same locality are −19.5 per mil for combined 290.0 to 291.5 and 750.0 to 751.5-cm depths, and −23.3 per mil for the 1350.0 to 1351.5-cm interval. These relations imply that C-13 content of organic matter in Pacific deep-sea sediments decreases with depth.

Although the data obtained from Recent sediment extracts do not conclusively establish the relative importance of direct accumulation of lipid materials as compared with transformation after burial, they indicate that, if transformation and isotopic fractionation are involved, they occur at an early stage in the history of petroleum formation. Until additional data are obtained, it appears that at least two different processes can be involved in petroleum formation. One process involves transformation of lipid substances without carbon isotope fractionation; the other involves isotope fractionation and transformation of non-lipid materials.

ORGANIC MATTER IN ANCIENT SEDIMENTS

The organic matter contained in ancient shales was analyzed to determine the relation between their isotopic compositions and those of associated petroleums. Two California shales, one of Miocene and the other of Pliocene age, were selected for this study. The organic material contained in these shales is not homogeneous. Only about 2 per cent of the total organic content is soluble in a 4:1 chloroform-methanol mixture. This soluble fraction, however, is isotopically similar to the insoluble organic fraction. A Pliocene shale from 11,100 feet in General Petroleum's McNally well No. 1, Los Angeles County, California, was treated with hydrochloric acid to remove carbonate carbon, and the residue was converted to carbon dioxide and analyzed. This carbon dioxide gave a δ-value of −22.3 per mil. The chloroform-methanol extract of another portion of the same sample analyzed −22.2 per mil. These values are similar to the −22.4 per mil δ-value reported by Craig (1953) for a Pliocene oil from South Belridge field in California.

The total organic matter in a Miocene shale from Standard of California's Rowland Estate well No. 1 (4,200-foot depth) has a δ-value of −22.8 per mil. This value also falls into the range of C-13/C-12 δ-values noted for California petroleums (Table I).

Chloroform-methanol extracts of Green River shale (Eocene age), Utah, analyzed −30.5 per mil. Oil produced by retorting Green River shale at 130°C. and reduced pressures for extended periods of time has a similar C-13/C-12 ratio (−31.0 per mil). The ratio of total organic matter in the shale was found to be slightly lower (−31.9 per mil). All of these values, however, are within the range of δ-values reported (Table II) for Eocene oils from the Uinta basin, Utah. These analyses indicate that carbon isotopic compositions of the various types of organic matter in ancient shales are similar to those of related petroleums.

This observation is in accord with the previous conclusion that either petroleum generation is not accompanied by isotope fractionation or that such fractionation occurs at an early stage of the formation process.

SUMMARY

C-13/C-12 analyses of petroliferous materials permit certain conclusions concerning the origin and evolution of petroleum. The fact that the magnitude of C-13/C-12 ratio differences between marine and terrestrial organisms is similar to the differences between petroleums geologically related to marine deposits and those related to non-marine sediments provides additional confirmation that petroleum can be derived from non-marine as well as marine organic sources. The lower C-13 content of petroleums, relative to that of organisms, implies that not all available biologic carbon is converted to petroleum. This requires that C-13-rich material must be lost in some manner. The change in isotopic composition may result from isotope fractionation during the conversion process. This apparent C-13 depletion can also be explained by selective preservation of the lipid fraction of plants which is isotopically lighter than plant cellulose. Analyses of soluble organic matter in Recent sediments do not resolve this question, but imply that both mechanisms may be involved. Perhaps a thorough study of the nature and distribution of organic matter in modern sediments will provide definitive answers to these important questions.

Analyses of chromatographic fractions of petroleum and soluble and insoluble organic constituents of ancient shales indicate that changes in chemical composition do not produce marked changes in isotopic composition. Gases occurring in petroliferous areas, however, are considerably lighter (lower C-13/C-12 ratios) than liquid petroleums. Thus a pronounced isotopic fractionation occurs if the gas is derived from petroleum. In view of the relatively narrow range of C-13/C-12 ratios in petroleums of widely different chemical compositions, it appears unlikely that any of the petroleums analyzed in this survey, or those analyzed by Craig (1953) and Wickman (1956), have released large amounts of gas. This implies that the conversion of heavier hydrocarbons into gas and subsequent loss of gas during geologic time is not an important mechanism for destroying oil accumulations. These interpretations, based on a brief exploratory survey of stable carbon isotope abundances in petroliferous materials, indicate that much valuable organic geochemical information can be derived from further carbon isotope studies.

REFERENCES

BARTON, D. C., 1937, "Evolution of Gulf Coast Crude Oil," *Bull. Amer. Assoc. Petrol. Geol.*, Vol. 21' pp. 914–46.

BRADLEY, W. H., 1931, "Origin and Microfossils of the Oil Shale of the Green River Formation of Colorado and Utah," *U. S. Geol. Survey Prof. Paper 168*, pp. 1–58.

———, 1948, "Limnology and the Eocene Lakes of the Rocky Mountain Region," *Bull. Geol. Soc. America*, Vol. 57, pp. 635–48.

BROOKS, B. T., 1950, "Catalysis and Carbonium Ions in Petroleum Formation," *Science*, Vol. 111, pp. 648–50.

———, 1952, "Evidence of Catalytic Action in Petroleum Formation," *Indus. Eng. Chem.*, Vol. 44, pp. 2570–77.

Craig, H., 1953, "The Geochemistry of the Stable Carbon Isotopes," *Geochimica et Cosmochimica Acta*, Vol. 3, pp. 53–92.

Emery, K. O., and Rittenberg, S. C., 1952, "Early Diagenesis of California Basin Sediments in Relation to Origin of Oil," *Bull. Amer. Assoc. Petrol. Geol.*, Vol. 36, pp. 735–806.

Landergren, S., 1954, "On the Relative Abundance of the Stable Carbon Isotopes in Marine Sediments," *Deep-Sea Research*, Vol. 1, pp. 98–120.

McKinney, C. R., McCrea, J. M., Epstein, S., Allen, H. A., and Urey, H. C., 1950, "Improvements in Mass Spectrometers for Measurement of Small Differences in Isotope Abundance Ratios," *Rev. Sci. Instruments*, Vol. 21, pp. 724–30.

McNab, J. G., Smith, Jr., P. V., and Betts, R. L., 1952, "The Evolution of Petroleum," *Indus. Eng. Chem.*, Vol. 44, pp. 2557–63.

Murphy, B. F., and Nier, A. O., 1941, "Variations in the Relative Abundance of Carbon Isotopes," *Phys. Rev.*, Vol. 59, pp. 771–72.

Nier, A. O., and Gulbransen, E. A., 1939, "Variations in the Relative Abundance of the Carbon Isotopes," *Jour. Amer. Chem. Soc.*, Vol. 61, pp. 697–98.

Smith, Jr., P. V., 1954, "Studies on Origin of Petroleum: Occurrence of Hydrocarbons in Recent Sediments," *Bull. Amer. Assn. Petrol. Geol.*, Vol. 38, pp. 377–404.

Urey, H. C., 1947, "The Thermodynamic Properties of Isotopic Substances," *Jour. Chem. Soc. London*, pp. 562–81.

———, Lowenstam, H. A., Epstein, S., and McKinney, C. R., 1951, "Measurement of Paleotemperatures and Temperatures of the Upper Cretaceous of England, Denmark, and the Southeastern United States," *Bull. Geol. Soc. America*, Vol. 62, pp. 399–416.

West, S. S., 1945, "The Relative Abundance of the Carbon Isotopes in Petroleum," *Geophysics*, Vol. 10, pp. 406–20.

Wickman, F. E., 1952, "Variations in the Relative Abundance of the Carbon Isotopes in Plants," *Geochimica et Cosmochimica Acta*, Vol. 2, pp. 243–54.

———, 1956, "The Cycle of Carbon and the Stable Carbon Isotopes," *ibid.*, Vol. 9, pp. 136–53.

Part III

MOLECULAR ARCHITECTURE

Editor's Comments on Papers 23 Through 28

In Part II emphasis was placed on the various organic molecules. The differences between these molecules were due to differences in elemental compositions, as shown by their molecular formulas. Many organic molecules, however, have the same molecular formula and are called *isomers.* Isomers may differ in the nature or sequence of bonding of their atoms or in the arrangement of their atoms in space. The former are constitutional isomers and the latter are stereoisomers. Stereoisomers that are nonsuperimposable mirror-images are called enantiomers; stereoisomers not related in this manner are called diastereoisomers. Stereochemistry, the study of stereoisomers, is very important in explaining many of the properties and features of organic molecules, and the application of stereochemistry to organic geochemistry has added a new dimension to this subject. The six papers that follow deal with the stereochemistry of different kinds of organic molecules. The papers illustrate various conventions used to describe stereochemical relationships. Stereoisomerism in biologically derived organic compounds usually results from the presence in these molecules of asymmetric carbon atoms that have four different atoms or groups linked tetrahedrally to them. The geometry of this atomic arrangement is such that nonsuperimposable, mirror-image relationships can exist. Where more than one asymmetric center exists in a molecule, several configurational stereoisomers are possible. Maclean et al. (Paper 23) were the first scientists to establish the relative stereochemistry of isoprenoid fatty acids in geological environments. The acids studied contain two or three asymmetric carbon atoms. Two conventions are used to describe the stereochemical geometry of asymmetric carbon atoms. The older, D and L convention, relates the relative positions in space of the substituents bonded to the asymmetric carbon atom to a reference compound whose D and L stereochemistry has been arbitrarily assigned. The newer convention utilizes R (rectus = right) and S (sinister = left) to designate unequivocally the absolute configuration at an asymmetric carbon atom. In Paper 23 the former convention is used, and the stereochemistry, for example, of pristanic acid in Green River Formation was found to be a 1:1 mixture of 2(L), 6(D), 10(D), 14-tetramethylpentadecanoic acid and 2(D), 6(D), 10(D), 14-tetramethylpentadecanoic acid. Follow-up papers (Cox et al., 1970; Maxwell et al., 1973) on the stereochemistry of isoprenoid acids from the Green River Formation used the R and S convention. The absolute stereochemistry of pristanic acid was determined to be a 1:1 mixture of 2(S), 6(R), 10(R), 14-tetramethylpentadecanoic acid and 2(R), 6(R), 10(R), 14-tetramethylpentadecanoic acid. Phytol (3, 7(R), 11(R), 15-tetramethyl-hexadec-2-en-1-ol), has the appropriate stereochemistry to serve as the probable source for the isoprenoid acids.

Structural formulas can be used to show bond orientations of asymmetric carbon atoms. Thick bonds can be visualized as being directed above the plane of the paper and dotted lines indicate bonds directed below the plane of the paper. The stereoisomers of pristanic acid are illustrated.

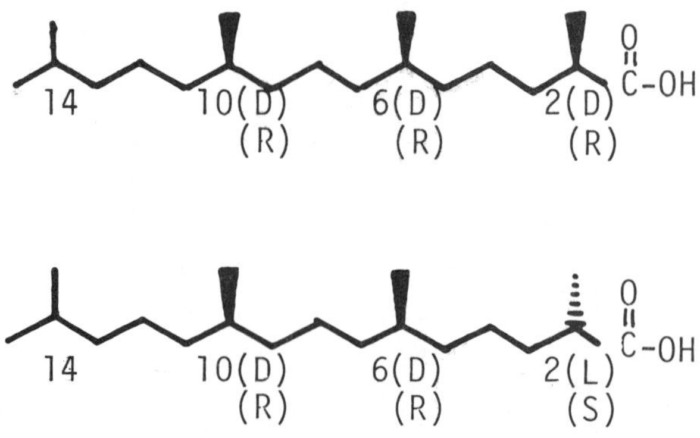

PRISTANIC ACID

In addition to isoprenoid acids, the absolute stereochemistry of the isoprenoid hydrocarbons, pristane (Maxwell et al., 1972; Cox et al., 1972) and farnesane (Brooks et al., 1977) have been determined.

Probably the first paper in the organic geochemical literature to consider the stereochemistry of a molecule is one by Hills et al. (Paper 24). They isolated the optically active pentacyclic triterpane, gammacerane, from the Green River Formation and were able to prove this triterpane's stereochemistry by the application of a gamut of techniques. The structural formula below shows the orientation of the bonds at the ring system carbon atoms to which substitutents (methyl groups or hydrogen) are attached. As with the structural formulas for isoprenoid acids, thick bonds are directed above the plane of the paper and dashed bonds below.

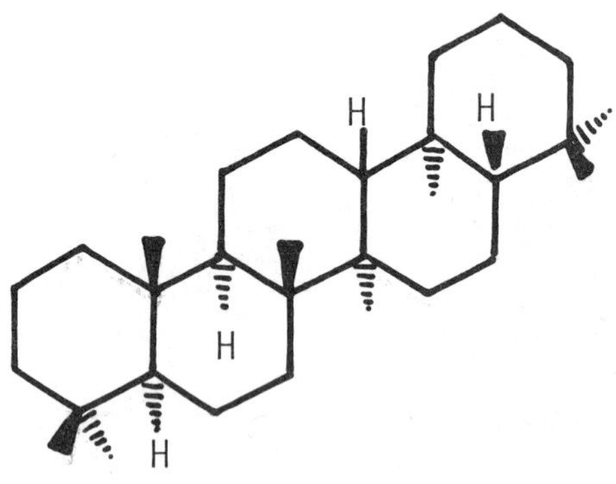

GAMMACERANE

Closely related to gammacerane are the hopanes, whose presence in the Messel oil shale has been described by Kimble et al.

(Paper 25), by Ensminger et al. (1972), and by Ensminger et al. (1974). In Paper 25, two hopane series of molecules are identified. These series differ in their stereochemistry at carbon number 17. Here another convention is used for these cyclic compounds to describe the directions of bonds. The major series of hopanes in the shale have this bond at carbon number 17 pointing above the plane of the paper. This bond direction is called β. In the minor series of hopanes the bond is located below the plane of the paper and is designated α.

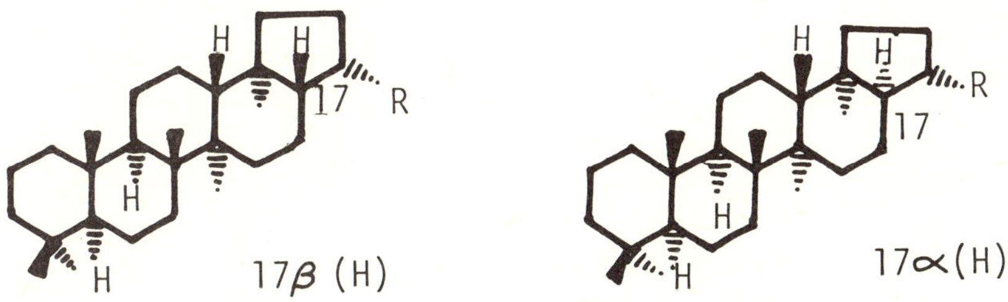

HOPANES

Because 17βH hopanes are found both in living organisms and in the shale, while 17αH hopanes have never been extracted from organisms, it is suggested that the 17αH hopanes in the shale form from 17βH hopanes through maturation processes (Ensminger et al., 1974).

The stereochemistry of steroid carboxylic acids in petroleum was determined by Seifert, Gallegos, and Teeter (Paper 26). In this paper, for example, the stereochemistry of cholanic acid isomers depends on the orientation of the carbon-hydrogen bond at carbon number 5. Both 5α- and 5β-cholanic acids are present with the 5α-acid about four to five times more abundant than the 5β-acid. In nature, however, the 5β acid is the most common. These authors point out that animal sources can account for all the steroid acids they found in petroleum but that plant sources also were likely contributors.

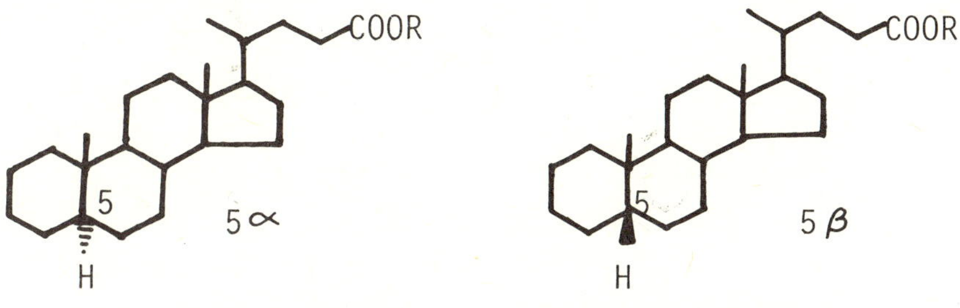

CHOLANIC ACID

Also the stereochemistry of steroid hydrocarbons (steranes) have been considered by Balogh, Wilson, and Burlingame (1971) and by Mulheirn and Ryback (1975, 1977). The stereochemistry of sterenes has been described by Rubenstein, Sieskind, and Albrecht (1975).

The importance to geochemistry of the stereochemistry of amino acids was first recognized by Hare and Mitterer (1967) and Hare and Abelson (Paper 27). In Paper 27 enzymatic techniques were used to follow the changes with geologic time in the relative abundance of D- and L-amino acid, ie., the extent of racemization in fossil shells. Hare and Abelson showed, for example, that L-isoleucine interconverts to D-alloisoleucine and that no measurable amounts of D-isoleucine or L-alloisoleucine are formed. They inferred that the extent of racemization (epimerization in the case of diastereomers such as isoleucine) of amino acids may be used to ascertain the age of shells. Enzymatic techniques for this kind of work have not been used extensively and now gas chromatographic methods are mainly employed to measure the extent of racemization of suites of amino acids. (For example, see Kvenvolden, Peterson, and Pollock, 1972.)

To show the stereochemistry of amino acids, Fischer projections are commonly used, and the D and L convention mentioned before designates the configuration of the amino acid relative to the standard of the series, serine. [(-)serine is arbitrarily designated L-serine.] In the Fischer projection, illustrated below by L-isoleucine and D-alloisoleucine, the main chain of an acyclic stereoisomer is drawn vertically and is imagined to be in or below the plane of the paper. The side chains are drawn horizontally and imagined to be projecting out of the plane of the paper.

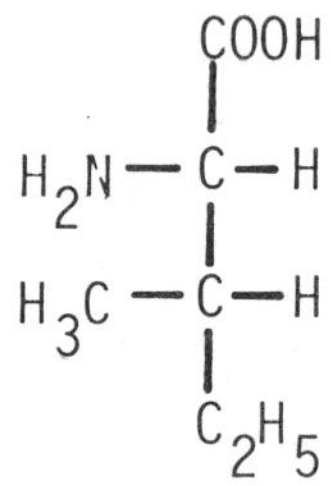

L-ISOLEUCINE

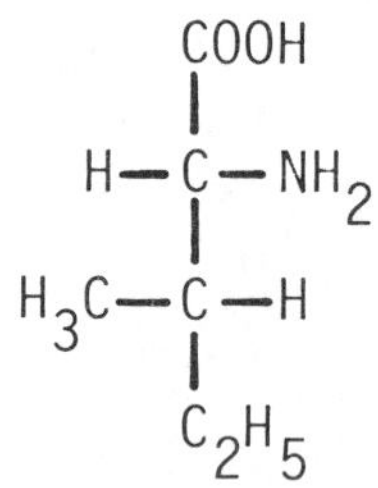

D-ALLOISOLEUCINE

The final paper in this section is concerned with stereochemistry that exists because of double bonds. Isomers that may be interconverted about double bonds are called geometric isomers or *cis-trans*-isomers. Atoms or groups at opposite ends of the double bond are considered to be *cis* to one another when they are on the same side of the plane of the molecule and *trans* when on the opposite sides. Van Vleet and Quinn (Paper 28) showed that the *cis* isomer accounted for about 92% and the *trans* isomer about 8% of the total palmitoleic (C16:1) and oleic (C18:1) acid present in a recent sediment from Narragansett Bay, Rhode Island. They reasoned that the *cis* isomers result directly from biological input; the *trans* isomers may also represent a direct contribution to the sediment, or the *trans* isomers may be a microbiological modification of the *cis* isomers. Additional work showing the stereochemical relationships of unsaturated fatty acids has also been done by Volkman and Johns (1977) and by Boon, de Leeuw, and Burlingame (1978).

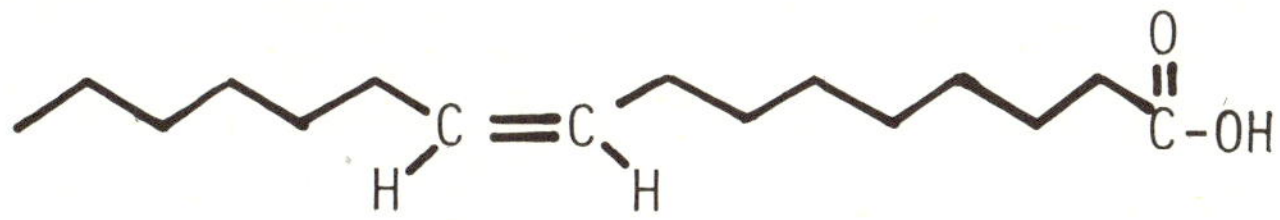

CIS-9-HEXADECENOIC ACID
(CIS-PALMITOLEIC ACID)

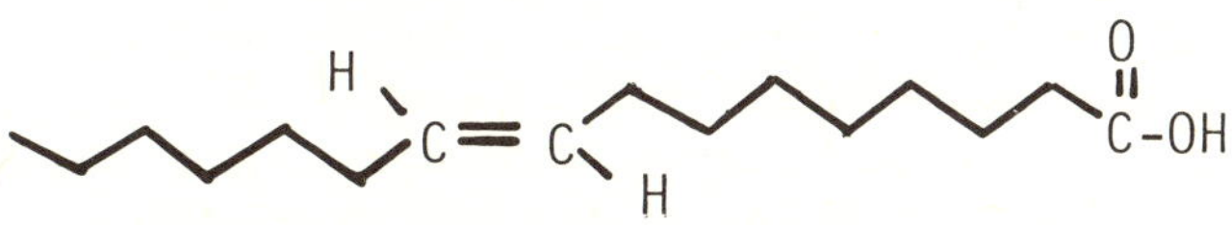

TRANS-9-HEXADECENOIC ACID
(TRANS-PALMITOLEIC ACID)

REFERENCES CITED

Balogh, B., D. M. Wilson, A. L. Burlingame. 1971. Carbon-13 NMR study of the stereochemistry of steranes from oil shale of the Green River Formation (Eocene). *Nature* **233**:261–263.

Boon, J. J., J. W. de Leeuw, and A. L. Burlingame. 1978. Organic geochemistry of Walvis Bay diatomaceous ooze—III. Structural analysis of the monoenoic and polycyclic fatty acids. *Geochim. et Cosmochim. Acta* **42**:631–644.

Brooks, P. W., J. R. Maxwell, J. W. Cornforth, A. G. Butlin, and C. B. Milne. 1977. Stereochemical studies of acyclic isoprenoid compounds—VI. The stereochemistry of farnesane from crude oil. In R. Campos and J. Goñi, eds., *Advances in Organic Geochemistry 1975.* ENADIMSA, Madrid, 81–97.

Cox, R. E., J. R. Maxwell, G. Eglinton, C. T. Pillinger, R. G. Ackman, and S. N. Hooper. 1970. The geological fate of chlorophyll: The absolute stereochemistries of a series of acyclic isoprenoid acids in a

50 million-year-old lacustrine sediment. *Chem. Commun.* **1970**: 1639–1641.

Cox, R. E., J. R. Maxwell, R. G. Ackman, and S. N. Hooper. 1972. Stereochemical studies of acylic isoprenoid compounds—III. The stereochemistry of naturally occurring (marine) 2,6,10,14-tetramethylpentadecane. *Canadian Jour. Biochem.* **50**:1238–1241.

Ensminger, A., P. Albrecht, G. Ourisson, B. J. Kimble, J. R. Maxwell, and G. Eglinton. 1972. Homohopane in Messel oil shale: First identification of a C_{31} pentacyclic triterpane in nature. *Tetrahedron Lett.* **36**: 3861–3864.

Ensminger, A., A. van Dorsselaer, Ch. Spyckerelle, P. Albrecht, and G. Ourisson. 1974. Pentacyclic triterpanes of the hopane type as ubiquitous geochemical markers: Origin and significance. In B. Tissot and F. Bienner, eds., *Advances in Organic Geochemistry 1973*. Editions Technip., Paris, 245–260.

Hare, P. E. and R. M. Mitterer (1967) Nonprotein amino acids in fossil shells: *Carnegie Inst. Washington Year Book* **65**:362–364.

Kvenvolden, K. A., E. Peterson, and G. E. Pollock. (1972). Geochemistry of amino acid enantiomers: Gas chromatography of their diasteromeric derivatives. In H. R. v. Gaertner and H. Wehner, eds., *Advances in Organic Geochemistry 1971*. Pergamon Press, Braunschweig, 387–401.

Maxwell, J. R., R. E. Cox, R. G. Ackman, and S. N. Hooper. 1972. The diagenesis and maturation of phytol. The stereochemistry of 2, 6, 10, 14-tetramethylpentadecane from an ancient sediment. In H. R. v. Gaertner and H. Wehner, eds., *Advances in Organic Geochemistry 1971*. Pergamon Press, Braunschweig, 277–291.

Maxwell, J. R., R. E. Cox, G. Eglinton, and C. T. Pillinger. 1973. Stereochemical studies of acyclic isoprenoid compounds—II. The role of chlorophyll in the derivation of isoprenoid-type acids in a lacustrine sediment. *Geochim. et Cosmochim. Acta* **37**:297–313.

Mulheirn, L. J., and G. Ryback. 1975. Stereochemistry of some steranes from geological sources. *Nature* **256**:301–302.

Mulheirn, L. J., and G. Ryback. 1977. Isolation and structural analysis of steranes from geological sources. In R. Campos and J. Goñi, eds., *Advances in Organic Geochemistry 1975*, ENADISMA, Madrid, 173–192.

Rubinstein, I., O. Sieskind, and P. Albrecht. 1975. Rearranged sterenes in a shale: Occurrence and simulated formation. *Jour. The Chem. Soc.* **19**:1833–1836.

Volkman, J. K., and R. B. Johns. 1977. The geochemical significance of positional isomers of unsaturated acids from an intertidal zone sediment. *Nature* **267**:693–694.

SUGGESTED ADDITIONAL READINGS

Eglinton, G. 1969. Organic geochemistry—The organic chemist's approach. In G. Eglinton and M. T. J. Murphy, eds., *Organic Geochemistry—Methods and Results*. Springer-Verlag, Berlin, 20–73.

Maxwell, J. R., C. T. Pillinger, and G. Eglinton. 1971. Organic geochemistry. *Quart. Rev., The Chem. Soc.* **25**:571–628.
Mislow, K. 1965. *Introduction to Stereochemistry.* W. A. Benjamin, Menlo Park, Calif., 193 p.
Patience, R. L., S. J. Rowland, and J. R. Maxwell. 1978. The effect of maturation on the configuration of pristane in sediments and petroleum. *Geochim. et Cosmochim. Acta* **42**:1871–1875.

23

Reprinted from *Nature* **218**:1019–1024 (1968)

Correlation of Stereoisomerism in Present Day and Geologically Ancient Isoprenoid Fatty Acids

by

IAIN MACLEAN
GEOFFREY EGLINTON*
K. DOURAGHI-ZADEH†
Department of Chemistry,
University of Glasgow

R. G. ACKMAN
S. N. HOOPER
Fisheries Research Board of Canada,
Halifax Laboratory,
Halifax, Nova Scotia

High resolution gas chromatography of the methyl esters of individual isoprenoid fatty acids isolated from a geological source of the Eocene period has shown a diastereoisomer composition compatible with a chlorophyll derivation for the acids. The L-menthyl esters give enhanced separation of isoprenoid acid diastereoisomers on an optically inactive GC substrate and could be employed on samples with more complex stereochemistry.

THE separation of diastereoisomers of the methyl ester of methyl branched fatty acids by gas chromatography[1] has recently been extended, using very efficient polar capillary columns, to isoprenoid fatty acids[2]. The relative stereochemistries (and in some cases, by supplementary use of rotation measurements, the absolute stereochemistry) of the asymmetric centres of such acids as phytanic (3,7,11,15-tetramethylhexadecanoic; I, $R=H$), pristanic (2,6,10,14-tetramethylpentadecanoic; II, $R=H$) and 4,8,12-trimethyltridecanoic (III, $R=H$) isolated from present day marine[2-6] and terrestrial animal[2,7-18] sources have been successfully determined. Speculation about the possible conversion of marine lipids to petroleums, bolstered by the discovery of pristanic and phytanic acids in a California petroleum[19] and in recent[20] and ancient (refs. 20–22 and unpublished results of Blumer, Douraghi-Zadeh, Douglas and Eglinton) sediments, has provoked interest in the source material and geological fate of these acids. We now report on the diastereoisomeric composition of four acids (I–IV, $R=H$) isolated from a sample of shale from the Green River Formation (Eocene, 50–60 million years). The results are consistent with the acids having arisen from the phytol (VI) portion of chlorophyll biosynthesized in Eocene times.

A 100 g sample of the shale from the 334 m depth of a core taken from Sulphur Creek, supplied by Dr W. E. Robinson of the US Bureau of Mines, was pulverized, demineralized with concentrated hydrochloric and hydrofluoric acids (1 : 1), and hydrolysed with methanolic potassium hydroxide. Extraction of the shale with benzene/methanol (1 : 1) gave a dark brown residue from which the acidic lipids were isolated by chromatography on a silica/potassium hydroxide column[23]. Esterification with methanol/benzene/concentrated sulphuric acid produced a crude ester fraction, from which we isolated the methyl esters of the monocarboxylic fatty acids by thin-layer chromatography on silica gel. The methyl esters (78 mg) were further purified by thin-layer chromatography on silica gel impregnated with 20 per cent silver nitrate, and the straight chain fatty acid methyl esters were removed from the branched chain esters by urea adduction (three times)[24]. The branched methyl esters (15 mg in 0·5 mg portions) were gas chromatographed preparatively on an 'Aerograph *A*90-*P*' fitted with a

* Present address: Organic Geochemistry Unit, School of Chemistry, University of Bristol.

† Present address: Department of Chemistry, Tehran University, Iran.

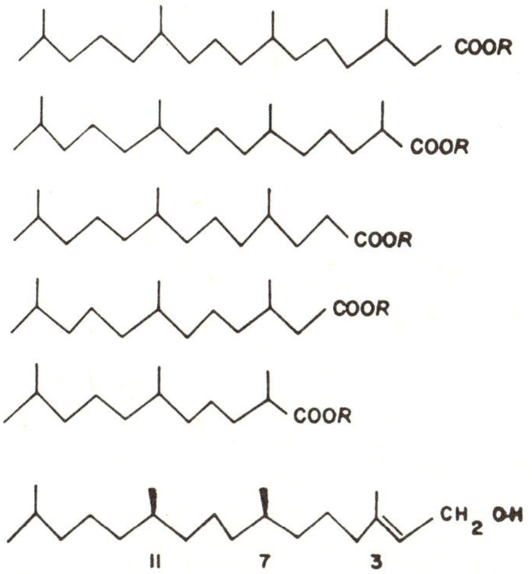

I, II, III, IV, V, VI

(*R* = H, methyl or L-menthyl)

column (7 m × 3 mm) of 3 per cent silicone gum rubber (*SE* 52) on 100–120 mesh 'Gaschrom *P*'. The rate of flow of helium was 60 ml./min and the oven temperature was programmed from 100°–300° C at 5°/min. The methyl esters of the C_{14}, C_{15}, C_{16}, C_{19} and C_{20} isoprenoid acids were collected and their structures were confirmed on an LKB 9000 gas chromatograph–mass spectrometer. The individual esters, which gave only single symmetrical peaks on packed column gas-liquid chromatography, were then analysed on highly efficient open tubular (capillary) columns coated with butanediol succinate polyester (BDS). Standard samples of the same esters, obtained synthetically or from biological sources, were studied similarly. Certain diastereoisomers of methyl pristanate, and of methyl phytanate, could be partially resolved on single columns (50 m × 0·25 mm) with an efficiency (calculated from methyl palmitate) of approximately 45,000 theoretical plates[2]. The lower isoprenoid fatty acids examined, specifically those possessing only two asymmetric carbon atoms, required two such columns operated in series (approximately 80,000 theoretical plates) for partial resolution of certain diastereoisomers[25]. The results are summarized in Table 1.

Natural phytol (VI), which is known[26,27] to have the D configuration at carbons 7 and 11, provides a convenient source of reference acids of established absolute configuration. Thus phytanic acid (I, *R* = H) was prepared from phytol (supplied by Stafford–Allen, Ltd) by catalytic hydrogenation (platinum oxide in methanol) and oxidation (chromic acid in acetic acid). The methyl phytanate (I, *R* = Me), obtained by treatment with diazomethane and purified by preparative gas–liquid chromatography, is a mixture of two diastereoisomers (3L, 7D, 11D and 3D, 7D, 11D) which appear as twin peaks of equal intensity (Fig. 2*A*) when run on a gas-liquid chromatograph with the BDS capillary column. The more slowly eluted isomer is known to be DDD by co-injection of a reference sample prepared from DDD dihydrophytyl glyceryl diethers which are produced by certain halophilic bacteria[28,29].

A mixture of all eight possible diastereoisomers of methyl phytanate (I, *R* = Me) might be expected to give four peaks corresponding to the four pairs of enantiomers. Synthetic methyl phytanate[30], however, which should have random stereochemistry, gives incomplete resolution resulting in a distorted triplet[25]. The extent of the inter-

Table 1. STEREOISOMERIC COMPOSITION OF ISOPRENOID ACIDS (I)–(V) BY GAS–LIQUID CHROMATOGRAPHY

No. of diastereo-isomers possible	Origin of sample	No. of peaks observed: Methyl ester	No. of peaks observed: L-Menthyl	Assignment†	Isomer ratios and comments
3,7,11,15-Tetramethylhexadecanoic (phytanic) (I, *R* = H)					
(8)	Fish and marine mammal oils[6]	2		LDD, DDD	LDD/DDD varies from ~1·0 to ~10
	Sheep[13], ox[12], butterfat[10,15]; Patient with Refsums syndrome[16]	2		LDD, DDD	LDD/DDD variable with DDD always in slight excess. Samples exclusively from New Zealand
	Halophilic bacteria[29]	1		DDD	
	Geological*	2		LDD, DDD	LDD/DDD ~0·7
	Synthetic (total synthesis)[30]	3		All	Gas–liquid chromatography shows a triplet due to incomplete resolution of the eight diastereoisomers which would occur as four pairs of antimers
	Synthetic (from phytol)*	2	2	LDD, DDD	Our sample gave a 1 : 1 ratio of diastereoisomers, whereas samples from other sources[25] all showed a slight DDD predominance. The L-menthyl esters increased the separation factor of the peaks
2,6,10-14-Trimethylpentadecanoic (pristanic) (II, *R* = H)					
(8)	Fish and marine mammal oils[6]	2		LDD, DDD	LDD/DDD varied from 0·5 to 3·5
	Sheep[14], butterfat[11]	2		LDD, DDD	LDD/DDD ratio variable with general DDD predominance. Samples exclusively from New Zealand
	Halophilic bacteria[29]	1		DDD	
	Geological*	2		LDD, DDD	Ratio of isomers ~1 : 1
	Fish oil (Kates/Sen Gupta)‖		4	All	An asymmetrical triplet was obtained §
5,9,13-Trimethyltetradecanoic					
(4)	Synthetic ‡ (phytol oxidation)*	1	1	DD	
4,8,12-Trimethyltridecanoic (III, *R* = H)					
(4)	Herring, seal oil, etc.[6,25]	1		DD	Suggests phytol origin
	Sheep fat[32]	1		DD	Suggests phytol origin
	Geological*	2		LD and DD	LD/DD ratio ~0·05
	Synthetic (from farnesol)[3]	2		All	1 : 1 ratio of peaks
	Synthetic ‡ (phytol oxidation)*	1	1	DD	
3,7,11-Trimethyldodecanoic (farnesanic) (IV, *R* = H)					
(4)	Geological*	2		LD and DD	LD/DD ratio ~0·05
	Synthetic (from farnesol)*	2	4	All	The methyl esters give an overlapping doublet which is poorly resolved. The L-menthyl esters resolve into an incompletely separated quartet as a result of overlap of two of the diastereoisomers (Fig. 3*D*)
	Synthetic ‡ (phytol oxidation)*	1	1	DD	
2,6,10-Trimethylundecanoic (V, *R* = H)					
(4)	Geological*				Insufficient material available for informative examination
	Synthetic (from farnesol)*	2	3	All	An ill-defined twin peak for the methyl ester was resolved into a triplet on conversion to the L-menthyl ester (Fig. 3*D*)
	Synthetic ‡ (phytol oxidation)*	1	1 ?	DD	

* Discussed in text.
† LDD phytanate refers to 3L,7D,11D-phytanate, which corresponds in the *R* and *S* nomenclature to 3*S*,7*R*,11*R*-phytanate.
‡ Phytol was oxidized with permanganate by the procedure of Murray[40].
§ Independently prepared by Drs A. K. Sen Gupta and M. Kates.
‖ Unpublished results of Ackman.

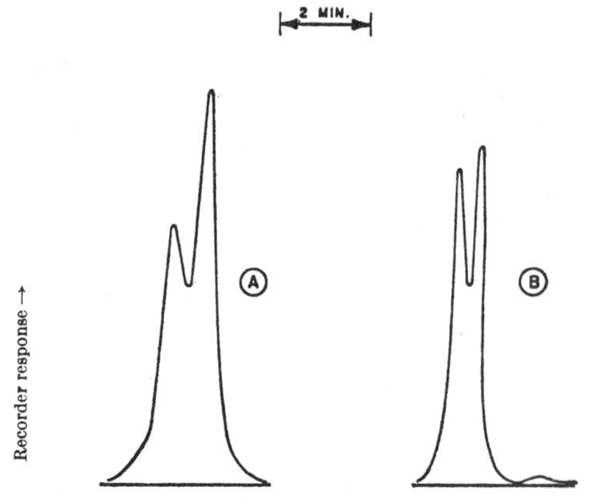

Fig. 1. Open-tubular (capillary) gas–liquid chromatography traces of the methyl esters of the branched chain fatty acids isolated from the Green River Shale. For details of chromatography see Table 2. *A*, Methyl ester of phytanic acid (I, *R* = Me). A small shoulder appears at the base of the leading edge. The largest component coincides with authentic 3D,7D,11D,15-tetramethylhexadecanoate. *B*, Methyl ester of pristanic acid (II, *R* = Me). The two major components do not show shoulders and the second coincides with authentic 2D,6D,10D,14-tetramethylpentadecanoate. *C*, Methyl esters of reference myristic (14 : 0) acid and of 4,8,12-trimethyltridecanoic acid (III, *R* = Me). A very small shoulder appears at the base of the leading edge of the major component; the latter coincides with authentic 4D,8D,12-trimethyltridecanoate. *D*, Methyl ester of 3,7,11-trimethyldodecanoic acid (IV, *R* = Me). A small shoulder is apparent on the leading edge. The major component coincides with the corresponding acid prepared by oxidation of phytol.

actions between the asymmetric centres is poorly understood.

Oxidation of phytol with potassium permanganate gave an acid fraction containing the DD isomers of the acids III, IV and V (*R* = H). The methyl esters of these acids were identified by combined gas chromatography–mass spectrometry (GC–MS) and on subsequent capillary gas chromatographic examination (Fig. 4) exhibited the expected sharp single peaks. The equivalent chain length (ECL) values of these esters were used in our comparative study of the corresponding geological samples. The separations of the diastereoisomeric pairs of methyl 4,8,12 trimethyltridecanoate (III, *R* = Me) and methyl 3,7,11-trimethyldodecanoate (IV, *R* = Me) prepared with mixed stereochemistry are shown (Fig. 2*C* and *D*) for comparison. It should be noted that on optically inactive gas–liquid chromatography liquid phases and supports, no resolution of optical antimers (that is, LD from DL or LL from DD) can be obtained. The preparation involves the non-specific hydrogenation of farnesol and gives samples showing two barely resolved peaks (respectively DL, LD and DD, LL).

Methyl phytanate (I, *R* = Me) and methyl pristanate (II, *R* = Me) from the Green River Shale were each resolved by capillary gas–liquid chromatography into two peaks (Fig. 1*A* and *B*), and in the case of the phytanate showed a predominance of the DDD (or LLL) isomer. Geological methyl 4,8,12-trimethyltridecanoate (III, *R* = Me) and methyl 3,7,11-trimethyldodecanoate (IV, *R* = Me), on the other hand, gave major peaks (Fig. 1*C* and *D*) with only about 5 per cent of the faster running LD (or DL) diastereoisomers, indicating that they consist essentially of the DD (or LL) isomers. Geological methyl 2,6,10-trimethylundecanoate (V, *R* = Me) was obtained in such a low yield that the chromatograms are not informative. Capillary gas–liquid chromatography examination of these methyl esters also does not resolve enantiomeric pairs into optical antimers and so assignments such as DDD for one peak in the chromatogram of the geological methyl phytanate are made only for simplicity: it may be correct, but the other (LLL) enantiomer remains a possibility. Further confirmation of the absolute configuration at asymmetric centres in the α or β positions can normally be obtained from optical rotations which are significant in sign and magnitude for these positions[25,29,31]. Results for other asymmetric centres provide useful supporting evidence when the amount of sample is adequate[31,32]. The $[M]_D$ of the isoprenoid acids, however, is so low that at the concentrations available from geological sources no definite conclusions can be drawn.

In the isoprenoid series of acids, methyl branching at all positions except that giving a terminal *iso* structure to the fatty acid chain gives rise to centres of asymmetry. Variations in the properties of the diastereoisomers will necessarily be small, for any two asymmetric centres along the chain are separated by four freely rotating C–C single bonds. The eight diastereoisomers of synthetic methyl 2,4,6-trimethyloctanoate, which have been separated by gas–liquid chromatography on an open-tubular column into four racemic pairs[1] with greater facility than isoprenoid acids, probably owe the greater differences in physical properties to the greater proximity of the asymmetric centres to the carboxyl group and to each other. The contribution of $[M]_D$ from the *anteiso* (ω-2) asymmetric centre is also significant[25,31].

Esterification with an optically active alcohol is the classical approach to resolution of racemates in the differing enantiomeric pairs of isoprenoid acids. In any event, the separation factor between individual diastereoisomers should be increased. A particularly appropriate

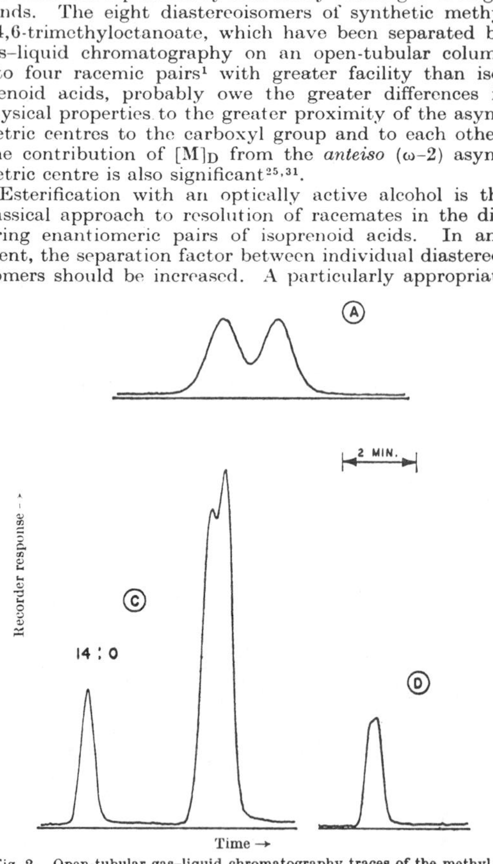

Fig. 2. Open-tubular gas–liquid chromatography traces of the methyl esters of synthetic branched chain fatty acids. For details of chromatography see Table 2. *A*, Methyl ester of phytanic acid (I, *R* = Me) prepared by oxidation of dihydrophytol of plant origin. The second component coincides with authentic 3D,7D,11D,15-tetramethylhexadecanoate. *C*, Methyl esters of reference myristic (14 : 0) acid and of 4,8,12-trimethyltridecanoic acid (III, *R* = Me) prepared by chain elongation from farnesol of plant origin. The second component coincides with authentic 4D,8D,12-trimethyltridecanoate and present-day samples of biological origin. *D*, Methyl ester of 3,7,11-trimethyldodecanoic acid (IV, *R* = Me) prepared from farnesol of plant origin. The second component coincides with the corresponding acid prepared by oxidation of phytol and therefore of DD structure.

ester or other derivative might permit the separation of all the possible diastereoisomers for that carbon skeleton. This approach to the analysis of mixtures of branched acids in nature is being investigated. Optical rotation measurements alone are useless when applied to a situation where a number of diastereoisomers differing little in rotation are possible. The gas–liquid chromatographic technique is in principle much more definitive and is applicable to smaller (μg) quantities.

We have accordingly prepared the L-menthyl esters of a number of isoprenoid acids by heating the acid chlorides under reflux in benzene containing L-menthol. The L-menthyl group should increase the conformational rigidity of the system, and its asymmetrical bulk would be expected to interfere with the diastereoisomeric interactions further along the chain from the carboxyl group. L-Menthyl phytanate (I, R = L-menthyl) prepared from phytol was more clearly resolved into two (LDD and DDD) gas–liquid chromatographic peaks than was the corresponding methyl ester (Fig. 3*A* and 2*A*, respectively), even though a double column was used for the methyl esters. In addition, the L-menthyl esters of DD acids III, IV and V, obtained from oxidation of phytol, again showed single sharp peaks (compare Fig. 4). In contrast, the L-menthyl ester of farnesanic acid (IV, R = H), prepared from farnesol by reduction, oxidation and esterification, gave a composite quartet (Fig. 3*D*), which must correspond to the four diastereoisomers (the L-menthyl esters of the DD, LL, DL and LD acids). The last peak to elute coincides exactly with that of the ester of IV prepared from phytol and is therefore DD. In the same chromatogram a triplet of lower retention time was assigned to the incompletely resolved diastereoisomers of the L-menthyl esters of 2,6,10-trimethylundecanoic acid (V) which is formed in small yield in the oxidation step. These results encourage us to prepare the L-menthyl esters of the geological isoprenoid acids, but results of this study are not yet available. We have also investigated

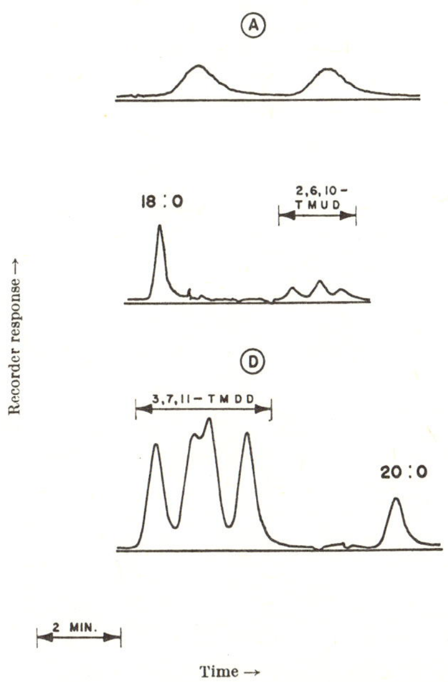

Fig. 3. Open-tubular gas–liquid chromatography traces of the L-menthyl esters of synthetic branched chain fatty acids. For details of chromatography see Table 2. *A*, L-Menthyl ester of phytanic acid sample illustrated in Fig. 2*A*. *D*. L-Menthyl ester of 3,7,11-trimethyldodecanoic acid illustrated in Fig. 2*D* with added reference methyl stearate (18 : 0) and arachidate (20 : 0). The minor triplet peak is believed to be the L-menthyl ester of a secondary oxidation product, 2,6,10-trimethylundecanoic acid (V).

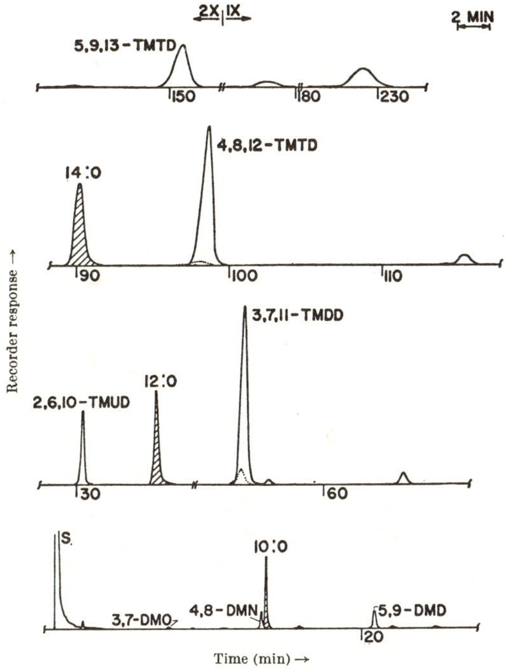

Fig. 4. Open-tubular gas–liquid chromatography trace (omitting lengthy blank sections) of an examination of the methyl esters of the branched chain fatty acids derived from permanganate oxidation of phytol. The sample was run on a BDS capillary (100 m × 0·25 mm inner diameter) of 80,000 theoretical plates at 130° with a helium pressure of 50 psi. Shaded peaks are added reference normal C_{10}, C_{12} and C_{14} acid methyl esters. Dotted lines reproduce peaks for III and IV of geological origin examined in identical GLC conditions. Identification of 3,7-dimethyloctanoate (DMO), 4,8-dimethylnonanoate (DMN) and 5,9-dimethyldecanoate (DMD) by calculation from GLC retention data[39]. Other labelled components, including 5,9,13-trimethyltetradecanoate (TMTD), identified also by combined GLC-MS.

the possibility of separating the diastereoisomers of dihydrophytol on gas–liquid chromatography by conversion to their L(+) lactates. Both the lactates and the acetoxypropionates (formed by acetylation of the lactates) gave resolution on the BDS or 'Apiezon *L*' capillary column inferior to that obtained with the menthyl esters.

Our interpretation of the stereochemical data (Table 1) in terms of the isoprenoid acids (I–IV, R = H) isolated from the shale may be related to the established occurrence[2–18] of these and other isoprenoid compounds in present day organisms and recent sediments. Three of the acids (I–III, R = H) are fairly widely distributed in marine animal lipids, although the actual quantities are seldom in excess of 1 per cent of the total fatty acid content. The ultimate source of the carbon skeletons of these acids is likely to be the phytol side chain of the chlorophyll molecule produced by the photosynthetic organisms such as phytoplankton at the beginning of the marine food web. Phytol (VI) obtained from chlorophyll is known[26,27] to have the D configuration at both asymmetric centres and it may be provisionally assumed that this applies to chlorophyll from all sources. Phytanic acid (I, R = H) derived from phytol by reduction and oxidation would therefore have the D configuration at carbons 7 and 11. Unless, however, a stereospecific reduction step is invoked the product will be a pair of acids enantiomeric at carbon 3 (that is, 3D, 7D, 11D, and 3L, 7D, 11D). Using the present technique of capillary column gas–liquid chromatography, surveys have revealed a range of values for the phytanate LDD/DDD ratio[2,6]. Pristanic acid (II, R = H) exhibits the same situation in regard to carbon atom 2. (Pristanic acid and

Table 2. GAS–LIQUID CHROMATOGRAPHY CONDITIONS EMPLOYED IN FIGURES (WITH RETENTION DATA)

Fig.	Sample and component*	Column No. (BDS)	Temperature (°C)	Pressure (psig)	Adjusted retention time (min) to leading edge	ECL value to leading edge (based on methyl esters)
1	*A* (phytanate)	7	150	40	61·7	16·98
	B (pristanate)	7	150	40	36·5	15·76
	C (14 : 0)	7+10	140	80	47·3	14·00
	C (4,8,12-TMTD)	7+10	140	80	50·3	14·17
	D (3,7,11-TMDD)	7+10	130	80	38·6	12·92
2	*A* (phytanate)	7+10	150	80	82	17·04
	C (14 : 0)	7+10	140	80	47·3	14·00
	C (4,8,12-TMTD)	7+10	140	80	50·3	14·17
	D (3,7,11-TMDD)	7+10	140	80	28·9	12·94
3	*A* (menthyl phytanate)	12	170	50	62·4	23·85
	D (above) (18 : 0)	12	170	20	27·1	18·00
	D (above) (menthyl 2,6,10-TMUD)	12	170	20	30·3	18·32
	D (below) (menthyl 3,7,11-TMDD)	12	170	20	51·5	19·72
	D (below) (20 : 0)	12	170	20	57·4	20·00
4	3,7-DMO	7+10	130	50	7·3	8·68
	4,8-DMN	7+10	130	50	13·3	9·94
	10 : 0	7+10	130	50	13·6	10·00
	5,9-DMD	7+10	130	50	20·7	10·87
	2,6,10-TMUD	7+10	130	50	30·2	11·69
	12 : 0	7+10	130	50	35·1	12·00
	3,7,11-TMDD	7+10	130	50	54·4	12·93
	14 : 0	7+10	130	50	89·7	14·00
	4,8,12-TMTD	7+10	130	50	97·5	14·18
	5,9,13-TMTD	7+10	130	50	150	15·09

* Methyl esters unless otherwise stipulated.

2,6,10-trimethylundecanoic acid (V) with an α-methyl group will be more susceptible to epimerization at this position.) The results to date for fat-rich tissues indicate that depot fats tend to reflect the ratios of diastereoisomers in dietary fats when preformed pristanic and phytanic acids are ingested, but it is not yet known with any certainty what selective stereochemical effects are involved in the conversion of ingested phytol to phytanic acid or in the eventual elimination or catabolism of these acids in the animal[2,6,33]. A sufficient variety of samples of marine origin have, however, been analysed to establish that in this milieu LDD phytanate normally exceeds DDD. This may be a result of preferential conversion of the DDD diastereoisomer to pristane in zooplankton where this hydrocarbon is widely observed[34,35]. In addition to biological processes leading to degradation of pristane and derivatives, there may be less specific chemical oxidation of isoprenoid acids and hydrocarbons in the surface layers of the oceans and products therefrom may be recycled back into the marine food web, complicating assessment of the origin of minor components.

The lower acids (III–V, $R = H$) have only two asymmetric centres which correspond to the D centres at carbons 7 and 11 of phytol. If racemization of these tertiary carbons is excluded, a phytol origin requires that each acid be present as a single diastereoisomer, DD. A comparison of acid III of geological origin may be made with present day samples isolated from biologically active systems. In suitable conditions of gas–liquid chromatography methyl esters of synthetic acids III and IV have been partially resolved into two components of nominally equal size (Fig. 2*C* and *D*)[25]. The second peak to elute has a retention time identical with III isolated from herring oil, seal blubber and sheep fat and also coincides with III prepared by the oxidation of phytol. In all three natural samples there was no evidence of the earlier eluting peak and the (+) optical rotation of the sheep fat sample[32] is believed to confirm a DD structure for the peaks which elute later. Further support for this view is found in observations indicating a general principle that "all D" (or "all L") isoprenoid fatty acid methyl esters will be the last diastereoisomers to elute from the gas–liquid chromatography columns used in this work[25]. An origin in phytol is therefore most likely for acid III of biological origin. Amounts of acid III are quite substantial in marine lipid systems[6], possibly because it may be treated to some extent by esterases and other depot fat biochemical processes as though it were palmitic (hexadecanoic) acid, the saturated fatty acid most prevalent in depot fat systems[4,5,36]. In contrast, acids IV and V are present in only trace amounts in marine lipid systems (unpublished results of Ackman), although myristic (tetradecanoic) acid is fairly plentiful. The reason for this differentiation, and the nature and details of the diastereoisomers of 3,7,11-trimethyldodecanoic (IV, $R = H$) and 2,6,10-trimethylundecanoic (V, $R = H$) acids in present day biological systems, remain to be studied.

For geological samples the findings for acids III and IV ($R = H$) are important in another respect, for they show that the asymmetric centres have not been racemized appreciably during the 50–60 million years of the entombment of the compounds themselves or their immediate precursors. Older rocks await study.

Many natural products based on the isoprenoid skeleton could conceivably have given rise to the isoprenoid fatty acids in the Green River Shale. Thus a dihydrophytyl glyceryl ether is an important lipid constituent of certain halophilic bacteria[28,29], but is known to be entirely DDD. Again, polyunsaturated compounds such as the carotenoids would give one isomer only by stereospecific reduction, or all possible stereoisomers by non-stereospecific reduction. Selective metabolic breakdown could be a further complicating factor, as already mentioned. The C_{15} isoprenoid acid (IV, $R = H$) could be produced by stereospecific reduction of farnesoic (3,7,11-trimethyldodeca-2,6,10-trienoic) acid, which is synthesized in the liver of mammals[37], or by oxidation of phyta-2,4-diene which occurs in zooplankton[38]. These results concerning the stereochemistry of the asymmetric centres of these acids from the shale conform best to a phytol derivation, however, presumably originated by the profuse phytoplanktonic growth in the shallow waters of the lakes of that period.

This work was supported in part by the US National Aeronautics and Space Administration and the Natural Environment Research Council. The GC–MS instrument was obtained through a grant from the Science Research Council to Drs G. Eglinton and C. J. W. Brooks at the University of Glasgow.

Received April 10, 1968.

1 Odham, G., *Fette Seifen Anstrichmittel*, **69**, 164 (1967).
2 Ackman, R. G., and Hansen, R. P., *Lipids*, **2**, 357 (1967).
3 Sen Gupta, A. K., and Peters, H., *Fette Seifen Anstrichmittel*, **68**, 349 (1966)
4 Ackman, R. G., and Sipos, J. C., *Comp. Biochem. Physiol.*, **15**, 445 (1965).
5 Ackman, R. G., Sipos, J. C., and Tocher, C. S., *J. Fish. Res. Board Canada*, **24**, 635 (1967).
6 Ackman, R. G., and Hooper, S., *Comp. Biochem. Physiol.*, **24**, 549 (1968).
7 Sonneveld, W., Haverkamp Begemann, P., van Beers, G. J., Keuning, R., and Schogt, J. C. M., *J. Lipid Res.*, **3**, 351 (1962).
8 Lough, A. K., *Biochem. J.*, **86**, 14 (1963).
9 Lough, A. K., *Biochem. J.*, **91**, 584 (1964).
10 Hansen, R. P., and Shorland, F. B., *Biochem. J.*, **50**, 358 (1951).
11 Hansen, R. P., and Morrison, J. D., *Biochem. J.*, **93**, 225 (1964).
12 Hansen, R. P., *Chem. and Ind.*, 303 (1965).
13 Hansen, R. P., *New Zealand J. Sci.*, **8**, 158 (1965).

[14] Hansen, R. P., *Chem. and Ind.*, 1258 (1965).
[15] Hansen, R. P., Shorland, F. B., and Morrison, J. D., *J. Dairy Res.*, **32**, 21 (1965).
[16] Hansen, R. P., *Biochim. Biophys. Acta*, **106**, 304 (1965).
[17] Hansen, R. P., *J. Dairy Res.*, **33**, 333 (1966).
[18] Avigan, J., *Biochim. Biophys. Acta*, **125**, 607 (1966).
[19] Cason, J., and Graham, D. W., *Tetrahedron*, **21**, 471 (1965).
[20] Blumer, M., and Cooper, W. J., *Science* (in the press).
[21] Eglinton, G., Douglas, A. G., Maxwell, J. R., Ramsay, J. N., and Ställberg-Stenhagen, S., *Science*, **153**, 1133 (1966).
[22] Haug, P., Schnoes, H. K., and Burlingame, A. L., *Science*, **158**, 772 (1967).
[23] McCarthy, R. D., and Duthie, A. H., *J. Lipid Res.*, **3**, 117 (1962).
[24] Nicolaides, N., and Ray, T., *J. Amer. Oil Chem. Soc.*, **42**, 702 (1965).
[25] Ackman, R. G., *J. Chromatog.* (in the press).
[26] Crabbé, P., Djerassi, C., Eisenbraun, E. J., and Shuyen-Liu, *Proc. Chem. Soc.*, 264 (1959).
[27] Burrell, J. W. K., Garwood, R. F., Jackman, L. M., Oskay, E., and Weedon, B. C. L., *J. Chem. Soc.*, 2144 (1966).
[28] Kates, M., Yengoyan, L. S., and Sastry, P. S., *Biochim. Biophys. Acta*, **98**, 252 (1965).
[29] Kates, M., Joo, C. N., Palameta, B., and Shier, T., *Biochemistry*, **6**, 3329 (1967).
[30] Sato, K., Mizuno, S., and Hirayama, M., *J. Org. Chem.*, **32**, 177 (1967).
[31] Abrahamsson, S., Ställberg-Stenhagen, S., and Stenhagen, E., *Progress in the Chemistry of Fats and Other Lipids*, Vol. VII, Part 1, pp. 98, 99 (Pergamon Press, Oxford, 1963).
[32] Hansen, R. P., *Biochim. Biophys. Acta* (in the press).
[33] Avigan, J., Steinberg, D., Gutman, A., Mize, C. E., and Milne, G. W. A., *Biochem. Biophys. Res. Commun.*, **24**, 838 (1966).
[34] Blumer, M., *Science*, **140**, 974 (1963).
[35] Avigan, J., and Blumer, M., *J. Lipid Res.* (in the press).
[36] Ackman, R. G., and Eaton, C. A., *Canad. J. Biochem.* (in the press).
[37] Christophe, J., and Popják, G., *J. Lipid Res.*, **2**, 244 (1961).
[38] Blumer, M., and Thomas, D. W., *Science*, **147**, 1148 (1965).
[39] Ackman, R. G., *J. Chromatog.*, **28**, 225 (1967).
[40] Murray, K. E., *Austral. J. Chem.*, **12**, 657 (1959).

24

Reprinted from *Chem. Commun.* **20**:752–754 (1966)

An Optically Active Triterpane, Gammacerane in Green River, Colorado, Oil Shale Bitumen

By I. R. Hills and E. V. Whitehead
(*The British Petroleum Company Limited, Sunbury-on-Thames, Middlesex*)

and

D. E. Anders, J. J. Cummins, and W. E. Robinson
(*Laramie Petroleum Research Centre, Laramie, Wyoming*)

In an earlier Paper,[1] the isolation of three pentacyclic triterpanes from an optically active petroleum distillate from Nigerian crude oil was reported. It was suggested there that the optically active hydrocarbons isolated by Oakwood *et al.*[2] and the hydrocarbon, m.p. 292—297°, separated from a Green River Colorado oil shale bitumen by Cummins and Robinson,[3] might also be pentacyclic triterpanes.

Upon request, after the work[1] had been reported, the group at Laramie Petroleum Research Centre kindly made available a sample of their hydrocarbon,[3] m.p. 292—297°. As received, it bore the reference KR 62—148 and for convenience will b referred to by this code number.

The melting points of KR 62—148 (292—297° and gammacerane (290°)[4,5] indicated that KR 62-148 might be identical with gammacerane (I), sinc gammacerane is the only triterpane of know structure which melts above 250°.

Professor F. B. Mallory[4] kindly provided us wit a few mg. of tetrahymane (gammacerane)[5] and w were able to compare this with the hydrocarbo KR 62—148.

The nuclear magnetic resonance spectra (6 Mc./sec.) of tetrahymane and KR 62—148 whic

TABLE

Relative retention volumes

0·01 in. i.d. glass capillary columns*		Relative retention volume n-C_{30} = 1·0		
Stationary phase	Length	Tetrahymane	KR 62–148	Mixture (1 : 1 wt)
SE 30 gum	58 ft.	1·49	1·49	1·47
SE 52 gum	148 ft.	1·69	1·68	1·70

* The approximate retention times from the methane peak to the $C_{30}H_{52}$ hydrocarbon on the SE 30 and SE 52 silicone gum columns, were respectively 1·75 and 2·5 hr.

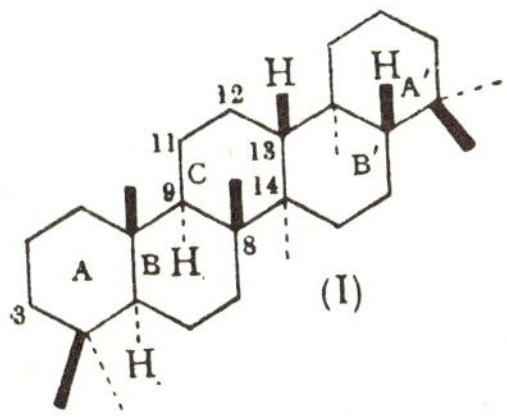

(I)

were virtually identical, showed that there were eight methyl groups in each molecule. These two spectra were different from, but showed certain similarities to the n.m.r. spectra of lupane and adiantane, a pentacyclic triterpane and nortriterpane respectively. All of these spectra also showed certain similarities to those of our triterpanes "E" and "F".[1]

Gas–liquid chromatograms were obtained from glass capillary columns coated with silicone gum. The carrier gas was hydrogen and the column was operated at 252° with a flame ionization detector. The relative retention volumes (R.R.V.) listed in Table 1 were determined on tetrahymane, KR 62—148 and on a 1 : 1 mixture of the two hydrocarbons. The mixed hydrocarbons showed a single symmetrical peak on each column and there was no evidence for any shoulder. The purity of tetrahymane on these columns appeared to be >99% whilst KR 62—148 was about 95%. Peak "O" in the gas–liquid chromatogram of the pentacyclic naphthene concentrate[1] has an R.R.V. of 1·49 (SE 30 gum 148 ft; n-C_{30} = 1·0) and may possibly be identical with tetrahymane and KR 62—148. Work in progress is directed towards the isolation from Nigerian crude oil of the hydrocarbon related to peak "O".

Tetrahymane had a specific optical rotation $[\alpha]^{21}_{546}$ + 29·4°±0·3° and the hydrocarbon KR 62—148 $[\alpha]^{21}_{546}$ + 31·9°±0·4°. These values were determined on as little as 1 mg. of sample using a new method developed by Mr. G. A. Speers for organic geochemical work at Sunbury. A full description of the method will appear in a future publication.

Uncorrected melting points determined in the laboratories at Sunbury were taken concurrently in open melting-point tubes in an oil bath. Tetrahymane, m.p. 301°; KR 62—148 m.p. 285—286°; a mixture (1 : 1) of tetrahymane with KR 62—148 melted at 288—292°. Each sample tended to sublime above 275° and showed signs of decomposition at the melting point. The slight differences in specific rotation and m.p. between these two hydrocarbons may be due to the presence in KR 62—148 of the impurities noted in its gas–liquid chromatogram.

The mass spectra of tetrahymane and KR 62—148 were recorded consecutively on an AEI MS-9 high-resolution mass spectrometer. These spectra were almost identical and relatively simple. The simplicity, no doubt, is due to the symmetry of the molecule which appears to favour fragmentation between C-8 and C-14, C-12 and C-13, C-9 and C-11, to yield identical fragments (II) at *m/e* 191 from rings A and B and from A′ and B′.

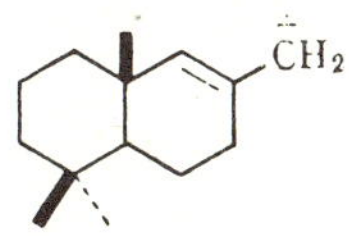

(II)

This fragment (II), containing rings A and B, is classed by Djerassi and his co-workers[6] as one of the most characteristic fragmentation products of pentacyclic triterpanes. The molecular ion at *m/e* 412 and the fragment at *m/e* 397 ($M-15$, methyl) as well as the fragments at *m/e* 205, 191, 137, 123, and 109 are all prominent in the mass spectra of pentacyclic triterpanes.[1] The accurate mass of the molecular ion from KR 62—148 was 412·409 and confirmed its molecular formula as $C_{30}H_{52}$. The mass spectrum of KR 62—148 indicated the presence of approximately 2·5% of a C_{31} homologue.

These experimental results confirm the identity of the hydrocarbon KR 62—148 as gammacerane. Derivatives of gammacerane, and derivatives of hopane closely related to gammacerane, are fairly widely distributed in the plant kingdom but pentacyclic triterpenes appear to occur only rarely in the animal kingdom. In fact, tetrahymanol (gammaceran-3β-ol), is the first pentacyclic triterpenoid alcohol isolated from an animal.[5] The ciliated protozoan *Tetrahymena pyriformis*, from which tetrahymanol was isolated[4] is widely distributed in Europe and America, usually in ponds, streams, stagnant pools, and thermal springs. It also occurs in the soil and probably in brackish water.[7]

Thus gammacerane (KR 62—148), isolated from the Mahogany-zone bed of the Green River Formation near Rifle, Colorado (Eocene *ca.* 52×10^{6}[b] years),[3] may be derived either from relatively simple plant forms or perhaps from protozoa both.

Other workers[8] have suggested the presence triterpenoid types in fossil fuels and sediments, b few triterpenoids of known structure[9] have yet be identified amongst the compounds isolated frc these sources. There appears to be less doubt nc that the pentacyclic triterpenoids and relat hydrocarbons had a biological origin, back geological time.

(*Received, August 22nd*, 1966; *Com.* 62

[1] I. R. Hills and E. V. Whitehead, *Nature*, 1966, **209**, 977.
[2] T. S. Oakwood, D. S. Shriver, H. H. Fall, W. J. McAleer, and P. R. Wunz, *Ind. Eng. Chem.*, 1952, **44**, 2568.
[3] J. J. Cummins and W. E. Robinson, *J. Chem. and Eng. Data*, 1964, **9**, 304.
[4] F. B. Mallory, J. T. Gordon, and R. L. Conner, *J. Amer. Chem. Soc.*, **1963**, **85**, 1362.
[5] Y. Tsuda, A. Morimoto, T. Sano, Y. Inubushi, F. B. Mallory, and J. T. Gordon, *Tetrahedron Letters*, 1965, 14
[6] H. Budzikiewicz, J. M. Wilson, and C. Djerassi, *J. Amer. Chem. Soc.*, **1963**, **85**, 3688.
[7] "An Introduction to the Study of Protozoa", D. L. Mackinnon and R. S. J. Hawkes, Clarendon Press, 1961, 2
[8] J. Marcusson, *Chem. Z.*, 1908, **32**, 377 and 391; W. G. Meinschein, *Bull. Amer. Assoc. Petrol. Geol.*, 1955, **43**, 92 *Geochim. Cosmochim. Acta*, 1961, **22**, 58; M. Louis, *Rev. Inst. franç. Pétrole*, 1964, **19**, 277; A. L. Burlingame, P. Ha T. Belsky, and M. Calvin, *Proc. Nat. Acad. Sci., U.S.A.*, 1965, **54**, 1406.
[9] D. H. R. Barton, K. H. Overton, and W. Carruthers, *J. Chem. Soc.*, 1954, 2047; K. Hejno, M. Horák, V. Jaroli M. Streibl, and F. Šorm, *Chem. and Ind*, 1958, 1142; *Coll. Czech. Chem. Comm.*, 1961, **26**, 459; 1963, **28**, 2318, 2443; 19 **30**, 873.

25

Reprinted from *Geochim. et Cosmochim. Acta* **38**:1165–1181 (1974)

Tri- and tetraterpenoid hydrocarbons in the Messel oil shale

B. J. Kimble, J. R. Maxwell, R. P. Philp and G. Eglinton
Organic Geochemistry Unit, School of Chemistry, University of Bristol, BS8 ITS, England

and

P. Albrecht, A. Ensminger, P. Arpino and G. Ourisson
Institut de Chimie, Université Louis Pasteur, Strasbourg, France

(*Received* 14 *November* 1973; *accepted in revised form* 28 *January* 1974)

Abstract—The high molecular weight constituents of the branched and cyclic hydrocarbon fraction of the Messel oil shale (Eocene) have been examined by high resolution gas chromatography and combined gas chromatography–mass spectrometry. The following compounds are present: perhydrolycopene (1; lycopane), together with one or more unsaturated analogues with the same skeleton; a series of 4-methylsteranes (2c) in higher abundance than their 4-desmethyl analogues; two series of pentacyclic triterpanes, one series (C_{27}–C_{32}) based on the hopane structure (3a–e), and the other (C_{27}–C_{31}) based on the 17α-H hopane structure (3a–d, 17αH); and an intact triterpene hop-17(21)-ene [3c, Δ17(21)]. Only two additional triterpanes were detected in minor concentrations, viz. 30-normoretane (3b, 21αH) and a C_{31} triterpane based on the hopane/lupane-type skeleton. The presence of these compounds suggests a significant microbial contribution to the forming sediment. Comparison of the tri- and tetraterpenoid hydrocarbons with those of the Green River Shale indicates differences in the organisms contributing to the two sediments.

Introduction

The Messel oil shale is an organic-rich Eocene sediment (~ 50 × 10^6 yr old) located 9 km north-east of Darmstadt, Germany. Geological studies indicate a particularly uneventful history and it is apparent that the sediment has experienced milder conditions (probably no more than 40°C) than those of the well-documented Green River Formation oil shale, U.S.A., of the same age. The shale is preserved in a basin 1000 m long and 700 m wide and the organic-rich layer has not been buried deeper than about 200 m (Matthes, 1968). It was probably deposited in a series of shallow swampy lakes linked by slow-moving fluvial systems; analyses of fossil plants and pollens (Matthes, 1968; Sittler, 1968) indicate that a hot, damp, tropical climate existed at the time of deposition. In the present study the sediment samples analysed comprise mainly gyttja (sapropelic black mud) with fine sandy clay intercalations (Matthes, 1968). The shale is composed of 25 per cent organic material, 35 per cent inorganic material and 40 per cent water (Matthes, 1968); the inorganic portion is mainly montmorillonite, and most of the organic carbon is present as kerogen.

Previous studies of the soluble fraction have identified isoarborinol (4a) (Albrecht and Ourisson, 1969), arborinone (4b) and friedelin (5), a series of 4α-methylstanols (2a) and 4α-methylstanones (2b), and very small quantities of phytosterols (6a,b,c) (Mattern *et al.*, 1970). Preliminary analysis of the hydrocarbons (Albrecht, 1969) indicated the absence of arborane, arborene and squalene; *n*-alkanes in the range *n*-C_{23} to *n*-C_{33} (C.P.I. > 6) are present, the main component being *n*-C_{27}. The lower molecular weight region of the branched and cyclic alkane fraction contains mainly isoprenoid alkanes with carbon numbers C_{15} (farnesane),

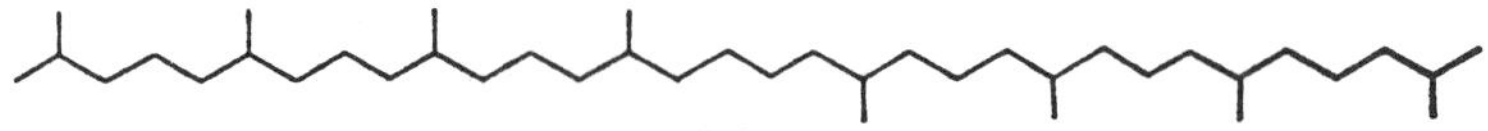

(1)

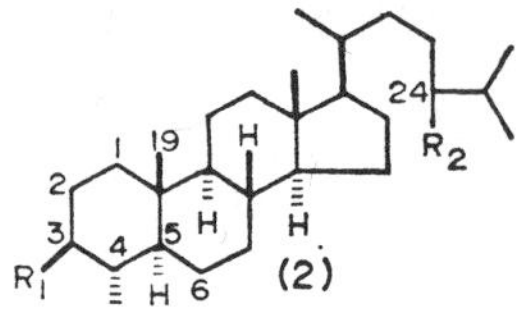

(2)

2a $R_1 = OH$, $R_2 = H, CH_3, C_2H_5$
b $R_1 = O$, $R_2 = H, CH_3, C_2H_5$
c $R_1 = H$, $R_2 = H, CH_3, C_2H_5$

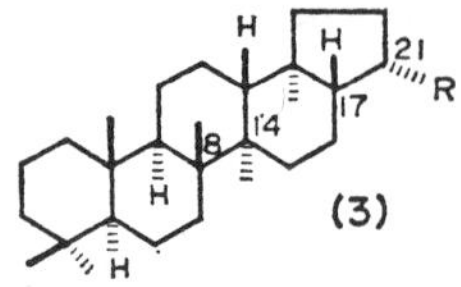

(3)

3a R = H
b $R = C_2H_5$
c $R = iso\,C_3H_7$
d $R = sec\,C_4H_9$
e $R = sec\,C_5H_{11}$

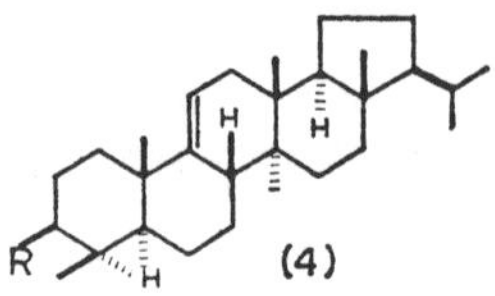

(4)

4a R = OH
b R = O

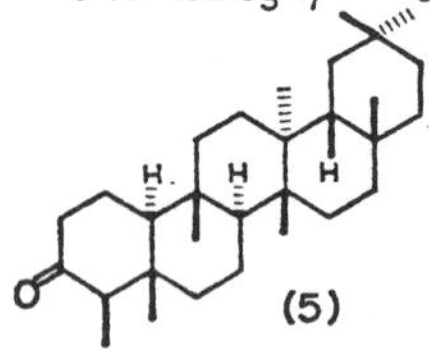

(5)

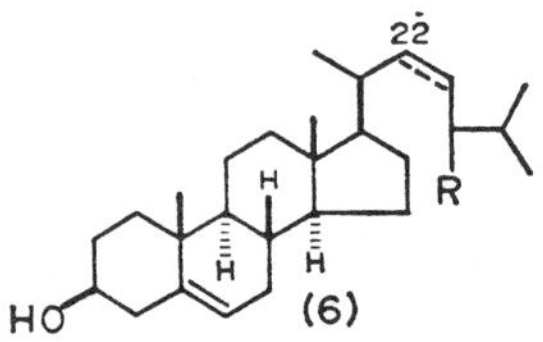

(6)

6a $R = CH_3$
b $R = C_2H_5$
c $R = C_2H_5$ and Δ-22

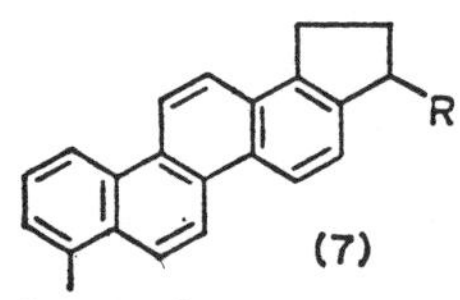

(7)

7a $R = iso\,C_3H_7$
b $R = C_2H_5$

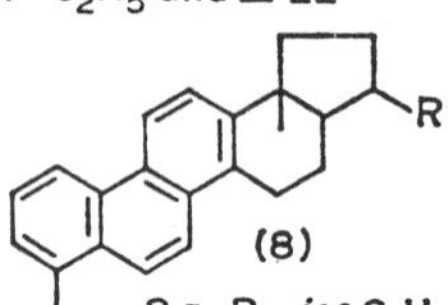

(8)

8a $R = iso\,C_3H_7$
b $R = C_2H_5$

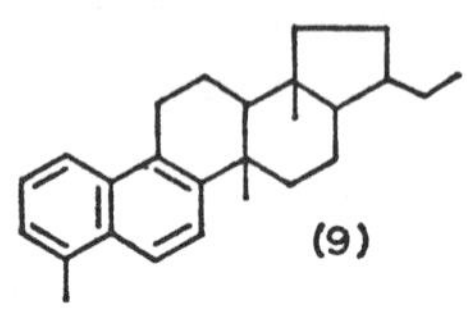

(9)

C_{16}, C_{18}, C_{19} (pristane) and C_{20} (phytane), in the relative abundances $C_{16} > C_{20} > C_{18} > C_{19} > C_{15}$ (ALBRECHT, 1969). Three types of aromatic hydrocarbons were isolated and mass spectrometric data indicated components of the formulae $C_{25}H_{24}$ (7a), $C_{24}H_{22}$ (7b), $C_{26}H_{30}$ (8a), $C_{25}H_{28}$ (8b) and $C_{26}H_{34}$ (9), compatible with the proposed structures indicated. The present study describes a detailed examination of the high molecular weight terpenoid components of the alkane and alkene fractions. A preliminary report of part of this work has appeared elsewhere (ENSMINGER *et al.*, 1972).

EXPERIMENTAL

The authentic standards used were: lycopane from catalytic hydrogenation of lycopene (Hoffmann La Roche) in ethyl acetate over palladium/charcoal; 1α-, 2α-, 3β- and 4α-methylcholestanes (KIMBLE *et al.*, in press); hopane, 17αH-hopane, moretane, 17αH-moretane from

Prof. R. E. Corbett; 22,29,30-trisnorhopane, 30-norhopane, homohopane and 17αH analogues (ENSMINGER *et al.*, 1974); cholestane; stigmastane from Dr. W. McCrae.

Gas–liquid chromatography (GLC)

GLC was initially carried out using a 10 ft × $\frac{1}{16}$ in. stainless steel packed column (3 per cent OV-17 on 100–120 mesh Chromosorb W; 270°C; nitrogen carrier gas, 10–12 ml/min), and a Perkin–Elmer Mark 1 F-11 gas chromatograph (injector block 290–310°C).

Preparative GLC conditions for lycopane were:—column 10 ft × $\frac{1}{4}$ in., 3 per cent SE 30 on 80–100 mesh Gas Chrom Q, 60 ml/min Ar, programmed from 150–300°C; effluent trapped in liquid nitrogen. Capillary GLC was carried out on three stainless steel open tubular capillary columns as follows:—

(i) 75 ft × 0·01 in. i.d. column coated with OV-101 in a Perkin–Elmer 226 gas chromatograph incorporating a pre-column carrier gas splitter. Carrier gas (He) flow rate was 2 or 3 ml/min; the injector block was maintained at 300°C and the column was operated isothermally at 250°C. Column efficiencies were calculated to be 25,000, 20,000 and 15,000 theoretical plates for n-C_{28}, n-C_{32} and n-C_{36} standard alkanes, respectively.

(ii) 150 ft × 0·01 in. i.d. column coated with Dexsil 300 (Perkin–Elmer Ltd.) was used in a Perkin–Elmer Mark II F-11 gas chromatograph. The flow rate (He) was 3 ml/min; the injector block was maintained at 320°C, and the column was operated isothermally at 280°C. Column efficiency was calculated to be 59,000 theoretical plates for 5α-cholestane.

(iii) 100 ft × 0·01 in. i.d. column coated with OV-101 (Perkin–Elmer Ltd.) was used as described in (ii), except for isothermal operation at 250°C. Column efficiency was calculated to be 25,000 theoretical plates for 5α-cholestane.

Combined gas chromatography–mass spectrometry (GC–MS)

Mass spectra were obtained using a Varian Aerograph 1200 gas chromatograph coupled by an all-glass single-stage Watson–Biemann He separator to a Varian MAT CH-7 single focussing mass spectrometer. The accelerating potential was 70 eV, the filament current 100 or 300 μA, and the instrument resolution 800–1000. The gas chromatograph was used initially with the packed column described above and subsequently with the capillary columns described in (ii) and (iii) above. Under these conditions the efficiency of column (ii) was *ca.* 55,000 theoretical plates for 5α-cholestane.

Isolation and analysis of branched and cyclic alkanes and alkenes

Analyses of the alkanes were carried out on two different solvent extracts of the shale.

(i) Extract I was obtained using a mixture of light petroleum (b.p. 60–80°C) and ethyl acetate (4:1) and the total hydrocarbons (0·25 per cent of 1 kg of dry rock) were isolated by column chromatography (SiO_2/light petroleum, b.p. 60–80°C). Additional column chromatography (SiO_2/light petroleum) and thin layer chromatography using silica gel impregnated with 10 per cent $AgNO_3$ (Ag^+—TLC) gave the total alkanes (226 ppm). Treatment of the total alkanes with 5 Å molecular sieve (O'CONNOR *et al.*, 1962), and urea adduction (COX, 1971) gave the total branched and cyclic alkanes (56 ppm) which were re-purified by Ag^+—TLC (hexane) and further simplified by thiourea adduction (MURPHY *et al.*, 1967). The resulting adduct and non-adduct fractions were analysed by packed column and capillary column (i) GLC and packed column GC–MS.

(ii) Extract II was obtained using cyclohexane; elemental sulphur was removed by the method of BLUMER (1957). The sulphur-free extract (0·2 per cent of 1·5 kg of rock) was chromatographed (neutral alumina column/hexane) to give the total hydrocarbons (0·04 per cent) from which the branched and cyclic hydrocarbons were isolated on adduction of the straight chain components with urea. The alkanes (*ca.* 20 ppm) and alkenes (*ca.* 270 ppm) were separated by Ag^+—TLC (hexane developer).

An aliquot of the branched and cyclic alkenes from Extract II was hydrogenated (PtO_2/C EtOAc/atmospheric pressure of H_2) and the resulting alkanes were isolated by Ag^+—TLC.

The alkane and alkene fractions obtained (Extract II) were analysed by capillary column [described under (ii) and (iii) above] GLC and by packed column and capillary GC–MS (columns ii and iii).

RESULTS

Identifications were based on high resolution gas chromatography on Dexsil 300 and OV-101 by coinjection with authentic standards and comparison of mass spectra with those of the standards. This method is similar to that used by HENDERSON *et al.* (1968a, b) for the identification of steranes and triterpanes in the Green River Formation oil shale. The chromatogram of the urea alkane non-adduct (Extract II) on Dexsil 300 is shown (Fig. 1), the peak numbers corresponding to the components summarized in Table 1.

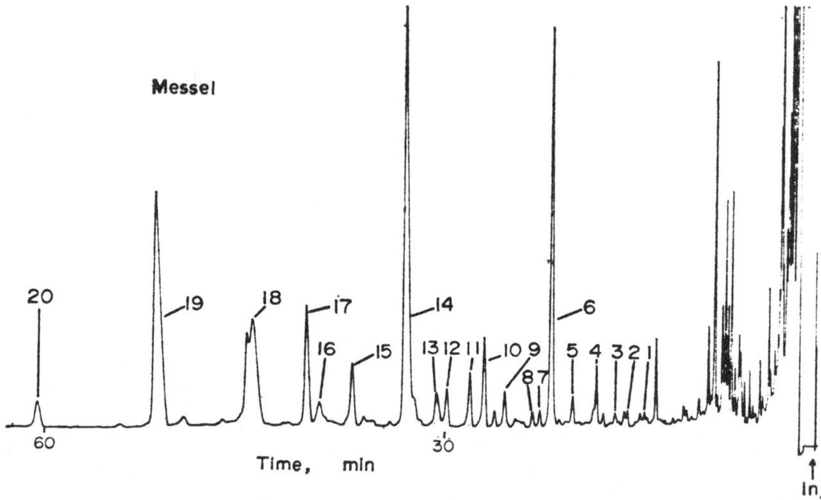

Fig. 1. Capillary gas chromatogram of Messel branched and cyclic alkanes (Extract II, urea non-adduct) on Dexsil 300 (isothermal 280°C).

Lycopane (*perhydrolycopene*, 1)

The mass spectrum (Fig. 2a) of the main component in the thiourea adduct fraction of Extract I (also peak 18 in Fig. 1) indicated an acyclic alkane with extensive branching. Detailed information could not be obtained from the GC–MS spectrum of the region above m/e 480 as the ions were of particularly low intensity. The GC–MS spectrum (Fig. 2b) of an authentic sample of lycopane from catalytic hydrogenation of lycopene strikingly resembled that of the Messel component but again it was not possible to determine the molecular ion exactly. More intense spectra of this region (incorporated into Figs. 2a, b) were obtained by direct insertion mass spectrometry (perfluorokerosene marker) of the Messel component isolated by preparative GLC and of authentic lycopane.

Further proof of identity was obtained by capillary column GLC. Coinjection of authentic lycopane (1; Kovats indices: 3471 on Dexsil 300, 3506 on OV-101) with the thiourea adduct fraction of Extract I and with the total branched and cyclic alkanes from Extract II (Fig. 1), produced peak enhancement. The similarities in

Table 1. Polyterpenoid branched and cyclic alkanes identified in Messel oil shale by capillary GC–MS
(Peak numbers refer to Fig. 1)

Peak	Formula	Assignment	Structure
1	$C_{27}H_{48}$	5β-Cholestane	10a; 5βH
2	$C_{27}H_{48}$	Cholestane	10a
3	$C_{28}H_{50}$	4-Methyl-5β-cholestane	2c; 5βH; R_2 = H
4	$C_{28}H_{50}$	4-Methylcholestane	2c; R_2 = H
5	$C_{27}H_{46}$	17αH-trisnorhopane	3a; 17αH
6	$C_{27}H_{46}$	Trisnorhopane	3a
7	$C_{29}H_{52}$	4-Methylergostane	2c; R_2 = CH_3
8	$C_{29}H_{52}$	Stigmastane	10c
9	$C_{29}H_{50}$	17αH-Norhopane	3b; 17αH
10	$C_{30}H_{54}$	4-Methylstigmastane	2c; R_2 = C_2H_5
11	$C_{29}H_{50}$	30-Normoretane	3b; R = 21αH
12	$C_{30}H_{52}$	17αH-Hopane	3c; 17αH
13	$C_{30}H_{50}$	Pentacyclic triterpene	
14	$C_{29}H_{50}$	Norhopane	3b
15	$C_{31}H_{54}$	17αH-Homohopane	3d; 17αH
16	$C_{31}H_{54}$	Pentacyclic triterpane	
17	$C_{30}H_{52}$	Hopane	3c
18	$C_{40}H_{82}$	Lycopane	1
19	$C_{31}H_{54}$	Homohopane	3d
20	$C_{32}H_{56}$	Bishomohopane	3e

the GC–MS spectrum of peak 18 (Fig. 1) and in the direct insertion spectra of authentic lycopane and the preparative GLC sample which included the shoulder on peak 18 suggest that the shoulder may be a compound similar to lycopane; indeed, its presence could indicate separation of diastereoisomers of lycopane.

In both spectra (Fig. 2) the ion at m/e 560 could represent the molecular ion for a mono-unsaturated or monocyclic alkane impurity of formula $C_{40}H_{80}$. However, this is unlikely because neither spectrum contains fragment ions (e.g. M-15 and

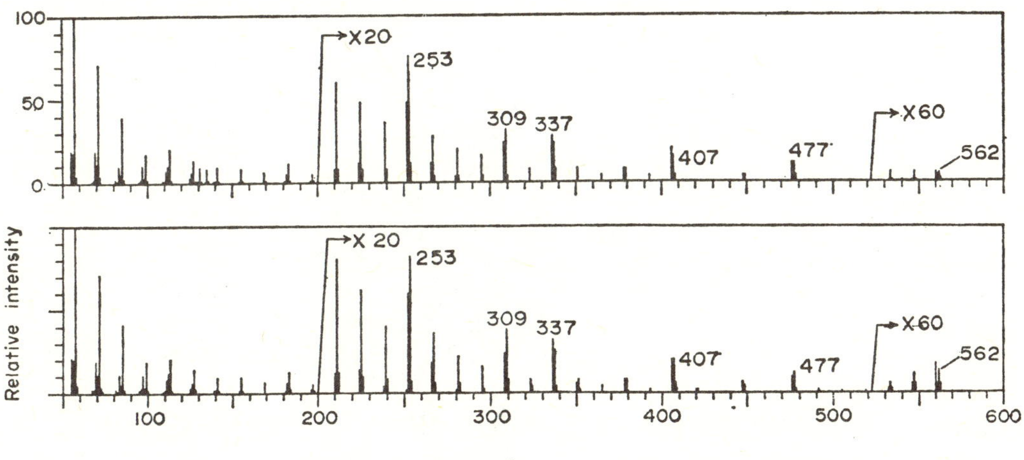

Fig. 2. Mass spectra from capillary GC–MS on Dexsil 300 of (a) Messel component 18 in Fig. 1 and (b) authentic lycopane ($C_{40}H_{82}$, 1), respectively.

C_nH_{2n-1}) expected from a monocyclic carotenoid alkane such as perhydro-γ-carotene. Also, bromination of authentic lycopane followed by Ag^+-TLC purification failed to produce any change in the spectrum. The M-2 ion at m/e 560 therefore appears to be a genuine fragment ion, as do those at m/e 252, 308, 336, 406 and 476, of the C_nH_{2n} series.

Methylsteranes and steranes (*Table* 1)

The spectra of four cycloalkanes ($C_{28}H_{50}$, $C_{28}H_{50}$, $C_{29}H_{52}$ and $C_{30}H_{54}$) present in the thiorea adduct (Extract I) and in the urea non-adduct (Extract II, Peaks 3, 4, 7 and 10, Fig. 1) showed fragmentation patterns analogous to those of the authentic sterane series cholestane ($C_{27}H_{48}$, 10a), ergostane ($C_{28}H_{50}$, 10b) and stigmastane ($C_{29}H_{52}$, 10c) (HENDERSON *et al.*, 1968a, b; ANDERSON *et al.*, 1969). The major features of the seven spectra are summarized in Table 2. Thus, for the Messel

Table 2. Structurally significant ions in the mass spectra of four Messel oil shale alkanes and those of authentic (ANDERSON *et al.*, 1969) cholestane, ergostane and stigmastane

Peak no. (Fig. 1)	Formula	Assignment	Structure	Significant ions					M-124	M-15	M
3	$C_{28}H_{50}$	4-Methyl-5β-cholestane	2c; 5βH; R_2 = H	163/165	217	231	232	246	262	371	386
4	$C_{28}H_{50}$	4-Methylcholestane	2c; R_2 = H	163	217	231	232	246	262	371	386
7	$C_{29}H_{52}$	4-Methylergostane	2c; R_2 = CH_3	163	217	231	232	246	276	385	400
10	$C_{30}H_{54}$	4-Methylstigmastane	2c; R_2 = C_2H_5	163	217	231	232	246	290	399	414
									M-110	M-15	M
—	$C_{27}H_{48}$	Cholestane	10a	149	203	217	218	232	262	357	372
—	$C_{28}H_{50}$	Ergostane	10b	149	203	217	218	232	276	371	386
—	$C_{29}H_{52}$	Stigmastane	10c	149	203	217	218	232	290	385	400

components the ions from ring D fragmentation occur at m/e 231 and 232 (cf. m/e 217 and 218 for the C_{27}–C_{29} steranes) and the fragment ions at m/e 163, 217 and 246 have similar relative intensities to the ions at m/e 149, 203 and 232, respectively, in the C_{27}–C_{29} steranes whose spectra have been previously reported in detail (HENDERSON *et al.*, 1968a, b; ANDERSON *et al.*, 1969). These features indicate that the Messel components are methyl-substituted analogues of cholestane, ergostane and stigmastane, the position of substitution being limited to one of the carbon atoms 1, 2, 3, 4, 6 and 19 (TOKÉS *et al.*, 1968). An authentic sample (KIMBLE *et al.*, in press) of 4α-methylcholestane (2c, R_2 = H) gave the expected spectrum and enhanced peak 4 (Fig. 1) when coinjected on Dexsil 300. Synthetic samples of 1α-methyl-, 2α-methyl-, 3β-methyl-, and 4α-methylcholestane which have similar mass spectra were all separable by capillary column GLC on Dexsil 300 (KIMBLE *et al.*, in press). The evidence suggests that peaks 3, 7 and 10 (Fig. 1) are also 4-methylsteranes. Comparison of the intensities of the ions at m/e 163 and 165 in the spectrum of peak 3 with those in the spectrum of the assigned 4-methylcholestane suggest that peak 3 is the 5β analogue of 4-methylcholestane, the relative intensities of the two ions paralleling those of the ions at m/e 149 and 151 in the spectra of cholestane and 5β-cholestane (GALLEGOS, 1971). Peaks 1, 2 and 8 were assigned in the usual way as 5β-cholestane (10a; 5βH), cholestane (10a) and stigmastane (10c) respectively, by, comparison with authentic standards.

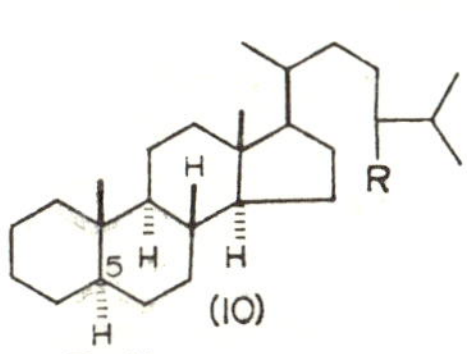

(10)

· 10a R = H
b R = CH_3
c R = C_2H_5

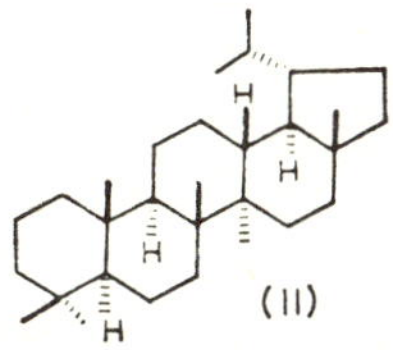

(11)

(12)

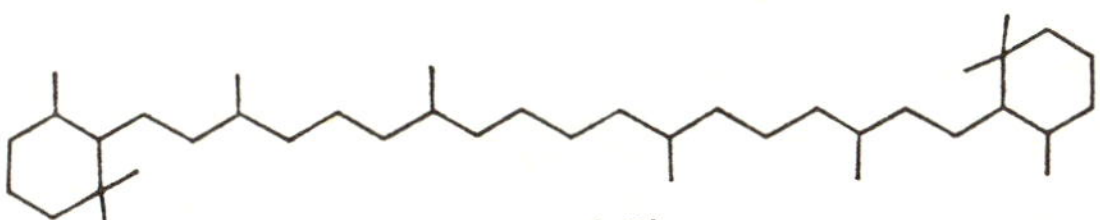

(13)

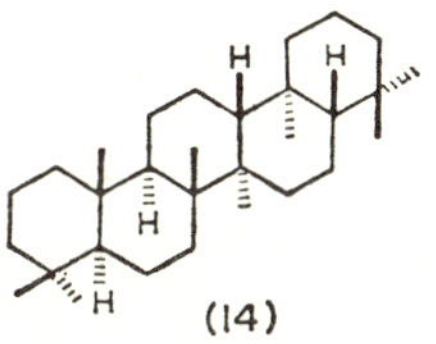

(14)

The stereochemistry of the C-4 methyl group remains unproven, as the 4α- and 4β-isomers may not be distinguishable under the GLC conditions used. Considerations of sterol biosynthesis and the relative abundance of the 4α-methylsterols compared to 4β-methylsterols in organisms (REES and GOODWIN, 1972, and references therein), together with the inherent stability of the 4α-isomers with respect to the 4β-methyl compounds (PYREK, 1969), suggest that these geological methyl steranes are probably the 4α-methyl series.

*Triterpanes** (*Table* 1)

The spectra of the major higher molecular weight components (listed in Tables 1 and 3) in the thiourea non-adduct fraction of Extract I all showed a very intense ion at m/e 191, typically found in pentacyclic triterpanes with two adjacent quaternary carbon atoms at C-8 and C-14 [for example, structures (3a–e)] or at C-8 and C-13 (KIMBLE *et al.*, in press). The molecular formulae (Table 3) range from $C_{27}H_{46}$ to $C_{32}H_{56}$ with the notable absence of $C_{28}H_{48}$; this range is related to the nature of the E ring substituent R which is also reflected in the masses of the intense ions at m/e 149, 177, 191, 205 and 219 (3a–e).

* In the present context, 'triterpane' is used to imply the saturated hydrocarbons presumably derived from triterpenoids, and to include their proximate higher (C_{31}, C_{32} . . .) and lower (C_{29}, C_{28}, C_{27}) homologues.

Table 3. C_{27}–C_{32} pentacyclic triterpanes in the Messel and Green River Shales

Formula	Messel triterpanes	Green River triterpanes*
$C_{27}H_{46}$	Trisnorhopane (3a, *ca.* 3·5%)† 17αH-Trisnorhopane (3a, 17αH, <1%)	$C_{27}H_{46}$ (trace)
$C_{28}H_{48}$	n.d.	$C_{28}H_{48}$ (trace)
$C_{29}H_{50}$	Norhopane (3b, *ca.* 3·5%) 17αH-Norhopane (3b, 17αH, <1%) Isoadiantane (3b, 21αH, <1%)	17αH-norhopane (3b, 17αH, *ca.* 2%)
$C_{30}H_{52}$	Hopane (3c, *ca.* 2%) 17αH-Hopane (3c, 17αH, <1%)	17αH-hopane (3c, 17αH, *ca.* 9%‡) gammacerane (14, *ca.* 3%) $C_{30}H_{52}$ (*ca.* 4%) $C_{30}H_{52}$ (<1%)
$C_{31}H_{54}$	Homohopane (3d, *ca.* 7%) 17αH-Homohopane (3d, 17αH, <1%) $C_{31}H_{54}$ (Hopane/lupane type <1%)	17αH-homohopane (3d, 17αH, <1%) $C_{31}H_{54}$ (<1%)
$C_{32}H_{56}$	Bishomohopane 3e (<1%)	$C_{32}H_{56}$ (<1%)

* Taken from data of BURLINGAME *et al.* (1965); HILLS *et al.* (1966); HENDERSON *et al.* (1968a, b); GALLEGOS (1971); ANDERS and ROBINSON (1971); BALOGH *et al.* (1973); VAN DORSSELAER *et al.* (in press).

† % of branched cyclic alkane fraction.

‡ Previously tentatively identified as hopane by HENDERSON *et al.* (1968b); shown to be 17α(H) isomer by WHITEHEAD (1971) and BALOGH *et al.* (1973).

n.d. not detected.

The major C_{31} triterpane (Fig. 1, peak 19) was identified (ENSMINGER *et al.*, 1972) as *one* of the two C-22 stereoisomers of homohopane (3d, R = *sec*-C_4H_9). In the present study the following authentic standards were coinjected with the branched and cyclic alkane fraction (Extract II, Fig. 1) on both Dexsil 300 and OV-101 phases: 22,29,30-trisnorhopane (3a); 30-norhopane (adiantane, 3b); hopane (3c); 22,29,30-trisnor-17αH-hopane (3a, 17αH); 30-nor-17αH-hopane (3b, 17αH); 17αH-hopane (3c); 17αH-homohopane (3d; 17αH); and 30-normoretane (isoadiantane 3b, 21αH). The resulting peak enhancements and the comparison of the capillary GC–MS mass spectra (Dexsil 300 and OV-101) of the standards with those of the Messel components (Fig. 3), allowed identification of components 6, 14 and 17 (Fig. 1) as the C_{27}, C_{29} and C_{30} members of the hopane series (3a–c, respectively), components 5, 9, 12 and 15 as the C_{27}, C_{29}, C_{30} and C_{31} members of the 17αH-hopane series (3a–d, 17αH, respectively). Similarly, component 11 (Fig. 1) was identified as 30-normoretane. The assignments are summarized in Table 1. Coinjection with authentic moretane (3c, 21αH), 17αH-moretane (3c, 21αH, 17αH) and arborane (cycloalkane from 4) showed that these compounds were either absent or present in very low abundances.

The Messel triterpanes, therefore, appear to form two related series, each series characterized by the stereochemistry at the D–E ring junction (i.e. carbon atom 17) and based on variations (H, C_2H_5, etc.) of the substituent R in the E ring. Since the

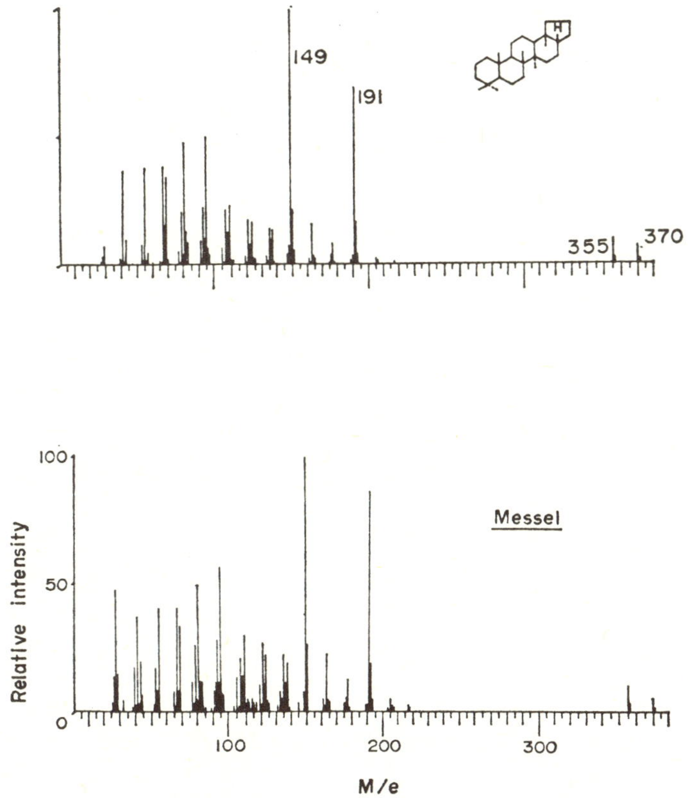

Fig. 3. Mass spectra from capillary GC–MS on Dexsil 300 of Messel component 6 in Fig. 1 and authentic 22, 29, 30-trisnorhopane (3a).

more abundant series by far is the hopane series (3a–d), it is inferred that the only C_{32} triterpane present, component 20 (Fig. 1, Table 1), is the C_{32} member of this series. Comparison of the mass spectrum of this component with those of the lower members of the hopane series is also in agreement with this assignment. Component 13 (Fig. 1) appears to be an unsaturated pentacyclic triterpene hydrocarbon, $C_{30}H_{50}$ (Table 1), with a spectrum corresponding to that of hopane (3) or lupane (11)-types of skeleton, i.e. a 5-membered E-ring; its presence in the branched/cyclic alkane fraction indicates that the double bond is in a hindered position. Component 16, the third C_{31} pentacyclic triterpane, also appears to be related to the hopane (3) or lupane (11) skeletons but its spectrum differs from that of homohopane (3d) and 17αH-homohopane (3d, 17αH) in the relative abundances of the ions at m/e 191 and 205.

Branched and cyclic alkenes

Preliminary information was obtained by catalytic hydrogenation of an aliquot followed by preparative Ag^+–TLC. An unsaturated fraction was obtained from the products and was dominated by a single component in the higher molecular weight region. The mass spectrum indicated a molecular formula $C_{30}H_{50}$, and was similar

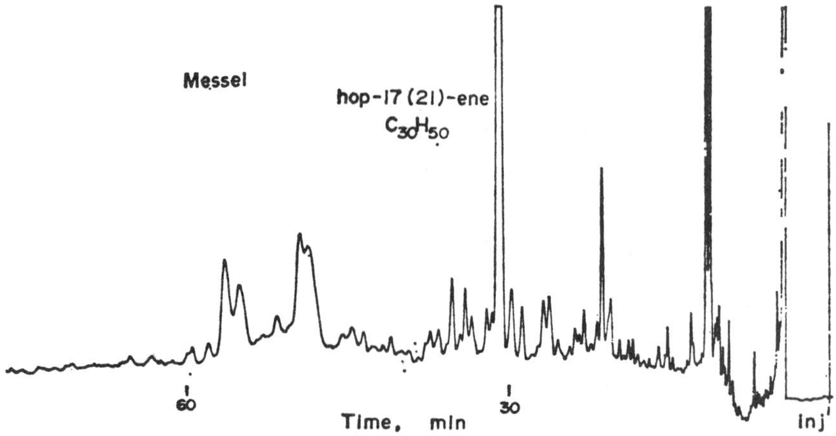

Fig. 4. Capillary gas chromatogram of Messel branched and cyclic alkenes (Extract II, urea non-adduct) on Dexsil 300 (isothermal 280°C).

to that obtained from an authentic sample of hop-17(21)-ene [3c, Δ17(21)]. This identification was confirmed by coinjection on both phases; furthermore, direct analysis of the branched and cyclic alkene fraction by GC–MS and GLC coinjection showed that the main component in Fig. 4 was hop-17(21)-ene, present intact in the shale. This component has a different mass spectrum from that of peak 13 in the branched/cyclic alkane fraction (Fig. 1, Table 1) which appears to be another hopene from its mass spectrum with a hindered double bond, since it is present in the alkanes after Ag^+–TLC.

A fraction, corresponding to alkanes in Ag^+–TLC R_f, was also isolated from the hydrogenation products. GLC of this fraction indicated four main components. The mass spectrum of the least volatile of these strongly resembled that obtained from authentic lycopane (Fig. 2), which produced peak enhancement on both phases. The identification of lycopane in this fraction indicates that one or more unsaturated compounds having the same carbon skeleton are present in the sediment.

DISCUSSION

Lycopane

The identification of lycopane (1) is the first reported occurrence in a geological material, although the structurally-related C_{30} compound squalane (12) has been identified in a Nigerian crude oil (GARDNER and WHITEHEAD, 1972). The bicyclic tetraterpane perhydro-β-carotene (13) has been identified in the Green River Formation oil shale (MURPHY *et al.*, 1967), and mass spectral data have also indicated the presence of another $C_{40}H_{82}$ component, components of molecular formulae $C_{40}H_{62}$ to $C_{40}H_{70}$ (GALLEGOS, 1971) and a monocyclic $C_{40}H_{80}$ alkane (ANDERS and ROBINSON, 1971). Similar data obtained from hydrocarbon samples isolated from several deep sea (JOIDES) cores indicated C_{40} compounds, including one of formula $C_{40}H_{74}$ (SIMONEIT and BURLINGAME, 1971).

Analysis of another deep sea (JOIDES) core (HARE and HOERING, 1972) indicated the presence of an acyclic C_{40} monohydroxy compound, apparently having the biosynthetically-interesting *head*-to-*head* coupling of two regular C_{20} units.

The presence of tetraterpanes in the Messel and Green River Formation oil shales indicates firstly a significant contribution of carotenoids to the original sediments, and secondly that conditions in these sediments permitted the reduction of carbon–carbon double bonds and the preservation of carotenoid carbon skeletons.

In contrast to the algal origin proposed for much of the organic matter in the Green River Formation oil shale (BRADLEY, 1970), the explanation of significant quantities of lycopane and the apparent absence of perhydro-β-carotene in the Messel shale is less obvious. Acyclic carotenoids have so far only been found as *major* carotenoids in photosynthetic bacteria (GOODWIN, 1965, 1966, 1970; JENSEN, 1963, 1967; WEEDON, 1965) and the distribution of cyclic and acyclic structures within the families indicates that non-sulphur purple bacteria (anaerobic heterotrophs found in soils, stagnant ponds, etc.; VAN NIEL, 1963) are likely to be the main producers of acyclic carotenoids to the exclusion of their cyclic analogues.

Methylsteranes

The series of 4-methylsteranes is of particular significance in view of the co-occurrence of their corresponding saturated 3-ketones and 3-alcohols (MATTERN *et al.*, 1970) and because cycloalkanes having these carbon skeletons have not so far been reported to occur in other geological materials.

4-Monomethylsterols having C_{28}, C_{29} or C_{30} carbon skeletons exist as *minor* sterol components in a variety of plants but relatively large amounts of unsaturated 3-hydroxy derivatives of 4α-methylcholestane have been reported in the bacterium *Methylococcus capsulatus* (BIRD *et al.*, 1971a, b).

The identification of 4-methylsteranes and their derivatives in a geological material is of interest, as the occurrence of large amounts of 4-methyl compounds relative to 4-desmethyl compounds contrasts with the relative abundances in present-day plant species. This may be explained by supposing that the original Messel flora was dominated by species synthesizing predominantly the 4-methyl structures, or that the 4-methyl compounds have survived diagenesis better than the 4-desmethylsterols, or that the common 4-desmethyl-phytosterols were initially incorporated into the sediment and were subsequently methylated at the 4-position. The established intermediacy of 4-methylsterols in the biosynthesis of 4-desmethyl-phytosterols (REES and GOODWIN, 1972, and references therein) supports the first explanation and further studies may identify other bacterial species that produce predominantly 4-methylsterols.

The co-occurrence of these skeletal types as alkanes, ketones and alcohols indicates that the diagenetic and/or maturation processes involved redox reactions producing ketones and hydrocarbons from biologically-occurring alcohols.

Triterpanes

The most abundant triterpanes present comprise a series (3a–e) based on the hopane ring structure. The presence of isoarborinol (4a) and arborinone (4b) as major constituents of the Messel shale, taken with the absence of the corresponding cycloalkane, arborane, in the hydrocarbon fraction, suggests that the geological hopane-type hydrocarbons derive from 3-desoxy-triterpanes with the same skeleton, rather than from 3-oxygenated precursors. Such 3-desoxycarbon skeletons occur in ferns (BERTI and BOTTARI, 1968; BOTTARI *et al.*, 1972). Also, recent studies have shown that hopane (3c), hop-22(29)-ene (3c, R-*iso* C_3H_5) and hop-17(21)-ene [3c, Δ17(21)] are present in the bacterium *Bacillus acidocaldarius*, along with two minor components ascribed to a C_{31} hopane-type alkane and alkene, respectively (DE ROSA *et al.*, 1971, 1973); hop-22(29)-ene has also been isolated from the bacterium *Methylococcus capsulatus* and other prokaryotes (BIRD *et al.*, 1971a, b). Re-examination of the mass spectrum of a polycyclic hydrocarbon isolated from each of three species of blue-green algae, *Lyngbya aestuarii*, *Nostoc* sp., and *Chroococcus turgidus* (GELPI *et al.*, 1970) suggests that this compound is probably also hop-22(29)-ene.

The second, minor triterpane series (3a–d, 17αH) present in the Messel shale has the 17αH stereochemistry; at present there is no report of the occurrence in living organisms of any compound with a carbon skeleton of this type.

Extended ($>C_{30}$) and degraded ($<C_{30}$) pentacyclic triterpanes have been detected by mass spectrometry in a variety of geological materials: oil shales (HENDERSON *et al.*, 1968a, b; DOUGLAS *et al.*, 1969; GALLEGOS, 1971; WSZOLEK *et al.*, 1971; ANDERS and ROBINSON, 1971; ARPINO *et al.*, 1971; VAN DORSSELAER *et al.*, in press), coal and lignite (MAXWELL, 1967; VAN DORSSELAER *et al.*, in press, and crude oil (DANIELLI *et al.*, 1968; HILLS *et al.*, 1970; WHITEHEAD, 1971; VAN DORSSELAER *et al.*, in press).

Such triterpane carbon skeletons may be considered to be of essentially a geochemical origin, i.e. formed by the alteration of C_{30} biolipids by the physical and chemical effects of the geological environment, a biogeochemical origin, i.e. formed by the action of organisms in the forming sediment, or a biochemical origin, i.e. formed by organisms before sedimentation.

(i) *Geochemical origin of triterpanes.* The formation of degraded triterpanes containing 27 and 29, but *not* 28, carbon atoms from a C_{30} precursor, is a problem similar to that of the formation of C_{16}, C_{18} and C_{19}, but *not* the C_{17}, isoprenoid alkanes from a C_{20} precursor. The C_{28} compound (cf. the C_{17} isoprenoid) can only be derived by cleavage of two C—C bonds located at the same carbon atom; only single carbon–carbon bond cleavages would be required for the formation of the C_{27} (cf. the C_{16} isoprenoid) and the C_{29} (cf. the C_{18} isoprenoid) structures. Participation of the insoluble organic kerogen matrix in the formation of these triterpanes has been proposed (ARPINO *et al.*, 1971). However, the subsequent identification (ENSMINGER *et al.*, 1972) of only *one* of the two C-22 stereoisomers of the C_{31} triterpane homohopane in the Messel shale is inconsistent with the particular route proposed by ARPINO *et al.* (1971). In addition, the occurrence of homohopane (3d) and other alkylated and dealkylated hopane-type hydrocarbons in recently-deposited lacustrine sediments (CRANWELL, 1973; EGLINTON *et al.*, in press) indicates that such compounds can be formed at the very earliest stages of diagenesis, although it is not yet certain that the compounds in these sediments are not derived from older geological mateials. However, if they are syngenetic with the contemporary sediments in which they occur, then it appears that thermal cracking (carbon–carbon bond cleavage) of the kerogen matrix is not a necessary prerequisite for their formation.

(ii) *Biogeochemical/biochemical origin of triterpanes.* Several related possibilities exist:—

(a) The alkylated, degraded and C_{30} triterpane skeletons were produced by species of the original Messel biota which were capable of synthesizing these skeletons directly (biogenesis), presumably from squalene or squalene epoxide followed by the gain or loss of carbon atoms. The occurrence of a C_{31} hopane-type alkene and alkane in a bacterium (DE ROSA *et al.*, 1971, 1973) indicates that a biosynthetic pathway exists in microorganisms for the formation of the C_{31} skeleton and its saturation.

(b) The triterpenoids originally deposited in the sediment were of the types found in present-day organisms (the C_{30} hopane and C_{29} adiantane skeletons) and carbon atoms were subsequently added and removed by microorganisms during early-stage diagenesis (bioalteration).

In cases (a) and (b), the additional carbon atoms would presumably be supplied by methionine. Mechanisms paralleling those involved in the biosynthesis of the C-24 alkylated sterols (LEDERER, 1969) can be envisaged which would be expected to form the observed single C_{22} stereoisomer (R or S; ENSMINGER *et al.*, 1972) of the C_{31} homohopane bearing a *sec*-butyl (1-methylpropyl) substituent.

Parallels for the specific removal of carbon atoms to form the C_{27} and C_{29}, but not the C_{28}, pentacylic structures are less obvious. The biosynthetic route for the formation of adiantane (C_{29}) derivatives in plants is not known, but the loss of carbon atoms from the C-4 and C-14 positions in sterols and certain tetracyclic triterpenoids proceeds by the sequential oxidation of a methyl group to a carboxyl group followed by decarboxylation (GOAD, 1969). Similar decarboxylation of a hopane skeleton would imply the stepwise removal of carbon atoms, which does not preclude formation of a C_{28} carbon skeleton. It appears, therefore, that, if route (b) is the one involved, the mechanism involved must be sensitive to the precise branched structure of the isopropyl side chain and specific reactive sites in the form of functional groups must be available at the 22–29 (C_{29} structure) and 21–22 (C_{27} structure) bonds. Both the biologically-occurring triterpenoids 22-hydroxyhopane and hop-22(29)-ene could provide such sites, the former by dehydration [to hop-22(29)-ene and hop-21(22)-ene] and the latter by reacting directly and by double bond isomerization to hop-21(22)-ene (which occurs very readily in the laboratory with mild acids).

(c) The triterpanes could be the degradation products of a biologically-produced C_{35} pentacyclic isoprenoid (WHITEHEAD, 1972). For example, the major Messel triterpane, homohopane (3d) could have arisen from decarboxylation of one of the C-22 stereoisomers of the homologous C_{32} acid (3d, R = $C_3H_6CO_2H$), itself formed by degradation of a pentacyclic C_{35} compound ENSMINGER *et al.*, in press). Possible precursor compounds have recently been isolated from a bacterium (FÖRSTER *et al.*, 1973).

Our present data do not allow distinction between possibilities (a), (b) and (c) but the evidence available points towards the involvement of microorganisms in the formulation of the extended and degraded hopanes in the Messel shale.

The origin of the 17αH-configuration in the minor Messel triterpane series (3a–d, 17αH) is also of interest because no carbon skeletons of this type have been found in living organisms. ENSMINGER *et al.* (in press) have provided evidence that the ratio of the 17αH alkane to the 17βH alkane in geological samples provides an indication of the extent of maturation. Thus, mature sediments and crude oils are characterized by a high abundance of the more stable 17αH alkanes relative to the 17βH alkanes, possibly as a result of epimerization of the 17βH alkanes. This may not be the situation with respect to all samples, however, because abundant hopanes with the 17αH configuration appear to be present in several recently-deposited lacustrine sediments (EGLINTON *et al.*, in press), suggesting that the 17αH configuration can be formed in some samples at the earliest stages of diagenesis, although their syngenetic origin with these contemporary sediments is not yet fully proved.

CONCLUSIONS

The carbon skeletons of the tri-and tetraterpenoid alkanes of the Messel shale provide an interesting comparison with those of the well-studied Green River Formation shale; the shales are of similar age ($\sim 50 \times 10^6$ yr) but the Messel appears to have experienced milder thermal conditions (~ 40°C, cf. 90–125°C for the Green River shale—BRADLEY, 1970). Both shales are derived from shallow lacustrine paleoenvironments with sub-tropical climates, but the Green River Formation was deposited by an extensive lake, contrasting with the much smaller area of the Messel environment which comprised a series of small swamp-lakes linked by slow-moving fluvial systems. The biological input to the Green River Formation shale is thought to have been mainly algae (Xanthophyceae, Chloropyceae, Cyanophyceae) together with water-borne or wind-blown pollens and spores (BRADLEY, 1970). Available data for the Messel shale (SITTLER, 1968; SCHULER, 1971) show that pollens from Pteridophyta, Myricaceae, Fagaceae (Castaneoideae) and Cupressaceae predominated in the paleoenvironment.

The alkane:alkene ratios of the two shales (*ca.* 2:1 and 1:4 for Green River and Messel, respectively) may reflect the milder diagenetic/maturation history of the Messel shale. Tri- and tetraterpenoid alkanes are present in high relative abundance in both shales; however, the only major Messel tetraterpane is the acyclic lycopane (*ca.* 3·5 per cent of branched and cyclic alkanes), whereas the bicyclic perhydro-β-carotene (*ca.* 16 per cent of branched and cyclic alkanes) is the major tetraterpane of the Green River shale. The major Messel steranes are the 4-methyl compounds which are less abundant than the triterpanes present, whereas those of the Green River shale are the 4-desmethyl analogues which are more abundant than the triterpanes (BURLINGAME *et al.*, 1965; HILLS *et al.*, 1966; HENDERSON *et al.*, 1968a, b; ANDERSON *et al.*, 1969; GALLEGOS, 1971; ANDERS and ROBINSON, 1971). These carbon skeleton differences are also evident in the corresponding stanol fractions (MATTERN *et al.*, 1970; STEEL and HENDERSON, 1972).

Differences are also apparent in the pentacyclic triterpane fractions when the Messel components are compared with the triterpanes of the Green River shale isolated by several investigators (Table 3). It is difficult to compare the data from several investigators who have used different techniques to study the Green River shale triterpanes, but the comparison in Table 3 appears to be valid because of the overall similarities in the different gas chromatographic distributions. An abundant Green River component is gammacerane (14) but no pentacyclic triterpanes with all six-membered rings could be detected in the Messel shale. The most abundant Messel triterpanes are the hopane series (3a–e), with the 17αH series (3a–d) present in lower concentrations. In the Green River shale the 17αH compounds (3b–d) predominate.

Taken together, the differences in tetraterpane, sterane and triterpane skeletons in the two sediments must reflect differences in the organisms contributing the precursor polyterpenoids, and therefore differences in the paleoenvironments. This also suggests that identification of the complete suites of the structurally-specific alkanes in a range of sediments may provide a chemical method of classification and characterization of paleoenvironments. 4-Methylsterols, acyclic carotenes and hopane-type triterpenoid hydrocarbons are abundant components of certain prokaryotic organisms, and identical or structurally-related compounds are found in the Messel shale suggesting a significant microbial contribution to the sediment. The hopane-type skeletons may also be derived in part from the hydrocarbons of Pteridophytes (ferns). These may be inferred to be one of the predominant species in the Messel paleoenvironment from the available data on fossil plants and pollens (SITTLER, 1968; SCHULER, 1971), and such compounds are relatively abundant in present-day Pteridophytes.

The reversal of the hopane: 17αH-hopane triterpane abundances in the two sediments may also reflect the milder diagenetic/maturation history of the Messel shale, as suggested by ENSMINGER *et al.* (in press).

Acknowledgements—We thank the Natural Environment Research Council (NERC GR/3/655), The National Aeronautics and Space Administration (subcontract from NGL 05-003-003) and ELF-ERAP for support. One of us (B. J. K.) is grateful to the Petroleum Research Fund (PRF 3286) for a Research Studentship. We also thank Professor R. E. CORBETT, Dunedin, New Zealand and Dr. W. McCRAE, formerly of Syntex, Palo Alto, California, for certain standards.

REFERENCES

ALBRECHT P. (1969) Constituents organiques de roches sédimentaires. Ph.D. Thesis, Université de Strasbourg, France.

ALBRECHT P. and OURISSON G. (1969) Triterpene alcohol isolated from oil shale. *Science* **163,** 1192–1193.

ANDERS D. E. and ROBINSON W. E. (1971) Cycloalkane constituents of the bitumen from Green River Shale. *Geochim. Cosmochim. Acta* **35,** 661–678.

ANDERSON P. C., GARDNER P. M., WHITEHEAD E. V., ANDERS D. E. and ROBINSON W. E. (1969) The isolation of steranes from Green River oil shale. *Geochim. Cosmochim. Acta* **33,** 1304–1307.

ARPINO P., ALBRECHT P. and OURISSON G. (1971) Studies on the organic constituents of lacustrine Eocene sediments. Possible mechanisms for the formation of some geolipids related to biologically occurring terpenoids. In *Advances in Organic Geochemistry, 1971,* (editors H. R. v. Gaertner and H. Wehner), pp. 173–187. Pergamon Press.

BALOGH B., WILSON D. M., CHRISTIANSEN P. C., and BURLINGAME A. L. (1973) 17αH-Hopane identified in oil shale of the Green River Formation (Eocene) by carbon-13 NMR. *Nature* **242,** 603–605.

BERTI G. and BOTTARI F. (1968) Constituents of ferns. *Progr. Phytochem.* **1,** 589–685.

BIRD C. W., LYNCH J. M., PIRT S. J. and REID W. W. (1971a) The identification of hop-22(29)-ene in prokaryotic organisms. *Tetrahedron Lett.* 3189–3190.

BIRD C. W., LYNCH J. M., PIRT S. J., REID W. W., BROOKS C. J. W. and MIDDLEDITCH B. S. (1971b) Steroids and squalene in *Methylococcus capsulatus* grown on methane. *Nature* **230,** 473–474.

BLUMER M. (1957) Removal of elemental sulphur from hydrocarbon fractions. *Anal. Chem.* **29,** 1039–1041.

BOTTARI F., MARSILI A., MORELLI I. and PACCHIANI M. (1972) Aliphatic and triterpenoid hydrocarbons from ferns. *Phytochemistry* **11,** 2519–2523.

BRADLEY W. H. (1970) Green River oil shale—concept of origin extended. *Bull. Amer. Geol. Soc.* **81,** 985–1000.

BURLINGAME A. L., HAUG P., BELSKY T. and CALVIN M. (1965) Occurrence of biogenic steranes and pentacyclic triterpanes in an Eocene shale (52 million years) and in an early Precambrian shale (2·7 billion years). *Proc. Nat. Acad. Sci.* **54,** 1406–1412.

COX R. E. (1971) Acyclic isoprenoids of geochemical significance. Ph.D. thesis, University of Bristol.

CRANWELL P. A. (1973) Personal communication.

DANIELLI N., GIL-AV E. and LOUIS M. (1968) Composition of the optically active fraction of a Texas petroleum. *Nature* **217,** 730–731.

DE ROSA M., GAMBACORTA A., MINALE L. and BU'LOCK J. D. (1971) Bacterial triterpenes. *Chem. Commun.* 619–620.

DE ROSA M., GAMBACORTA A., MINALE L. and BU'LOCK J. D. (1973) Isoprenoids of *Bacillus acidocaldarius*. *Phytochemistry* **12,** 1117–1123.

DOUGLAS A. G., EGLINTON G. and MAXWELL J. R. (1969) The organic geochemistry of certain samples from the Scottish Carboniferous Formation. *Geochim. Cosmochim. Acta* **33,** 579–590.

EGLINTON G., MAXWELL J. R. and PHILP R. P. (in press). Organic geochemistry of sediments from contemporary aquatic environments. In *Advances in Organic Geochemistry 1973.*

ENSMINGER A., ALBRECHT P., OURISSON G., KIMBLE B. J., MAXWELL J. R. and EGLINTON G. (1972) Homohopane in the Messel oil shale: first identification of a C_{31} pentacyclic triterpane in nature. *Tetrahedron Lett.* **36,** 3861–3864.

ENSMINGER A., VAN DORSSELAER A., SPYCKERELLE C., ALBRECHT P. and OURISSON G. (in press). Pentacyclic triterpanes of the hopane type as ubiquitous geochemical markers. Origin and significance. In *Advances in Organic Geochemistry 1973.*

ENSMINGER A., VAN DORSSELAER A., SIESKIND O., ALBRECHT P. and OURISSON G. (1974) Unpublished results.

FÖRSTER H. J., BIEMANN K., HAIGH W. G., TATTRIE N. H. and COLVIN J. R. (1973) The structure of novel C_{35} pentacyclic terpenes from *Acetobacter xylinum*. *Biochem. J.* **135,** 133–143.

GALLEGOS E. J. (1971) Identification of new steranes, terpanes and branched paraffins in Green River shale by combined capillary gas chromatography and mass spectrometry. *Anal. Chem.* **43,** 1151–1160.

GARDNER P. M. and WHITEHEAD E. V. (1972) The isolation of squalane from a Nigerian petroleum. *Geochim. Cosmochim. Acta* **36,** 259–263.

GELPI E., SCHNEIDER H., MANN J. and ORÓ J. (1970) Hydrocarbons of geochemical significance in microscopic algae. *Phytochemistry* **9,** 603–612.

GOAD L. J. (1969) Sterol biosynthesis. In *Natural Substances Formed Biologically from Mevalonic Acid,* (editor T. W. Goodwin), pp. 45–77. Biochemical Society Symposia No. 29.

GOODWIN T. W. (1965) Distribution of carotenoids. In *Chemistry and Biochemistry of Plant Pigments,* (editor T. W. Goodwin), pp. 127–142. Academic Press.

GOODWIN T. W. (1966) The carotenoids. In *Comparative Phytochemistry,* (editor T. Swain), pp. 121–137. Academic Press.

GOODWIN T. W. (1970) Algal carotenoids. In *Aspects of Terpenoid Chemistry and Biochemistry,* (editor T. W. Goodwin), pp. 314–356. Academic Press.

HARE P. E. and HOERING T. C. (1972) Organic geochemical studies on a long sediment core from the Cariaco Trench. *Carnegie Inst. Wash. Yearb.* **71,** 584–592.

HENDERSON W., WOLLRAB V. and EGLINTON G. (1968a) Identification of steroids and triterpenes from a geological source by capillary gas–liquid chromatography and mass spectrometry. *Chem. Commun.* 710–712.

HENDERSON W., WOLLRAB V. and EGLINTON G. (1968b) Identification of steranes and triterpanes from a geological source by capillary gas–liquid chromatography and mass spectrometry. In *Advances in Organic Geochemistry, 1968,* (editors P. A. Schenck and I. Havenaar), pp. 181–208. Pergamon Press.

HILLS I. R., WHITEHEAD E. V., ANDERS D. E., CUMMINS J. J. and ROBINSON W. E. (1966) An optically active triterpane, gammacerane, in Green River, Colorado, oil-shale bitumen. *Chem. Commun.* **20,** 752–754.

HILLS I. R., SMITH G. W. and WHITEHEAD E. V. (1970) Hydrocarbons from fossil fuels and their relationship with living organisms. *J. Inst. Petrol.* **56,** 127–137.

JENSEN S. L. (1963) Carotenoids of photosynthetic bacteria—distribution, structure and biosynthesis. In *Bacterial Photosynthesis*, (editors H. Gest, A. San Pietro and L. P. Vernon) pp. 19–34. Antioch Press.

JENSEN S. L. (1967) Recent advances in the chemistry of natural carotenoids. *Pure Appl. Chem.* **14,** 227–244.

KIMBLE B. J., MAXWELL J. R., PHILP R. P. and EGLINTON G. (in press) Identification of steranes and triterpanes in geolipid extracts by high resolution gas chromatography and mass spectrometry. Submitted to *Chem. Geol.*

LEDERER E. (1969) Some problems concerning biological C-alkylation reactions and phytosterol biosynthesis. *Quart Rev.* **23,** 453–481.

MATTERN G., ALBRECHT P. and OURISSON G. (1970) 4-Methyl-sterols and sterols in the Messel oil shale (Eocene). *Chem. Commun.* 1570–1571.

MATTHES G. (1968) Les couches Eocènes dans la région du fosse rhénan septentrional. *Mém. Bureau Rech. Geol. Min.* **58,** 327–337.

MAXWELL J. R. (1967) Studies in organic geochemistry. Ph.D. thesis, University of Glasgow.

MURPHY M. T. J., McCORMICK A. and EGLINTON G. (1967) Perhydro-β-carotene in the Green River shale. *Science* **157,** 1040–1042.

O'CONNOR J. G., BURROW F. H. and NORRIS M. S. (1962) Determination of normal paraffins in C_{20} to C_{32} paraffin waxes by molecular sieve adsorption. *Anal. Chem.* **34,** 82–85.

PYREK J. ST. (1969) Personal communication.

REES H. H. and GOODWIN T. W. (1972) Biosynthesis of triterpenes, steroids and carotenoids. In *Biosynthesis* (Chemical Society Specialist Periodical Report) **1,** 59–118.

SCHULER M. (1971) Personal communication.

SIMONEIT B. and BURLINGAME A. L. (1971) Preliminary organic analyses of the DSDP (JOIDES) cores, Legs V–IX. In *Advances in Organic Geochemistry, 1971*, (editors H. R. v. Gaertner and H. Wehner), pp. 189–228. Pergamon Press.

SITTLER C. (1968) L'analyse pollinique dans l'Est de la France. Etude des formations Eocènes ou rapportées a l'Eocène et des stratotypes palynologiques de Borken et de Messel. *Mem. Bur. Rech. Geol. Min.* **58,** 165–171.

STEEL G. and HENDERSON W. (1972) Isolation and characterization of a series of stanols from the Green River Shale. *Nature* **238,** 148–149.

TOKÉS L., JONES G. and DJERASSI C. (1968) Mass spectrometry in structural and stereochemical problems. CLXI Elucidation of the course of the characteristic ring D fragmentation of steroids. *J. Amer. Chem. Soc.* **90,** 5465–5477.

VAN DORSSELAER A., ENSMINGER A., SPYCKERELLE C., DASTILLUNG M., SIESKIND O., ARCPINO P., ALBRECHT P., OURISSON G., BROOKS P. W., GASKELL S. J., KIMBLE B. J., PHILP R. P., MAXWELL J. R. and EGLINTON G. (in press) Degraded and extended hopane derivatives (C_{27} to C_{35}) as ubiquitous geochemical markers. *Tetrahedron Lett.*

VAN NIEL C. B. (1963) A survey of the photosynthetic bacteria. In *Bacterial Photosynthesis*, (editors H. Gest, A. San Pietro and L. P. Vernon), pp. 459–467. Antioch Press.

WEEDON B. C. L. (1965) In *The Chemistry and Biochemistry of Plant Pigments*, (editor T. W. Goodwin), pp. 75–125. Academic Press.

WHITEHEAD E. V. (1972) Personal communication.

WHITEHEAD E. V. (1971) Chemical clues to petroleum origin. *Chem. Ind.* 1116–1118.

WSZOLEK P. C., GELPI E. and BURLINGAME A. L. (1971) A new approach to the isolation of milligram amounts of significant geochemical compounds. In *Advances in Organic Geochemistry, 1971*, (editors H. R. v. Gaertner and H. Wehner), pp. 229–247. Pergamon Press.

26

Reprinted from *Jour. Amer. Chem. Soc.* **94**:5880–5887 (1972)

Proof of Structure of Steroid Carboxylic Acids in a California Petroleum by Deuterium Labeling, Synthesis, and Mass Spectrometry[1]

Wolfgang K. Seifert,* Emilio J. Gallegos, and Richard M. Teeter

Contribution from the Chevron Oil Field Research Company and Chevron Research Company, Richmond, California 94802. *Received August 31, 1971*

Abstract: Discovery and proof of structure of four stereoisomeric C_{22}–C_{24} steroid acids in virgin petroleum in admixture with thousands of C_{16}–C_{31} petroleum carboxylic acids are described. The position of the carboxyl groups was fixed *via* reduction to alcohols and tosylation and reduction with lithium aluminum deuteride. The identity of synthetic deuterium-labeled and unlabeled 20-methyl-5α-pregnanes and 5α- and 5β-cholanes with the derivatives of the natural products is demonstrated by gas chromatography (gc) combined with mass spectrometry (ms). Final proof for the presence of two stereoisomers of 5α-pregnane-20ξ-carboxylic acid and of 5α- and 5β-cholanic acid was obtained by gcms of natural product and synthetic perfluoroalcohol esters. Both animal sources (bile acids) and plant sources (*i.e.*, unsaturated sterols) are proposed to explain the presence of the steroid acids in their observed ratios.

The structure of carboxylic acids in petroleum has been the subject of investigations throughout the century.[2] One incentive for scientists in this area of research lies in the proposal[3] that acids may be the precursors of petroleum hydrocarbons. The theory that petroleum is of biological origin had its beginning with the discovery of porphyrins[4] supplemented by recent work of Baker, *et al.*,[5] and others. It has now become a generally accepted concept supported by the presence[6] of a variety of compounds in sediments that appear to be related to similar ones occurring in the living organism (*e.g.*, amino acids, fatty acids, carbohydrates, sterols[7]). Similarly, biological marker hydrocarbons such as isoprenoids,[8] pentacyclic triterpanes,[9] tricyclic diterpanes,[10d] and steranes[10] in sediments[10a–e] and petroleum[10f,g] have become identifiable by modern means of instrumentation. The progress in petroleum carboxylic acid research has been retarded because of the enormous complexity of the mixtures and the difficulty of isolation and separation. Thus, with the exception of fatty acids, the C_{10} limit had not been exceeded until 1964 when Cason[11] and coworkers first discovered and proved the structure of four C_{14}–C_{20} isoprenoid[11a] and several[11b,c] cyclic and acyclic C_{10}–C_{11} carboxylic acids in petroleum. This finding was followed by a series of papers on isoprenoid acids in sediments, the most recent of which[12a–c] summarizes the literature.

We now wish to report the first discovery of *steroid* carboxylic acids in any petroleum source with structure proof for two C_{22} and two C_{24} stereoisomeric steroid acids. Our interest in the structure of individual carboxylic acids grew out of our work on classes[13a] of carboxylic acids in a virgin California petroleum of Pliocene age Field (Midway–Sunset, 10^7 years); it resulted in the addition of some 40 new compound classes of carboxylic acids[13b,c] from C_{16} to C_{31} to the existing knowledge of a few classes known to date and summarized in one of our papers.[13d]

Despite the extensive separation[13e] depicted by Scheme I, a rather "pure" thin-layer chromatographic subfraction of D-4, which is free of phenols,[13f] was estimated by mass spectrometry[13g] to contain at least 1500 compounds. Reduction of this fraction yielded hydrocarbons which could be more readily and extensively separated. The location of the carboxyl group was fixed by parallel reduction to deuteriomethyl and methyl (Scheme I). This approach resulted in working on a milligram scale at the hydrocarbon level; therefore, the only instrumental method of analysis applicable to the nanogram quantities of individual steroid acid derivatives present in the mix-

(1) Partial preliminary communication: W. K. Seifert, E. J. Gallegos, and R. M. Teeter, *Angew. Chem., Int. Ed. Engl.*, **10**, 747 (1971); *Angew. Chem.*, **83**, 805 (1971).
(2) H. L. Lochte and E. R. Littman, "Petroleum Acids and Bases," Chemical Publishing Co., New York, N. Y., 1955.
(3) J. E. Cooper and E. E. Bray, *Geochim. Cosmochim. Acta*, **27**, 1113 (1963).
(4) A. Treibs, *Angew. Chem.*, **49**, 682 (1936), and references therein.
(5) (a) E. W. Baker, *J. Amer. Chem. Soc.*, **88**, 2311 (1966); (b) E. W. Baker, T. F. Yen, J. P. Dickie, R. E. Rhodes, and L. F. Clark, *ibid.*, **89**, 3631 (1967).
(6) (a) G. Eglinton and M. T. J. Murphy, Ed., "Organic Geochemistry," Springer Verlag, New York, N. Y., 1969; (b) P. Albrecht and G. Ourisson, *Angew. Chem., Int. Ed. Engl.*, **10**, 209 (1971).
(7) (a) G. Mattern, P. Albrecht, and G. Ourisson, *Chem. Commun.*, 1570 (1970); (b) W. Henderson, W. E. Reed, G. Steel, and M. Calvin, *Nature (London)*, **231**, 308 (1971).
(8) (a) J. G. Bendoraitis, B. L. Brown, and L. S. Hepner, *Anal. Chem.*, **34**, 49 (1962); (b) R. A. Dean and E. V. Whitehead, *Tetrahedron Lett.*, **21**, 768 (1961).
(9) I. R. Hills, E. V. Whitehead, D. E. Anders, J. J. Cummins, and W. E. Robinson, *Chem. Commun.*, **20**, 752 (1966).
(10) (a) A. L. Burlingame, P. Haug, T. Belsky, and M. Calvin, *Proc. Nat. Acad. Sci. U. S.*, **54**, 1406 (1965); (b) W. Henderson, V. Wollrab, and G. Eglinton, *Chem. Commun.*, 710 (1968); P. A. Schenck and I. Havenaar, *Proc. Int. Meet. 4th, Org. Geochem. 1968*, 181 (1969); (c) P. C. Anderson, P. M. Gardner, E. V. Whitehead, D. E. Anders, and W. E. Robinson, *Geochim. Cosmochim. Acta*, **33**, 1304 (1969); (d) E. J. Gallegos, *Anal. Chem.*, **43**, 1151 (1971); (e) D. E. Anders and W. E. Robinson, *Geochim. Cosmochim. Acta*, **35**, 661 (1971); (f) I. R. Hills and E. V. Whitehead, *Nature (London)*, **209**, 977 (1966); (g) T. C. Hoering, personal communication.
(11) (a) J. Cason and D. W. Graham, *Tetrahedron*, **21**, 471 (1965); (b) J. Cason and K. L. Liauw, *J. Org. Chem.*, **30**, 1763 (1965); (c) J. Cason and A. I. A. Khodair, *ibid.*, **31**, 3618 (1966); **32**, 3430 (1967).
(12) (a) A. E. Douglas, M. Blumer, G. Eglinton, and K. Douraghi-Zadeh, *Tetrahedron*, **27**, 1071 (1971); (b) R. E. Cox, J. R. Maxwell, G. Eglinton, and C. T. Pillinger, *Chem. Commun.*, 1639 (1970); (c) R. C. Murphy, M. V. Djuricic, S. P. Markey, and K. Bieman, *Science*, **165**, 695 (1969).
(13) (a) W. K. Seifert and R. M. Teeter, *Chem. Ind. (London)*, 1464 (1969); (b) W. K. Seifert and R. M. Teeter, *Anal. Chem.*, **42**, 750 (1970); (c) W. K. Seifert and R. M. Teeter, *ibid.*, **42**, 180 (1970); (d) W. K. Seifert, R. M. Teeter, W. G. Howells, and M. J. R. Cantow, *ibid.*, **41**, 1638 (1969); (e) W. K. Seifert and W. G. Howells, *ibid.*, **41**, 554 (1969); (f) W. K. Seifert, *ibid.*, **41**, 562 (1969); (g) W. K. Seifert and R. M. Teeter, *ibid.*, **41**, 786 (1969).

Scheme I

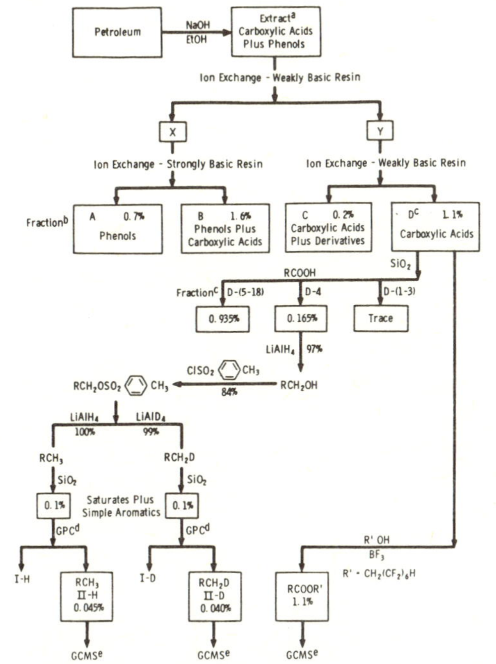

[a] This extract represents 3.54% of the total crude oil; 2.5% based on crude oil is RCOOH, see Tables I–III of reference 13e. [b] Percentages are based on total crude oil. [c] Represents 40% of all RCOOH present in crude oil. [d] Gel permeation chromatography, see Figure 1. [e] Gas chromatography plus mass spectrometry.

tures was capillary gas chromatography (gc) combined with mass spectrometry (ms).

More specifically, acid fraction D-4, representing 6% of all acids and equal to 0.17% of the petroleum, was investigated (Scheme I). Fractions of predominantly saturated hydrocarbons isolated from both the labeled and unlabeled portions in identical fashion by silica gel chromatography were further separated by gel permeation chromatography (Scheme I). The deviation of the refractive index from that of the solvent is given in Figure 1. It illustrates the duplication between labeled and unlabeled fractions for the three steps of reduction, silica gel chromatography, and gel permeation chromatography. The investigation was focused on the pair II-H and II-D (Figure 1) because they showed the highest concentration of steroid hydrocarbons. The gas chromatogram of the unlabeled hydrocarbon is given in Figure 2A and is identical with that of the labeled sample. The two largest peaks were recognized as terpanes from their mass spectra.[14]

(14) A complication which interfered with the interpretation of the gcms data was the continual emergence, from the chromatograph, of a mixture of unresolved components which resulted in a significant background to each steroid mass spectrum. Correction for these background contributions was made by averaging backgrounds taken before and after the steroid peak and deducting the average from the observed mass spectrum of the steroid. The error inherent in this method is due to the variation of the background during the emergence of the individual steroid compounds. Therefore, the significance of the mass spectra depends upon the general shape of the pattern rather than the exact intensities of the fragment peaks.

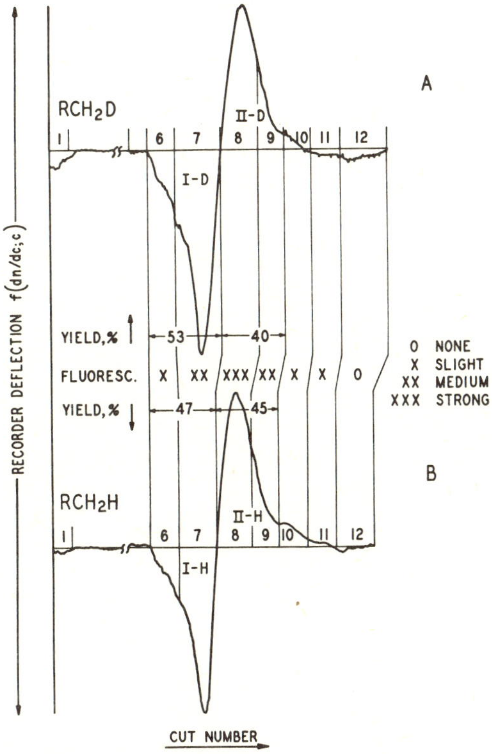

Figure 1. Gel permeation chromatograms of deuterium-labeled hydrocarbons (silica gel fractions I of hydrocarbons derived from carboxylic acid fraction D-4) and hydrogen analogs.

C_{22} Acids. The mass spectrum of the first significant steroid hydrocarbon emerging after the C_{24}-terpane is shown in Figure 3A and that of its deuterated counterpart in Figure 3C. The dominant fragments in steroids at *m/e* 218, 217, and 232 have been shown to arise from cleavage of ring D[15] (see formula I for numbering). Because the deuterated steroid (Figure 3C) shows no increase in mass at these positions, the carboxylic acid group cannot be attached to rings A, B, or C, but must be at ring D. Moreover, the presence of the fragment at *m/e* 232 shows C_{15} to be unsubstituted because this particular fragment arises from cleavage of the 13–17 and 15–16 bonds.[15] Furthermore, the appearance of the spectrum (Figure 3A) as a whole resembles closely that of a naturally occurring steroid with substitution on C_{17}, and on that basis substitution at C_{16} can be excluded.[16] The parent minus methyl fragments in the labeled and unlabeled compounds at *m/e* 288 and 287, respectively, must represent loss of the 18- and 19-methyl groups because no loss of deuteriomethyl (mass 16) is observed.

Reported spectra of 5α- and 5β-pregnane[15] show that the stereochemistry of rings A and B can be recognized as trans by the near absence of a fragment at *m/e* 151

(15) L. Toekes, G. Jones, and C. Djerassi, *J. Amer. Chem. Soc.*, **90**, 5465 (1968).
(16) C. Djerassi, personal communication, 1971.

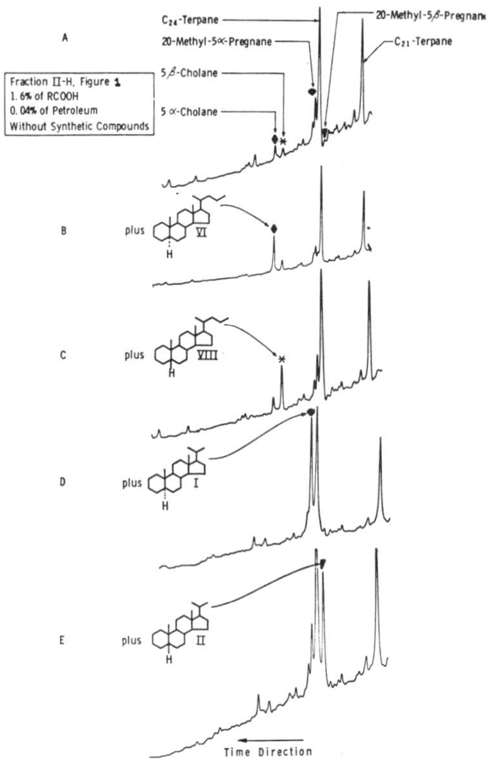

Figure 2. Gas chromatogram of carboxylic acid derived hydrocarbons coinjected with synthetic steroids. Time is measured on a variable scale.

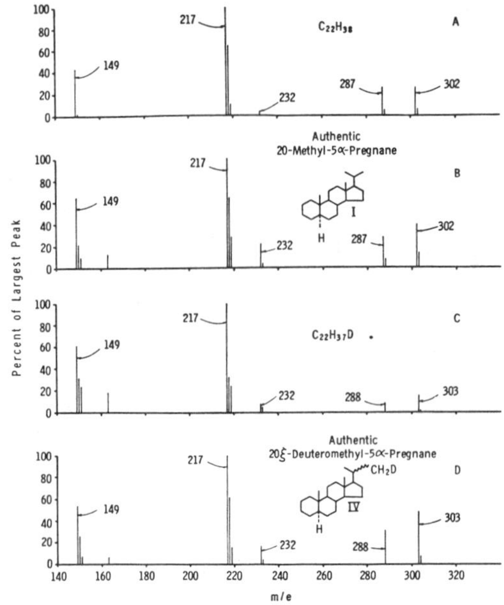

Figure 3. Mass spectra of 5α-pregnane-20ξ-carboxylic acid derived petroleum hydrocarbons (gcms; after background deduction in A and C) *vs.* synthetic compounds.

Scheme II

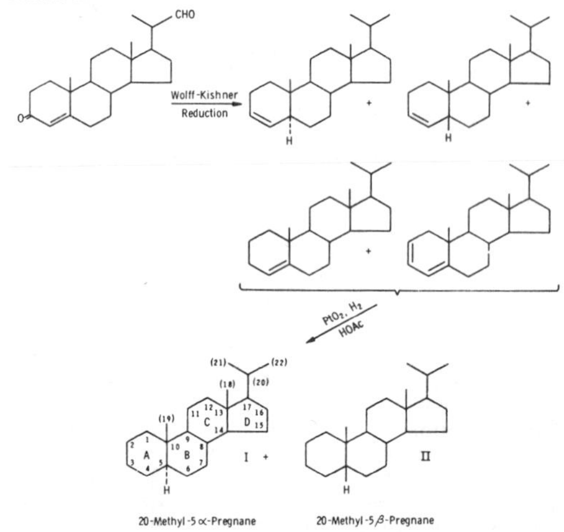

and as cis by a fragment at *m/e* 151 of intensity about equal to or greater than that at *m/e* 149; this observation was confirmed in this laboratory by comparison of the spectra of 5α-cholestane and 5β-cholestane. Figures 3A and 3C indicate the trans A/B configuration for this naturally occurring steroid carboxylic acid.

The unlabeled trans (I) and cis (II) hydrocarbons were synthesized as a mixture as shown in Scheme II, and separated by preparative gas chromatography.

The mass spectral fragmentation pattern of I was obtained from gcms data collected under the same conditions as was that of the petroleum-derived hydrocarbon, and the two are shown in Figures 3A and 3B. Differences in relative intensities are due to uncertainties in the subtraction of background. Gas chromatographic coinjection of the synthetic 5α-trans compound I with the petroleum carboxylic acid derived hydrocarbons illustrates the enhancement of the natural product peak of interest (Figure 2D). The synthetic 5β-cis compound shows a fragment at *m/e* 151 of approximately the same intensity as its *m/e* 149 peak. Coinjection with the petroleum-derived hydrocarbon (Figure 2E) showed it to have a retention time identical with that of a very small peak emerging just prior to the C_{21} terpane. A weak steroid mass spectrum was detected at this position in the deuterium-labeled mixture with *m/e* 151 equal to *m/e* 149. Our conclusion is that cis isomeric pregnanecarboxylic acid is present in this petroleum in much smaller (trace) quantity than the trans isomer.

In order to confirm the proposed position of the carboxyl group at C_{20} in this C_{22}-steroid acid, the required deuterium-labeled hydrocarbon IV was synthesized *via* the authentic carboxylic acid (III) as shown in Scheme III.

The general mass spectral pattern of this 20ξ-deuteriomethyl-5α-pregnane (Figure 3D) obtained under gcms conditions is very similar to that derived from the nat-

Scheme III

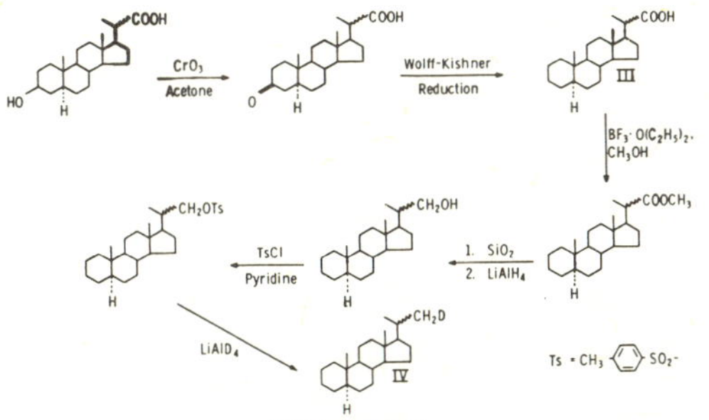

22ξ-Deuteromethyl-5α-Pregnane

ural product (Figure 3C); gas chromatographic coinjection of IV with the deuterium-labeled hydrocarbon mixture results in peak enhancement the same as that shown in Figure 2D for the unlabeled hydrocarbon I.

A comparison of the intensities of the fragments at *m/e* 44 of the labeled and unlabeled pairs additionally supports the proposed structure. This peak, in the unlabeled species, is due to $C_2{}^{13}CH_7$. In the labeled species, *m/e* 44 is enhanced by the presence of C_3H_6D from the side chain at C_{17} of the steroid nucleus. Again, background corrections reduce the accuracy of the figures, but the height increase (relative to the height of the *m/e* 217 peak) is the same (12%) for the synthetic pair (4–16%) and the petroleum-based pair (1–13%). This indicates that the side-chain structures are probably the same. Proof would require a demonstration that the *n*-propyl group yields a different increase.

To further prove the presence of the isopropyl group and to determine whether the carboxyl group at C_{20} is stereochemically α or β or both, an ester of crude carboxylic acid fraction D (Scheme I, representing 40% of all carboxylic acids in this petroleum) with 1,1,7-trihydroperfluoroheptanol[13g] was investigated by gcms. Figure 4 shows the mass chromatogram (see Experimental Section) using the intensities of the parent peak of the ester at *m/e* 646 to monitor the occurrence of the various isomers of the C_{22} acids. Three isomers, A, B, and C, of very similar mass spectral fragmentation pattern were observed (Table I). The ratio of the retention times of B/A was 1.023 and of C/A, 1.042. To clarify this situation, the fluoroalcohol ester of synthetic acid III was prepared; it showed the presence of two isomers (gcms) whose mass spectral fragmentation patterns are very similar to one another (Table I). Their retention time ratio was 1.023, indicating that A and B of the petroleum sample are the same as A and B of the authentic sample and that isomer C occurs in the petroleum only. When the mixture of these two synthetic perfluoroalcohol ester 5α isomers was coinjected with the natural product ester mixture, compounds A and B (Figure 4) were enhanced, not compound C. The agreement of the mass spectral fragmentation patterns of compounds A[17] and B with their synthetic counterparts of equal retention time is seen from Table I. The height ratio of the peaks at *m/e* 149

(17) The spectrum of compound A was not taken at the maximum intensity of the *m/e* 646 peak (see Figure 4) because at the time of scan 328, an unidentified component of the mixture was starting to elute and its fragments were interfering. This was revealed when the plots of peaks other than *m/e* 646 failed to change in synchronism with the 646 peak.

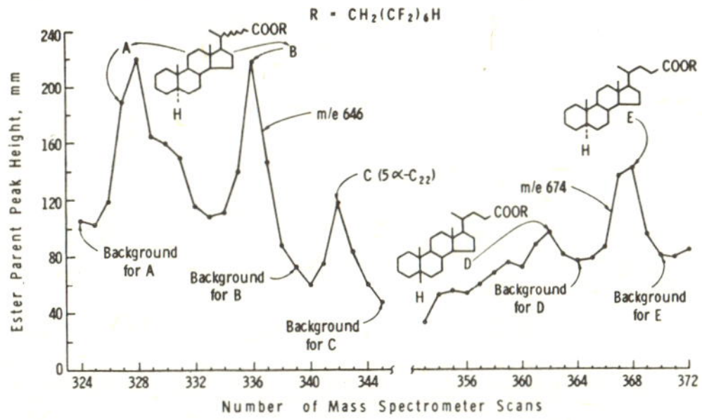

Figure 4. Molecular ion mass chromatogram of esters (with 1,1,7-trihydroperfluoroheptanol) of petroleum steroid carboxylic acids. The number of mass spectrometer scans is from sample injection; The scan cycle is 8.7 sec. The entire spectrum is used to generate Tables I and II.

Table I. Mass Spectra[a] of 5α-Pregnane-20ξ-carboxylic Acid Fluoroalcohol Esters by Gas Chromatography–Mass Spectrometry (R = $CH_2(CF_2)_6H$)

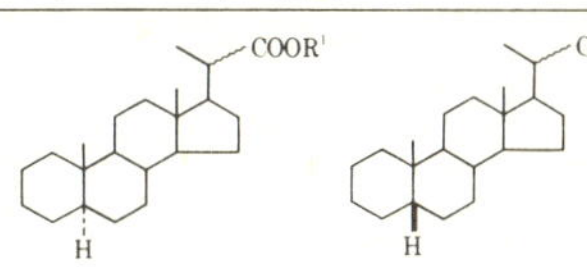

m/e	Synthetic	A, from petroleum[b]	Synthetic	B, from petroleum[b]	C, from petroleum[b]
149	66	98	62	97	74
151	14	20	12	41	19
217	100	100	100	100	100
218	30	33	34	35	43
232	35	33	42	37	20
374	0	7	0	0	4
388	1	2	2	4	1
631	21	17	23	39	22
646	32	26	32	41	26
Retention time, min	47.3		48.4		49.3

[a] Peak intensities relative to *m/e* 217 [b] After background deduction.

and 151 indicates both A and B to be 5α (Table I). They are believed to be the isomers with the carboxyl group at C_{20}, either α or β. The latter distinction is no longer detectable after reduction to deuterium-labeled hydrocarbon IV which gives the appearance of one single compound by gcms. Our conclusion is that both 5α-pregnane-20α-carboxylic acid and its 20β isomer are present in this petroleum.

The mass spectrum of C (Table I) indicates an A/B trans 5α isomer. It may differ from A and B by having either a branched chain at C_{17} in the α configuration or a straight chain in either α or β configuration at C_{17}. From Figure 4 the amount of the unidentified C_{22}-steroid carboxylic acid (isomer C) is estimated to be about one-tenth that of the sum of the identified acids (isomers A and B).

C_{24} Acids. In addition to the above-described C_{22}-steroid acids, work on the acid-derived, labeled, and unlabeled hydrocarbons revealed two additional stereoisomeric C_{24}-steroid acids. Comparison of their mass spectra (Figures 5A and 5C) and the same mass

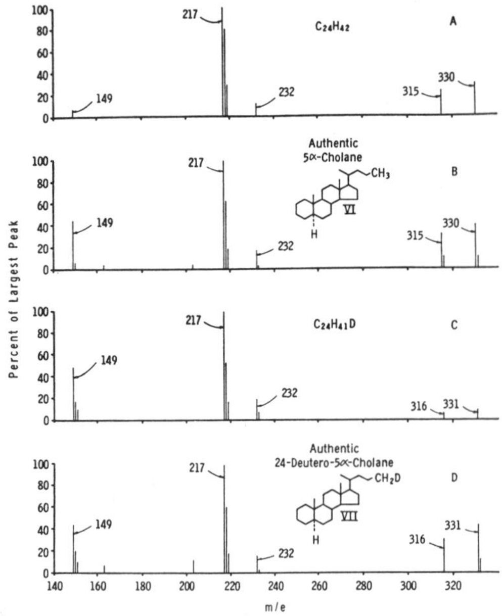

Figure 5. Mass spectra of 5α-cholanic acid derived petroleum hydrocarbons (gcms; after background deduction in A and C) *vs.* synthetic compounds.

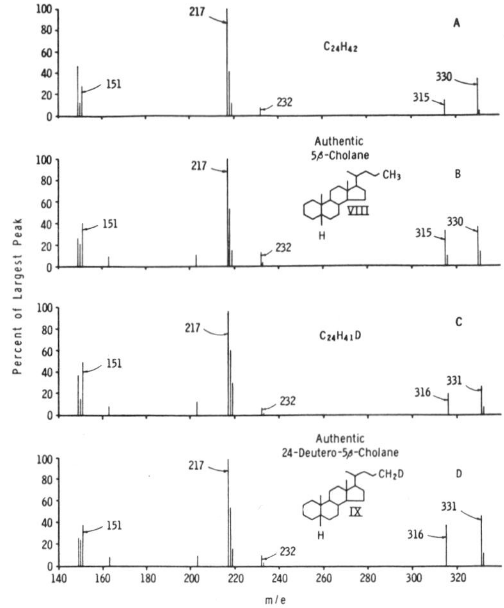

Figure 6. Mass spectra of 5β-cholanic acid derived petroleum hydrocarbons (gcms; after background deduction in A and C) *vs* synthetic compounds.

spectral arguments as described above for the deuterium-labeled and unlabeled C_{22} compounds led us to synthesize 5α-cholane (VI) and 24-deuterio-5α-cholane (VII) *via* 5α-cholanic acid (V) as shown in Scheme IV.

Scheme IV

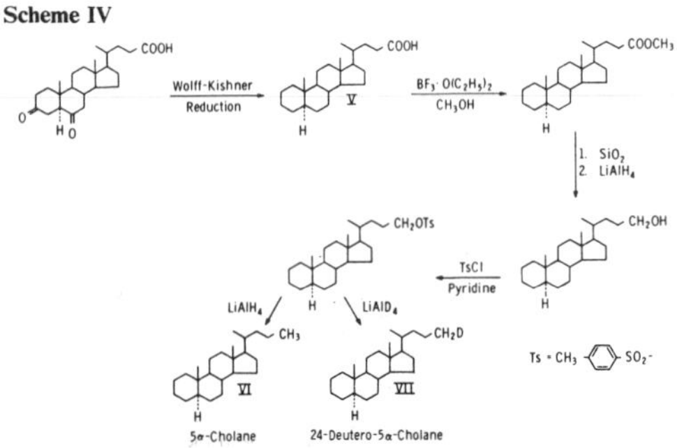

Analogously 5β-cholane (VIII) and 24-deuterio-5β-cholane were synthesized by the above shown route from 5β-cholanic acid (X). It is noteworthy that the experimental conditions for converting the authentic steroid acids to the corresponding hydrocarbons of known stereochemistry were identical with those applied to the natural product acids, thus demonstrating the stability of the hydrocarbon skeletons of the latter during this transformation. The mass spectral similarity of the four synthetic labeled and unlabeled hydrocarbons with their natural product counterparts is seen in Figures 5 and 6. Gc coinjection is illustrated in Figures 2B and 2C showing the shorter retention time of the A/B cis isomer.

Final proof of the presence of 5α- and 5β-cholanic acids in this petroleum was obtained from the gcms data of the fluoroalcohol ester of fraction D (Scheme I). The mass chromatogram (Figure 4) of the petroleum acid fluoroalcohol ester monitored by the intensity of the parent peak at *m/e* 674 for the C_{24} acids demonstrates the presence of two isomers, D and E, whose mass spectral fragmentation patterns are compared with those of the esters from synthetic 5α-cholanic acid (V) and 5β-cholanic acid (X) in Table II. Gc coinjection with the natural product esters enhanced compounds E and D (Figure 4), respectively, the 5β isomer having the slightly shorter retention time. Additional evidence comes from the retention times of synthetic and petroleum-derived 5α-cholanic acid esters when measured relative to the most abundant of the pregnane carboxylic acid esters. Retention time ratios of 1.122 for the ester of acid V/A and 1.123 for E/A were measured.

As in the mass spectra of hydrocarbons, the 5α configuration of the synthetic (V) and naturally occurring 5α-cholanic acid E is illustrated by the predominance of *m/e* 149 over *m/e* 151 in the fluoroalcohol ester spectra (Table II); analogously, the esters of naturally occurring 5β-cholanic acid D and of synthetic acid X show the increase of the intensity at *m/e* 151 required by the 5β configuration. The absolute values of some of the fragment intensities of the ester of the naturally occurring 5β-cholanic acid D are distorted by background which has a particularly large effect on this isomer because its amount is the smallest of all of the steroid acid esters of Figure 4. Fragments due to McLafferty rearrangement at *m/e* 374 in 5α- and 5β-

Table II. Mass Spectra[a] of 5α- and 5β-Cholanic Acid Fluoroalcohol Esters by Gas Chromatography–Mass Spectrometry ($R = CH_2(CF_2)_6H$)

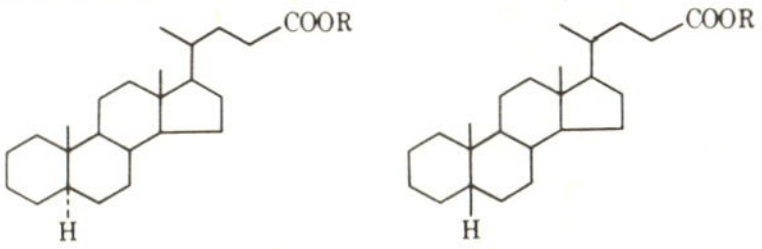

m/e	Synthetic	E, from petroleum[b]	Synthetic	D, from petroleum[b]
149	42	51	26	60
151	8	11	34	80
217	100	100	100	100
218	51	38	55	40
232	37	36	19	20
373	3	0	4	8
374	3	4	4	16
564	10	11	9	12
659	34	30	44	76
674	54	28	85	80
Retention time, min	53.1[c]		52.4[c]	

[a] Peak intensities relative to *m/e* 217. [b] After background deduction. [c] Of the petroleum-derived esters.

cholanic acids are small in synthetic and natural product esters and even smaller at *m/e* 388 in pregnane carboxylic acid ester. The reason for the latter is probably the rigidity of ring D of the steroid system[16] coupled with interference by the C_{19} methyl with the planarity of the six-membered ring intermediate.[18]

Amounts of Acids and Their Significance. An estimate of the relative amounts of the individual steroid acids in this petroleum can be based on the intensities (relative to *m/e* 217) of the parent peaks of the four synthetic acid esters given in Tables I and II combined with the relative heights of parent peaks of the derivatives A–E of the naturally occurring acids in the mass chromatogram (Figure 4). The sum of the two 5α-pregnane-20ξ-carboxylic acids, A and B, is about seven times the quantity of 5α-cholanic acid E; the latter occurs in four–five times greater amount than 5β-cholanic acid D. From the hydrocarbon work described above, 5β-pregnane-20ξ-carboxylic acid appears to be present in an amount smaller than 5β-cholanic acid.

The sum of the two predominant 5α-pregnane-20ξ-carboxylic acids A and B, both of which become 20-methyl-5α-pregnane after reduction, is estimated from the occurrence of the latter in reduced fraction D-4 (Scheme I and Figure 2A) and in reduced fraction D (Scheme I) to be about 6 ppm of the petroleum. Therefore, from the ratios described above, 5α-and 5β-cholanic acids are estimated at about 1 and 0.2 ppm of the petroleum, respectively. Because only 40% of all acids in this petroleum were investigated and since the complexity of the mixture prevented in-hand isolation, these analytical values allow an estimate of the order of magnitude of all steroid acids identified as about 0.1–10 ppm of the petroleum. It is therefore obvious that proof of structure, including stereochemical details of such polar compounds in the enormously complex mixtures in which they naturally occur, requires the elaborate separation sequences combined with labeling techniques (which exclude contamination) and modern instrumental analysis plus organic synthesis described here. The fortuitous factors in this work were good sensitivity of steroid mass spectra and the ease of interpretation of the labeled compounds due to the extensive mass spectral work done on steroid systems by others.[15]

Because of the presence of functional groups and the recognition of stereochemical features, the detailed information which can be derived biogeochemically from steroid acids in petroleum may exceed that of any biological marker hydrocarbon and that of any carboxylic acid except those isoprenoid acids recently shown[12b] to be linked to the phytyl side chain of chlorophyll. While these latter acids supplement the abundant evidence for the predominant plant genesis of petroleum, the identification of the steroid acids reported here justifies postulation of some animal contribution. Because the observed abundance ratio of 5α-C_{22} to 5β-C_{22} (30:≪1) is different from the corresponding ratio of 5α-C_{24} to 5β-C_{24} (4:1), thermodynamic equilibrium between 5α and 5β in both pairs is most unlikely. In addition, the greater amount of 5β-C_{24} over 5β-C_{22}, which is inverse to the ratio in the 5α series (5α-C_{22}'s > 5α-C_{24}), warrants some speculation.

Most C_{24} bile acids presently known[19] possess A/B cis configuration (*i.e.*, 5β), and no abundant 5β-C_{24} steroids of plant origin are known. Therefore, the most likely sources for the 5β-C_{24} acid are 5β-C_{24} bile acids which have various numbers of hydroxyl groups (*i.e.*, cholic acid). Upon dehydration and hydrogenation, transformations generally accepted by petroleum chemists, all these 5β bile acids would yield the observed 5β-cholanic acid. The 5α-C_{24} acid, being more abundant than the 5β-C_{24} isomer, could be derived by isomerization of the latter, because the trans-5α isomer is the thermodynamically more stable one,[20] or in part from the less abundant 5α-C_{24} bile acids. Allowing β oxidation of 5α-C_{24} to 5α-C_{22}, the latter can also be explained from bile acids.

However, plants also have to be considered as possible sources, *i.e.*, the 5α-C_{22} steroid acid isomers are derivable by oxidative C_{22}–C_{23} double bond cleavage of sterols which occur in large variety in the plant kingdom (*i.e.*, dihydroergosterol, algae) or, less likely, oxidation of steroidal sapogenins. Thus, the larger quantity and number of 5α-C_{22} steroid acid isomers may reflect a greater variety of sources. Analogously, 5α-C_{24} can be derived by oxidative C_{24}–C_{25} double bond cleavage of zymosterol.

In summary, animal sources alone can explain all steroid acid isomers found. Plant sources, although they alone cannot explain all the isomers found in their observed ratios, are very likely contributors. Consequently, both, which existed at Pliocene age, are proposed to explain the results, unless recent bacterial activity has occurred voiding the above conclusions.

Experimental Section

Instrumentation. For gcms analysis of hydrocarbons, an F&M Model 810 gas chromatograph without a splitter was used in

(18) J. A. Gilpin and F. W. McLafferty, *Anal. Chem.*, **29**, 990 (1957).

(19) E. Heftmann, "Steroid Biochemistry," Academic Press, New York, N. Y., 1971, pp 55–59.

(20) M. N. Mitra and W. H. Elliot, *J. Org. Chem.*, **34**, 2170 (1969).

conjunction with an AEI MS-9 mass spectrometer. A 45 m × 0.05 mm i.d. capillary column coated with OV-17 operated at a He flow rate of about 2 ml/min was coupled directly to the source of the MS-9 without the use of an interface. This achieves maximum sample sensitivity with minimum loss of chromatographic resolution. The source pressure ion gauge read 1–2 × 10^{-5} Torr. A temperature of 330° was maintained at both the injection port and the column exit to the MS-9 source. The source was held at 240° using a temperature controller; the chromatograph was programmed from 100 to 340° at a rate of 4°/min using the total ion monitor of the MS-9 to obtain the chromatograph traces. Mass spectra were recorded at a magnet scan rate of 5 sec/decade at about 1000 resolution. A typical sample size injected using a 1-μl syringe was 0.3 μl. Peaks in the mass spectra of Figures 3, 5, and 6 smaller than 10% of the largest peak were not included unless of particular interest.

Gcms of the fluoroalcohol esters was performed using the same chromatograph but with a Nuclide 12-90-G mass spectrometer. The gcms conditions were similar to those described above for the MS-9 with the following differences: column length, 60 m; He flow rate, 5 ml/min, and pressure at the source ion gauge, 5 × 10^{-6} Torr. The mass spectrometer was set to cycle in the linear magnet current mode to gather spectra from *m/e* 140 to 750 by repetitive scanning at intervals of 8.7 sec. The typical peak width for the steroid esters in question is about 14.0 sec. Mass spectra were gathered at an electron accelerating voltage of 50 eV. All mass spectral data were recorded on a Honeywell Model 7600 tape recorder at a tape speed of 9.5 cm/sec. Selected spectra were then reproduced on a Honeywell Model 1508A Visicorder.

Both unresolved components of the mixture and column bleed interfered with the steroid mass spectra. A plot of molecular ion peak height as a function of the number of mass spectrometer scans for the petroleum fluoroalcohol ester gcms, defined as a mass chromatogram, is given in Figure 4. Scans 327, 336, 342, 362, and 368 were read and from them were subtracted scans 324, 339, 345, 364, and 370, respectively, to yield, after normalization, the partial spectra presented in Tables I and II. Average standard deviation of gas chromatographic retention time ratios of fluoroalcohol esters is 0.3% on duplicate runs. The partial spectra of the synthetic compounds were also obtained from gcms data but are more accurate because only column bleed background was present, and its subtraction could be done with better assurance that it was constant.

Exact mass measurements of synthetic labeled and unlabeled hydrocarbons and the intermediates leading to their synthesis were preferred to conventional elemental analyses. They were performed by the peak-matching technique referred to appropriate peaks of perfluorotributylamine using a quartz direct injection probe[21] for sample introduction. An average standard deviation of 4.6 ppm was observed. For preparative gas chromatography, a Wilkens Aerograph Model A-350 instrument was operated isothermally at 270°. The column was 6 mm o.d. × 6 m stainless steel packed with 3% OV-17 on Gas Chrom Q; He flow was 40 ml/min. The traps were 3 mm o.d. × 10 cm glass, air-cooled. Gas chromatographic coinjection of synthetic steroid hydrocarbons with samples derived from the petroleum (Figure 2) was made in a Model 204-1B Aerograph on a 45 m × 0.05 mm i.d. OV-17 column programmed from 180 to 300° at 6°/min, at 3–4 ml of He flow using 0.1 mg of sample.

The instrumentation for gel permeation chromatography was described previously.[13d] Optical rotations were measured in chloroform on a Cary Model 60 spectropolarimeter. Melting points were obtained on a Fisher Scientific Instrument melting point block and are uncorrected.

Isolation, Conversion, and Separation of Natural Product Acids (Scheme I). The isolation of carboxylic acid fraction D-4 from Midway–Sunset 31E, California crude, has been described previously.[13e] The earlier reported[13d] three-step procedure for conversion (for yields, see Scheme I) of carboxylic acids to hydrocarbons *via* alcohols was applied to this fraction by reducing half of the *p*-toluenesulfonate with lithium aluminum hydride and the other half with lithium aluminum deuteride.

Solvents for chromatography were purified by distillation and chromatography on alumina. Milligram quantities of both labeled and unlabeled hydrocarbon fractions were chromatographed on purified[22] Grace Davison Chemical Co. 100–200 mesh activated Grade 923 desiccant silica gel using large columns (2.5 cm i.d. × 60 cm). A 1-l. petroleum ether eluate containing 60% of the charge was worked up and used for subsequent gel permeation chromatography. It was later found that a much smaller column (1.25 cm i.d. × 130 cm) operated at 14 psi of pressure with hexane suffices to separate charges of several hundred milligrams and allows the use of silica gel without prepurification simply by purging the column with 200 ml of hexane before charging. (In this case, the saturated hydrocarbons are eluted free of aromatics in the first 140 ml of eluate in 15–20 min.) Deuterium labeled and unlabeled samples contained significant amounts of aromatics and were both further separated by preparative (30-mg quantities) gel permeation chromatography[13d] in identical yields, avoiding contamination by cleaning the pump and extended purging of the columns with distilled toluene (Figure 1). The unlabeled fraction II-H and the labeled fraction II-D were subjected to gcms.

20-Methyl-5α- and -5β-pregnane. 3-Oxapregn-4-ene-20β-carboxaldehyde (the Upjohn Co., 3.29 g, 10 mmol) was reduced to hydrocarbon by a modified Wolff–Kishner reduction[23] in triethanolamine solvent. The reaction mixture was cooled to room temperature and mixed thoroughly with water (30 ml) and dichloromethane (30 ml). After separation, the aqueous phase was extracted twice with dichloromethane (15 ml each); the organic solutions were combined and washed twice with dilute aqueous HCl, once with dilute aqueous sodium bicarbonate, and once with water and dried with magnesium sulfate; evaporation of the solvent yielded 2.79 g (92% of theory) of a dark brown oil. Gas chromatography (3% OV-17 on Gas Chrom Q) showed the presence of three major species totaling about 90–95% of the sample. The mass spectrum indicated about 92% of mol wt 300 material (theory 300) and about 8% of mol wt 298 material.

This crude mixture of isomeric monoolefins (presumably contaminated by a small amount of diene) was hydrogenated. A 0.5-g (1.66 mmol) solution of the olefin in 5 ml of glacial acetic acid with 60 mg of suspended PtO_2 (84%) was subjected to 80 psi of hydrogen pressure for 3 hr at 22° and an additional 3 hr at 50° with theoretical hydrogen consumption. After filtering off the catalyst and azeotroping off the solvent with excess *n*-heptane, 0.5 g of a dark oil was obtained. Chromatography on silica gel using the above-described small column yielded 0.35 g of a mixture of 72% of 20-methyl-5β-pregnane and 28% of 20-methyl-5α-pregnane as determined by gas chromatography; the retention time ratio was found to be 5α/5β = 1.08 (lit.[24] 1.12). Isolation of each pure isomer was effected by repeated preparative gas chromatography and retrapping (to remove traces of the other isomer) and subsequent microsublimation[25] to obtain samples free from traces of silicone due to gc column bleed.

20-Methyl-5α-pregnane had the following physical characteristics: mp 113° (lit.[26] mp 111–112°); mol wt calcd for $C_{22}H_{38}$, 302.2973; found, 302.2966 (all molecular weights determined by high-resolution mass spectrometry); $[\alpha]_D$ +7.3 ± 0.8° (lit.[26] $[\alpha]_D$ +8.9°); partial mass spectrum in Figure 3B.

20-Methyl-5β-pregnane showed the following characteristics: mp 77–78°; mol wt calcd for $C_{22}H_{38}$, 302.2973; found, 302.2971; $[\alpha]_D$ +8.3 ± 0.8°; *m/e* 151 equals 47% of *m/e* 217 at a source temperature of 124°; *m/e* 151:*m/e* 149 = 1.15. For gc coinjection with petroleum acid derived hydrocarbons, see Figures 2D and 2E.

5α-Pregnane-20ξ-carboxylic Acid Methyl Ester. Chromium trioxide (78 mg) was dissolved in 3.3 ml of an acid mixture prepared from 0.87 ml of concentrated sulfuric acid and 3 ml of water. This oxidant solution[27] was added at 22° over the course of 1 hr to a suspension of 267 mg (0.77 mmol) of 5α-pregnan-3β-ol-20ξ-carboxylic acid, mp 278–279° (lit.[28] 274–276° (Steraloids, Inc.)), in 1.5 ml of acetone. The reaction mixture was diluted with 15 ml of water and extracted with dichloromethane which was washed with water, dried over magnesium sulfate, and evaporated to dryness. A second run with 471 mg (1.35 mmol) was made, and the combined products from the two runs were recrystallized from

(21) E. J. Gallegos, Invited Paper No. 30, Seventh World Petroleum Congress, Mexico City, Mexico, April 2–8, 1967.

(22) R. Hernandez, R. Hernandez, and L. R. Axelrod, *Anal. Chem.*, **33**, 371 (1961).

(23) L. F. Fieser and M. Fieser, "Reagents for Organic Synthesis," Wiley, New York, N. Y., 1967, p 435.

(24) C. J. W. Brooks and L. Hanaineh, *Biochem. J.*, **87**, 151 (1963).

(25) W. H. Melhuish, *Nature (London)*, **184**, 2933 (1959).

(26) J. P. Dusza and W. Bergmann, *J. Org. Chem.*, **25**, 79 (1960).

(27) See ref 23, p 142.

(28) W. Bergmann, D. H. Gould, and E. M. Low, *J. Org. Chem.*, **10**, 570 (1945).

chloroform–cyclohexane to yield 466 mg (63% of theory) of 3-oxo-5α-pregnane-20ξ-carboxylic acid: mp 252°; mol wt calcd for $C_{22}H_{34}O_3$, 346.2508; found, 346.2518. The keto acid was reduced to the acid and converted to its methyl ester by the same techniques as used for the preparation of 5α-cholanic acid methyl ester (below). The crude ester (200 mg) was purified by silica gel chromatography on the above-described small column by eluting with cyclohexane followed by benzene. The cyclohexane eluate (700 ml) and the first 200 ml of benzene eluate were discarded. Pure ester product was isolated from the subsequent 400 ml of benzene eluate: mp 87.5–88°; mol wt calcd for $C_{23}H_{38}O_2$, 346.2872; found, 346.2891; ir carbonyl absorption at 1740 cm^{-1}. The gas chromatogram indicated the presence of more than 90% of one isomer.

5α-Pregnane-20ξ-carboxylic Acid and 1,1,7-Trihydroperfluoroheptyl Ester. A portion (6.6 mg) of the purified methyl ester (above) was treated with 5 ml of 2 *N* ethanolic KOH at 95° for 27 hr, neutralized with 1 *N* HCl at 0°, and extracted with ether six times. The ether solution was washed twice with water and evaporated to dryness. The 5.9 mg of crude acid was chromatographed on SiO_2 with hexane and benzene. Recovery was 4.0 mg of pure acid: mol wt calcd for $C_{22}H_{36}O_2$, 332.2775; found, 332.2728; ir carbonyl absorption at 1710 cm^{-1}. The acid was esterified with 1,1,7-trihydroperfluoroheptanol by a previously described method.[13g] The mass spectral fragmentation pattern of this ester as obtained by gcms is given in Table I.

20ξ-Deuteriomethyl-5α-pregnane. 5α-Pregnane-20ξ-carboxylic acid methyl ester, whose preparation is described above, was reduced with lithium aluminum hydride to its corresponding alcohol:[13d] mol wt calcd for $C_{22}H_{38}O$, 318.2922; found, 318.2946. The alcohol was converted to the tosylate (mol wt calcd for $C_{29}H_{44}O_3S$, 472.3009; found, 472.3017), and the latter was reduced with lithium aluminum deuteride to give 20ξ-deuteriomethyl-5α-pregnane, mp 112–113°; the corresponding unlabeled hydrocarbon synthesized by a different route (above) has mp 113°; mol wt calcd for $C_{22}H_{37}D$, 303.3036; found, 303.3046; partial mass spectrum in Figure 3D.

5α-Cholanic Acid Methyl Ester. 3,6-Dioxo-5α-cholanic acid (Mann Research Laboratories, mp 203–206° (lit.[29] 208–210°), 0.377 g, 0.97 mmol) was reduced by a modified Wolff–Kishner reaction[23] using triethanolamine as the solvent. A solution of the reaction mixture in dichloromethane was washed three times with dilute aqueous hydrochloric acid and with water and dried (magnesium sulfate). Evaporation of the solvent yielded crude 5α-cholanic acid (318 mg, 91% of theory). The acid was dissolved in a mixture of methanol (15 ml) and boron trifluoride etherate (3 ml) and heated under reflux for 2 hr. A dichloromethane solution of the reaction mixture was washed with aqueous sodium bicarbonate and water and dried over magnesium sulfate, and the solvent evaporated, yielding 5α-cholanic acid methyl ester (336 mg, 101% of theory); the latter was purified by chromatography on silica gel as described for the 5α-pregnane-20ξ-carboxylic acid methyl ester (above). Gas chromatographic coinjection of this "5α" ester with authentic 5β-cholanic acid methyl ester revealed that this product consisted of 77% of the 5α and 23% of the 5β isomer (retention time 5α/5β = 1.04); mol wt calcd for $C_{25}H_{42}O_2$, 374.3185; found, 374.3157. The 5β isomer probably was already present in the starting diketo acid, as indicated by a gas chromatogram of its methyl ester. This would not be surprising because the 5α-diketo acid is commercially prepared from 5β bile acids. An alternative would be isomerization of some 5α to 5β during the Wolff–Kishner reaction.

5α-Cholanic Acid and 1,1,7-Trihydroperfluoroheptyl Ester. The acid (mol wt calcd for $C_{24}H_{40}O_2$, 360.3028; found, 360.3035) was prepared by hydrolysis of the methyl ester. The fluoroalcohol ester was prepared from the acid as described previously.[13g] Gcms gave the fragmentation pattern shown in Table II; gc retention times: synthetic 5α- and 5β-fluoroalcohol esters 1.019 *vs.* 1.013 for natural product derivatives E/D.

5α-Cholane and 24-Deuterio-5α-cholane. The mixture of 5α- and 5β-cholanic acid methyl esters (described above) was converted to the corresponding mixture of stereoisomeric cholanols and tosylates by previously described methods;[13d] lithium aluminum deuteride was used for part of the tosylate to give (after preparative gc and sublimation) pure 24-deuterio-5α-cholane. 5α-Cholane showed the following characteristics: mp 73–74°; mol wt calcd for $C_{24}H_{42}$, 330.3286; found, 330.3296; $[\alpha]_D$ 22.0 ± 2.0°; partial mass spectrum in Figure 5B; gc coinjection with petroleum-based hydrocarbon in Figure 2B. 24-Deuterio-5α-cholane showed the following characteristics: mp 73–74°; mol wt calcd for $C_{24}H_{41}D$, 331.3349; found, 331.3359; partial mass spectrum in Figure 5D.

24-Deuterio-5β-cholane. 24-Deuterio-5β-cholane was prepared by reduction of the previously described[13d] corresponding tosylate with lithium aluminum deuteride and purified as above: mp 92–92.5° (lit.[30] mp of 5β-cholane, 90°); mol wt calcd for $C_{24}H_{41}O$, 331.3349; found, 331.3299; partial mass spectrum in Figure 6D; gc coinjection of 5β-cholane with petroleum acid derived hydrocarbon in Figure 2C; gc retention time for 5α/5β cholane, 1.045.

Acknowledgment. We gratefully acknowledge Professor C. Djerassi and Dr. R. Hale, Stanford University, and also Professor M. Calvin, University of California, Berkeley, for stimulating discussions. We thank Dr. W. D. Redfield for optical rotation measurements, Dr. M. J. R. Cantow for gel permeation chromatography, Dr. W. G. Howells for some preliminary separations, and Mr. R. G. Prince for technical assistance. Grateful acknowledgment for support of this work and for the permission to publish is made to the management of Chevron Oil Field Research Co., La Habra, and Chevron Research Co., Richmond, Calif. This paper was presented in part at the Gordon Research Conference on Organic Geochemistry, Holderness, N. H., Aug 1970.

(29) H. Ishii, Y. Nozaki, T. Okumura, and D. Satoh, *Yakugaku Zasshi*, **81**, 805 (1961).

(30) R. T. Blickenstaff and F. C. Chang, *J. Amer. Chem. Soc.*, **80**, 2726 (1958).

ERRATA

Page 5880, column 2, line 15 should read: "Pliocene age (Midway–Sunset Field, 10^7 years); it resulted . . ."

Page 5887, column 2, line 32 should read: "92.5° (lit. [30] mp of 5β-cholane, 90°); mol wt calcd for $C_{24}H_{41}D$, . . ."

27

Reprinted from *Carnegie Inst. Washington Year Book* **66**:526–528 (1968)

RACEMIZATION OF AMINO ACIDS IN FOSSIL SHELLS

P. E. Hare and P. H. Abelson

The proteins of all living forms consist principally of *L*-amino acids. The universal use of the one stereoisomer is a phenomenon of special concern to persons interested in the origin and evolution of life. Why was one isomer chosen, and what factors determined the choice?

Present views of processes involved in the origin of life emphasize a period of abiologic synthesis during which amino acids would accumulate in a primitive ocean. It is therefore of interest to study changes in amino acids with passage of time. The present study was devoted to determining the optical configuration of the amino acids in Recent and fossil *Mercenaria* shells.

The shells were dissolved in HCl, and proteins or peptides were converted to free amino acids by hydrolysis in 6 *N* HCl. Calcium was precipitated with HF, and the supernatant solution was dried in vacuo. After adjustment of the *p*H, solutions of the amino acids were treated with enzymes that specifically attack either the *L* or *D* forms, and destroy them. The enzymes employed included *D*-amino oxidase, *L*-amino oxidase, and *L*-glutamic decarboxylase. Behavior of the enzymes was thoroughly tested with standard solutions of known content and of known amino acid configuration. After digestion of the samples with enzymes, aliquots of the mixture were treated with enough 25% trichloracetic acid to precipitate the enzymes and to give a final *p*H of about 2. The mixture was centrifuged, and the supernatant solution was placed in the amino acid analyzer.

This instrument is based on conventional design, but it has been modified to attain high sensitivity for small samples. Output of the spectrophotometer is monitored by a recorder and digital integrator. In the present experiments the apparatus was operated at somewhat reduced sensitivity. Accordingly, 10^{-8} moles of amino acid resulted in 50,000 counts on the integrator. It is possible to measure a peak of 200 counts with a probable error of 50 counts. Because of this sensitivity and because of the specificity and effectiveness of the *L*-amino oxidase in destroying completely the *L* isomer of a number of the amino acids, the appearance of as little as a few per cent of *D* isomer could be detected. In comparing the amino acid content of solutions before and after enzyme treatment, glycine constituted an ideal internal standard, since it is inert to the various enzymes used in this study. Overall, the combination of amino acid analyzer and enzyme digestion provides a sensitivity for detection of racemization that is at least three orders of magnitude better than a typical electronic polarimeter.

The total amount of amino acids present in fossil shells decreases with age of the fossil. In Table 37 the amino acid composition of each sample is indicated in residues per thousand as well as the total micromoles per gram of shell. The distribution of amino acids changes with time, reflecting different reaction rates for the disappearance of each amino acid. As pointed out in last year's report, some amino acids not found in Recent shells are present in fossil shells.

In Table 38 data are summarized on the optical configuration of the amino acids, which are present in significant quantities and which react with the specific enzymes used in this

TABLE 37. Amino Acid Composition in Residues per Thousand for Inner Layer of *Mercenaria* Shells

	Recent, total	Upper Pleistocene		Miocene, total
		Total	Insoluble	
Aspartic acid	122	133	142	35
Threonine	73	47	38	<1
Serine	89	38	29	<1
Glutamic acid	87	109	105	139
Proline	111	118	111	120
Glycine	93	88	67	104
Alanine	75	110	56	166
Cystine	25	4	25	. . .
Valine	46	65	62	127
Methionine	14	13	15	10
Alloisoleucine	. . .	8	1	42
Isoleucine	23	25	33	34
Leucine	32	43	47	78
Tyrosine	32	44	56	38
Phenylalanine	22	31	41	52
Ornithine	. . .	8	5	9
Lysine	71	68	77	46
Histidine	21	10	14	. . .
Arginine	38	37	79	. . .
Ammonia*	(62)	(108)	(60)	(780)
μM/g shell	16.1	6.5	. . .	1.1

* Ammonia values not included in total residues.

study. The results show that complete racemization has occurred in the Miocene specimen, and appreciable racemization has occurred even in the Upper Pleistocene from Wailes Bluff. Data from the insoluble fraction from Wailes Bluff indicate that some racemization occurred while the amino acids were still bound in peptide linkage in an insoluble state. As expected, all the amino acids in the Recent shell are of the *L* configuration.

From the experimental results with specific enzymes it appears that part of the *L*-isoleucine originally present is converted to *D*-alloisoleucine and that no appreciable amounts of *D*-isoleucine or *L*-alloisoleucine are formed.

Alloisoleucine is not a protein amino acid. It is not present in Re-

TABLE 38. Percentage of *D*-amino Acid Isomer in Amino Acids from Shells of *Mercenaria*

	Recent, total	Upper Pleistocene, total	Upper Pleistocene, insoluble	Miocene, total
Glutamic acid	<5	22	11	47
Proline	0	31	19	52
Alanine	0	40	10	51
Valine	0	28	5	52
Alloisoleucine	. . .	98	100	95
Isoleucine	0	0	0	<5
Leucine	0	26	5	48
Tyrosine	0	25	16	50
Phenylalanine	0	25	8	49

cent shells but is present in fossils. The relationship of *L*-isoleucine and *D*-alloisoleucine is intriguing: Isoleucine has two asymmetric carbon atoms, and as a result there are four possible isomers, *D*- and *L*-isoleucine and *D*- and *L*-alloisoleucine. Racemization of *L*-isoleucine involves the amino carbon and yields *D*-alloisoleucine. Racemization involving the second asymmetric carbon atom proceeds very slowly, if at all. We have not detected significant formation of *D*-isoleucine or *L*-alloisoleucine from *L*-isoleucine or *D*-alloisoleucine under circumstances in which most amino acids are completely racemized.

Isoleucine and alloisoleucine behave differently on cation-ion-exchange columns; alloisoleucine is eluted more readily than isoleucine. This indicates that isoleucine is more hydrophobic. Thus, the attempted racemization of isoleucine results in products easily separable by nonbiologic processes. In contrast, when amino acids such as alanine racemize, the physical-chemical behavior of the products is almost identical. They are equally hydrophilic, and they cannot be separated by ion-exchange methods. An adsorption surface possessing asymmetric properties is required to separate them. In a prebiotic world it would be much more feasible to separate alloisoleucine and isoleucine than to separate racemic pairs of amino acids such as alanine.

The apparent characteristic time for racemization is of the order of 10^5 years. Enough racemization should occur in shorter periods so that it should be possible to study the process in specimens 30,000 years old that can be dated by C^{14}. The rate of racemization is affected by temperature. Laboratory tests conducted at 160°C for 1 week showed that amino acids in shells were racemized in that period.

These results indicate that for most of the amino acids the mixture of stereo forms accumulating in a primitive ocean would be racemic. If some unknown process presented a slight advantage to one form or another, racemization would tend to vitiate the advantage. On the other hand, once a decision was made to use the *L* form, supplies of it would be replenished by racemization from the *D* form as the *L* isomer was used up.

28

Reprinted from *Nature* **262**:126–128 (1976)

Characterisation of monounsaturated fatty acids from an estuarine sediment

PREVIOUS investigations of fatty acids in marine sediments have primarily been concerned with the concentrations of the individual acids rather than their chemical configuration[1–4]. Brown and Swidler[5] have demonstrated the ability of clay to catalyse the conversion of oleic acid (18:1Δ^9 *cis*) to elaidic acid (18:1Δ^9 *trans*) accompanied by extensive rearrangement of the double bond position. (18:1 = normal carbon chain length: number of double bonds). As most biologically produced olefinic acids are believed to be in the *cis* configuration, diagenetic changes in the sediment may alter the *cis* isomers to the more stable *trans* isomers, as well as cause migration of the double bond. Consequently, investigations of the monounsaturated fatty acid *cis*/*trans* ratio and position of the double bonds, with depth in the sediment, may yield significant information on diagenetic pathways of these acids. We have determined the presence of small amounts of palmitelaidic acid (16:1Δ^9 *trans*) and elaidic acid (18:1Δ^9 *trans*), as well as other positional isomers of these trans acids, in a Recent estuarine sediment. In addition, we have also confirmed that the major monounsaturated fatty acids are palmitoleic acid (16:1Δ^9 *cis*) and oleic acid (18:1Δ^9 *cis*). We give here the methods for isolating these acids and confirming their structures, and their geochemical significance is discussed.

The general collection and analytical procedures have previously been reported in detail[1–3]. Surface sediment (upper 8–10 cm) was collected from station D in Narragansett Bay, Rhode Island[2]. The sediment was saponified (under reflux) for 2 h using 2:1 (*v*/*v*) 0.5 *N* KOH in 90% methanol-benzene, filtered through a Whatman No. 1 filter, and the filtrate transferred to a separatory funnel. After addition of distilled water to give two phases, the non-saponifiable fraction was removed and backwashed with *p*H 10 distilled water. The resulting aqueous phase and the initial saponifiable fraction were then both acidified to *p*H 2 and extracted three times with petroleum ether. The combined ether phases were evaporated to near dryness at room temperature under reduced pressure using a rotary evaporator. Subsamples were then taken for each analysis.

To subsamples I and II were added 17:1Δ^{10} *cis* or 17:1Δ^{10} *trans* fatty acid internal standards. Each subsample was then methylated using BF_3–methanol and the methyl esters isolated by thin-layer chromatography (TLC)[1]. The methyl esters were further separated on $AgNO_3$–TLC to isolate both *cis* and *trans* monounsaturated esters[6,7].

The *cis* and *trans* fractions were individually analysed by flame ionisation gas–liquid chromatography (GLC) using a Hewlett–Packard model 5750 gas chromatograph. The resulting chromatogram for the *trans* methyl esters is shown in Fig. 1. The proper configurational assignments were made on the basis of relative retention times and coinjection with authentic standards on a stainless steel column (6m × 2.2mm inside diameter) containing GP–15% SP-2340 on 100/120 Chromosorb P, AW-DMCS (Supelco), resolution[8] = 0.83 for 16:1Δ^9 *cis* and *trans*. Each sample was also hydrogenated and co-injected with standard saturated esters

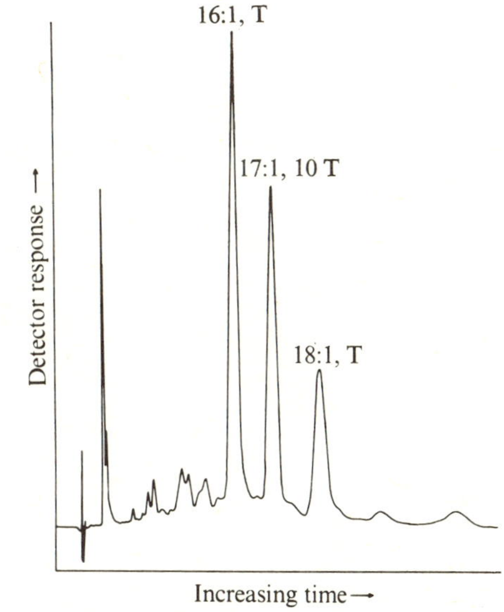

Fig. 1 Gas chromatogram of 16:1 and 18:1 *trans* methyl esters, in presence of 17:1Δ^{10} *trans* methyl ester internal standard, recovered from station D surface sediment.

To determine the quantitative amount of each acid, the areas of the peaks of the naturally occurring acids (before and after hydrogenation) were compared with the areas of the internal standard. (No natural 17:1 *cis* or *trans* was detected in the sediments.) The GLC analysis of I and II indicated the quantities of fatty acids in the sediment to be as follows (μg acid per g dry weight of sediment): for 16:1 *cis*, 17.1 μg g^{-1}, for 16:1 *trans*, 1.7 μg g^{-1}, for 18:1 *cis*, 9.6 μg g^{-1}, and for 18:1 *trans*, 0.8 μg g^{-1}. Analytical uncertainty was found to be ± 10% or less. Thus in both cases, the *cis* isomer accounted for ~ 92% and the trans isomer ~ 8% of the total 16:1 and 18:1 acids.

To investigate the possibility of the *trans* isomer being generated from the *cis* isomer during the saponification, methylation, or TLC procedures, a series of blanks was carried through the entire analysis. Each blank consisted of a pre-extracted sediment plus various combinations of 16:1, 17:1 and 18:1 *cis* standard acids. No *trans* acids were generated during this procedure

The configurational assignments were verified by infrared spectroscopy and combined gas chromatography–mass spectrometry (GC–MS) analyses. Two separate subsamples, III and IV, were processed by the above methods without internal

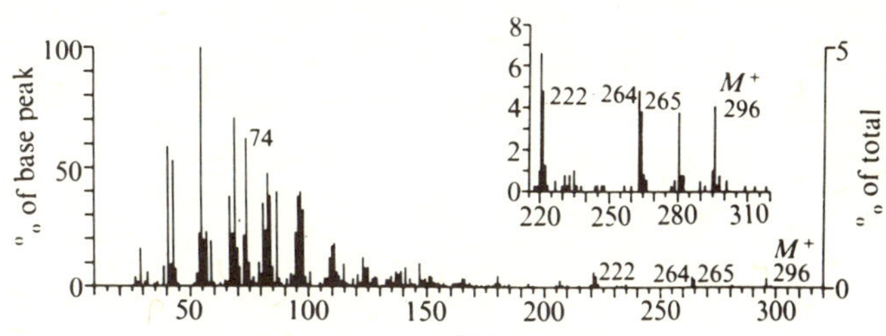

Fig. 2 GC–MS spectrum of 18:1 *trans* methyl ester recovered from station D surface sediment; no internal standard added (*m*/*e* 220–310 expanded in upper right).

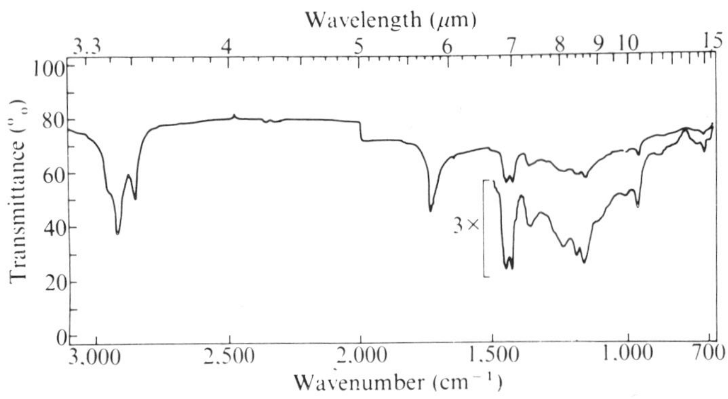

Fig. 3 Infrared spectrum of 16:1 and 18:1 *trans* methyl esters recovered from station D surface sediment; no internal standard added.

standards. Portions of the purified *cis* and *trans* methyl esters were each analysed by GC–MS (Finnigan 1015C). The resulting GC–MS spectra for these esters were compared with spectra for authentic standards and agreed well in all cases. The GC–MS spectrum for the *trans* methyl ester of the eighteen carbon fatty acid is shown in Fig. 2. This spectrum shows the molecular ion peak at m/e 296 (M^+). Some characteristic fragmentation peaks are m/e 265 ($M^+ - 31$), m/e 264 ($M^+ - 32$), m/e 222 ($M^+ - 74$), and m/e 74 (McLafferty rearrangement)[9,10].

Separate portions of III and IV were analysed by infrared spectroscopy (Perkin Elmer model 521) and compared with authentic standard infrared spectra. The infrared spectrum of the *trans* methyl esters is presented in Fig. 3 and shows the characteristic *trans* absorption at 965 cm^{-1} (10.3 μm). The infrared spectrum of the *cis* methyl esters showed both the absence of the *trans* band at 965 cm^{-1} and the presence of a stronger absorption in the region of 3,015 cm^{-1} (3.32 μm). The infrared and GC–MS spectra together with the GLC analyses confirm the presence of both 16:1 and 18.1 *cis* and *trans* acids.

Portions of III and IV were oxidised using $KMnO_4$ in a modification of Downing and Greene's[11] procedure to determine the position of the double bond. The sixteen and eighteen carbon *cis* oxidation products yielded primarily the C_7 and C_9 methyl esters and the C_9 dimethyl ester, indicating the double bond to be predominantly in the 9–10 position for both fatty acids. The ratios of the dimethyl esters recovered indicate that ~ 62% of the double bonds in the *cis* isomers are in the 9–10 position, 19% are in the 11–12 position, and 19% are in other possible positions from C_5 to C_8, and C_{10}.

Analysis of the sixteen and eighteen carbon *trans* oxidation products again yielded primarily the C_9 dimethyl ester and the C_7 and C_9 methyl esters, indicating the double bonds in both *trans* isomers to be predominantly in the 9–10 position. The ratios of the dimethyl esters recovered from the oxidation of the *trans* isomers indicate that ~ 70% of the double bonds are in the 9–10 position, 23% are in the 11–12 position, and 7% are in other positions as above. Within the accuracy of the method, there is no significant difference between the positions of the double bonds in the *cis* and *trans* isomers.

The three most common monounsaturated fatty acids found in organisms are oleic acid, palmitoleic acid, and vaccenic acid ($18:1\Delta^{11}$ *cis*)[12]. It appears that the major monounsaturated acids present in the examined surface sediment are the $16:1\Delta^9$, $18:1\Delta^9$ and $18:1\Delta^{11}$. These are present as ~ 92% in the *cis* configuration and 8% in the *trans* configuration. Since the double bonds are primarily in two positions, the 9–10 and 11–12 positions, this would seem to indicate that the *cis* isomers present in this surface sediment result from a direct biological input. The *trans* isomers result from either a direct input or from microbial modification of the *cis* isomers in the sediments. The *trans* isomers apparently do not result from the clay catalysed isomerisation of the *cis* isomer. According to Brown and Swidler[5] the clay catalysed isomerisation of the *cis* octadecenoic acid yields a mixture of the *trans* isomers with the double bond ranging from the 2–17 positions, and the 9–10 *trans* isomer making up only 23% of the total.

Although the absolute quantities of the monounsaturated fatty acids normally decrease with depth in the sediment[2], secondary effects may result from clay catalysis. As depth increases, the clay catalysed conversion of the *cis* to *trans* isomers may occur until an equilibrium mixture of *cis* and *trans* results (for the 18:1 acid, the equilibrium mixture is ~ 33% *cis*, 67% *trans*[5,13]). As the *cis* isomer is isomerised to the *trans* isomer, it may also result in a migration of the double bond throughout the molecule. Thus with increasing depth in the sediments, there exist two potential biogeochemical markers based on monounsaturated fatty acids, the ratio of *cis*/*trans* and the position of the double bond.

This work was supported by the National Sea Grant Program. We thank Dr Chris Brown for the infrared analyses and Dr Peter Rogerson and Curt Norwood for the GC–MS analyses.

EDWARD S. VAN VLEET
JAMES G. QUINN

Graduate School of Oceanography,
University of Rhode Island,
Kingston, Rhode Island 02881

Received February 5; accepted May 18, 1976.

1 Farrington, J. W., and Quinn, J. G., *Geochim. cosmochim. Acta*, **35**, 735–741 (1971).
2 Farrington, J. W., and Quinn, J. G., *Nature*, **230**, 67–69 (1971).
3 Farrington, J. W., and Quinn, J. G., *Geochim. cosmochim. Acta*, **37**, 259–268 (1973).
4 Johnson, R. W., and Calder, J. A., *Geochim. cosmochim. Acta*, **37**, 1943–1955 (1973).
5 Brown, L. H., and Swidler, R., *US Pat.* 2,065,248, *Nov.* 20, 1962; *Chem. Abstr.* **58**, 11221F (1963).
6 Morris, L. J., in *Metabolism and Physiological Significance of Lipids* (edit. by Dawson, M. C., and Rhodes, D. N.) (Wiley, New York, 1963).
7 Cubero, J. M., and Mangold, H. K., *Microchem. J.*, **9**, 227–236 (1965).
8 McNair, H. M., and Bonelli, E. J., *Basic Gas Chromatography* (Varian Aerograph, Walnut Creek, 1968).
9 McCloskey, J. A., in *Topics in Lipid Chemistry* (edit. by Gunstone, F.D.) (Wiley, New York, 1970).
10 Silverstein, R. M., Bassler, G. C., and Morrill, T. C., *Spectrometric Identification of Organic Compounds* (Wiley, New York, 1974).
11 Downing, D. T., and Greene, R. S., *Lipids*, 3, 96–100 (1968).
12 White, A., Handler, P., and Smith, E. L., *Principles of Biochemistry* (McGraw-Hill, New York, 1968).
13 Swern, D., and Scanlan, J. T., *Biochem. Prep.*, 3, 119–120 (1953).

Part IV

MOLECULAR EXPERIMENTATION

Editor's Comments on Papers 29 Through 34

In the preceding sections of this book, emphasis has been placed on an analytical approach to the geochemistry of organic molecules. That is, classes of organic compounds have been sought and discovered through the analysis of many different kinds of geological materials. From the kinds of organic compounds present, interpretations have been made regarding geochemical processes. Another approach to investigating organic geochemical processes is through laboratory simulation. Two distinct methods of laboratory simulation are possible: (1) Individual organic compounds or classes of compounds can be added to a geological material like clay or sediment. This mixture is then subjected to certain experimental conditions such as elevated temperatures. An analysis then seeks to find the resulting products from the added compounds. Comparisons are made between the resulting products and naturally occurring molecules; or (2) Materials like sediments and shells can be subjected directly to experimental conditions involving perhaps elevated temperatures and pressures, without the introduction of known compounds. The experimental conditions induce geochemical changes in the indigenous organic compounds of the geologic material. Comparisons are made between the chemical compositions before and after the experiment. Changes in chemical compositions can lead to inferences regarding diagenesis of organic molecules.

Four papers have been selected for this part to illustrate experiments in which individual organic compounds have been added to sedimentary materials and subjected to high temperatures, and two papers demonstrate laboratory induced diagenesis of organic matter in sediments and shells. Many of the laboratory simulations involving added compounds have sought to explain various processes leading to petroleum genesis. One of the first investigations that attempted to simulate in the laboratory the possible role of fatty acid in the formation of hydrocarbons of petroleum was by Jurg and Eisma (Paper 29). In these experiments *n*-docosanoic acid (behenic acid, C_{22}) was added to the clay, bentonite, and the mixtures, with and without water, were heated in sealed tubes to 200°C for 89 and 760 hours. The major product was the *n*-alkane, *n*-heneicosane (C_{21}). Lower molecular weight *n*-alkanes were also produced along with *n*-alkanes having up to 34 carbon atoms. The experiment showed that the process of decarboxylation can produce alkanes but other processes also are required. An expanded version of this kind of work is given by Eisma and Jurg (1969).

Starting with a hydrocarbon rather than a fatty acid, Henderson et al. (Paper 30) pyrolyzed *n*-octacosane (C_{28}) at 375°C for varying periods of time with and without bentonite. They observed that this single hydrocarbon altered to an homologous series of *n*-alkanes having a smooth distribution of even- and odd-carbon-numbered molecules. Other products included branched-cyclic alkanes, alkenes, and aromatic hydrocarbons. These experiments indicate that thermal and catalytic processes may account in part for the hydrocarbons in petroleum.

In the two experiments just mentioned, two processes were operative: decarboxylation and thermal alteration. Shimoyama and Johns (Paper 31) simulated both of these processes in a series of experiments in which *n*-docasanoic acid (the same compound used by Jurg and Eisma in Paper 29) was heated along with the clay, montmorillonite (bentonite), to 200 and 250°C for 50–500 hours. The resulting compounds led to the interpretation that decarboxylation followed by thermal cracking is a reasonable process in petroleum genesis. Other information pertaining to this work can be found in Shimoyama and Johns (1972) and Waples (1972).

The experiments discussed thus far have tried to account for geochemical processes involving fatty acids and hydrocarbons. Laboratory simulation experiments have been done on other classes of compounds including, for example, amino acids (Vallentyne, 1964), sterols (Rhead, Eglinton, and Draffan, 1971), and porphyrins (Casagrande and Hodgson, Paper 32). Paper 32 is not the first to simulate porphyrin diagenesis in the laboratory; Treibs in Paper 1 appears to have priority. Paper 32, however, is important because it demonstrates that in laboratory simulations a single porphyrin can be altered to yield an homologous series of porphyrins as observed by Thomas and Blumer (1964)., Morandi and Jensen (1966), and Baker (see Paper 18) in natural geological materials. The experiment consisted of heating mesoporphyrin IX along with various mixtures including clay and water in sealed tubes at 250°C for 280 hours. The resulting homologous series of porphyrins supports the transalkylatron mechanims first proposed by Baker in Paper 18.

The use of radiolabeled compounds in these kinds of laboratory simulation studies has been emphasized by Eglinton (1972). A number of compounds have been investigated including oleic acid (Rhead et al., Paper 11; Gaskell et al., 1976), sterols (Gaskell and Eglinton, 1975), and phytol (Brooks and Maxwell, 1974). The use of radiolabels has the advantage that only small quantities are needed, and the products can be detected whether expected or not.

The second approach to simulating organic geochemical processes in the laboratory takes advantage of the compounds already present in the geological material. Two papers show how this approach can be used. Mitterer and Hoering (Paper 33) investigated the compounds that are released upon heating the insoluble organic material in a recent marine sediment. These experiments produced *n*-alkanes, isoprenoid hydrocarbons, and porphyrins. The results of the work demonstrated that kerogen can be a major source of hydrocarbons. A follow-up of this kind of line of investigation has been given in a related series of papers, an example of which is Ishiwatari et al. (1977) and references therein.

Laboratory simulation of the diagenetic reactions of amino acids in shells was accomplished by Hare and Mitterer (Paper 34) who incubated under a variety of conditions fragments of modern shells. The changes taking place in the amino acids upon heating, in the presence of water, of modern shell fragments are analogous to and duplicate many of the effects found in fossil shells. Thus laboratory studies of chemical diagenesis have provided new insights into possible organic geochemical processes.

REFERENCES

Brooks, P. W. , and J. R. Maxwell. 1974. Early stage fate of phytol in a recently deposited lacustrine sediment. In B. Tissot and F. Bienner, eds., *Advances in Organic Geochemistry 1973*. Editions Technip., Paris, 977–991.

Eglinton, G. 1972. Laboratory simulation of organic geochemical processes. In H. R. v. Gaertner and H. Wehner, eds., *Advances in Organic Geochemistry 1971*. Pergamon Press, Braunschweig, 29–48.

Esima, E., and J. W. Jurg. 1969. Fundamental aspects of the generation of petroleum. In G. Eglinton and M. T. J. Murphy, eds., *Organic Geochemistry—Methods and Results*. Springer-Verlag, Berlin, 676–698.

Gaskell, S. J., and G. Eglinton. 1975. Rapid hydrogenation of sterols in a contemporary lacustrine sediment. *Nature* **254**:209–211.

Gaskell, S. J., M. M. Rhead, P. W. Brooks, and G. Eglinton. 1976. Diagenesis of oleic acid in an estuarine sediment. *Chem. Geology* **17**:319–324.

Ishiwatari, R., M. Ishiwatari, B. G. Rohrback, and I. R. Kaplan. 1977. Thermal alteration experiments on organic matter from recent marine sediments in relation to petroleum genesis. *Geochim. et Cosmochim. Acta* **41**:815–828.

Morandi, J. R., and H. B. Jensen. 1966. Comparison of porphyrins from shale oil, oil shale, and petroleum by absorption and mass spectroscopy. *Jour. Chem. and Eng. Data* **11**:81–88.

Rhead, M. M., G. Eglinton, and G. H. Draffan. 1971. Hydrocarbons produced by the thermal alteration of cholesterol under conditions simulating the maturation of sediments. *Chem. Geology* **8**:277–297.

Shimoyama, A., and W. D. Johns. 1972. Formation of alkanes from fatty acids in the presence of $CaCO_3$. *Geochim. et Cosmochim. Acta* **36**: 87–91.

Thomas, D. W., and M. Blumer. 1964. Porphyrin pigments of a Triassic sediment. *Geochim. et Cosmochim. Acta* **28**:1147–1154.

Vallentyne, J. R. 1964. Biogeochemistry of organic matter—II. Thermal reaction kinetics and transformation products of amino compounds. *Geochim. et Cosmochim. Acta* **28**:157–188.

Waples, D. W. 1972. Catalytic formation of hydrocarbons from fatty acids. *Nature Phys. Sci.* **237**:63–64.

SUGGESTED ADDITIONAL READINGS

Didyk, B. M., Y. I. A. Alturki, C. T. Pillinger, and G. Eglinton. 1975. Petroporphyrins as indicators of geothermal maturation. *Nature* **256**: 563–565.

Ikan, R., Z. Aizenshtat, M. J. Baedecker, and I. R. Kaplan. 1975. Thermal alteration experiments on organic matter in recent marine sediments—I. Pigments. *Geochim. et Cosmochim. Acta* **39**:173–185.

Ikan, R., M. J. Baedecker, and I. R. Kaplan. 1975. Thermal alteration experiments on organic matter in recent marine sediments—II. Isoprenoids. *Geochim. et Cosmochim. Acta* **39**:187–194.

Ikan, R., M. J. Baedecker, and I. R. Kaplan. 1975. Thermal alteration experiments on organic matter in recent marine sediments—III. Aliphatic and steroidal alcohols. *Geochim. et Cosmochim. Acta* **39**:195–203.

Leeuw, J. W. de, V. A. Correia, and P. A. Schenck. 1974. On the decomposition of phytol under simulated geological conditions and in the top-layer of natural sediments. In B. Tissot and F. Bienner, eds., *Advances in Organic Geochemistry 1973*. Editions Technip., Paris, 993–1004.

29

Reprinted from *Science* **144**:1451–1452 (1964)

PETROLEUM HYDROCARBONS: GENERATION FROM FATTY ACID

J. W. Jurg and E. Eisma

In connection with the generation of petroleum, several investigators have published data on the formation of hydrocarbons from various types of organic matter in sediments (*1*). Other authors (*2*) have performed model experiments with pure organic compounds. The present investigation also deals with a pure organic compound, a saturated fatty acid.

In many young sediments that have never suffered great depth of burial, the amount of normal (*n*-)alkanes with an odd number of carbon atoms greatly exceeds the amount of those with an even number in the range of about 27 to 37 carbon atoms (*3*). This "odd predominance" depends, among other things, on the depth of burial; with increasing depth the odd predominance decreases and generally disappears. The fact that depth plays such an important role suggests that temperature has a great influence on the formation of *n*-alkanes. It is often thought that these *n*-alkanes are formed from straight-chain fatty acids.

We have studied the formation of hydrocarbons from behenic acid (normal $C_{21}H_{43}COOH$). One gram of this acid was kept at 200°C in a sealed glass tube, with 3 g of bentonite clay as a catalyst, for 89 and 760 hours. The tube was opened in an atmosphere of hydrogen, and the volatile reaction products were stripped off with a stream of hydrogen at 150°C and collected in a vessel placed in liquid nitrogen. This low-molecular-weight fraction was analyzed by gas chromatography. The residue containing the nonvolatile reaction products was subjected to extraction with *n*-pentane. The extract containing the hydrocarbons with a higher molecular weight was passed over a silica-alumina column to obtain the saturated hydrocarbons, which were then separated as the urea adducts into a fraction rich in *n*-alkanes and a fraction containing the branched-chain and cycloalkanes. The *n*-alkane fraction was analyzed by means of gas chromatography.

The amounts of the various components generated after heating are listed in the table.

The total amount of hydrocarbons with 3, 4, or 5 carbon atoms increases in proportion to the duration of the heating. With increasing heating time the amount of saturated hydrocarbons increases very strongly, but the amount of unsaturated hydrocarbons decreases. This suggests that the unsaturates are intermediates in the formation of the saturated hydrocarbons.

Gas chromatography was used to detect the *n*-alkanes from C_{14} up to C_{34} in the fraction with a higher molecular weight. There was a strong predominance of the C_{21} *n*-alkane, which can be attributed to the decarboxylation of the fatty acid. Figure 1 is a distribution diagram of this fraction. The total amount of *n*-alkanes with a higher molecular weight was about 2 mg.

A remarkable thing, however, is that *n*-alkanes with a longer chain than that of the original acid are also present. This suggests that the *n*-alkanes present in crude oil need not necessarily be derived from fatty acids with longer chain lengths.

A factor that may be significant in the generation of petroleum is the presence of water. In order to gain an idea of its possible effect, we heated the behenic acid and the bentonite clay in a sealed tube with enough water to ensure its presence as a liquid at 200°C. In the low-molecular-weight fraction resulting from the reaction mixture, only very small quantities of the hydrocarbons from C_1 up to C_6 were detected, whereas in the higher molecular-weight fraction the *n*-alkanes up to C_{20}

were detected. These experiments with and without water revealed a marked difference in the ratio of the various components to one another. This difference was the most striking with respect to the normal and branched chain C_4 and C_5 hydrocarbons. In the experiments with water the ratio was about 0.1, whereas in the experiments without water this ratio was about 4. The same figures were obtained for the ratio of iso- to normal-pentane. We have tentatively concluded that, because isomerization is very strong in the experiments without water, carbonium ions are the intermediates (*4*). Since isomerization is much less pronounced in the experiments with water, it is possible that radicals (*5*) act as intermediates in this case.

The behenic acid did not yield hydrocarbons after heating at 200°C without the bentonite clay. The bentonite contained only about 0.1 percent of organic carbon, but a blank run with it yielded only negligible quantities of hydrocarbons.

Table 1. Amounts of low-molecular-weight hydrocarbons generated during heating at 200°C of 1 g of behenic acid with bentonite clay.

Hydrocarbons	Heating period (hr): 89 (µmole)	760 (µmole)
Groups		
C_3	0.18	1.12
C_4	0.29	2.20
C_5	0.33	2.69
Individual compounds		
Propane	0.03	0.87
Propene	0.15	0.25
n-Butane	0.02	0.54
Iso-butane	0.08	1.50
Butenes	0.19	0.16
n-Pentane	0.02	0.38
Iso-pentane	0.07	2.20
Pentenes	0.24	0.11
n-Hexane	0.02	0.29

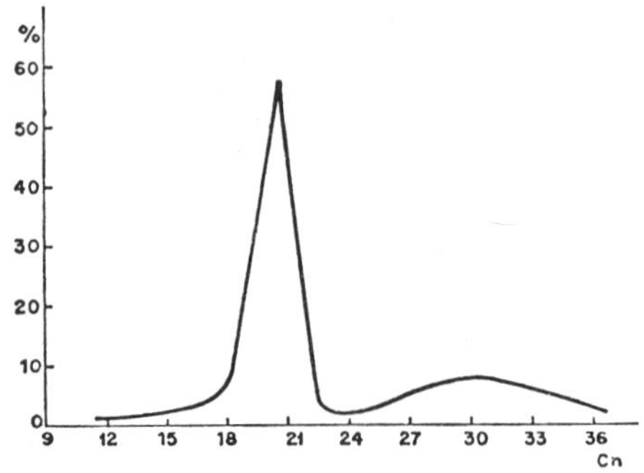

Fig. 1. Distribution of the *n*-alkanes with a carbon number greater than C_{14}. The total amount of this fraction was about 2 mg.

We fully realize that the conditions and the substances are quite different from those occurring in sediments. The results suggest that in the study of the processes of oil generation it should be realized that the presence of water may influence the mechanism by which the hydrocarbons are produced and that *n*-alkanes are not solely produced by the decarboxylation of fatty acids.

References and Notes

1. J. M. Hunt, Geochemical data on organic matter in sediments, *Proc. Intern. Scientific Oil Conference*, October 8-13, 1962, Budapest; P. H. Abelson, Organic geochemistry and the formation of petroleum, *World Petroleum Cong. 1963*, Section 1, paper 41; P. H. Abelson and T. C. Hoering, Abstracts for the meeting of the Geochem. Soc., New York, November 17-20, 1963; J. D. Mulik and J. G. Erdman, *Science* **141**, 806 (1963); R. Robinson, *Nature* **199**, no. 4889, 113 (1963).
2. J. E. Cooper and E. E. Bray, A postulated role of fatty acids in the formation of petroleum, Am. Chem. Soc. Spring Meeting, 1963, Los Angeles; J. A. Bedov, S. D. Pustilnikova, L. V. Ratnikova, Al A. Petrov, *Neftekhimija* **2**, 213 (1962); J. G. Erdman, *Geochim. Cosmochim. Acta* **22**, 16–36 (1961).
3. N. P. Stevens, E. E. Bray, E. D. Evans, *Bull. Am. Assoc. Petrol. Geolog.* **40**, 975 (1956).
4. B. S. Greensfelder, H. H. Voge and G. M. Good, *Ind. Eng. Chem.* **41**, 2573 (1949).
5. A. Kossiakoff and F. O. Rice, *J. Am. Chem. Soc.* **65**, 590 (1943).

6. We are indebted to Professor Dr. G. C. A. Schuit of Eindhoven Technological University for valuable discussions.

10 April 1964

Reprinted from *Nature* **219**:1012–1016 (1968)

Thermal Alteration as a Contributory Process to the Genesis of Petroleum

by

W. HENDERSON*
G. EGLINTON

Organic Geochemistry Unit,
School of Chemistry,
University of Bristol

P. SIMMONDS

Jet Propulsion Laboratory,
Los Angeles,
California

J. E. LOVELOCK

Bowerchalke,
Near Salisbury,
Wiltshire

The products of pyrolysis of *n*-octacosane ($C_{28}H_{58}$) have been identified by capillary gas-liquid chromatography. The reaction is affected by clay minerals. The experiments provide further support for the view that thermal and catalytic alteration of the organic material present in sediments in part explains the genesis of petroleum and other hydrocarbon minerals.

Many theories have been proposed for the origin of petroleum, ranging from purely abiogenic[1] and biogenic[2] to bimodal[3]. It is now generally accepted that a predominantly biogenic origin, with subsequent alteration of the organic debris which accumulates in sediments, is the most likely explanation. Any theories proposed for the alteration processes have to account, among others, for the following generalizations. (i) The concentrations of fatty acids are higher in modern sediments than they are in ancient sediments and petroleums[4]. (ii) The concentrations of paraffins in modern sediments are usually lower than in ancient sediments and petroleums[5]. (iii) The even/odd carbon number preference index (c.p.i.) values for fatty acids are higher in modern than in ancient sediments[4]. (iv) The odd/even c.p.i. values for paraffins are higher in modern than in ancient sediments[6].

The principal processes occurring in a sediment are usually held to be bacterial action[7–9], radioactive bombardment[10–12] and thermal alteration[13–18]. Thermal alteration, which may include thermally induced catalytic alteration of the organic debris, is probably the single most important factor. The differences found by several workers[17–21] for the distribution patterns of *n*-paraffins, obtained from similar sediments at increasing depths in a single formation, may be caused largely by thermal alteration. Temperatures of around 200° C have been reported for deeply buried sediments[21].

Thermal and Catalytic Cracking of *n*-Octacosane

In this paper the thermal and catalytic alteration of an *n*-paraffin (*n*-octacosane) is studied at different temperatures for varying periods of time. From the results we try to assess the importance of the related geochemical processes which may bring about the alteration of paraffins and other biologically derived lipids in a sediment.

Samples of *n*-octacosane were heated at 375° C for varying periods of time. The appearance of the contents changed markedly with increasing time; initially the *n*-octacosane was colourless and crystalline, but as the heat treatment progressed the products became liquid and increasingly yellow coloured and carbonized films were formed on the inside surface of the glass. Accompanying the deepening colour were marked pressure rises due to the formation of increasing quantities of gaseous products, probably low molecular weight hydrocarbons, such as methane, ethane and propane.

A typical gas–liquid chromatography (GLC) analysis of two of the four fractions isolated from the 75 h pyrolysis is illustrated in Fig. 1. The alkane fraction consists predominantly of *n*-alkanes with a relatively smooth carbon number distribution centred around *n*-C_{21}. A striking feature is the complexity of the two alkene fractions. Each carbon position is a complex mixture of peaks with as many as seven components distinguishable ranging from C_{28} to C_{14} in carbon number. The carbon number distribution patterns of the alkenes are relatively smooth with maxima at C_{22} and C_{18}. The complexity is due to the presence of geometric and positional double bond isomers and possibly, to a small extent, of branched chain or cyclic isomers. Work is in progress to resolve these complex mixtures and to determine the structures of the individual alkenes which probably play an important part in the pyrolysis reaction.

The yields and relative abundances of the hydrocarbons obtained are shown in Table 1. The general trends with increasing time are as follows: (i) the yield of alkanes decreases, (ii) the other *n*-alkanes are formed in increasing amounts relative to the *n*-C_{28} alkane, (iii) the proportion of branched and cyclic alkanes increases, (iv) the yield of alkenes and aromatics increases, (v) the carbon number maxima for the alkanes (Fig. 2) and the alkenes shift to lower values.

The *n*-alkane distributions for the 75 and 150 h pyrolyses

* On leave from the Department of Chemistry, University of Glasgow.

Table 1. PERCENTAGE YIELDS OF THE PER 75, 100 AND 150H DUCTS AFTRO PYROLYSIS OF n-C_{28} ALKANE AT 375° C

Time (h)	Yield* of products† (%)	Alkanes‡ / products† (%)	n-C_{28} Alkane‡ / n-alkanes (%)	n-Alkanes‡ / Alkanes (%)	Branched cyclic‡ / Alkanes (%)	Alkenes* / Products† (%)	Aromatics* / Products† (%)
0	—	—	99·9	—	—	—	—
75	90	80	89	95	5	5	1
100	82	76	41	77	23	8	2
150	15	70	8	73	27	12	12

* Yields calculated by weighing (accuracy ±0·1 mg).
† Products soluble in benzene.
‡ Yields calculated by measuring peak areas.

are shown in bargraph form in Fig. 2. These do not show the lower molecular weight hydrocarbons which are formed by the pyrolysis and subsequently lost in the work-up procedure. Certain apparently preferential fragmentations of the n-octacosane carbon chain occur; for example, the higher abundance of the n-C_{26} alkane in the 75 h pyrolysis suggests that the loss of a two carbon fragment is slightly preferred. Similarly, the initial abundance of the n-C_{14} alkane indicates a slight preferential cleavage at the midpoint, that is, between carbons 14 and 15 in the n-C_{28} carbon chain. The loss of a methyl radical has a low probability. The overall trend towards generation of lower molecular weight alkanes gives the n-alkane carbon number distribution patterns an increasingly skew appearance with time, but the 150 h pattern resembles those for certain crude petroleums.

These experiments conducted at 375° C have thus revealed considerable conversion of n-octacosane into other hydrocarbons. Lower temperatures would be more acceptable as a parallel for geological experience and we have examined the effect of heating n-octacosane at 200° C for 1,000 h in a vacuum. Less than 0·1 per cent alteration occurred, because the activation energy for the fission

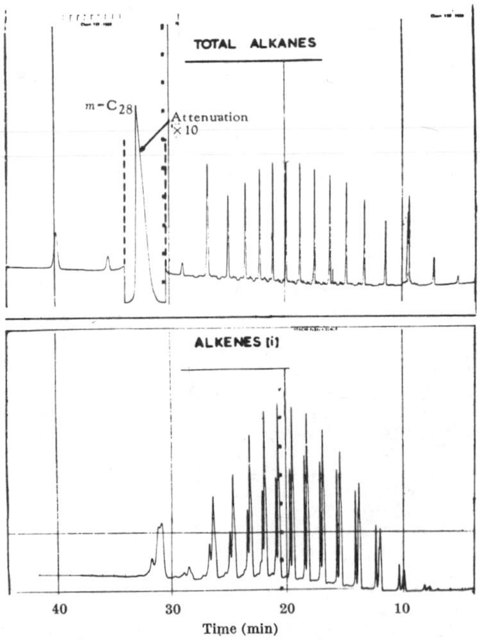

Fig. 1. Gas-liquid chromatograms of the total alkane fraction and the alkene (i) fraction from the 75 h pyrolysis of n-octacosane at 375° C. Conditions (Perkin-Elmer F 11 (Mk II), modified to reduce dead volume and the temperature gradient between the detector and the oven): 175 foot × 0·01 inch stainless steel capillary column coated with 'Apiezon L' grease, temperature programmed from 100° C to 275° C at 8°/min, H_2 flow rate 3 ml./min. The peak by n-octacosane was attenuated ten-fold; the two peaks of longer retention time are given by n-C_{29} and n-C_{30} impurities present in the original n-C_{28} to the extent of 0·1 per cent.

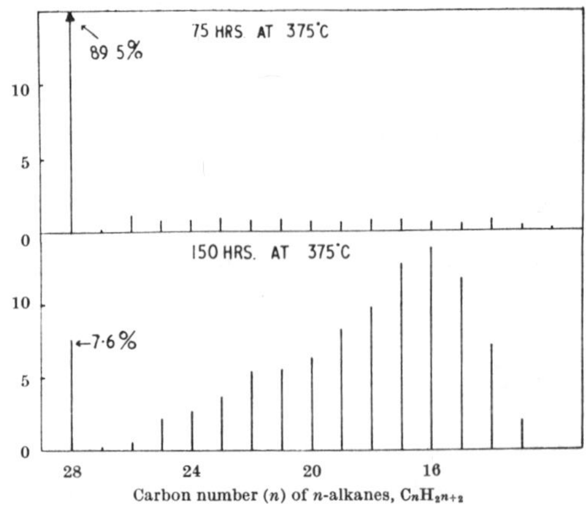

Fig. 2. The percentage abundance of the individual n-alkanes in the pyrolysis products from the n-C_{28} alkane, calculated by measuring GLC peak areas.

of carbon–carbon bonds is high and the reaction consequently slow. Thermally induced cracking of carbon compounds is accelerated by numerous catalytic materials[19,22,23]. Indeed, the carbonized film forming on the wall of the glass tube during the experiment may itself act as a catalyst[22].

We have heated n-octacosane in the presence of bentonite, which is a commercially important clay (Table 2). With bentonite alone, at 200° C for 170 h, approximately 1 per cent of the n-octacosane had been converted into other alkanes and olefins and aromatic materials, as indicated by their behaviour on TLC and GLC. Thus the bentonite had brought about a more than ten-fold increase in the extent of alteration compared with the previous experiment, which had been conducted over a much longer period. Two experiments, one with added water and one without, have been conducted at 375° C. The effect of the bentonite at this temperature was to bring about the conversion of more than 90 per cent of the n-octacosane to insoluble black carbonaceous material. Only a small proportion of the n-octacosane remained in the solvent–soluble products (Table 2), which contained much larger proportions of alkenes and aromatics than in the previous experiments (Table 1).

Perhaps the most interesting result is the high concentration of branched or cyclic alkanes apparent from the gas chromatographic record obtained for the alkane fraction from the treatment involving bentonite alone (Fig. 3A). The repeating pattern of peaks is very reminiscent of that found for the alkane fractions of crude petroleums. Gas chromatograms for the alkane fractions derived from two

Table 2. PERCENTAGE YIELDS OF THE PRODUCTS AFTER PYROLYSIS OF (a) n-C_{28} ALKANE IN THE PRESENCE OF BENTONITE AND (b) THE GREEN RIVER SHALE

Starting material	Temp. (° C)	Time (h)	Yield* of products† (%)	Alkanes / Products (%)	Alkenes‡ / Products (%)	Aromatics‡ / Products (%)
(a) n-C_{28} alkane/bentonite§	200	170	98	98	1	1
n-C_{28} alkane/bentonite§	375	60	8	55	10	35
n-C_{28} alkane/bentonite§/ H_2O (1 : 10 : 1)	375	60	2	35	20	45
(b) Green River shale (3 g)	375	150	0·07	5	55	40
Green River shale (10 g)	200	250	0·7	70	24	5
Green River shale (10 g)	200	1,000	1·0	53	17	30
Green River shale (4 g) (untreated)	Ambient	—	0·6	56	26	17

* Yields calculated by weighing (accuracy ± 0·1 mg).
† Products soluble in benzene/methanol (3 : 1).
‡ Class of compound identified by behaviour on $AgNO_3/SiO_2$ TLC only. Yields are approximate.
§ Bentonite (B.D.H. Technical Grade), exhaustively pre-extracted with benzene/methanol, heated to 500° C (4 h) and stored in a desiccator over silica gel.

crude oils of very different ages are shown in Fig. 3*C* and 3*D*. The Nonesuch seep oil exudes in very small quantities from fissures in the Parting shale stratum of the Nonesuch formation which is of Late Pre-Cambrian age (*c.* 1×10^9 years). This shale has been shown to contain various biogenic residues including compounds of isoprenoid type and porphyrins[24,25]. The Boscan crude oil is produced commercially from Zulia, Venezuela (Cretaceous, *c.* 1×10^8 yr). The much older Nonesuch oil gives a chromatogram which in certain aspects closely resembles that of the *n*-octacosane pyrolysis products. The Boscan oil, on the other hand, shows many divergences from a regularly repeating pattern, which in our view implies a less extensive alteration. The slight predominance of even carbon number *n*-alkanes in this and other oils may be the result of hydrogenation in the sediment of corresponding long chain acids and alcohols[15,26,27].

Bimodal carbon number distributions could conceivably arise by thermal alteration of biological source material of two rather different but narrow carbon number groupings, such as plant waxes (*c.* C_{30}) and glyceride lipids (*c.* C_{16}).

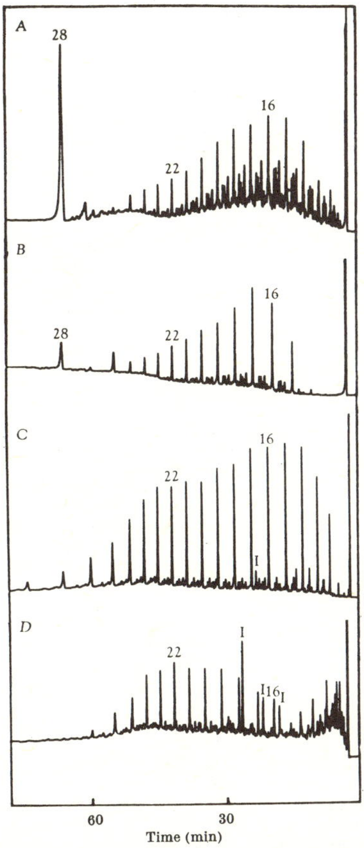

Fig. 3. Gas chromatograms of the alkane fractions from heat treatment of *n*-octacosane and from certain crude oils. *A* and *B*: fractions formed by the heat treatment (375° C for 60 h) of *n*-octacosane in the presence of: *A*, bentonite, and *B*, bentonite and water. Alkane fractions from *C*, Nonesuch seep oil, and *D*, Boscan crude oil. Conditions (see Fig. 1): 200 foot × 0·01 inch stainless steel capillary column coated with 7-*PPE* and 'Gas Quat *L*' (from Armour Industrial Chemical Co., Illinois), temperature programmed from 100° C to 250° C at 3°/min, helium flow rate 3 ml./min. *I* indicates prominent peaks corresponding to isoprenoid hydrocarbons.

Laboratory Heat Treatment of a Sediment

A catalytic clay such as bentonite can only be a poor approximation to the complex mineral matrix comprising most sediments. It is commonly assumed that such matrices possess marked catalytic properties and to some extent these have been demonstrated by laboratory experiments involving heat treatment of sediments[13–18]. In a previous paper[16], we showed that a homologous series of *n*-alkanes had been formed during the pyrolysis of an Eocene lacustrine sediment (Green River shale). Where the complex mixture of hydrocarbons liberated can distil out of the shale, the yield of distillate (shale oil) is high (500° C; 10 per cent by weight for our sample of the Green River shale and up to 40 per cent by weight for the Scottish Torbanite (Carboniferous)); the geological parallel would be an igneous intrusion into or near to a sedimentary formation, the liberated hydrocarbons possibly appearing as veins of hydrocarbon minerals[28,29]. Hunt[30] has shown that igneous intrusions into sedimentary formations lead to thermal decomposition of the organic matter, yielding low molecular weight hydrocarbons.

When the laboratory pyrolysis is conducted in a closed system, the hydrocarbons initially formed suffer further breakdown and eventually carbonization. After 150 h at 375° C the Green River shale (powdered to pass 100 mesh) yielded only a small quantity of solvent–soluble material (Table 2), which contained major proportions of fractions resembling olefinic and aromatic hydrocarbons in their chromatographic behaviour. These conditions, presumably involving both thermal and catalytic processes, had extensively altered the organic content of the shale and had resulted in almost total destruction of the alkane fraction, including the steranes and triterpanes present in the untreated shale[31–33]. By contrast, relatively mild thermal treatments at 200° C for 250 and 1,000 h brought about an increase in the solvent–soluble extract compared with unheated shale. The increase in lower molecular weight compounds arises from thermal and catalytic cracking of the kerogen—the insoluble polymeric organic debris—in the shale[17,18]. The kerogen is presumably formed during early diagenesis and represents an accumulation of many types of compounds interlinked and cross-linked by a variety of covalent bonds. We infer from the thermally induced production of *n*-alkanes that the Green River kerogen contains polymeric networks bearing long chain substituents. Isoprenoid hydrocarbons such as pristane and phytane are major constituents of some crude oils, and sediment extracts of Pre-Cambrian age. These compounds can be generated from kerogen by thermal cracking, their production thereby counteracting their more rapid destruction.

Most carbon skeletons formed biologically range in carbon number from C_{10} to C_{40}, and Mair[34] has pointed out that probably well over 95 per cent of the hydrocarbons in petroleums have carbon numbers below 40. Welte[17,18] and Mitterer and Hoering[15], using kerogen alone and in the original sediment, have arrived at similar conclusions while Brooks and Smith[19] have related the *n*-alkane patterns of different coals to their rank and the presumed extent of thermal alteration. In an earlier survey, Mueller[28] related the composition of hydrocarbon minerals to the presence of igneous veins. Recent studies[35,36] involving the oxidative and hydrolytic degradation of Green River kerogen demonstrate the abundance of long alkyl side chains.

Considering the extremely complex nature of the organic debris in a sediment, it is obvious that no one class of organic compound such as the alkanes can alone be responsible for the production of petroleum. There is experimental evidence relevant to the thermal and catalytic alteration of other types of compound, such as the fatty acids[37–39], alcohols[15,40–42] and isoprenoid hydrocarbons[43–45]. In effect, thermal alteration in the laboratory of a complete sediment such as the Green River shale can provide information on the alteration of a very wide range of compounds, but the situation is experimentally complex. The Green River shale formation itself has had a mild thermal history[46,47] and the extractable alkanes have carbon numbers and carbon skeletons very suggestive of a plant origin[31-33,48,49] typified by the cuticular waxes of land plants[50].

Genesis and Maturation of Petroleums

Our experiments with *n*-octacosane and the Green River shale show that even brief exposures, in geological terms, to temperatures of or above 200° C must result in extensive alteration of organic debris in sediments. This explains the absence of odd/even carbon number preference in the *n*-alkane fractions isolated from crude oils derived from deep wells[18], where temperatures can reach 200° C. Certainly a lack of odd/even predominance cannot be taken as definite evidence of either abiotic synthesis[51-55] or an unusual biological origin (in the sense of a smooth carbon number distribution for the constituent *n*-alkanes, *n*-fatty acids, etc.[56,57]) because thermal alteration will quickly eliminate any carbon number alternation.

Our experiments indicate that an *n*-alkane (C_nH_{2n+2}) degrades thermally to the complete homologous series of *n*-alkanes, C_mH_{2m+2} (where $m<n$), and the corresponding *n*-alkenes, $C_{n-m}H_{2(n-m)}$ (Fig. 4). Other products are formed, but we believe that the significant point which bears on the observed distribution of *n*-alkanes in crude petroleums and sediments is the formation of the homologous series of *n*-alkanes. Each *n*-alkane in the mixture undergoing thermal alteration (maturation) will itself be generating a new series of *n*-alkanes of lower carbon number. This situation merits mathematical treatment[58]. The pattern of abundance of the carbon number is affected by the laboratory work-up procedure, because the lower molecular weight alkanes are preferentially lost by volatilization, resulting in a maximum. In geological conditions similar losses would become apparent as natural gas accumulations. Cyclic or branched alkanes, such as the isoprenoid hydrocarbon squalane, undergo more rapid thermal cracking, but only very small amounts of *n*-alkane are formed. We therefore believe that straight chain lipids must be the main source of the *n*-alkanes found in petroleum and other ancient sediments.

The thermal cracking processes require the formation of equal amounts of alkanes and alkenes. Alkenes are normally trace constituents of petroleums and other sediments, so that if these mechanisms apply in geological conditions, then the first formed alkenes must be removed, possibly by conversion to other products such as cyclic and aromatic hydrocarbons[23,59]. Other processes which may operate in a sediment include hydrogenation[15], oxygenation and polymerization[13,60]. Olefins and alkanes undergo thermal cracking at about the same rate, but in the presence of an acidic catalyst such as bentonite the olefins form carbonium ions more readily and crack at a much higher rate[22]. Olefins formed by thermal cracking might well undergo isomerization, rearrangement, cyclization and hydrogen transfer in the presence of such catalysts[13,15,60]. While thermal cracking produces high concentrations of *n*-alkanes, catalytic cracking therefore produces mostly branched and cyclic alkanes[22]. A combination of both processes could explain the relative proportions of normal, branched and cyclic alkanes found in sediments, particularly when the supposed biological source material is deficient in cyclic and branched carbon compounds. Water is present in sedimentary deposits during diagenesis, but it is evident from the laboratory experiments that water does not inhibit these alteration processes, though it does affect the composition of the products[37-39].

Thermal alteration of most sediments and crude oils does not appear to have proceeded to anywhere near completion, which is in accord with the observation that sedimentary formations generally experience temperatures in the range 20°–200° C. For this range we have no laboratory data. But it seems reasonable to assume that processes bringing about alteration of organic materials at 200° C and above in the laboratory may also be effective in sedimentary formation at much lower temperatures over geological time measured in millions of years. The organic debris is trapped in close contact with the silicate matrix, and the varying proportions and type of the two materials will affect the extent of the thermal and catalytic cracking. Furthermore, extensive physical and chemical data from the Los Angeles and Ventura Basins have led Philippi[21] to conclude that the bulk of oil generation in these basins takes place at temperatures above 115° C. Estimates of the extent of alteration which could conceivably take place during a given period of geological time and within a given range of temperature have been made[61] and are said to preclude extensive thermal (radical) cracking. But kinetic data are not available for the catalytic cracking which may occur in sedimentary conditions, and we agree with previous workers (for example, refs. 61 and 62) who have held that cracking must be at least a contributory factor in the genesis of petroleum, and one which becomes of increasing importance with increasing age, depth of burial and temperature.

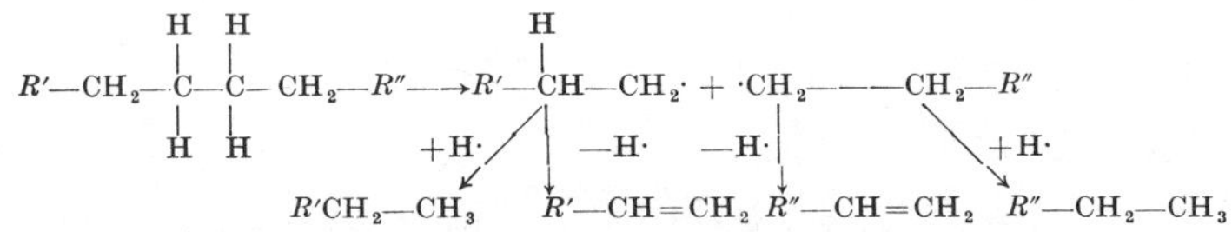

Disproportionation by homolytic fission.

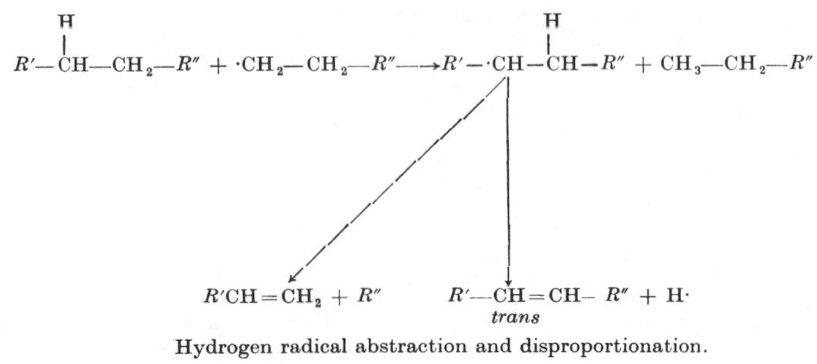

Hydrogen radical abstraction and disproportionation.

Fig. 4. Mechanisms involved in the thermal cracking of alkanes.

This work was supported in part by the US National Aeronautics and Space Administration and the Natural Environment and Science Research Councils (SRC research studentship for W. H.).

ANALYTICAL METHODS

Samples of *n*-octacosane (0·078 g each, 99 per cent pure by GLC) were sealed in clean 'Pyrex' glass tubes and heated at 375° C for periods of 75, 100 and 150 h. Extraction with benzene followed by evaporation furnished the "products" (Tables 1 and 2). These were examined by analytical thin-layer chromatography (TLC) using a silver nitrate impregnated silicic acid layer eluted with *n*-hexane. This indicated, by R_F values and comparison with standard compounds, that the products contained alkanes, two different classes of alkene ((i) and (ii)) and aromatic hydrocarbons. Infrared spectrometry confirmed the presence of the two types of olefinic double bond: (i) chiefly *trans*-disubstituted (ν_{max}, 965 cm^{-1}), and (ii) chiefly terminal alkenes (ν_{max}, 990 and 910 cm^{-1}).

Separation into the alkane, alkene and aromatic categories was carried out by preparative TLC and the fractions analysed by capillary GLC on 'Apiezon *L*' grease (pre-treated to remove the polar material by passing a hexane solution of the grease through neutral alumina). Combination gas chromatography—mass spectrometry (GC–MS) confirmed the carbon number data for the *n*-alkanes and the alkenes.

Received June 18; revised July 22, 1968.

[1] Rudakov, G., *Chem. Geol.*, **2**, 179 (1967).
[2] Meinschein, W. G., *Bull. Amer. Assoc. Petrol. Geol.*, **43**, 925 (1959).
[3] Robinson, R., *Nature*, **199**, 113 (1963).
[4] Cooper, J. E., *Nature*, **193**, 744 (1962).
[5] Bray, E. E., and Evans, E. D., *Geochim. Cosmochim. Acta*, **22**, 2 (1961).
[6] Bray, E. E., and Evans, E. D., *Erdol u. Köhle*, **18**, 421 (1965); *Chem. Abst.*, **63**, 8043 h.
[7] Stone, R. W., and Zobell, C. E., *Ind. Eng. Chem.*, **44**, 2564 (1952).
[8] Kusnetsov, S. I., Ivanov, M. U., and Lyalikova, N. N., *Introduction to Geological Microbiology* (McGraw-Hill, New York, 1963).
[9] Davis, J. B., *Geol. Soc. Amer. Ann. Meet.* (New Orleans, 1967).
[10] Lind, S. C., and Bardwell, D. C., *J. Amer. Chem. Soc.*, **48**, 2335 (1926).
[11] Colombo, U., Denti, E., and Sironi, G., *J. Inst. Petrol.*, **50**, 228 (1964).
[12] Sheppard, C. W., and Burton, V. L., *J. Amer. Chem. Soc.*, **68**, 1636 (1946).
[13] Brooks, B. T., *Ind. Eng. Chem.*, **44**, 2570 (1952).
[14] Hoering, T. G., and Mitterer, R. M., *Geol. Soc. Amer. Ann. Meet.* (New Orleans, 1967).
[15] Mitterer, R. M., and Hoering, T. G., *Carnegie Inst. Wash. Year Book*, **65**, 510 (1966).
[16] Douglas, A. G., Eglinton, G., and Henderson, W., in *Organic Geochemistry* (edit. by Hobson, G. D., and Speers, G. C.) (Pergamon Press, London, in the press, 1968).
[17] Welte, D. H., *Geologischen Rundschau*, **55**, 131 (1965).
[18] Welte, D. H., *Bull. Amer. Assoc. Petrol. Geol.*, **49**, 2246 (1965).
[19] Brooks, J. D., and Smith, J. W., *Geochim. Cosmochim. Acta*, **31**, 2389 (1967).
[20] Robinson, W. E., Cummins, J. J., and Dineen, G. U., *Geochim. Cosmochim. Acta*, **29**, 249 (1965).
[21] Philippi, G. T., *Geochim. Cosmochim. Acta*, **29**, 1021 (1965).
[22] Greensfelder, B. S., Voge, H. H., and Good, G. M., *Ind. Eng. Chem.*, **41**, 2573 (1949).
[23] Thomas, C. L., *Ind. Eng. Chem.*, **41**, 2564 (1949).
[24] Eglinton, G., Scott, P. M., Belsky, T., Burlingame, A. L., and Calvin, M., *Science*, **145**, 263 (1964).
[25] Barghoorn, E. S., Meinschein, W. G., and Schopf, J. W., *Science*, **148**, 461 (1965).
[26] Kvenvolden, K. A., *Nature*, **209**, 573 (1966).
[27] Kvenvolden, K. A., *J. Amer. Oil Chem. Soc.*, **44**, 628 (1967).
[28] Mueller, G., *Cong. Geol. Intern.*, *CR* (19th Session, Algiers, 1952), No. 12, 279 (1954).
[29] Geissman, T. A., Sim, K. Y., and Murdoch, J., *Experientia*, **23**, 793 (1967).
[30] Hunt, J. M., in *Proc. Intern. Sci. Oil Conf.* (Budapest, October 8–13, 1962).
[31] Hills, I. R., Whitehead, E. V., Anders, D. E., Cummins, J. J., and Robinson, W. E., *Chem. Commun.*, **20**, 752 (1966).
[32] Burlingame, A. L., Haug, P., Belsky, T., and Calvin, M., *Proc. US Nat. Acad. Sci.*, **54**, 1406 (1965).
[33] Henderson, W., Wollrab, V., and Eglinton, G., *Chem. Commun.*, **13**, 710 (1968).
[34] Mair, B. J., *Geochim. Cosmochim. Acta*, **28**, 1303 (1964).
[35] Robinson, W. E., and Lawlor, D. L., *Fuel*, **40**, 375 (1961).
[36] Burlingame, A. L., and Simoneit, B. R., *Science*, **160**, 531 (1968).
[37] Jurg, J. W., and Eisma, E., *Science*, **144**, 1451 (1964).
[38] Jurg, J. W., thesis, Univ. Eindhoven (1967).
[39] Jurg, J. W., and Eisma, E., in *Organic Geochemistry: Methods and Results* (edit. by Eglinton, G., and Murphy, Sister M. T. J.) (Springer-Verlag, New York, in the press, 1968).
[40] Martin, R. L., Winters, J. C., and Williams, J. A., *Nature*, **199**, 110 (1963).
[41] Smith, H. M., *J. Amer. Oil Chem. Soc.*, **44**, 680 (1967).
[42] Degens, E. T., *Geochemistry of Sediments* (Prentice Hall, New Jersey, 1965).
[43] Douglas, A. G., and Mair, B. J., *Science*, **147**, 499 (1965).
[44] Erdman, J. G., *Geochim. Cosmochim. Acta*, **22**, 16 (1961).
[45] Day, W. C., and Erdman, J. G., *Science*, **141**, 808 (1963).
[46] Milton, C., and Eugstar, H. P., in *Researches in Geochemistry* (edit. by Abelson, P. H.), 118 (John Wiley, New York, 1959).
[47] Jones, J. D., and Vallentyne, J. R., *Geochim. Cosmochim. Acta*, **21**, 1 (1960).
[48] Murphy, Sister M. T. J., McCormick, A., and Eglinton, G., *Science*, **157**, 1040 (1967).
[49] Cummins, J. J., and Robinson, W. E., *Chem. Eng. Data*, **9**, 304 (1964).
[50] Eglinton, G., and Hamilton, R. J., *Science*, **156**, 1322 (1967).
[51] Meinschein, W. G., *Space Sci. Rev.*, **2**, 665 (1963).
[52] Friedel, R. A., and Sharkey, A. G., *Science*, **139**, 1203 (1963).
[53] Lovelock, J. E., *Nature*, **207**, 568 (1965).
[54] McCarthy, E. D., and Calvin, M., *Nature*, **216**, 642 (1967).
[55] Studier, M. H., Hayatsu, R., and Anders, E., *Geochim. Cosmochim. Acta*, **32**, 151 (1968).
[56] Oró, J., Tornabene, T. G., Nooner, D. W., and Gelpi, E., *J. Bacteriol.*, **93**, 1811 (1967).
[57] Stransky, K., Streibl, M., and Herout, V., *Coll. Czech. Chem. Commun.*, **32**, 3213 (1967).
[58] Kvenvolden, K. A., and Weiser, D., *Geochim. Cosmochim. Acta*, **31**, 1281 (1967).
[59] Badger, G. M., Donnelly, J. K., and Spotswood, T. M., *Austral. J. Chem.*, **19**, 1023 (1966).
[60] Frost, A. V., *Uspekhi Khim.*, **14**, 501 (1945).
[61] McNab, J. G., Smith, P. V., and Betts, R. L., *Ind. Eng. Chem.*, **44**, 2556 (1952).
[62] Silverman, S. R., *J. Amer. Oil Chem. Soc.*, **44**, 691 (1967).

31

Reprinted from *Nature Phys. Sci.* **232**:140–144 (1971)

Catalytic Conversion of Fatty Acids to Petroleum-like Paraffins and their Maturation

AKIRA SHIMOYAMA & WILLIAM D. JOHNS

Department of Geology, University of Missouri, Columbia, Missouri 65201

Experiments on the decarboxylation of n-docasanoic acid and the subsequent cracking of the n-paraffins are described and related to petroleum genesis.

IT has been proposed that the transformation of naturally occurring fatty acids to paraffins during burial of sediments plays an important role in petroleum genesis. This idea is based extensively on studies of the amounts and molecular distributions of these compounds in sediments[1,2]. Hydrocarbon contents are in general much lower in modern sediments than in ancient sediments[3]; the ratios of odd to even carbon-number n-paraffin molecules are higher in modern than in ancient sediments and near unity in petroleum[4]. On the other hand, modern sediments contain greater amounts of fatty acids than ancient sediments. The ratios of even to odd carbon-number n-fatty acids decrease from high values to near unity from modern to ancient sediments and petroleum reservoir waters[5].

The transformation processes proposed involve the formation by decarboxylation of n-paraffins with one less carbon atom in their skeletons than their precursor acids. In laboratory thermal experiments the decarboxylation of fatty acids has been shown to produce n-paraffins with one less carbon atom in the presence of a clay mineral catalyst[6,7]. The formation of smaller amounts of other n-paraffins with both shorter and longer chains than the fatty acid precursor has also been observed[6,7].

The well balanced distribution of odd and even carbon-number molecules characteristic of n-paraffins in ancient sediments and petroleum cannot be obtained by assuming that decarboxylation is the only important step involved in starting with fatty acid distributions found in recent sediments[1]. Mathematical models have, for example, assumed generation of shorter from longer chain fatty acids so as to match those computed with paraffin and fatty acid carbon preference index (c.p.i.) values found in ancient sediments[1,8]. Such fatty acid conversions have not been observed as important reactions in experimental thermal studies.

Clays also catalyse the cracking of n-paraffins, producing a spread of n-paraffins under thermal treatments[9]. This article deals experimentally with two important steps in the generation of petroleum-like n-paraffins—the decarboxylation of a fatty acid and the subsequent cracking of n-paraffins produced from the acid as a result of the thermal-catalytic effects of montmorillonite. Combinations of these steps provide a possible conversion path from fatty acids to petroleum-like hydrocarbons with well balanced molecular distributions of the odd and even carbon numbers, that is c.p.i. values approaching unity.

Decarboxylation of n-Docasanoic Acid

We have studied the n-paraffins between C_{16} and C_{28} generated from n-docasanoic acid ($C_{21}H_{43}COOH$) as well as the acid recovered among the reaction products. The range of paraffins studied was limited chiefly by our analytical capability for quantitative estimation. A sample of 50 mg of the C_{22} n-fatty acid was heated with 100 mg of Ca-montmorillonite catalyst at 200° C and 250° C for 50–500 h. The n-paraffins and residual n-fatty acid (as methyl ester) were isolated from the reaction products and analysed by gas–liquid chromatography (GLC). A typical GLC analysis of the n-paraffin fraction for a 50 h run at 200° C is shown in Fig. 1. Each component was estimated quantitatively from peak areas on the chromatograms by means of internal standards. The molecular distribution of n-paraffins estimated for these conditions is shown in Fig. 2. The C_{21} n-paraffin—with one less carbon atom than the precursor acid—is the most abundant n-alkane produced. The relative amounts and molecular distribution of these paraffins are very similar to those produced in similar conditions by previous investigators[6]. The loss of one carbon fragment has been attributed to

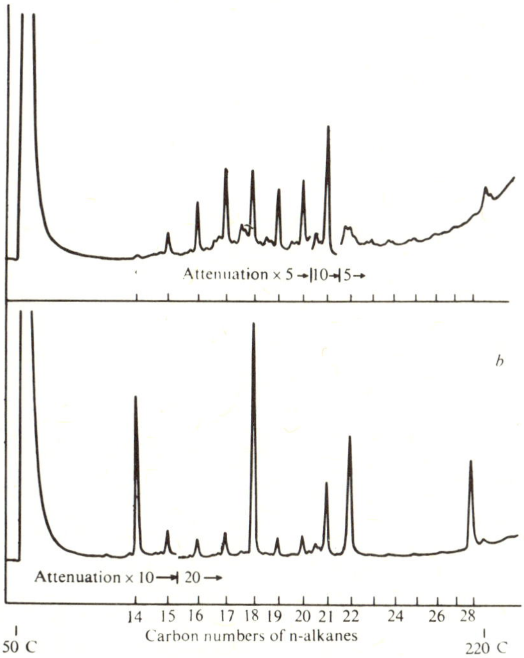

Fig. 1 Gas–liquid chromatograms of n-paraffins generated from C_{22} n-fatty acid (50 mg) in the presence of Ca-montmorillonite catalyst at 200° C for 50 h. *a*, Peaks of an n-paraffin fraction; *b*, peaks of the n-paraffin fraction with internal standards C_{14}, C_{18}, C_{22}, and C_{28}. Stainless steel column (122 cm by 0.63 cm) packed with 1% DEGS on 'Chromosorb W (AW)', temperature programmed from 50° C to 230° C at 5° min^{-1}.

decarboxylation of the acid. The small double peaks in Fig. 1 (upper curve) are believed to represent alkenes, but this has not been confirmed.

The C_{21} n-paraffin was the most abundant alkane produced in a series of heating experiments with the montmorillonite catalyst. The extent of decarboxylation varied, however, with the duration and temperature of the run. The amounts of n-paraffins generated between C_{16} and C_{28} are given in Tables 1 and 2. At 200° C (Table 1) the amount of C_{21} n-paraffin produced shows a maximum of about 60 μg after 150 to 300 h and decreases slightly after 500 h. At 250° C (Table 2) the quantity reaches about 110 μg after 50 h; a similar amount is produced after 150 h. The amount is much smaller after 300 and 500 h runs. This indicates that the decarboxylation has been promoted to the greatest extent during the first 50 h of heating.

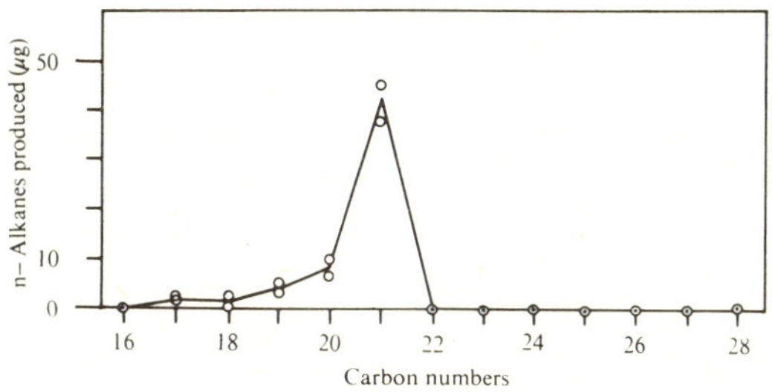

Fig. 2 Distribution of n-paraffins from n-docasanoic acid with Ca-montmorillonite at 200° C for 50 h.

This levelling off of the decarboxylation process with time can be correlated with the amount of precursor acid recovered from the reaction products. Variations in the amount of recovered fatty acid with the duration of runs at 200° C and 250° C are shown in Fig. 3. At 200° C the maximum consumption of the acid occurs during the first 150 h of heating. There is no significant change in the amount of acid consumed on further heating, even though more than half of the initial charge is still available for conversion. This can probably be explained in terms of the effective life of the catalyst. The catalyst probably became inert after about 150 h because of deposition of carbonaceous organic matter on its active sites. At 250° C acid consumption is both greater and faster than at 200° C. More than 80% of the initial charge was converted during the first 50 h. The catalyst is evidently more active at 250° C than at 200° C. The branching ratio of the C_{21} n-paraffin production to the acid consumption is 0.2 to 0.3% at both temperatures for 50 h runs. This similarity suggests that the reactions at both temperatures are essentially the same as far as paraffin formation and acid consumption are concerned. Therefore the curve for the change in amount of acid recovered at 250° C is drawn in a manner analogous to that at 200° C (Fig. 3), instead of connecting point by point. At present we

Table 1 Amounts of n-Paraffins (μg) produced after Heating n-Docasanoic Acid (50 mg) with Ca-Montmorillonite at 200° C for Various Times

Carbon No.	50 h a	50 h b	50 h Mean	150 h a	150 h b	150 h Mean	300 h a	300 h b	300 h Mean	500 h a	500 h b	500 h Mean
16	—	—	—	5.7	4.3	5.0	11.6	9.3	10.5	11.4	11.1	11.3
17	1.5	2.3	1.9	5.9	3.1	4.5	6.7	7.6	7.2	7.7	7.1	7.4
18	0.5	2.3	1.4	7.8	5.7	6.8	6.5	7.3	6.9	7.1	5.1	6.1
19	3.2	5.5	4.4	9.4	8.5	9.0	7.8	11.1	9.5	9.8	8.2	9.0
20	6.7	10.2	8.5	14.2	13.7	14.0	12.3	14.3	13.3	13.9	12.8	13.4
21	37.2	45.1	41.2	62.5	59.2	60.9	59.3	60.4	59.9	53.5	49.1	51.3
22	—	—	—	—	—	—	—	—	—	—	—	—
23	—	—	—	0.6	0.5	0.6	—	—	—	—	—	—
24	—	—	—	2.2	2.4	2.3	—	—	—	0.8	—	0.4
25	—	—	—	0.5	0.8	0.7	—	—	—	—	—	—
26	—	—	—	1.9	1.1	1.5	—	—	—	—	—	—
27	—	—	—	0.9	1.1	1.0	—	—	—	0.1	—	0.1
28	—	—	—	1.5	0.7	1.1	—	—	—	0.4	—	0.2
Total	49.1	65.4	57.4	113.1	101.1	107.4	104.2	110.0	107.3	104.7	93.4	99.2

a and b are duplicate runs. —, Not detected.

can only explain the scatter of data points as experimental error, although each duplicated run shows excellent reproducibility.

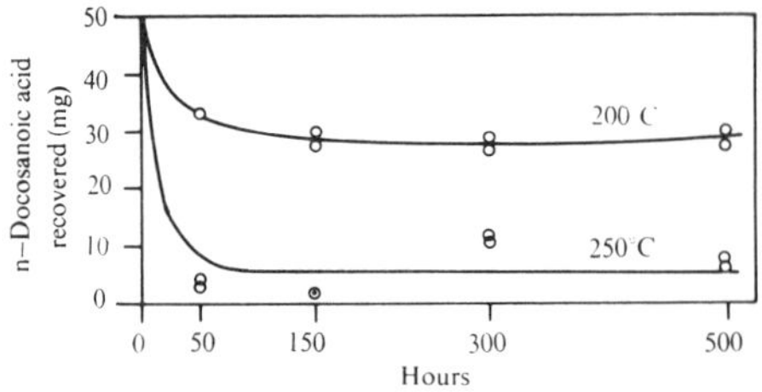

Fig. 3 Amounts of n-docasanoic acid recovered after heating of 50 mg with Ca-montmorillonite at 200° C and 250° C.

Generation of n-paraffins in the range C_{16} to C_{28} accounts for less than 1% of the acid consumed. Formation of alkenes and alkanes shorter than C_{16}, which we have not yet investigated, must account to some extent for the amount of acid consumed. Fatty acids other than residual precursor acid were not detected even in trace amounts. It seems that a large proportion of the acid consumed has been converted to kerogen-like matter. This conversion is suggested by the formation of an organic insoluble substance which is deposited on the catalyst. This kerogen-like material can probably become a further source of hydrocarbons at higher temperatures[10,11].

Cracking of n-Paraffins

The n-paraffins in the range C_{16} to C_{20} are also produced in amounts sufficient to describe processes involved in their formation. Variations in the amounts of these n-paraffins with the duration of runs at 250° C are shown in Fig. 4. The curve for C_{21} paraffin indicates that this component has been produced in greatest quantity between 50 and 150 h as mentioned previously. The curves for C_{16} and C_{17} paraffins demonstrate an increase in production between 150 and 300 h, whereas a marked decrease in the quantity of C_{21} paraffin is evident. These two contrasting changes show that the C_{16} and C_{17} components have been generated from the C_{21} paraffin by cracking of C–C bonds. The curves for C_{18}, C_{19} and C_{20} are composites of those for C_{21} and C_{16} or C_{17}, that is, they follow the same pattern of increase as C_{21} until about 150 h and then change to the pattern of C_{16} or C_{17} at times

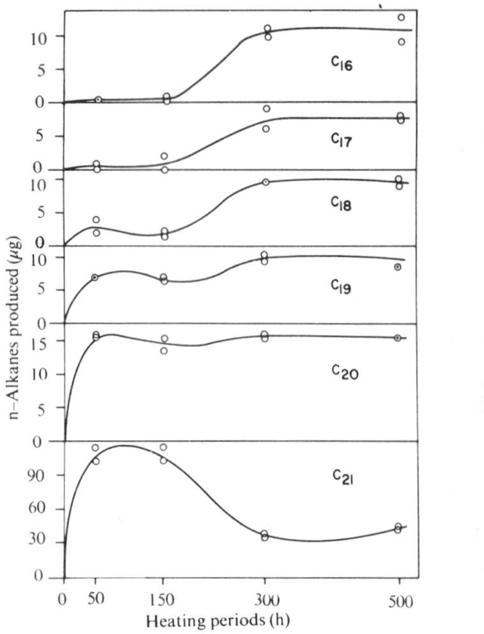

Fig. 4 Amounts of n-paraffins generated from n-docasanoic acid with Ca-montmorillonite at 250° C.

approaching 500 h. The similarity of the curves for C_{18}, C_{19} and C_{20} and that for C_{21} indicates that these n-paraffins, like the C_{21} component, have been generated from the fatty acid. The cracking of C–C bonds of the acid molecules has probably been promoted to some extent as decarboxylation of the acid has taken place. Concordance of the latter portion of the C_{18}, C_{19} and C_{20} curves with the curves of C_{16} and C_{17} suggests that these paraffins were also formed from the C_{21} paraffin between 150 and 300 h, as were the C_{16} and C_{17} components.

The paraffins C_{16} to C_{20} generated at 200° C point to the cracking of C–C bonds of the fatty acid molecules to generate components with less carbons than the acid. The cracking of the C_{21} paraffin at 200° C to generate the other n-paraffins

Table 2 Amounts of n-Paraffins (μg) produced after Heating n-Docasanoic Acid (50 mg) with Ca-Montmorillonite at 250° C for Various Times

Carbon	50 h			150 h			300 h			500 h		
No.	a	b	Mean	a	b	Mean	a	b	Mean	a	b	Mean
16	0.5	—	0.3	1.1	—	0.6	11.6	9.9	10.8	13.0	9.2	11.1
17	1.1	—	0.6	2.3	0.1	1.2	6.6	9.4	8.0	8.2	7.5	7.9
18	3.8	1.7	2.8	2.2	1.5	1.9	9.4	9.6	9.5	9.8	9.0	9.4
19	6.9	7.1	7.0	6.9	6.4	6.7	10.5	9.4	10.0	8.5	8.4	8.5
20	15.9	15.5	15.7	15.3	13.6	14.5	15.3	16.0	15.7	15.3	15.4	15.4
21	102.1	113.4	107.8	106.3	115.9	111.1	35.7	38.3	37.0	45.2	42.1	43.7
22	1.6	—	0.8	—	—	—	1.3	—	0.7	1.1	—	0.6
23	—	—	—	—	0.3	0.2	0.8	0.4	0.6	0.9	—	0.5
24	0.9	—	0.5	0.1	—	0.1	2.8	1.6	2.2	2.6	0.9	1.8
25	0.1	—	0.1	—	—	—	0.4	—	0.2	0.2	—	0.1
26	0.9	—	0.5	0.3	—	0.2	2.2	0.4	1.3	1.6	0.1	0.9
27	0.5	—	0.3	—	—	—	1.1	0.4	0.8	0.3	—	0.2
28	0.4	—	0.2	1.0	0.1	0.6	2.3	1.1	1.7	1.1	0.9	1.0
Total	134.7	137.7	136.6	135.5	137.9	137.1	100.0	96.5	98.5	107.8	93.5	101.1

a and b are duplicate runs. —, Not detected.

is, however, not clearly indicated because the temperature is too low. To confirm that the reaction is indeed too slow to observe at 200° C, we estimated the reaction kinetics constants using data obtained at 250° C. We assumed that (1) the degradation follows a first order reaction; (2) the reaction started at 150 h with 111 μg of C_{21} n-paraffin and continued to 300 h at which time the amount of the n-paraffin was 37 μg; (3) additional C_{21} n-paraffin was not produced from the fatty acid during the degradation; (4) the Arrhenius factor, A, is 5×10^{13} s^{-1} as suggested by previous investigators[11]; (5) the effect of the montmorillonite catalyst was constant during the degradation. The equation for the first order reaction is

$$\frac{dN}{dt} = -kN$$

or

$$\ln \frac{N}{N_0} = -kt$$

and

$$k = Ae^{-E/RT}$$

where N is the amount of C_{21} n-paraffin at time t, N_0 is the initial amount of the n-paraffin, k is the reaction rate, A is the Arrhenius factor, E the activation energy, R the gas constant, and T the absolute temperature. Using the experimentally determined amounts formed after 150 and 300 h at 250° C, the unknowns at this temperature were calculated as

$$E = 46.5 \text{ kcalories mol}^{-1}$$
$$k = 2.0 \times 10^{-6}\ s^{-1}$$

Previous studies of the thermal (non-catalytic) cracking of hydrocarbons in petroleum gave a calculated activation energy of 49 kcalories mol^{-1} and a reaction rate of 2.7×10^{-9} s^{-1} at 267° C (ref. 12). Our calculations show a somewhat lower activation energy and an appreciably greater reaction rate because of the catalytic effect of the clay. Thus the more favourable reaction kinetics in our experiments suggest the importance of clay mineral catalytic cracking.

Assuming that the activation energy and Arrhenius factor are constant at both 200° C and 250° C, the reaction rate and relative amount of C_{21} n-paraffin formed at 200° C were calculated as

$$k = 1.8 \times 10^{-8}\ s^{-1}$$

$$\frac{N}{N_0} = 0.99 \text{ for 150 h degradation}$$

$$\frac{N}{N_0} = 0.98 \text{ for 350 h degradation}$$

As the formation of the C_{21} n-paraffin and the consumption of the acid reached a maximum at about 150 h at 200° C, the cracking of the n-paraffin was apparently initiated at 150 h. The calculated 1% loss of C_{21} paraffin after 150 h is in agreement with the experimental data. In a similar way other investigators[9] working with C_{28} n-paraffin have shown a 1% loss by degradation in the presence of a bentonite (montmorillonite) catalyst, in that case after heating 170 h at 200° C. There is a discrepancy, however, between observed and calculated degradation after 350 h of heating at 200° C; the calculated loss of C_{21} is 2% and the observed loss 17%. Thus the observed data indicate even more rapid reaction than calculated. This discrepancy highlights the shortcomings of our attempt at kinetic analysis based on such limited data. We realize that our values are broad approximations, but feel that they give us at least a qualitative picture of the catalytic effect of the montmorillonite in the cracking reaction.

Genesis and Maturation

In living matter, and in modern sediments containing residues of living matter, even carbon-number molecules of fatty acids predominate over odd carbon-number molecules. In petroleum and ancient sediments n-paraffin molecules do not show marked odd or even carbon-number predominance. In our heating experiments, decarboxylation of the C_{22} n-fatty acid chiefly produced the C_{21} member of the group of C_{16} to C_{28} n-paraffins. The acid also generated smaller amounts of n-paraffins with chains shorter than C_{21} and trace amounts with chains longer than the original fatty acid. Some of these paraffins underwent subsequent degradation on further heating. This type of maturation was demonstrated by the cracking of C_{21} paraffin at 250° C.

Table 3 Values of the c.p.i. for n-Paraffins Generated in the Range C_{16} to C_{21} from C_{22} n-Fatty Acid in the Presence of Montmorillonite Catalyst at 200° C and 250° C for Various Times

200° C							
50 h		150 h		300 h		500 h	
a	b	a	b	a	b	a	b
5.8	4.2	2.8	3.0	2.4	2.6	2.2	2.2
250° C							
50 h		150 h		300 h		500 h	
a	b	a	b	a	b	a	b
5.5	7.0	6.2	8.1	1.5	1.6	1.6	1.7

a and b refer to the same duplicate runs as in Tables 1 and 2.

A c.p.i. index has been proposed[4] as a means of evaluating the progress of diagenesis and maturation of n-paraffins in sediments. The c.p.i. has been defined as the mean of two ratios which are determined by dividing the sum of the concentrations of odd carbon-number n-paraffins by the sum of the even carbon-number components over a given carbon-number range. We have estimated approximate c.p.i. values for the n-paraffins produced in our experiments (Table 3). Our c.p.i. values simply represent the ratio of the sum of odd carbon-number components to the sum of the even carbon-number molecules between C_{16} and C_{21}.

$$\text{c.p.i.} = \frac{C_{17} + C_{19} + C_{21}}{C_{16} + C_{18} + C_{20}}$$

The higher c.p.i. values for the shorter heating periods at 200° C and 250° C are due, of course, to the greater amounts of C_{21} produced. These values decrease considerably over longer heating periods because of the cracking of the C_{21} component; the decrease is more rapid at 250° C. The resultant values of 1.5 to 1.7 fall within the range reported for ancient shales[4]. The significant implication to the genesis and maturation of n-paraffins is that the C_{22} n-fatty acid has been converted to n-paraffins with c.p.i. values similar to those of paraffins found in ancient sediments. We also see that initially the paraffins formed have the high c.p.i. values characteristic of modern sediments.

There are differing opinions about the likelihood of extensive thermal cracking in suitable geological conditions and temperatures in the sedimentary column[8,12]. The ages of source rocks of petroleum are known to extend from the Pre-Cambrian to the Pleistocene[13], and the maximum temperatures to which most of the source rocks have been subjected are probably 100° C to 150° C (ref. 14), although temperatures as high as 200° C have been reported[15]. Our experiments show that the cracking of n-paraffins is a much slower reaction than the initial conversion of the fatty acid. Therefore the time necessary for the formation of petroleum-like n-paraffins from n-fatty acids depends principally on the cracking reaction. The change in c.p.i. value between 150 and 300 h of heating at 250° C is due to the loss of two-thirds of the C_{21} n-paraffin. Extrapolation of this change to geological time and temperature parameters can be carried out using the reaction kinetics constants we have estimated.

Assuming conditions in nature analogous to our experiments, we find that the maturation of petroleum-like paraffins requires times of the order of 6,500 yr at 150° C, 0.21 m.y. at 125° C, 11 m.y. at 100° C and 960 m.y. at 75° C. Obviously these time and temperature relations cannot be applied rigorously to the natural situation. Nevertheless, they suggest that catalytic cracking can be a significant factor in petroleum genesis within reasonable geological times in the temperature interval 100° to 150° C. We are able to relate genesis and maturation of n-paraffins, as expressed by c.p.i. values.

Analytical Methods

A sample of 50 mg of C_{22} n-fatty acid was sealed in 'Pyrex' glass tubes with 100 mg of Ca-montmorillonite. The tubes were heated at 200° C and 250° C for 50, 150, 300 and 500 h. The reaction products were collected by Soxhlet extraction with benzene–methanol solution with dissolved potassium hydroxide. Two aliquots were taken from each extracted solution. Internal standards of C_{14}, C_{18}, C_{22} and C_{28} n-paraffins and C_{18} n-fatty acid were added to one of a pair of aliquots at this stage. The aliquots were subjected to acidification of the potassium salts of the fatty acid, methyl esterification of the acid and urea adduction for isolation of n-paraffins and esters of residual fatty acid or any acids produced. Silica gel column chromatography was used to separate n-paraffin and ester fractions which were analysed by a gas–liquid chromatograph with a stainless steel column (122 cm by 0.63 cm) packed with 1% DEGS on 'Chromosorb W (AW)'. Identification of each compound was based on the isolation procedure and retention times of the internal standards on the gas–liquid chromatograms. Estimates of amounts were made—by means of the internal standards—from the peak areas on the chromatograms. Because of the small amounts and small volumes (30 to 40 μl.) of the fractions isolated, blank runs with the C_{22} fatty acid were carried out by the same procedures. Only trace amounts of contaminants were introduced during the isolation procedures and these were subtracted where appropriate. The C_{18} and C_{22} fatty acids and the C_{14}, C_{18}, C_{22} and C_{28} n-paraffins utilized were more than 99% pure by thin-layer chromatography.

This study was supported by a grant from the US National Science Foundation.

Received July 13, 1971.

1 Cooper, J. E., and Bray, E. E., *Geochim. Cosmochim. Acta*, **27**, 1113 (1963).
2 Kvenvolden, K. A., *Nature*, **209**, 573 (1966).
3 Hunt, J. M., *Geochim. Cosmochim. Acta*, **22**, 37 (1961).
4 Bray, E. E., and Evans, E. D., *Geochim. Cosmochim. Acta*, **22**, 2 (1961).
5 Cooper, J. E., *Nature*, **193**, 744 (1962).
6 Jurg, J. W., and Eisma, E., *Science*, **144**, 1451 (1964).
7 Eisma, E., and Jurg, J. W., in *Organic Geochemistry: Methods and Results* (edit. by Eglinton, G., and Murphy, M. T. J.) (Springer, New York, 1969).
8 Kvenvolden, K. A., and Weiser, D., *Geochim. Cosmochim. Acta*, **31**, 1281 (1967).
9 Henderson, W., Eglinton, G., Simmonds, P., and Lovelock, J. E., *Nature*, **219**, 1012 (1968).
10 Hunt, J. M., in *Proc. Intern. Sci. Oil Conf.* (Budapest, October 1962).
11 Hoering, T. C., and Abelson, P. H., *Carnegie Inst. Washington Yearbook*, **62**, 229 (1963).
12 McNab, J. G., Smith, P. V., and Betts, R. L., *Ind. Eng. Chem.*, **44**, 2556 (1952).
13 Hedberg, H. D., *Bull. Amer. Assoc. Petrol. Geol.*, **48**, 1755 (1964).
14 Leverson, A. I., in *Geology of Petroleum* (Freeman, San Francisco, 1954).
15 Philippi, G. T., *Geochim. Cosmochim. Acta*, **29**, 1021 (1965).

32

Reprinted from *Nature Phys. Sci.* **233**:123–124 (1971)

GEOCHEMICAL SIMULATION EVIDENCE FOR THE GENERATION OF HOMOLOGOUS DECARBOXYLATED PORPHYRINS

Daniel J. Casagrande and Gordon W. Hodgson

PORPHYRINS, which are believed to be degradation products of chlorophyll, are found in petroleum and ancient sediments. The porphyrins observed in petroleum as etioporphyrin (etio) and deoxophylloerythroetioporphyrin (DPEP) have been shown not to be single compounds but members of two homologous series[1-4], one consisting of those derived from chlorophyll in which the isocyclic ring is intact and the other of those in which the ring is absent. The relative abundance of the homologues in each series is usually nearly symmetrical about the median member, which is usually the direct daughter of the presumed parent, chlorophyll *a*, although sometimes lowered by one or two CH_2 groups. The puzzling feature of the distribution of the porphyrins is the occurrence of daughter members with molecular weights greater than that of the parent skeleton. Transalkylation has been suggested as a mechanism[5-7], and preliminary experiments at high temperatures have shown that it is possible[8]. We now have been able to demonstrate, in geochemically plausible conditions, the conversion of a single porphyrin to a homologous series similar to that of the naturally occurring porphyrins.

Not only is it necessary to demonstrate the evolution of porphyrins into homologous series, but it is equally important to illustrate conditions in which biogenic carboxylated pigments are decarboxylated, as this conversion is required to account for the structure of essentially all porphyrins of ancient sediments and petroleum. The results of our investigations show indeed that such decarboxylation reactions take place concomitantly with the transalkylation reactions.

An authentic porphyrin was heated in controlled conditions with various substances which have geochemical relevance, including water, fatty acids, benzene and/or clay. Thus a typical mixture included the porphyrin (mesoporphyrin IX dimethyl ester), clay (montmorillonite) and water in proportions of 3 mg : 1 g : 2 ml. The clay and porphyrin were ground in a mortar and put into a heavy-walled glass tube; water was added and the tube was immediately sealed under vacuum. The reaction conditions were such that the water remained in the liquid state. At the end of the reaction time the tube was opened and the constituents extracted with benzene–methanol (9 : 1). The material soluble in the benzene–methanol solvent was evaporated *in vacuo* to dryness and the residue partitioned between hydrochloric acid (6 M) and diethyl ether (1 : 1 v/v). Free-base porphyrins remained in acid solution whereas any porphyrins chelated with metals during the experiment entered the ether phase. After the phases were separated, the acid fraction was neutralized with sodium acetate and the associated porphyrins transferred into ether. Mesoporphyrin IX dimethyl ester was used because it is a carboxylated porphyrin that is readily obtainable from commercial sources.

Mesoporphyrin IX dimethyl ester was purchased from K and K Laboratories. All solvents were of reagent grade. A Beckman 'DB' spectrophotometer and Turner '210' spectrofluorometer were used for monitoring purposes. A Finnigan mass spectrometer component of a '1015 GC-MS' system was used for porphyrin mass spectral analysis. After evaporation into capillary tubes, each sample was put into a solid probe attachment. The temperature of the probe was usually between 200° and 225° C (to minimize fragmentation a 15 eV ionization potential was used).

Mesoporphyrin IX dimethyl ester was used to illustrate the mass spectral analytical procedure. A single ion at 594 a.m.u. was observed and the fragmentation pattern agreed with data published by Jackson *et al.*[8]. The primary fragments included ions due to the removal of the following groups: $—CH_2CO_2CH_3$, $—CO_2CH_3$, $—OCH_3$ and $—CH_3$. Porphyrins from a Boscan crude oil were demetallated and extracted using methanesulphonic acid to illustrate the method further. The mass spectrum from the Boscan porphyrins showed the etio and phyllo series reported by Baker[4]. Petroleum from Lloydminster (Alberta, Canada) was extracted by the same technique, and it seems that at least an etio series exists in that oil.

Table 1 lists the various geochemical simulations comprising the initial study. When mesoporphyrin IX dimethyl ester (MPDME) was heated alone at 250° C, a homologous

Table 1 Typical Simulation Experiments

(1) Mesoporphyrin IX dimethyl ester
(2) Mesoporphyrin IX dimethyl ester + clay
(3) Mesoporphyrin IX dimethyl ester + water
(4) Mesoporphyrin IX dimethyl ester + water + montmorillonite
(5) Mesoporphyrin IX dimethyl ester + acetic acid
(6) Mesoporphyrin IX dimethyl ester + acetic acid + montmorillonite
(7) Mesoporphyrin IX dimethyl ester + propionic acid
(8) Mesoporphyrin IX dimethyl ester + propionic acid + montmorillonite

Experimental conditions were 250° C in sealed tubes for 280 h.

series of porphyrins was found in the molecular weight range of 436 to 534. The principal member was mass 478 which corresponds to a double decarboxylation; mass 464 had a peak height of one half that of mass 478, whereas all other members of the series were very much less abundant. The significant observations were (*a*) a homologous series of porphyrins was generated, and (*b*) the porphyrin molecule was completely decarboxylated.

The free-base porphyrin (acid-soluble fraction) recovered from a system which included montmorillonite showed that the clay system gave a mass spectrum with a marked increase in the range (below mass 478) and intensity of peaks corresponding to those in the non-clay system. The series started at mass 408 and peaks at 450 and 464 were almost as intense as the 478 mass peak. Some degree of symmetry in the envelope of peaks was also observed; in petroleum and oil shale, porphyrins have a marked degree of symmetry. Metal complexed porphyrins were also recovered from the same system and these showed a similar homologous series. The mass spectrometer displacement of the parent ion suggests that iron may have been the complexing metal. From the limited number of experiments with MPDME with and without clay, it seems that clays have significant effect on the products that are observed apart from the obvious production of metal complexes, namely that of increasing the length of the homologous series.

The following reaction systems involving water also yielded homologous series: MPDME+water; MPDME+acetic acid; MPDME+propionic acid; MPDME+propionic acid+clay; MPDME+benzene. Histograms of the mass spectra for the MPDME+clay system (free-base) are shown in Fig. 1 together with corresponding data for porphyrins extracted from the Lloydminster crude oil. Further investigations using the above systems are being carried out.

The data are in good agreement with suggested transalkylation mechanisms for the origin of homologous porphyrins. The significant aspect of the work is that the reaction takes place in mild conditions in the presence of water and clay minerals. The geochemical plausibility of the system is equally notable because not only does the transalkylation seem to take place under these conditions but so also does the decarboxylation of the precursor porphyrins, another required geochemical conversion whose occurrence has not been demonstrated previously.

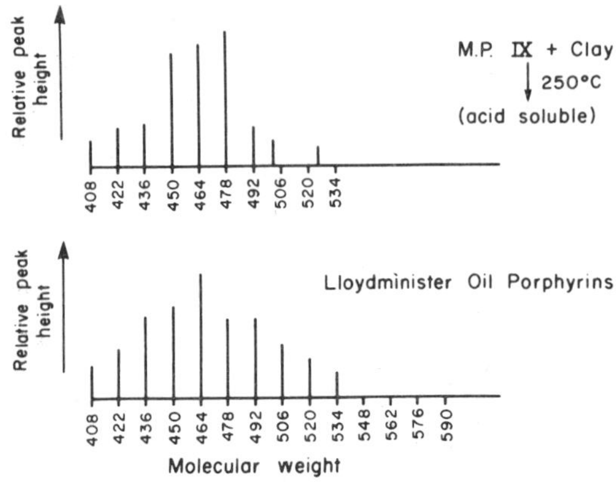

Fig. 1 Histogram of mass spectrum of homologous series of porphyrins generated from mesoporphyrin IX dimethyl ester+clay, compared with a similar histogram obtained from an asphaltic petroleum, Lloydminster crude oil.

[1] Dean, R. A., and Whitehead, E. V., *Proc. Sixth World Petrol. Cong.*, Section 5, Paper 9 (1963).
[2] Thomas, D. W., and Blumer, M., *Geochim. Cosmochim. Acta*, **28**, 1147 (1964).
[3] Morandi, J. R., and Jensen, H. B., *J. Chem. Eng. Data*, **11**, 81 (1966).
[4] Baker, E. W., *J. Amer. Chem. Soc.*, **88**, 2311 (1966).
[5] Baker, E. W., Yen, T. F., Dickie, J. P., Rhodes, R. E., and Clark, L. F., *J. Amer. Chem. Soc.*, **89**, 3631 (1967).
[6] Hodgson, G. W., Baker, B. L., and Peake, E., *Proc. Seventh World Petrol. Cong.*, **2**, 117 (1967).
[7] Yen, T. F., Boucher, L. J., Dickie, J. P., Tynan, E. C., and Vaughan, G. B., *J. Inst. Petrol.*, **55**, 87 (1969).
[8] Jackson, A. H., Kenner, G. W., Smith, K. M., Aplin, R. T., Budzikiewicz, H., and Djerassi, C., *Tetrahedron*, **21**, 2913 (1965).

33

Reprinted from *Carnegie Inst. Washington Year Book* **66**:510–514 (1968)

PRODUCTION OF HYDROCARBONS FROM THE ORGANIC MATTER IN A RECENT SEDIMENT

R. M. Mitterer and T. C. Hoering

An instructive approach to the diagenesis of organic matter in sediments is to heat a Recent sediment in the laboratory and analyze the products formed. This procedure gives clues to the types of organic reactions occurring under natural conditions. The formation of low-molecular-weight hydrocarbons from mild heating of Recent and ancient sediments was reported in *Year Book 62* and *Year Book 63*. This year we have examined the higher-molecular-weight normal hydrocarbons produced in similar experiments. The formation of isoprenoid hydrocarbons and porphyrins has also been noted. Experiments in which octadecene and octadecanol have been heated with sedimentary organic matter have yielded octadecane.

About 1% to 4% of the total organic matter in a Recent sediment is soluble in organic solvents and consists of hydrocarbons and organic compounds containing such elements as oxygen, nitrogen, and perhaps sulfur. The remainder, kerogen, is insoluble in organic solvents and is isolated by destroying the inorganic minerals with hydrofluoric acid.

Recent marine sediment from the San Nicolas Basin, described in the following section of this report, was extracted twice in an ultrasonic tank with a mixture of benzene and methanol (7:3 vol) to remove preexisting soluble material. The insoluble kerogen remains in the sediment after this treatment. The usual procedures for isolation and identification of hydrocarbons—liquid-liquid extraction (heptane: methanol), silica-gel column chromatography, urea adductination, and gas-liquid chromatography—were followed.

About 30 ppm of soluble paraffinic hydrocarbons on a dry sediment weight basis was recovered from unheated San Nicolas sediments. Values from nearby basins range from 10 to 130 ppm (Emery, 1960). The distribution of normal hydrocarbons is typical of that for Recent sediments, molecules with 27, 29, and 31 carbon atoms predominating.

To investigate the results of exposure of kerogen to elevated temperatures about 250-g portions of the dried, extracted sediment, containing kerogen, were sealed into a stainless steel bomb, pressurized to about 200 psig with nitrogen, and heated in a constant-temperature oven to 162°, 184°, 225°, and 250°C. The experiments lasted from 3 to 14 days, with the shorter runs at the higher temperatures. The heated sediment was extracted, and the hydrocarbons were separated as previously described. Table 31 lists the amounts of paraffinic hydrocarbons generated at the indicated temperatures and times.

The newly generated hydrocarbons are different from those of the original sediment extract, for little or no odd-carbon preference is seen, and significant amounts of compounds with less than 27 carbon atoms are found. The C-27, C-29, and C-31 *n*-hydrocarbons, prevalent in Recent sediment extract, are present in small amounts in the heated samples, indicating that the new hydrocarbons could not be a result of incomplete extraction. Figures 90 and 91 show the gas-liquid chromatograms of the extractable hydrocarbons from unheated Recent sediment and hydrocarbons extracted from sediment heated to 225°C.

The newly formed hydrocarbons had little optical activity. The specific rotation of the unheated sediment extract was $[\alpha]_D = +13.62$, whereas the extract of the sediment heated to 225°C had a value of $[\alpha]_D = +0.71$. The specific rotation is measured in degrees of rotation/($dm \times$ g/ml).

TABLE 31. Amount of Paraffinic Hydrocarbons Generated by Mild Heating of Recent Sediment

Temperature °C	Time, days	Percentage of Total Organic Matter
Original Sediment		0.04
162	14	0.04
184	8	0.04
225	8	0.16
225	3	0.05
250	3	0.24

Certain organic compounds diagnostic of life processes, such as isoprenoid hydrocarbons, are found in ancient sediments and petroleum. Pristane and phytane, two of the isoprenoid hydrocarbons, can be derived from the phytol side-chain of the chlorophyll molecule. Pristane, but not phytane, has been reported in Recent sediments. Because of their branched nature, pristane and phytane do not form adducts with urea and thus can be separated from the normal paraffins. The nonadducted fractions of the extracts from heated sediment contained substances with similar gas chromatographic retention times of pristane and phytane. The compounds were separated by preparative gas-liquid chromatography, and the individual peaks were analyzed on two different column subtrates, together with pure pristane and phytane. The retention times of the unknown compounds were identical with those of pristane and phytane on each column. The infrared spectra and the mass spectra of the compounds agreed with those of pure pristane and phytane. In a typical experiment, e.g., 184°C, pristane and phytane each comprised about 1% of the hydrocarbons formed, about the same concentration as *n*-C-17 and *n*-C-18 hydrocarbons. Experiments at other temperatures gave similar results.

Another group of compounds commonly found in petroleum is the porphyrins. They have not been found in Recent sediments, though their possible precursors, pheophytins, have been reported. Porphyrins were generated by heating Recent sediment at 225° and 250°C.

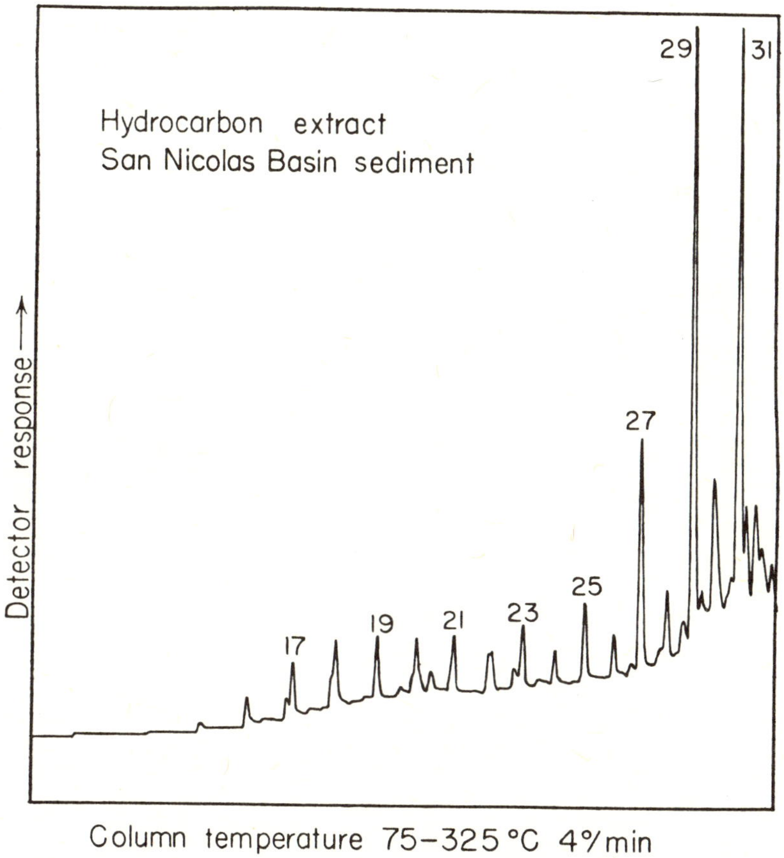

Fig. 90. Gas chromatogram of total hydrocarbons extracted with benzene-methanol from San Nicolas Basin sediment. Column: 10-foot by ⅛-inch 7% Apiezon substrate on Chromosorb W. Helium flow 40 ml/minute. Temperature programmed from 75° to 325°C at 4°/minute. The numbers at the peaks indicate the number of carbon atoms in the *n*-alkane.

The porphyrins were eluted from a column of silica gel with benzene and then purified by thin-layer chromatography. Visible absorption spectra were obtained on a Beckman DB spectrophotometer. Figure 92 shows the spectrum obtained after demetallation with methane sulfonic acid. The large Soret peak at 390 mμ is characteristic of porphyrins. When excited with light of wavelength 492 mμ the porphyrins fluoresced strongly at 628 mμ.

The chemical steps leading to porphyrins from chlorophyll or pheophytin are hydrogenation, hydrogenolysis, and decarboxylation. Different combinations of these reactions will result in different porphyrins. The presence of porphyrin indicates that at least some of these reactions have occurred.

The catalytic effect of the inorganic minerals on organic compounds has been suggested repeatedly and apparently demonstrated in the instance of a fatty acid and montmorillonite (Jurg and Eisma, 1964). Hoering and Abelson (*Year Book 62*), however, have previously shown that heating of Green River shale and kerogen derived from it resulted in qualita-

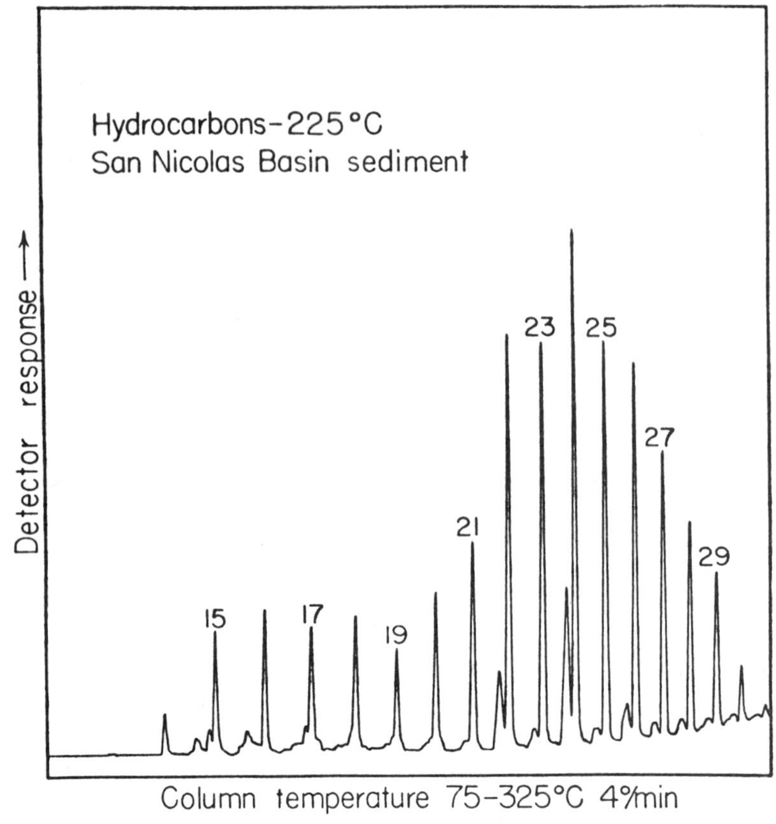

Fig. 91. Gas chromatogram of normal hydrocarbons extracted with benzene-methanol from San Nicolas Basin sediment heated to 225°C. Column: 10-foot by ⅛-inch 7% Apiezon substrate on Chromosorb W. Helium flow 40 ml/minute. Temperature programmed from 75° to 325°C at 4°/minute. Peak numbers indicate the number of carbon atoms in the *n*-alkane.

tively and quantitatively similar amounts of low molecular weight hydrocarbons.

In the present work, Recent sediment and kerogen derived from it by demineralization with HF were heated to 250°C in a similar experiment to study the higher molecular weight hydrocarbons formed. The new results support the findings of Hoering and Abelson; the yield and distribution of hydrocarbons from the sediment and kerogen were about the same. The results of both experiments show that mild heating is capable of releasing saturated hydrocarbons from kerogen.

The probable precursors of the normal hydrocarbons are normal alkyl chains, chemically bonded into the kerogen, and their release involves bond breakage and hydrogenolysis. The formation of normal hydrocarbons, isoprenoids, and porphyrins in the heated sediment indicates that hydrogenolysis has occurred and implies the presence of a hydrogen donor in the sediment that is capable of acting as a reducing agent. The following experiments were performed to explore further the reducing power of sediments.

A sample of sediment containing organic matter was treated with deuterium oxide at 100°C for several weeks in a sealed flask to replace some

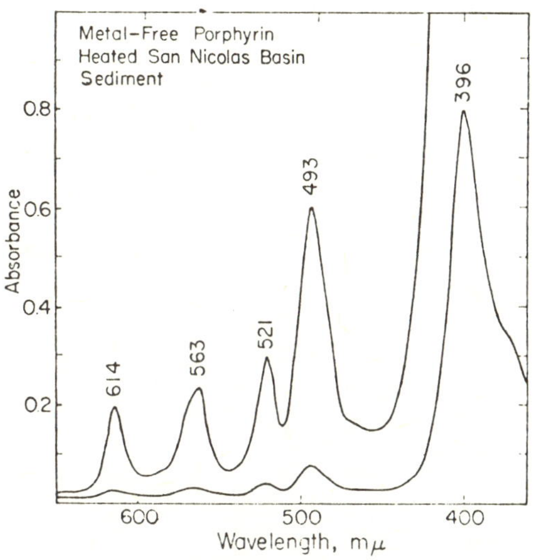

Fig. 92. Absorption spectra in heptane of metal-free porphyrin extracted from San Nicolas Basin sediment heated to 225°C for 8 days.

of the easily exchangeable hydrogen atoms with deuterium. After drying, the sample was heated at 225°C for 3 days in a sealed bomb. The infrared spectrum of the newly formed hydrocarbons had a new band at 4.3 μ, due to *C-D* vibrations, indicating that deuteration had occurred. Hoering and Abelson had previously investigated the deuteration of volatile hydrocarbons in a similar experiment.

In another series of experiments various compounds capable of being reduced were added to 100 g of dried, unextracted sediment. The sediment was suspended in pentane or heptane, and about 1 g of 1-octadecanol, 1-octadecene, stearic acid, and laurone were added to the mixture, which was then evaporated at reduced pressure. The sediment and the pure organic substance were heated at 225°C for about 3 days.

Approximately 140 mg of octadecane, the C-18 saturated hydrocarbon, was formed from about 1 g of 1-octadecanol, the C-18 alcohol. Octadecene, a C-18 hydrocarbon with one double bond, was converted to the saturated hydrocarbon in about 9% yield. About 40% was recovered as unreacted octadecene and about 50% was apparently incorporated into the kerogen. The experiment was also performed with separated kerogen concentrate as a possible hydrogen donor. Octadecane was formed from octadecene in about 17% yield. Again, about 50% of the octadecene was not recovered.

Less conclusive results were obtained by heating stearic acid and laurone with sediment. Only 2% to 3% of the stearic acid and laurone was recovered as hydrocarbon. At this level the indigenous hydrocarbons of the unextracted sediment and those generated through heating become prominent.

Identification of octadecane was made by gas-liquid chromatographic analysis, after paraffinic hydrocarbons were isolated by the standard chemical procedures. No octadecane impurity was detected in either the starting octadecene or octadecanol although the alcohol was found to contain about 3% of other paraffinic impurities.

These results provide further support for the view that kerogen in Recent sediments is a major source of hydrocarbons. Mild heating releases much larger amounts of hydrocarbons than are present in the natural sediment.

The present experiments have given insight into the nature of some of the reactions occurring in Recent sediment. To form free hydrocarbons and porphyrins, bond breakage and hydrogenolysis or hydrogenation must occur. The model experiments with olefin and octadecanol and the deuterium experiment illustrate the reactivity and reducing power of Recent sediments. Reduction is an important reaction of the sedimentary organic matter accompanying the release of alkyl chains from kerogen.

34

Reprinted from *Carnegie Inst. Washington Year Book* **67**:205–208 (1969)

LABORATORY SIMULATION OF AMINO-ACID DIAGENESIS IN FOSSILS

P. E. Hare and R. M. Mitterer

Changes in the amino-acid content of fossil shells with passage of time have been found to be consistent with deductions based on the order of stability of free amino acids heated in dilute aqueous solutions, as shown first by Abelson (*Year Book 53*, pp. 97–101; and 1959) and more recently by Vallentyne (1964).

Amino acids found stable in such heating tests included alanine, glycine, glutamic acid, valine, and the leucines (Abelson, 1959); these are the amino acids that are primarily found in fossils older than Pleistocene. Amino acids unstable to heating, such as serine and threonine, and therefore not expected in older fossils, have been reported (e.g. Degens, 1965, pp. 202–280) in older sediments. Amino-acid stabilization by surrounding matrix has been suggested (Prashnowsky and Schidlowski, 1967) to explain the apparent anomaly of unstable amino acids in ancient deposits.

In addition, experiments at this laboratory on heating free amino acids in solutions indicate that *p*H can be a significant factor in the stability of some amino acids. Samples of a standard mixture of amino acids buffered at *p*H 2.0, 4.6, and 9.4 (as measured at 25°C) were heated in sealed tubes under nitrogen for 4 days at 140°C. Arginine disappeared completely at *p*H 9.4 to form ornithine, whereas at *p*H 2.0 and 4.6 no appreciable amount of arginine was converted to ornithine. Cystine was stable only at *p*H 2.0, virtually disappearing at both other *p*H values. The alloisoleucine-to-isoleucine ratio observed after incubation was 0.05 at *p*H 2.0 but 0.2 at *p*H 9.4. Aspartic acid was less stable at *p*H 4.6 than at *p*H 2.0 or 9.4. At *p*H 9.4 the serine-to-threonine ratio was 10 after heating. At *p*H 4.6 the serine-to-threonine ratio was 1.5, and at *p*H 2.0 it was reversed at 0.3.

In view of this *p*H effect and the possibility of a matrix effect on amino-acid stability, we have attempted to simulate more closely the diagenesis of amino acids in fossils by subjecting fragments of modern shells to incubation under a variety of temperatures and environments. A first series of experiments tested the role of water in the environment, allowing for the fact that in nature water is usually present. Modern shell fragments of *Mercenaria* were heated to 160°C for 10 hours in varying amounts of water or water vapor. The sample heated dry showed little or no reaction, in contrast to the samples heated in the presence of water or water vapor.

In subsequent experiments shell fragments of about 100 mg were heated in 1 ml of water sealed in a tube under nitrogen. The changes observed upon heating Recent shell fragments in the presence of water are analogous to and duplicate those changes found in the natural series of fossils. With continued heating the unstable amino acids virtually disappeared, the remaining amino acids racemized, and the mixture of amino acids approached that found in the oldest fossils examined.

Evidence that we are dealing with the same mechanisms and changes that take place in the natural environment is found in the data from a series of Recent and fossil shells heated for 1 day at 185°C. The results are shown in Table 18.

In the oldest shell, a sample from the Miocene St. Mary's formation, relatively few changes took place. Serine and threonine were already absent; the ratio of alloisoleucine to isoleucine was already the equilibrium ratio, so that heating for 1 day at 185°C in the presence of water changed the pattern very little. Similar treatment of a specimen from Wailes Bluff, of Upper Pleistocene age, caused much more extensive changes, but the final mixture resembled that ob-

TABLE 18. Amino-Acid Composition for Inner Layer of *Mercenaria* Shells Before and After Heating to 185°C for 24 Hours, nM/g shell

	Recent		Upper Pleistocene		Miocene	
	Unheated	Heated	Unheated	Heated	Unheated	Heated
Aspartic acid	1960	80	865	78	35	12
Threonine	1170	0	306	0	0	0
Serine	1430	0	247	0	0	0
Glutamic acid	1400	575	709	363	140	140
Proline	1790	570	767	375	120	120
Glycine	1500	735	572	325	104	99
Alanine	1210	850	715	523	166	165
Cystine	403	0	26	0	0	0
Valine	740	290	423	198	127	114
Methionine	225	77	85	35	10	10
Alloisoleucine	0	100	52	69	42	42
Isoleucine	370	80	163	55	34	34
Leucine	515	190	280	115	78	74
Tyrosine	515	134	286	94	38	38
Phenylalanine	354	147	202	84	52	47
Ornithine	0	97	52	62	9	5
Lysine	1143	155	442	81	46	20
Histidine	338	0	65	0	0	0
Arginine	612	0	241	0	0	0
Ratio allo/iso	0	1.25	0.32	1.25	1.25	1.25

tained in the previous experiment. When a Recent shell was heated, profound changes were observed, the final pattern again approaching that obtained in the other two heated samples.

To study the kinetics of the reactions in more detail, fragments of modern shells were heated at temperatures of 165°, 140°, 125°, 105°, and 90°C for periods of time ranging from 1 hour to over 3 months. The shell fragments were heated in water in tubes sealed under nitrogen. The total amount of amino acids decreased with time of heating. Amino-acid ratios changed systematically with time of heating, in a fashion analogous to that found in fossil shells.

Comparison of data on free amino acids and shell fragments showed that free amino acids reacted more slowly. For example, in 4 days at 140°C, serine decreased only to 80% of its original value (at *p*H 9.4), whereas in the shell fragments it decreased to less than 10%. This apparent decrease in stability in the shell matrix seemed to be true for all the amino acids.

Isoleucine racemizes to alloisoleucine by a first-order reaction. This reaction is reversible, and by heating the two separately we have shown that the isoleucine-to-alloisoleucine reaction is approximately 25% faster than the reverse reaction, so that at equilibrium the ratio of alloisoleucine to isoleucine is about 1.25.

The Arrhenius equation used in Fig. 83 relates the rate constant of a reaction to the absolute temperature. A plot of the time necessary for a given amount of isoleucine to diminish to $1/e$ of its initial value at a particular temperature is shown. It is seen that extrapolation of data based on the laboratory treatment checks well with data from a series of radiocarbon-dated fossil-shell material from areas with average temperatures of 23° to 24°C. These data were obtained from the middle shell layer of *Mercenaria*. There is evidence that other shell structures in *Mercenaria* react at somewhat different rates.

Point *A* represents data from *Foraminifera* tests from an Antarctic deep-sea core (core 15–16 of Goodell and Watkins, 1968) supplied by Florida State University. Sedimentation rates are based on recognized magnetic reversals. Agree-

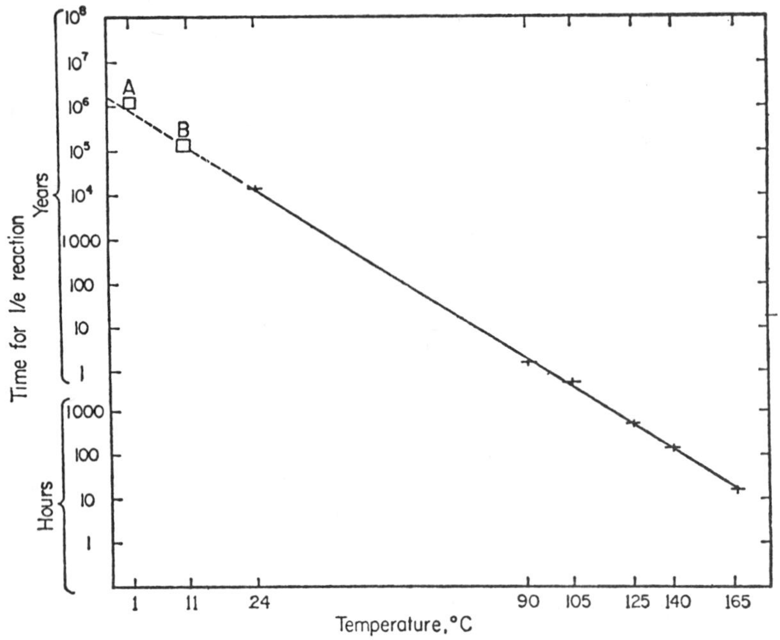

Fig. 83. Arrhenius plot of time (1/*e* of reaction) as a function of temperature^{-1}.

ment with extrapolated values is good and gives us confidence that the reactions occurring in the laboratory treatment are similar to those occurring in fossils.

Point *B* represents data from the Upper Pleistocene Wailes Bluff locality. Radiocarbon dating shows that this deposit is older than 40,000 years. Amino-acid ratios suggest that it is probably around 60,000 years old if the temperature has been around 11° to 12°C.

The agreement between the values found in these Pleistocene samples and the Arrhenius plot suggests potential uses of amino-acid ratios in fossils. If the temperature is known precisely, an approximate age can be calculated. Alternatively, effective incubation temperature can be inferred on well-dated material. Still another potential use is for stratigraphic correlation, which should be particularly useful in deep-sea cores, since the temperatures are uniform and constant.

In addition to providing quantitative information on reaction rates, the heating experiments afforded an opportunity to study contamination of the shell and loss of amino acids from it. Amino acids were found in the water surrounding the heated shell fragment. The concentration of amino acids in the water varied directly with time of heating and temperature. With prolonged heating at higher temperatures, the amounts of amino acids outside the shell were approximately equal to those within the shell.

In another experiment, to test for the possibility of amino acids migrating into the shell, two shell fragments were placed in a 1-ml solution containing 10 μM of the nonprotein amino acid norleucine, which is not found in modern or fossil shells. The sealed tube containing the shell fragments was heated for 4½ days at 165°C under the usual conditions. After the tube was opened, one fragment was removed, thoroughly washed, and prepared for amino-acid analysis. The other fragment was washed and dried, and surface material was removed with a dental tool. The fragment was then acid-washed to remove more of the outer surface and analyzed. In the water-washed sample, norleucine amounted to 1.3% of the total amino acids. In the sample that was abraded and acid-washed, norleucine was about 0.5% of the total acids.

The results suggest that young intact fossil shells are probably not contaminated to a significant degree by amino acids in groundwater and sediment. On the other hand, older fossil specimens, having lost much of their amino acids and having been continuously subjected to contamination over a long period of time, might contain some adventitious amino acids.

The results of this study indicate that it is possible to simulate the diagenetic reactions occurring in nature by simple heating of shell fragments in water. Furthermore, when the data from calcareous fossils were compared with those from free amino acids heated in aqueous solutions, it was found that there is indeed a matrix effect. The matrix effect resulted in faster reaction in the shell than for free amino acids, however, in contrast to a stabilizing matrix effect that has been postulated for sediments.

Part V

MOLECULAR APPLICATIONS

Editor's Comments on Papers 35 Through 42

35 **BREGER**
Diagenesis of Metabolites and a Discussion of the Origin of Petroleum Hydrocarbons

36 **PHILIPPI**
On the Depth, Time and Mechanism of Petroleum Generation

37 **TISSOT et al.**
Influence of Nature and Diagenesis of Organic Matter in Formation of Petroleum

38 **WINTERS and WILLIAMS**
Microbiological Alteration of Crude Oil in the Reservoir

39 **BLUMER and SASS**
Oil Pollution: Persistence and Degradation of Spilled Fuel Oil

40 **BADA and PROTSCH**
Racemization Reaction of Aspartic Acid and Its Use in Dating Fossil Bones

41 **MITTERER**
Ages and Diagenetic Temperatures of Pleistocene Deposits of Florida Based on Isoleucine Epimerization in Mercenaria

42 **KVENVOLDEN, LAWLESS, and PONNAMPERUMA**
Nonprotein Amino Acids in the Murchison Meteorite

This final part attempts to show, using eight examples, how studies of the geochemistry of organic molecules can be applied in a number of special problem areas. One of these areas is petroleum geochemistry, which constitutes a significant portion of the broader field of organic geochemistry. To illustrate petroleum geochemistry, three classic papers dealing with aspects of the origin of petroleum are included here. These papers span a period of about fifteen years during which time ideas about petroleum genesis evolved considerably. Breger (Paper 35) emphasizes the possible role of lipids, particularly fatty acids, in the generation of hydrocarbons in petroleum. He postulates reactions that can produce certain compounds found in petroleum. Of particular interest is the Diels-Alder reaction, which, in concert with other mechanisms, can account for some straight-chain, branched-chain, cyclic, and aromatic hydrocarbons.

In one of the most cited papers in petroleum geochemistry, Philippi (Paper 36) describes in great detail the processes of petroleum generation in the Los Angeles and Ventura Basins of California. By comparing hydrocarbons in recent and ancient sediments and by measuring the total hydrocarbon content and the compositions of paraffins and naphthenes, he concluded that in source beds the petroleum "is formed from, and within, the vastly more abundant nonhydrocarbon organic matter." Temperature was postulated to be one of the most important factors in the petroleum generation process.

In contrast to the approach of Philippi (Paper 36), who examined the hydrocarbons of sediments, Tissot et al. (Paper 37) stressed the direct investigation of the kerogen component of sediments. They were able to define the paths of evolution during maturation of various types of kerogen. From these studies, predictions could be made regarding sources and thermal maturity of buried organic matter and the products, oil and/or gas, that could be expected. The general scheme for petroleum generation that Tissot et al. describe is now widely accepted. This scheme includes incorporation of hydrocarbons derived from original biological precursors, but the amount of hydrocarbon from this source is small. Some nonhydrocarbon molecules in the original biological mixture are transformed into hydrocarbons, but the bulk of petroleum comes from the thermal breakdown of kerogen.

An area of overlap between organic geochemistry and microbiology is illustrated by the often cited paper by Winters and Williams (Paper 38). These authors were the first to document the role of microorganisms in the alteration of petroleum. They noted several lines of evidence that implicated microorganisms—evidence such as reduction or loss of *n*-alkanes, increase in nitrogen content, and increase in optical activity. About 10% of all crude oils analyzed in their laboratories showed evidence for microbial alteration. This paper stimulated additional research on the alteration of petroleum in reservoirs. (See suggested additional readings.)

Environmental organic geochemistry is a field that has received accelerated attention since about 1969. Concern about the fate and effects of petroleum in the environment has led to numerous studies of petroleum in sediments and organisms. As a consequence of this intense interest, improved analytical techniques have been developed, and a new appreciation has been gained about the background of hydrocarbons that exists in nature. One of the first papers to address the question of what happens to petroleum products spilled into the environment was by Blumer and Sass (Paper 39). They followed for two years the chemical changes that took place in fuel oil spilled in Buzzard's Bay, Massachusetts. By applying techniques of organic geochemistry they measured the content and degradation of oil in the sediments of the Bay, determined criteria by which to identify fossil fuel pollutants, and gained new knowledge on the persistence of pollutants in the environment. Numerous investigations concerned with the environmental consequences of spilled petroleum have followed. Some of these are noted in "Suggested Additional Readings."

The last three papers deal with the use of amino acids as parameters in three different research areas—archaeology, geochronology-paleothermometry, and cosmochemistry. The ubiquity and structural properties of these compounds make them so useful in these three areas. An important property is the fact that amino acids in protein of living systems are optically active and exist in only one of two possible structural configurations called enantiomers. In fossils, the enantiomers interconvert by a process called racemization (epimerization when diastereomers are involved). This process is dependent on both time and temperature. The extent of racemization of amino acids in fossils is used to date fossil bones in archaeological sites and to establish correlations, ages, and average temperatures of Pleistocene deposits. The paper by Bada and Protsch (Paper 40) illustrates the use of racemization of amino acids in dating fossil bones. The method can be used to date fossil bones in the range from about five thousand years old to perhaps several hundred thousand years old depending on the amino acid measured. Amino acid dating is a potentially valuable tool for archaeology and may be very useful in helping to unravel the history of human evolution.

To illustrate the use of amino acids for dating geologic deposits, a paper by Mitterer (Paper 41) is included. In this paper, the extent of epimerization of isoleucine in the clam *Mercenaria* is measured, and this parameter is used to assign ages to Pleistocene marine deposits of Florida. In addition he was able to calculate average diagenetic temperature for the Late Pleistocene. Thus, this paper shows that both geochronologic and paleothermometric information can be obtained through studies of amino acids.

The last paper by Kvenvolden, Lawless, and Ponnamperuma (Paper 42) describes the occurrence of amino acids in the Murchison meteorite. These amino acids are believed to have originated through abiotic, cosmochemical processes. The abiotic nature of these processes is suggested by the fact that most of the amino acids discovered in the meteorite are not found in terrestrial protein and by the observation that the amino acids with asymmetric carbon atoms are present as racemic mixtures.

Other references to the use of amino acids in the three areas of investigation outlined above are given in "Suggested Additional Readings."

SUGGESTED ADDITIONAL READINGS

Petroleum Geochemistry

Abelson, P. H. 1963. Organic geochemistry and the formation of petroleum. *6th World Petrol. Cong. Proc.*, Sec. 1. 397–407.

Albrecht, P., M. Vanderbroucke, and M. Mandengue. 1976. Geochemical studies on the organic matter from the Douala Basin (Cameroon)—I. Evolution of the extractable organic matter and the formation of petroleum. *Geochim. et Cosmochim. Acta* **40**:791–799.

Baker, D. R. 1962. Organic geochemistry of Cherokee Group in southestern Kansas and northeastern Oklahoma. *Am. Assoc. Petrol. Geol. Bull.* **46**:1621–1642.

Bray, E. E., and E. D. Evans. 1965. Hydrocarbons in non-reservoir-rock source beds. *Am. Assoc. Petrol. Geol. Bull.* **49**:248–257.

Brooks, J. D., and J. W. Smith. 1967. The diagenesis of plant lipids during the formation of coal, petroleum and natural gas—I. Changes in the *n*-paraffin hydrocarbons. *Geochim. et Cosmochim. Acta* **31**:2389–2397.

Brooks, J. D., and J. W. Smith. 1969. The diagenesis of plant lipids during the formation of coal, petroleum and natural gas—II. Coalification and the formation of oil and gas in the Gippsland Basin. *Geochim. et Cosmochim. Acta* **33**:1183–1194.

Cordell, R. J. 1972. Depths of oil origin and primary migration: A review and critique. *Am. Assoc. Petrol. Geol. Bull.* **56**:2029–2067.

Dow, W. G. 1977. Kerogen studies and geological interpretations. *Jour. Geochem. Explor.* **7**:79–99.

Durand, B., and J. Espitalie. 1976. Geochemical studies on the organic matter from the Douala Basin (Cameroon)—II. Evolution of kerogen. *Geochim. et Cosmochim. Acta* **40**:801–808.

Erdman, J. G. 1965. Petroleum—Its origin in the earth. In *Fluids in Subsurface Environments—A Symposium,* Mem. 4. Am. Assoc. Petrol. Geol., Tulsa, Okla.:20–52.

Fuex, A. N. 1977. The use of stable isotopes in hydrocarbon exploration. *Jour. Geochem. Explor.* **7**:155–188.

Haun, J. D. 1971. Origin of Petroleum. *Am. Assoc. Petrol. Geol. Repr. Ser.* **1**. 192 p.

Haun, J. D. 1973. Origin of Petroleum II. *Am. Assoc. Petrol. Geol. Repr. Ser.* **9**. 210 p.

Hedberg, H. D. 1964. Geologic aspects of origin of petroleum. *Am. Assoc. Petrol. Geol. Bull.* **48**:1755–1803.

Hunt, J. M. 1979. *Petroleum Geochemistry and Geology*. Freeman, San Francisco, 617 pp.

McIver, R. D. 1967. Composition of kerogen—Clue to its role in the origin of petroleum. *7th World Petrol. Cong. Proc.* **2**:25–36.

Meinschein, W. G. 1961. Significance of hydrocarbons in sediments and petroleum. *Geochim. et Cosmochim. Acta* **22**:58–64.

Ronov, A. B. 1958. Organic carbon in sedimentary rocks (in relation to the presence of petroleum). *Geochem.* **5**:510–536.

Seifert, W. K., and J. M. Moldowan. 1978. Application of steranes, terpanes and monoaromatics to the materation, migration and source of crude oils. *Geochim. et Cosmochim. Acta* **42**:77–95.

Staplin, F. L. 1969. Sedimentary organic matter, organic metamorphism and oil and gas occurrence. *Bull. Canadian Petrol. Geol.* **17**:47–66.

Stevens, N. P. 1956. Origin of petroleum—A review. *Am. Assoc. Petrol. Geol. Bull.* **40**:51–61.

Tissot, B., J. Espitalie, G. Deroo, C. Tempere, and D. Jonathan. 1974. Origine et migration des hydrocarbures dans le Sahara Oriental (Algérie). In B. Tissot and F. Bienner, eds., *Advances in Organic Geochemistry 1973.* Editions Technip., Paris, 315–334.

Tissot, B. P., and D. H. Welte. 1978. *Petroleum Formation and Occurrence*. Springer-Verlag, Berlin, 538 p.

Vandenbroucke, M., P. Albrecht, and B. Durand. 1976. Geochemical studies on the organic matter from the Douala Basin (Cameroon)—III. Comparison with the early Toarcian shales, Paris Basin, France. *Geochim. et Cosmochim. Acta* **40**:1241–1249.

Welte, D. 1965. Relation between petroleum and source rock. *Am. Assoc. Petrol. Geol. Bull.* **49**:2246–2268.

Welte, D. H. 1972. Petroleum exploration and organic geochemistry. *Jour. Geochem. Explor.* **1**:117–136.

Williams, J. A. 1974. Characterization of oil types in Williston Basin. *Am. Assoc. Petrol. Geol. Bull.* **58**:1243–1252.

Geochemical Microbiology

Bailey, N. J. L., H. H. Krouse, C. R. Evans, and M. A. Rogers. 1973. Alteration of crude oil by waters and bacteria—Evidence from geochemical and isotope studies. *Am. Assoc. Petrol. Geol. Bull.* **57**:1276–1290.

Bailey, N. J. L., A. M. Jobson, and M. A. Rogers. 1973. Bacterial degradation of crude oil: Comparison of field and experimental data. *Chem. Geology* **11**:203–221.

Evans, C. R., M. A. Rogers, and N. J. L. Bailey. 1971. Evolution and alteration of petroleum in western Canada. *Chem. Geology* **8**:147–170.

Environmental Geochemistry

Gearing, P., J. N. Gearing, T. F. Lytle, and J. S. Lytle. 1976. Hydrocarbons in 60 northeast Gulf of Mexico shelf sediments: A preliminary survey. *Geochim. et Cosmochim. Acta* **40**:1005–1017.

Farrington, J. W., and J. G. Quinn. 1973. Petroleum hydrocarbons in Narragansett Bay—I. Survey of hydrocarbons in sediments and clams *(Mercenaria mercenaria). Estuarine and Coastal Marine Sci.* **1**:71–79.

Farrington, J. W., and P. A. Meyers. 1975. Hydrocarbons in the marine environment. In G. Eglinton, ed., *Environmental Chemistry* **1**. The Chemical Soc., London, 109–136.

Palacas, J. G., P. M. Gerrild, A. H. Love, and A. A. Roberts. 1976. Baseline concentrations of hydrocarbons in barrier-island quartz sand, northeastern Gulf of Mexico. *Geology* **4**:81–84.

Wakeham, S. G., and R. Carpenter. 1976. Aliphatic hydrocarbons in sediments of Lake Washington. *Limnology and Oceanography* **21**: 711–723.

Amino Acids—Archaeology

Bada, J. L., R. A. Schroeder, and G. F. Carter. 1974. New evidence for the antiquity of man in North America deduced from aspartic acid racemization. *Science* **184**:791–793.

Bada, J. L. 1972. The dating of fossil bones using the racemization of isoleucine: *Earth and Planetary Sci. Letters* **15**:223–231.

Bada, J. L., and P. M. Helfman. 1977. Amino acid racemization dating of fossil bones. *World Archaeology* **7**:160–173.

Masters, P. M., and J. L. Bada. 1975. Racemization of isoleucine in fossil molluscs from Indian middens and interglacial terraces in southern California. *Earth and Planetary Sci. Letters* **37**:173–183.

Wehmiller, J. F. 1977. Amino acid studies of the Del Mar, California, Midden Site: Apparent rate constants, ground temperature models, and chronological implications. *Earth and Planetary Sci. Letters* **37**: 184–196.

Amino Acids—Geochronology, Paleothermometry

Bada, J. L., R. Protsch, and R. A. Schroeder. 1973. The racemization reaction of isoleucine used as a paleotemperature indicator. *Nature* **241**:349–395.

Miller, G. H., and P. E. Hare. 1975. Use of amino acid reactions in some Arctic marine fossils as stratigraphic and geochronological indicators. *Carnegie Inst. Washington Year Book* **74**:612–617.

Mitterer, R. M. 1974. Pleistocene stratigraphy in southern Florida based on amino acid diagenesis in fossil *Mercenaria. Geology* **2**:425–428.

Amino Acid—Cosmochemistry

Cronin, J. R., and C. B. Moore. 1971. Amino acid analyses of the Murchison, Murray and Allende carbonaceous chondrites. *Science* **172**: 1327–1329.

Kvenvolden, K. A., J. G. Lawless, K. Pering, E. Peterson. J. Flores, C. Ponnamperuma, I. R. Kaplan, and C. Moore. 1970. Evidence for extraterrestrial amino acids and hydrocarbons in the Murchison meteorite. *Nature* **228**:923–926.

Lawless, J. G., K. A. Kvenvolden, E. Peterson, C. Ponnamperuma, and C. Moore. 1971. Amino acids indigenous to the Murray meteorite. *Science* **173**:626–627.

Lawless, J. G., K. A. Kvenvolden, E. Peterson, C. Ponnamperuma, E. Jarosewich. 1972. Evidence for amino-acids of extraterrestrial origin in the Orgueil meteorite. *Nature* **236**:66–67.

Pollock, G. E., C-N. Cheng, S. E. Cronin, and K. A. Kvenvolden. 1975. Stereoisomers of isovaline in the Murchison meteorite. *Geochim. et Cosmochim. Acta* **39**:1571–1573.

35

Diagenesis of metabolites and a discussion of the origin of petroleum hydrocarbons

I. A. Breger
U.S. Geological Survey, Washington 25, D.C.

Abstract—Proteins and carbohydrates are rapidly degraded to compounds of no direct interest in the problem of the origin of petroleum. Lignin, if carried into marine basins in the form of humic substances, is probably the major progenitor of kerogen rather than the precursor of petroleum. Pigments are but minor contributors to petroleum.

The fate of fatty acids in a marine environment is not completely understood. Although they may not be directly decarboxylated biochemically, it is shown how they can be converted into oxygenated or dehydrogenated acids more reactive than the parent compounds. Illustrations are also given for Diels–Alder reactions that could account for the formation from these compounds of the alicyclic and aromatic hydrocarbons in petroleum.

It is most likely that crude oil is generated in sediments containing concentrations of lipids, the character of which governs the nature of the oil that is formed.

There is no longer any argument regarding the organic origin of crude oil. Knowledge of the mechanisms whereby complex oxygen-, nitrogen-, and sulphur-bearing biological progenitors are converted into a complex mixture of hydrocarbons is, however, still limited. In an attempt to achieve a better basis for understanding how such conversion may have occurred, an effort has been made to consider the fate of various biological substances in diagenetic environments. In particular, it was of considerable interest to evaluate any interactions that could possibly occur between initial degradation products under the geochemical conditions of petroleum formation.

Crude oils are generally thought to be formed from complex organic materials in marine muds. Of necessity, initial consideration must be given to the nature of this organic matter in any study of the origin of petroleum.

Proteins, carbohydrates, and lipids make up the three major classes of substances produced in life processes; these three types of compounds along with lignin and pigments could, therefore, be expected to contribute to a marine sediment in which oil could later be formed.

Progenitors

Proteins

Proteins, both the water soluble and water insoluble varieties, are composed of fewer that thirty different α-amino acids tied together in peptide linkages. Most proteins are sensitive to attack by proteolytic enzymes, such as pepsin or trypsin, which hydrolyse the peptide bonds to form partially decomposed proteins. The latter are then attacked and degraded by other known enzymes. Proteolytic processes, which in effect hydrolyse the peptide bond, operate with but a slight energy requirement and serve to prepare the proteins for eventual utilization as energy sources on other oxidative or reductive biochemical processes. Although these processes and enzymes are those normally associated with human biochemical processes, there can be little doubt that analogous degradations occur in other animal systems. In general, these processes would be expected to result in

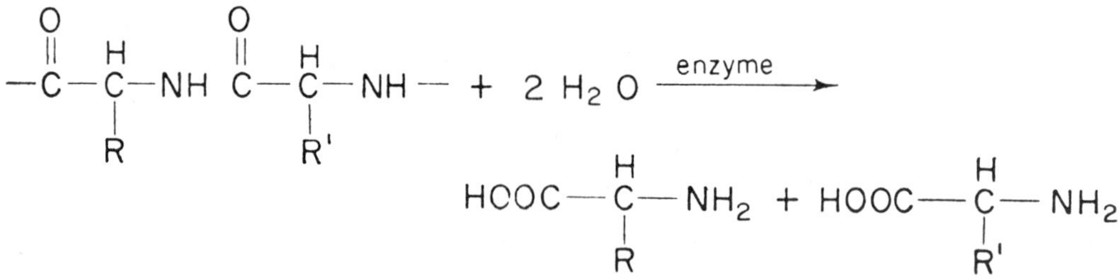

Hydrolysis of a peptide bond

the elimination of proteinaceous matter as carbon dioxide, water, ammonia, nitrogen and hydrogen sulphide. Other labile products, such as urea, would, if produced, eventually yield similar degradation products.

Proteins, therefore, appear to undergo biochemical degradations leading to compounds of no direct interest in the problem of the origin of petroleum. Only in certain cases of putrefaction can biological processes result in the formation of stable nitrogen bases that might be incorporated into sediments. Examples of these processes are the formation of indole and skatole from tryptophane, cadaverine from lysine, and putrescine from arginine.

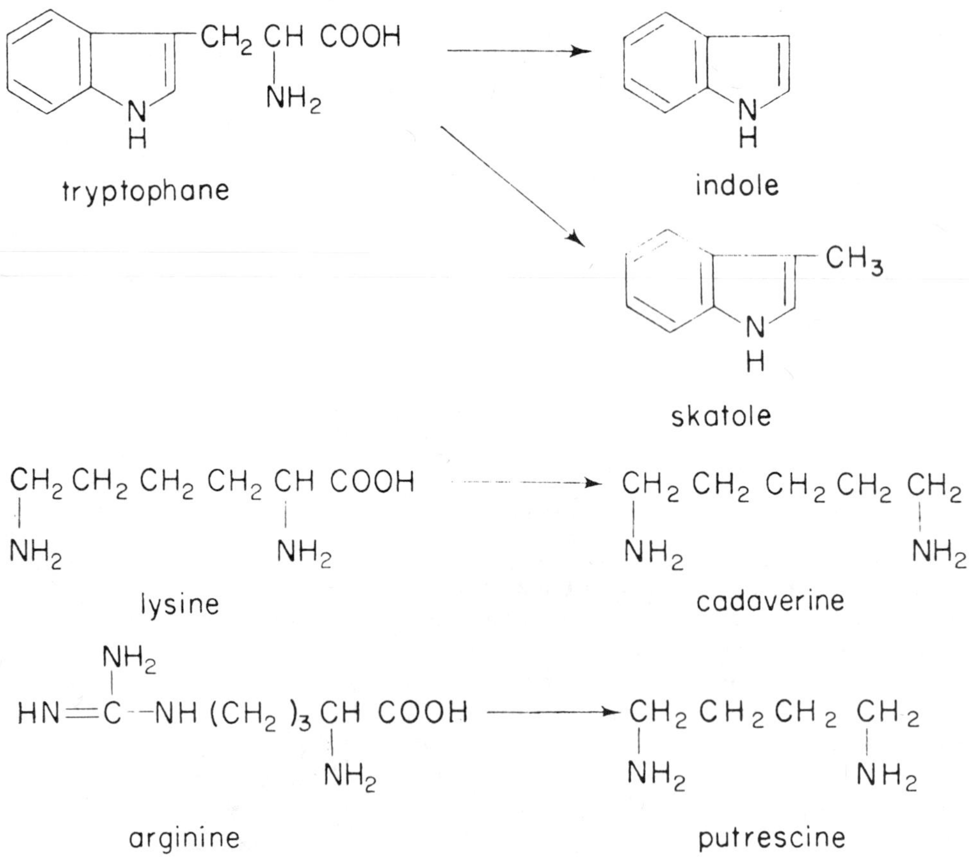

Certain proteins of biological origin are associated with other compounds and are known as conjugated proteins. An example of such a compound is haemoglobin, in which the globin, a protein, is bound through imidazole nitrogen to protoporphyrin-III containing an atom of reduced iron (HARPER, 1957, p. 60).

It is evident that degradation of the protein will lead to a residue of porphyrin which, as evidenced by its occurrence in petroleum, is resistant to biochemical decomposition.

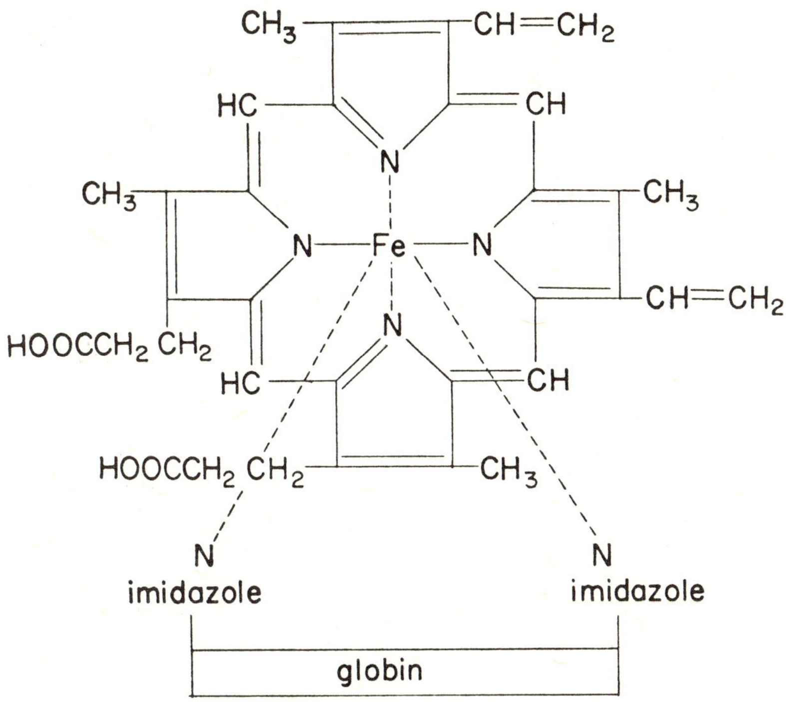

Reduced haemoglobin

Recent studies have shown that some amino acids survive in certain environments. ABELSON (1957) for example, has been able to identify free amino acids in calcareous fossil shells. The minute quantities of such residual amino acids, however, attest only to a successful entrapment mechanism, but do not nullify the conclusion that proteins and their amino acids are labile, are biologically decomposed, and probably contribute very little to the overall picture of petroleum genesis.

Carbohydrates

Carbohydrates account for a major portion of both plant and animal residues. They are classified as saccharides or disaccharides, both of which are water soluble, some tri- and tetrasaccharides, and finally as essentially insoluble polysaccharides such as starch, cellulose, and glycogen. Monosaccharides are characterized as either polyhydroxy aldehydes or ketones; the formation of polysaccharides is accomplished by the elimination of water between the monomers to form oxygen bridges as shown below for the disaccharide, cellobiose:

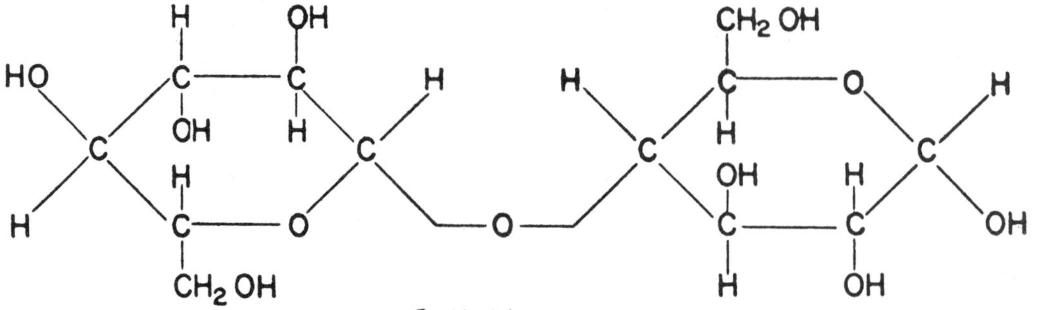

Cellobiose

Carbohydrates are subject to a great variety of enzymatic reactions that degrade them into simpler substances. Under anaerobic conditions fermentation can occur in which a six-carbon monosaccharide, glucose, is converted into alcohol without the addition of oxygen. Other enzymatic processes are able to convert sugars into fat. As is well known, the cellulose

$$\underset{\text{glucose}}{C_6H_{12}O_6} \xrightarrow{\text{enzyme}} \underset{\text{ethyl alcohol}}{2C_2H_5OH} + 2CO_2$$

is bacteriologically ingested by herbivores depolymerized into mono- and di-saccharides that the animal can then metabolize as a source of energy. The enzymatic hydrolysis of carbohydrates is similar to the proteolytic degradation of proteins; in both cases the polymers are broken down into individual units or shorter chains with a minimum transfer of energy. The products of these reactions are subject to oxidation or reduction and are, therefore, sources of energy for life processes.

In general, the final products of carbohydrate metabolism are carbon dioxide and water. Although the widespread disappearance of appreciable residues of carbohydrates is evidence of their sensitivity to biochemical degradation, there is some evidence that under certain conditions carbohydrates can be protected and can, therefore, survive through geological time. VALLENTYNE and WHITTAKER (1956) have recently found free sugars in lake waters and sediments, and THEANDER (1952) reported the presence of galactose, glucose, mannose, arabinose, xylose and rhamnose in hydrolysates from sphagnum peat. The presence of small quantities of cellulose in the Eocene brown coals of Germany has also long been known. The very fact that a careful search for these residual minute quantities of sugars has been necessary is in itself significant in demonstrating the rapid and nearly total degradation of carbohydrates.

Lignin

Lignin, although not a normal constituent of marine organisms, can nevertheless be a major contributor to potential source beds for petroleum if it is carried into marine basins by rivers in the form of humic substances. Lignin has a phenylpropane structure and is an aromatic compound. It also has a flavonoid structure which, when opened, develops acidic properties. Using the Russell formula for lignin (RUSSELL, 1948) not as an actual representation of the structure but merely as a guide for illustration of potential degradative reactions, it is possible to demonstrate such rupture as shown below.

The product illustrated is demethylated because the methoxyl content of lignin is known to decrease rapidly on degradation. Double bonds may also be formed by the dehydration of β-hydroxy ketones, and the resultant derivative of catechol is extremely reactive towards condensation reactions. Ring opening with attendant formation of a chalcone, as shown below, has been postulated to occur rather rapidly in the presence of dilute (0·12 N) alkali (FUJISE and SASAKI, 1938). Products of lignin degradation such as those shown must be considered in any conclusions regarding the fate of metabolic products in diagenetic environments. However, it is felt that these substances are the progenitors of kerogen and not precursors of petroleum.

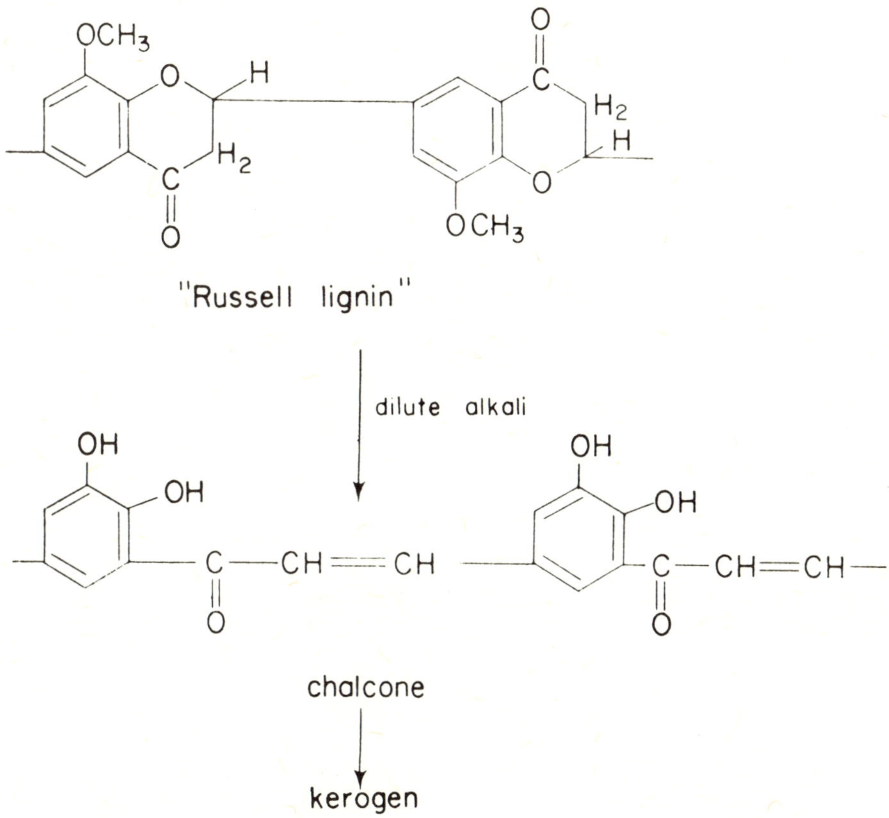

Pigments

Pigments of various types are produced by plants and animals and should be expected to contribute to sediments in which petroleum is formed. These pigments are numerous in nature and variety but for the sake of brevity, only flavonoids, carotenes, and porphyrins will be considered. Examples of the first two types are shown below; the structure of protoporphyrin-III has already been given.

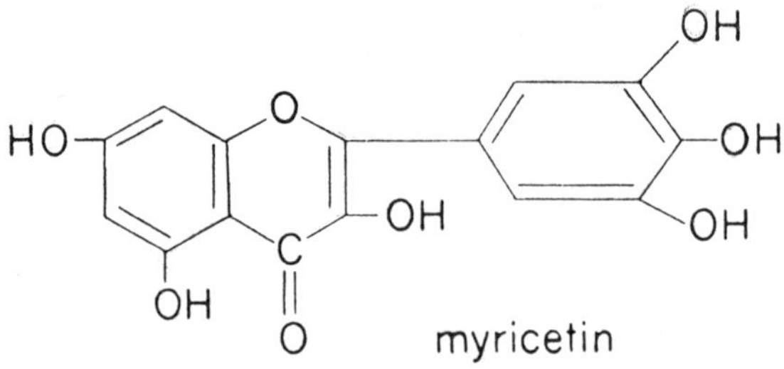

myricetin

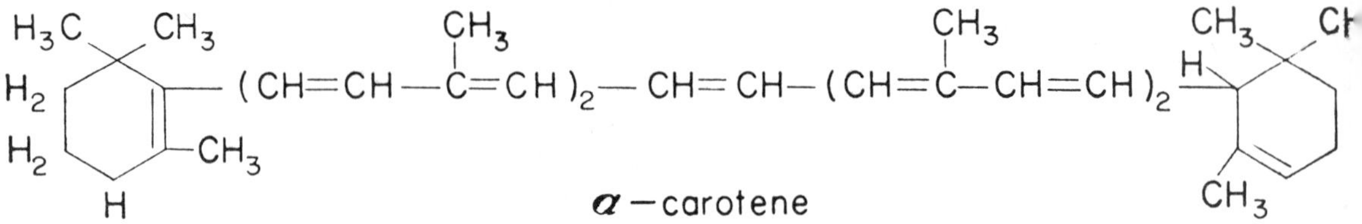

α-carotene

In general, the percentage of these pigments in biological substances are quite low and it may well be for this reason that flavonoids and carotenes have not been identified in marine sediments or crude oil even if they survive. Porphyrins, because of their high stability compared to other types of pigments, do survive and have been used successfully to demonstrate the low-temperature history of crude oil. Flavonoids, having the same basic structure as lignin, would be expected to hydrolyse to open-chain structures similar to those already discussed. Polyenes such as carotene, however, are extremely reactive and may well disappear by interaction with other substances or by intra- or intermolecular conversion. It may be noted in passing that β-carotene yields toluene, *m*-xylene, and 2:6-dimethylnaphthalene on heating at only 110°C. Degradation becomes understandable if the formula for β-carotene is written as shown on page 303 (KUHN and WINTERSTEIN, 1932).

Considering that minute quantities of amino acids and proteins have been found as residues of major biological products, it seems likely that residues of these pigments, which represent only minor products, will eventually also be identified.

Lipids

To this point the discussion has covered proteins, carbohydrates, lignin and pigments; a consideration of lipids, which represents a third major biological product, has been left until last because of their particular importance.

Biochemically the term "lipid" refers to those substances that are insoluble in water but soluble in ether, benzene or chloroform. They may be simple lipids, such as esters of fatty acids and alcohols, or of fatty acids with glycerol; compound lipids containing nitrogen, phosphorus, or sulphur; or derived lipids such

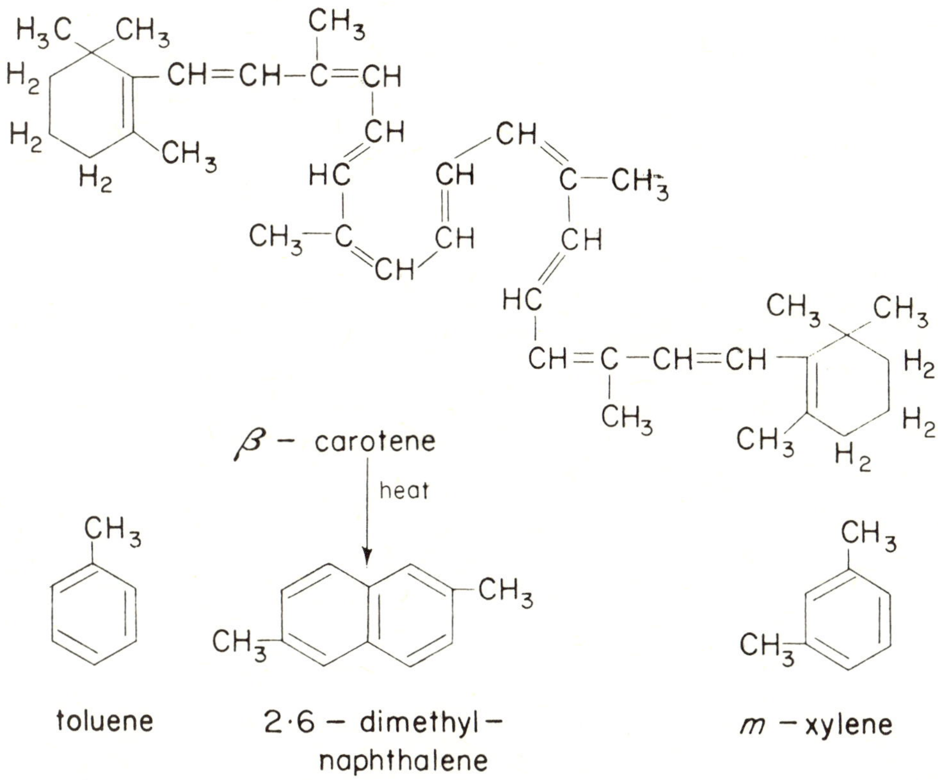

Thermal degradation of β-carotene

as fatty acids, sterols or alcohols. For purposes of this discussion it is necessary to consider only the derived lipids inasmuch as many of the other varieties eventually break down to these simpler forms.

Natural glycerides, triesters of glycerol with various saturated and unsaturated fatty acids, disappear rapidly from depositional environments, presumably as a result of biological processes. Again, as in the case of proteins and polysaccharides, the glycerides are readily hydrolysable compounds yielding glycerol and fatty acids. The glycerol then serves as an energy source for biological processes, but the fate of the fatty acids, both saturated and unsaturated, is less well understood.

Fats and fatty acids are derived biologically from carbohydrates and amino acids. These reactions are dependent upon a two-carbon radical mechanism and, normally, give rise to even-numbered carbon chains. On degradation of a lipid, then there should be a preponderance of even-numbered fatty acids over those with odd numbers of carbon atoms in the chain. Saturated acids containing from four to twenty-six carbon atoms have been found in fats, but those with twelve, sixteen or eighteen carbon atoms are in the majority. Many unsaturated acids containing from one to six double bonds in unconjugated or conjugated positions are also known (Cowan, 1953). A number of fatty acids, ranging from formic to nonanoic, have been isolated from crude oil (Hancock and Lochte, 1939).

The fate of fatty acids in a marine environment is not well understood. Rosenfeld (1948) found that unsaturation decreased with increase in depth of burial, but that bacterial dehydrogenation also was quite general. No clear-cut evidence for biological decarboxylation of fatty acids has been noted.

Chemical Conversion of Fatty Acids to Hydrocarbons

Inasmuch as fatty acids have basic hydrocarbon-like structures and apparently a relatively high degree of stability compared to other metabolic products, they present a class of substances of particular interest to the origin of petroleum. For this reason, it may be of some value to consider possible mechanisms by which such acids might be converted into the mixture of aliphatic, alicyclic, and aromatic hydrocarbons known to occur in crude oil. In any such consideration it is essential to recognize that biological degradation of metabolic products probably occurs very rapidly at the upper surface of uncompacted recent sediments. Following this initial decomposition, further reactions will most likely be chemical in nature. Because of the paucity of available information, much of what is to be suggested is of necessity speculative in nature.

The direct biochemical conversion of fatty acids to hydrocarbons in oil-forming quantities has not yet been reported. For this reason it is of considerable interest to note a number of potentially important enzymatically catalysed conversions that are known to occur in fatty acids and which may serve as important intermediate steps in petroleum synthesis.

Drawing an analogy from oxidative processes known to occur in the human liver, fatty acids might be converted by bacterial mechanisms into the following acids (Harper, 1957, pp. 196–198).

$RCH{=}CHCOOH$	α:β-unsaturated acid
$RCHOHCH_2COOH$	β-hydroxy acid
$RCOCH_2COOH$	β-keto acid

Human physiological reactions may also lead to the formation of keto or hydroxy acids, or acetone (Harper, 1957, pp. 196–198):

CH_3COCH_2COOH	acetoacetic acid
$CH_3CHOHCH_2COOH$	β-hydroxybutyric acid
CH_3COCH_3	acetone

These compounds, which represent intermediates in the oxidative degradation of fatty acids, are all more reactive compounds that the initial fatty acids and therefore represent substances which could enter into secondary reactions. Of particular interest in the potential ketonic cleavage of the β-keto acid in either weakly acidic or alkaline medium:

$$\underset{\beta\text{-keto acid}}{RCOCH_2COOH} \longrightarrow \underset{\text{ketone}}{RCOCH_3} + CO_2$$

This reaction could lead to decarboxylation of a fatty acid and its conversion to an active ketone containing less oxygen and, moreover, an odd number of carbon atoms, assuming the initial acid to have had the usual even-numbered chain. An additional example of decarboxylation is that known to occur anaerobically through the medium of bacterial decarboxylase:

$$\underset{\text{pyruvic acid}}{2CH_3COCOOH} + H_2O \rightarrow \underset{\text{acetic acid}}{CH_3COOH} + CO_2 + \underset{\text{lactic acid}}{CH_3CHOHCOOH}$$

In this case dismutation occurs leading to the formation of the α-hydroxy acid, lactic acid (PORTER, 1946).

To summarize then, although fatty acids may not be directly decarboxylated by bacteria, a number of changes are considered possible all of which result in oxygenated or dehydrogenated acids more reactive chemically that the parent substances. In addition to these new products, we must still consider those unsaturated acids initially available from the decomposition of glycerides within the overall picture of potential reactions leading to hydrocarbons.

One of the most powerful mechanisms for the interaction of the above substances is the Diels–Alder reaction in which products may be formed from the following type of system:

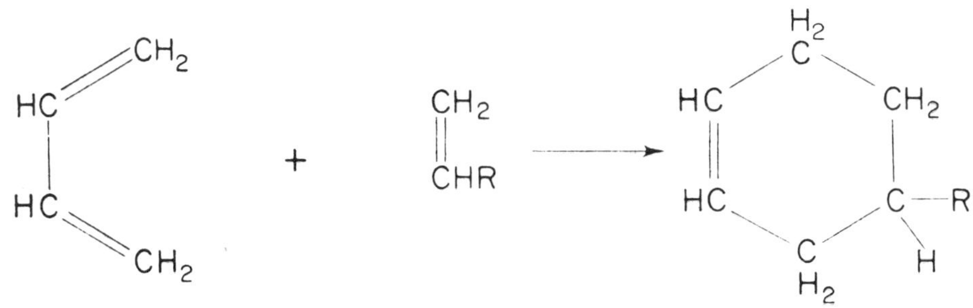

The R in this case usually represents an activating carbonyl (>C=O) group. As an example, the interaction between eleostearic acid from tung oil, and crotonic acid, would be expected to yield cyclic products. See top of page 306.

The Diels–Alder reaction is a purely chemical process operative at normal or elevated temperatures. This reaction can take place in hydrous media as shown by SOWA (1951), who found that α:β-unsaturated acids react readily with conjugated dienes at pH values below 7. See page 306.

Under a wide variety of conditions a broad spectrum of partially unsaturated ring compounds might be expected to result from the Diels–Alder reaction. Hydrogenation to alicyclic compounds or dehydrogenation to aromatic compounds is not inconceivable. There still remains, however, the problem of decarboxylating the products of this reaction to yield hydrocarbons. If the Diels–Alder reaction occurs rapidly in the zone of biological activity, then it is possible that the resultant acids are decomposed to ketones or aldehydes, in which case a mechanism for the reduction of the carbonyl group to a methylene (CH_2) or methyl (CH_3) group is necessary. No specific information on such a mechanism is available. Perhaps a more likely mechanism would be one in which the carbonyl

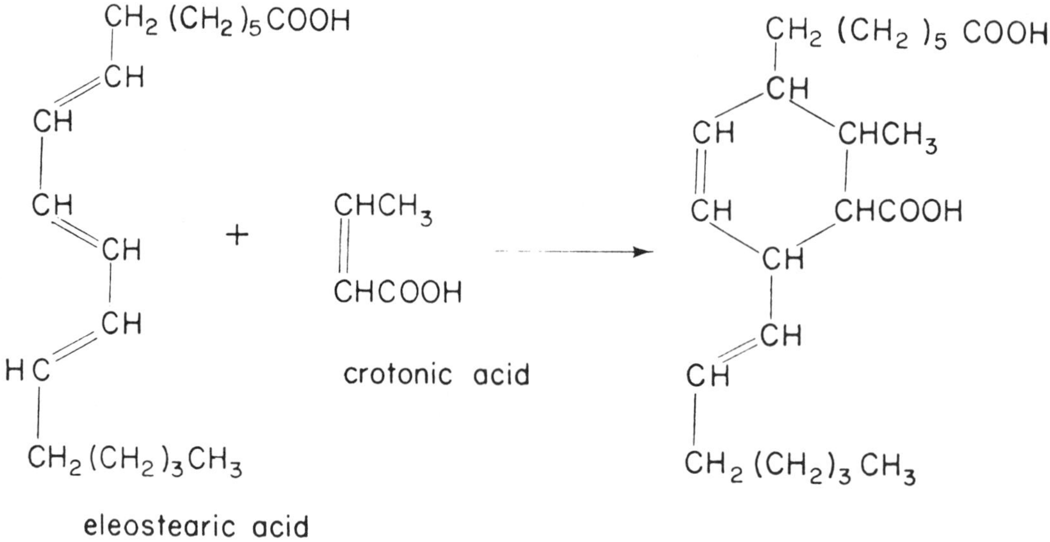

Example of a Diels–Alder reaction showing one of several possible products obtainable from eleostearic and crotonic acids.

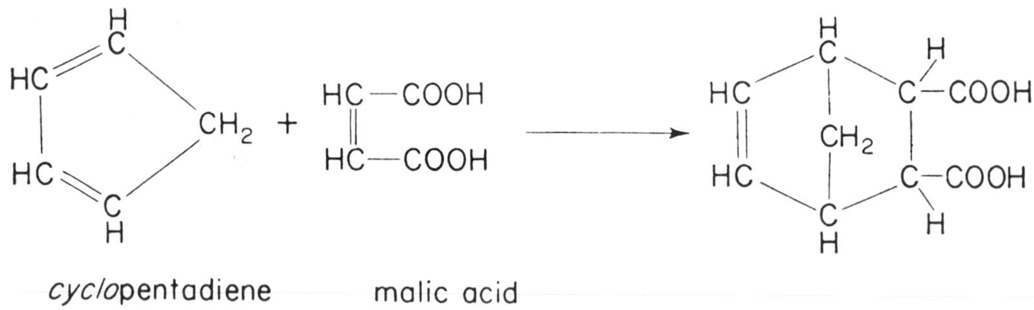

group is reduced to an alcoholic hydroxyl which then splits out as water to leave as unsaturated bond:

$$RCH_2CH_2OH \rightarrow RCH{=}CH_2 + H_2O$$

The availability of hydrogen for saturation of this bond is probable in biologically active Recent sediments.

$$RCH{=}CH_2 + H_2 \rightarrow RCH_2CH_3$$

Many studies of the polymerization of drying oils by the Diels–Alder reaction have been reported. As an example, linoleic and linolenic acids, components of vegetable oils, have been found to dimerize by the Diels–Alder reaction following which unreacted double bonds isomerize into reactive, conjugated positions (CHAMPETIER and PETIT, 1945).

DISCUSSION

Although the preceding discussion is highly speculative, it does lead to a number of suggestions regarding the origin of petroleum.

(1) Current consideration that the complex organic constituents of marine muds and shales are converted to hydrocarbons by unknown processes and sources of energy is subject to revision. The contributions of proteins and carbohydrates to marine sediments are probably very small in view of the labile nature of these compounds; the major contribution to shales may actually be of terrestrial origin, namely humic substances derived from lignin. Such substances probably have little oil-forming potential thus explaining why some highly organic shales, such as the Chattanooga shale, are not major source beds.

(2) The most likely contributors to the formation of crude oil are the fatty acids and related substances. Initial biological degradation and associated Diels–Alder reactions may convert such substances into straight-chain compounds with both odd and even numbers of carbon atoms, to alicyclic compounds, and even, upon dehydrogenation, into aromatic substances. The numerous possibilities inherent in interaction of diverse saturated and unsaturated compounds could easily lead to a complex mixture of structures such as is found in crude oil.

(3) Many biochemical and chemical reactions, such as those that have been noted, result in products that still contain oxygen in the form of carbonyl, hydroxyl or carboxyl groups. Degradation of carboxyl groups to keto or hydroxy acids may be followed by decarboxylation. In view of these considerations, it would seem more profitable to search for those biochemical or chemical mechanisms able to reduce carbonyl or hydroxyl groups rather than for those that decarboxylate. According to BEERSTECHER (1954), bacterial hydrogenation of ketones, dehydrations, and the reduction of double bonds are all common reactions, but no specific examples are cited.

(4) Occurence of small quantities of hydrocarbons in recent sediments may be related to the fact that bacteria are able to produce traces of such material as part of their protoplasm (ZOBELL, 1951). The predominance of odd-numbered hydrocarbons over those with even-numbered chains may indicate a metabolic process in which a β-keto acid has been decarboxylated and the resulting ketone has been reduced. In general, however, reports of bacterial production of hydrocarbons are difficult to evaluate and not too convincing as far as the origin of petroleum is concerned (BEERSTECHER, 1954).

(5) Small amounts of condensed aromatic hydrocarbons can be formed by thermal degradation of carotenoid pigments, but it is considered more plausible that aromatics are formed by Diels–Alder mechanisms. The thermal history of a sediment and its accompanying oil cannot, however, be disregarded, for the rates of chemical reactions double for every 10°C rise in temperature.

(6) Based on these considerations it seems most likely that crude oil is generated in sediments containing high concentrations of lipids. If this is so, then petroleum could be formed in marine, brackish, or fresh-water environments, and the variety of hydrocarbons produced would be to a great extent dependent upon the contributory lipids. In this respect perhaps the paraffinic character of old oils and the naphthenic and aromatic character of young oils are related to variations in saturated and unsaturated lipids.

Numerous degradation products other than those mentioned in this paper have at one time or another been suggested as intermediates. Although some intermediates may merit consideration, in general it is felt that those structures discussed above are by far the most important. Limitations in the scope of this paper have also prevented a discussion of sources of energy, such as heat and pressure, or of catalytic effects that may have influenced the course and speed of the reaction.

Acknowledgment—Many thanks are due to A. M. Pommer of the U.S. Geological Survey for his helpful comments and discussions. This report concerns work done by the U.S. Geological Survey on behalf of the Division of Research, U.S. Atomic Energy Commission.

References

Abelson P. H. (1957) Some aspects of paleobiochemistry. *Ann. N.Y. Acad. Sci.* **69,** 276–285.

Beerstecher Ernest Jr. (1954) Petroleum microbiology pp. 109–112. Elsevier, Houston.

Champetier G. and Petit J. (1945) Thermal polymerization of drying oils. Study of the products of saponification and ethanolysis of linseed stand oils. *Bull. Soc. chim. France.* **12,** 680–688.

Cowan J. C. (1953) Lipids in Organic Chemistry, an Advanced Treatise, (Edited by Gilman) (2nd Ed.) Vol. 3, pp. 178–217. John Wiley, New York.

Fujise S. and Sasaki H. (1938) Optically active flavanones—111. Asymmetric synthesis of a hydroxyflavone from a hydroxychalcone. *Ber. dsch. chem. Ges.* **71B,** 314–344.

Hancock K. and Lochte H. L. (1939) Acidic constituents of a California straight-run gasoline distillate. *J. Amer. Chem. Soc.* **61,** 2448–2452.

Harper H. A. (1957) *Review of physiological chemistry*, (6th Ed). Lange Medical Publications, Los Altos, California.

Kuhn R. and Winterstein A. (1932) Thermal decomposition of carotenoid pigments. Communication XXV on conjugated double bonds. *Ber. Dsch. chem. Ges.* **65,** 1873–1880.

Porter J. R. (1946) Bacterial chemistry and physiology. John Wiley, New York.

Rosenfeld W. D. (1948) Fatty acid transformation by anaerobic bacteria. *Arch. Biochem.* **16,** 263–273.

Russell A. (1948) Interpretation of lignin—I. The synthesis of gymnosperm lignin. *J. Amer. Chem. Soc.* **70,** 1060–1064.

Sowa F. J. (1951) Diels–Alder reactions in aqueous media. Brit. pat. 657911.

Theander Olof (1952) Sphagnum peat—I. Preliminary studies on the carbohydrate constituents. *Svensk Kem. Tidskr.* **64,** 197–199.

Vallentyne J. R. and Whittaker J. R. (1956) On the presence of free sugars in filtered lake water. *Science*, **124,** 1026–1027.

Zobell C. E. (1951) Contributions of bacteria to the origin of oil. *Proceedings of the Third World Petroleum Congress, The Hague*, 1951. Sec. 1. pp. 414–420.

36

Reprinted from *Geochim. et Cosmochim. Acta* **29**:1021–1049 (1965), with the permission of Microforms International Marketing Corporation as exclusive copyright licensee of Pergamon Press journal back files

On the depth, time and mechanism of petroleum generation*

G. T. Philippi
Shell Development Company, Houston, Texas

(*Received* 4 *January* 1965; *in revised form* 5 *April* 1965)

Abstract—The subsurface petroleum generation process has been studied. In the Los Angeles and Ventura basins, California, hydrocarbon content and hydrocarbon/noncarbonate carbon ratio increase slowly with depth and age in upper and lower Pliocene shales and more rapidly in the deeper and warmer upper Miocene shales. During the oil generation process, and simultaneous with the increase of the hydrocarbon/noncarbonate carbon ratio, the composition of the shale normal paraffins and naphthenes boiling above 325°C gradually changes and eventually becomes very similar to that of the waxy oils of the basin.

In the Los Angeles basin as well as in the Ventura basin the bulk of petroleum is generated at depths where the subsurface temperature is above 115°C and where, therefore, the shales are sterile. The conclusion that petroleum is formed from sediment organic matter essentially by thermal, nonbiological processes is inescapable. Because petroleum generation is a chemical process, it is strongly temperature-dependent. This explains the strong increase of the hydrocarbon/noncarbonate carbon ratio at greater depths, where the temperature is higher. It also explains the greater depth of oil generation and of equality of shale and oil hydrocarbon composition in the Ventura basin compared with the Los Angeles basin. This is caused by the smaller temperature gradient in the Ventura basin. The importance of the temperature history of potential oil source rocks is discussed. Appreciable amounts of oil are formed only if the temperature history of the source sediments is sufficiently intense and if the nature and the amount of the organic source material are appropriate.

In the Los Angeles and Ventura basins most producing horizons are much shallower than the oil source beds, and vertical migration of petroleum is very common. Evidence is presented that vertical migration of petroleum takes place through faults and fractures, not through the bulk of the shales.

Introduction

For many years controversies have existed in the literature about the mechanism of petroleum generation. Is petroleum formed biochemically? Is it derived partially from the cell material of the organisms which contributed organic matter to the source sediments and partially from the cells of bacteria which may live in the pores of the sediments to appreciable depths? Is petroleum formed from the organic matter of sediments by nonbiochemical chemical reactions? Is petroleum formed relatively early or late in the geologic sequence, and how long does it take? Questions such as these persist in the literature and have not been satisfactorily answered because of lack of adequate geochemical data. This paper is a report of a study which aids the solution of such questions.

Since the mid-1940's we have observed that certain shales and argillaceous limestones from oil basins have appreciable indigenous hydrocarbon contents up to several thousand parts per million (Philippi, 1956). A method to determine whether the hydrocarbons extracted from a rock sample are indigenous or not was described in this publication. The indigenous hydrocarbon content of sedimentary

* EPR Publication No. 387.

rocks from oil basins with a normal noncarbonate carbon content in the range of 0·5–5 per cent varies considerably, but an upper limit exists of the order of magnitude of a few thousand parts per million by weight for hydrocarbons boiling above 325°C. Others have also demonstrated the presence of small quantities of hydrocarbons in shaly sediments (HUNT and JAMIESON, 1956; HUNT, 1961).

Table 1. Average noncarbonate carbon content, content of indigenous hydrocarbons boiling above 325°C, and hydrocarbon/noncarbonate carbon ratio of Recent sediments (after BLUMER, 1958)

Environment (location)	Average noncarbonate carbon content (%)	Average hydrocarbon content (ppm)	Hydrocarbons/ noncarbonate carbon
Soils			
(Texas)	1·2	60	0·0050
Peats			
(Florida)	37·0	350	0·0009
Fresh-Water Lake			
(Grand Lake, Louisiana)	0·3	41	0·0137
Near-Shore Mud Flat			
(Sabine Pass, Texas)	0·9	23	0·0026
Lagoonal			
(Laguna Madre, Texas)	0·4	20	0·0050
Carbonate Muds			
(Florida)	1·2	20	0·0017
Interdeltaic			
(Louisiana)	0·8	20	0·0025
Deltaic			
(Mississippi Delta)	1·2	80	0·0067
Prodeltaic			
(Mississippi Delta)	0·8	80	0·0100
Open Marine Shelf			
(Gulf of Mexico)	0·9	30	0·0033
Deep Marine Basin			
(Chubasco Trench, Mexico)	1·7	23	0·0014
(Channel Islands Area, California)	2·1	100	0·0048
(Cariaco Trench, Venezuela)	3·4	140	0·0041

From a study of Recent sediments for API Project 4, entitled "Origin and Environment of Source Sediments", TRASK and WU (1930) concluded that "liquid petroleum probably does not occur in sediments at the time of their deposition, and that if it is present, it is in very small amounts, certainly less than 3 parts per 100,000". Although no accurate hydrocarbon analyses of sediments were feasible in 1930, the inference of TRASK and WU, in a general way, was correct and has since been confirmed. SMITH (1954) was the first to recognize and measure quantitatively the occurrence of very small quantities of hydrocarbons in Recent sediments. For three Orinoco Delta Recent sediments, SMITH found hydrocarbon values of 25–82 parts per million by weight, based on dry sediment. For four deep-water marine sediments off the coast of West Africa, hydrocarbon values of 28–36 parts per million were found. Similar low hydrocarbon contents of Recent sediments have been observed in this laboratory by BLUMER (1958) (see Table 1).

In Table 2 examples are presented of several ancient sediments with a moderately high to high indigenous hydrocarbon content. Comparison of the data of Table 2 and Table 1 shows that the indigenous hydrocarbon content of several ancient sediments is much higher than that of Recent sediments. Even more significant is the fact that for many ancient sediments the values of the hydrocarbon/noncarbonate carbon ratio are much higher. The large difference in

Table 2. Noncarbonate carbon content, content of indigenous hydrocarbons boiling above 325°C, and hydrocarbon/noncarbonate carbon ratio of some ancient sediments

Age and formation (location)	Average noncarbonate carbon content (%)	Average hydrocarbon content (ppm)	Hydrocarbons/ noncarbonate carbon
U. Mio.—Div. D & E (Ventura and Los Angeles Basins, California)	3·12	1577	0·050
Mio.—Telissa (Sungei Taham Field, S. Sumatra)	0·89	896	0·101
Cret.—2nd White Specks (Alberta, Canada)	2·02	1572	0·078
Cret.—Graneros (Northeast Colorado)	3·08	1239	0·040
Cret.—LaLuna (La Paz Field, Venezuela, S.A.)	3·67	2360	0·064
Cret.—LaLuna Equivalent (Casabe Field, Colombia, S.A.)	1·54	1146	0·074
Perm.—Leonard (Delaware Basin, Texas)	1·66	875	0·053
Miss.—Lodgepole (Eastern Montana)	0·79	267	0·034
Ord. —Winnipeg (Eastern Montana)	0·53	394	0·074

hydrocarbon/noncarbonate carbon ratio of certain ancient sediments and of Recent sediments strongly suggests that the generation of petroleum is essentially a subsurface process. The depth at which this process takes place and its mechanism and duration have never been clearly understood. The following sections contribute to the solution of these problems.

The Composition of the High-Boiling Normal Paraffins from Recent Sediments

Smith (1954) has shown that the hydrocarbons from Recent sediments contain the three major types—aromatics, naphthenes and paraffins—as do crude oils. Although *gross* composition of Recent sediment hydrocarbons resembles that of crude oils, a closer study of their detailed composition reveals very significant differences. Stevens, Bray and Evans (1956) and Bray and Evans (1961) were the first to point out the extreme difference in composition of the normal paraffins of the C_{22}–C_{34} carbon-number range from Recent sediments on the one hand and of those from crude oils and ancient sediments on the other hand. They observed that in Recent sediments normal paraffin molecules with an odd number of carbon atoms strongly predominate over those with even numbers, whereas, in contrast, the normal paraffins from crude oils and several ancient sediments have approximately equal concentrations of odd- and even-carbon-number molecules.

The explanation of the strong odd-carbon-number predominance of the normal paraffins from Recent sediments is that the normal paraffins synthesized by many living organisms contain practically only normal paraffin molecules with odd carbon numbers. As an example, the normal paraffins from practically all natural waxes which have been analyzed show a strong odd-carbon-number predominance (WARTH, 1956). Because of their biologic origin, the normal paraffins from Recent sediments, which are a mixture derived from many species of living organisms, also exhibit a strong odd-carbon-number predominance. The fact that the normal paraffins from many ancient sediments and from crude oils are so very different from Recent sediment paraffins and exhibit smooth carbon-number distribution curves is due primarily to the oil generation process, as will be explained in a later section.

THE RING-NUMBER DISTRIBUTION OF THE HIGH-BOILING NAPHTHENES FROM RECENT SEDIMENTS

In addition to normal paraffins, the high-boiling naphthenes from Recent through upper Miocene sediments have been studied because of their importance in the problem of the origin of oil. For this purpose first an isoparaffin-naphthene concentrate was prepared by carefully removing the aromatics by liquid chromatography from the total hydrocarbons extracted from the sediment, and then the normal paraffins were removed by urea adduction. The isoparaffin-naphthene concentrates obtained in this way were distilled in an efficient microstill, designed by BLUMER (1962), into the following fractions:

Fraction 1	*Fraction 2*	*Fraction 3*	*Fraction 4*
–325°C	325–370°C	370–420°C	420–470°C

Because fraction 1 suffers from evaporation losses as a result of the experimental procedure, only fractions 2, 3 and 4 have been investigated in detail. They represent the C_{19}–C_{32} carbon-number interval, which is within the light-to-medium lubricating oil range. Isoparaffin-naphthene fractions 2, 3, and 4 were analyzed by mass spectrometric parent peak methods (O'NEAL, 1954; O'NEAL *et al.*, 1955; SCHISSLER *et al.*, 1957; HOOD and O'NEAL, 1959). Thus, the percentages of isoparaffins (zero rings) and of naphthenes with one ring, two rings, three rings etc., up to eight rings per molecule have been determined. Rings per naphthenic molecule were determined with the methods used.

Results of the study of four different Recent environments are presented in Fig. 1. Numerical data are shown in Table 3. These results indicate that the composition of the naphthene-isoparaffin concentrates varies considerably with environment of deposition. It is now known that the protoplasm of many (probably most) organisms contains extremely small quantities of hydrocarbons which differ in composition for different species (SIRAHANA, 1938; SMITH, 1954; LARSEN and HAUG, 1956; STEVENS, BRAY and EVANS, 1956; BLUMER, MULLIN and THOMAS, 1963). Thus, the hydrocarbons from Recent sediments are derived from

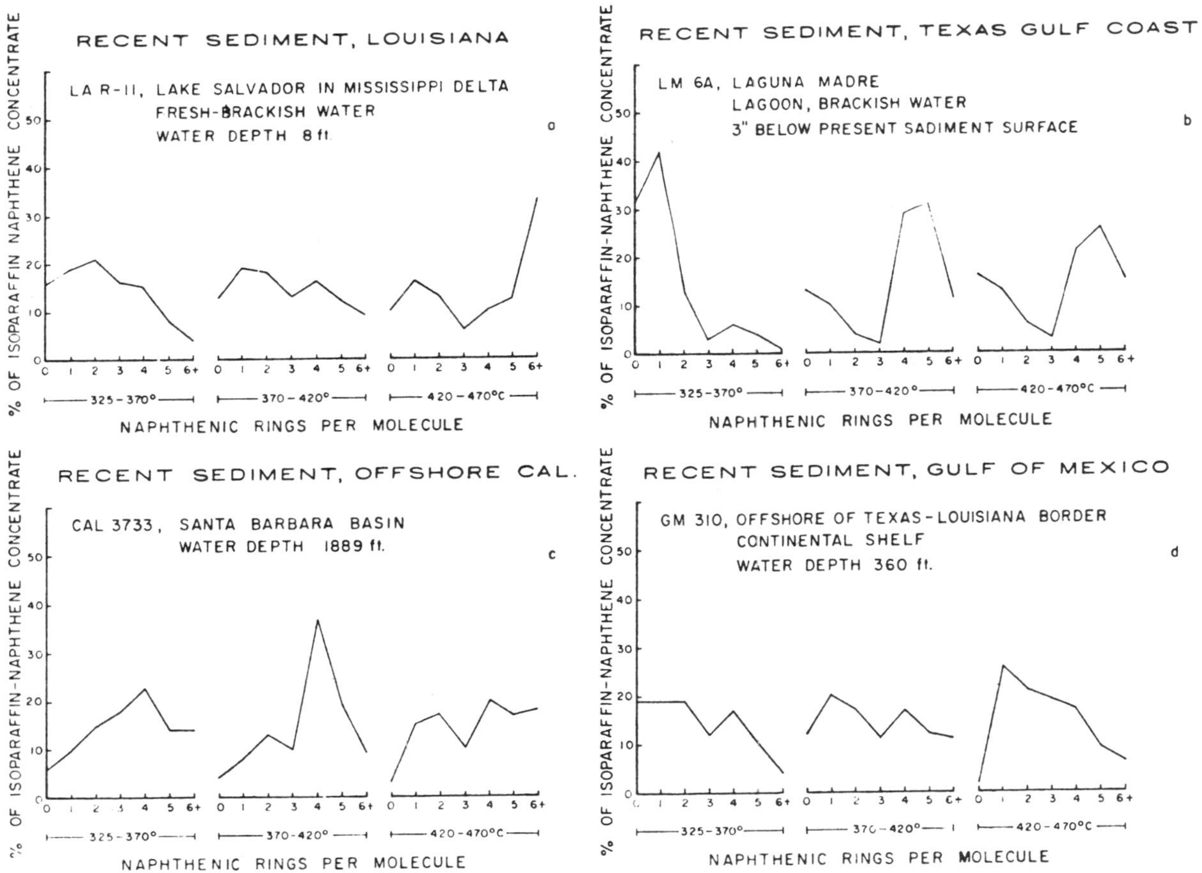

Fig. 1. Per cent of isoparaffins (zero rings) and of naphthenes with one ring, two rings, three rings etc., per molecule in isoparaffin-naphthene concentrates from Recent sediments.

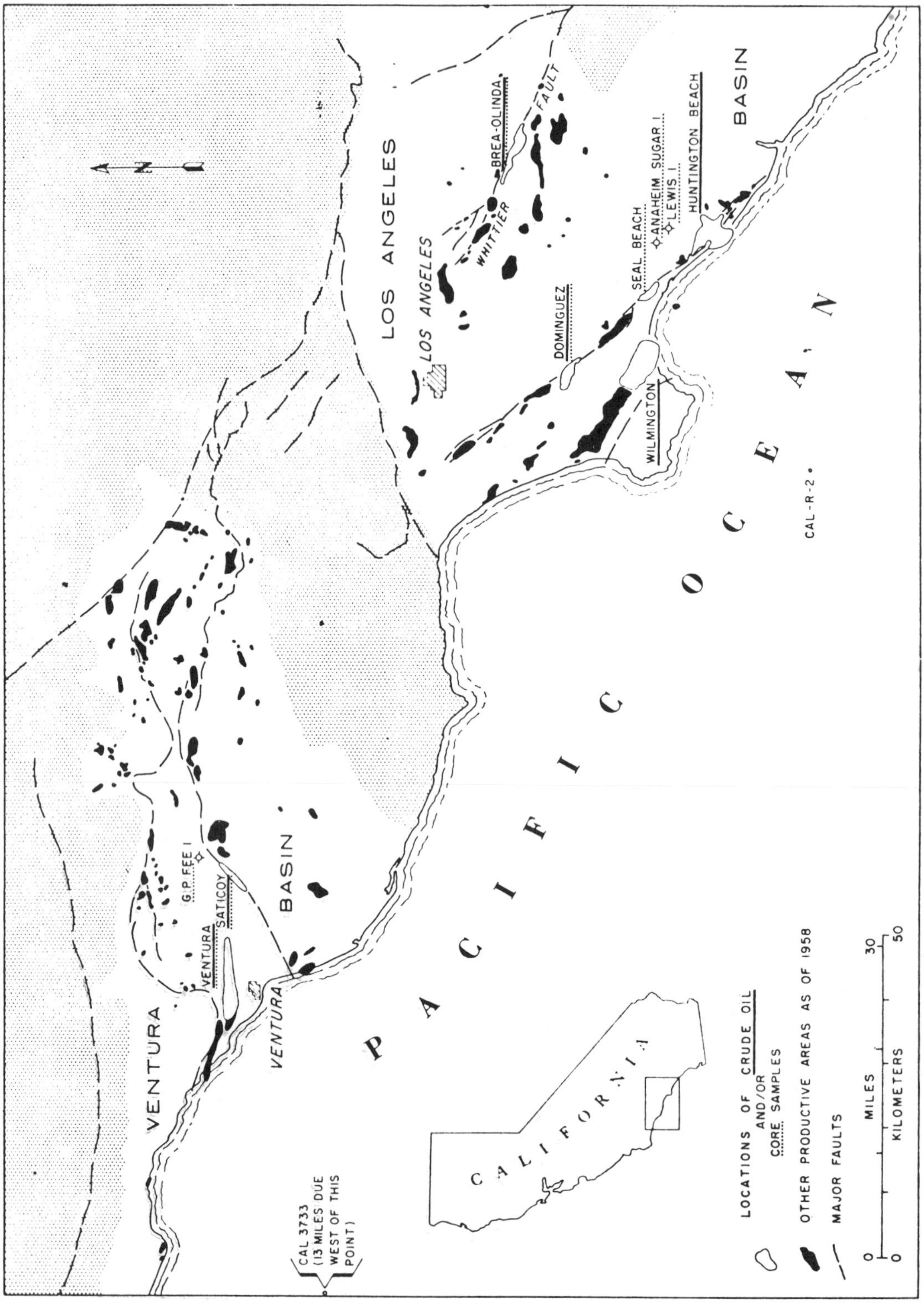

Fig. 2. Map of the oil fields of the Los Angeles and Ventura basins, California.

the protoplasm of the many types of organisms, including bacteria, which live in the particular environment of deposition and from terrestrial plant matter which has been washed in. Therefore, the considerable differences in hydrocarbon composition of the four Recent environments shown in Fig. 1 are caused basically by different environmental conditions and ecology.

Subsurface Generation of Petroleum in the Los Angeles and Ventura Basins of California

Figure 2 is a map of the oil fields of the Los Angeles and Ventura basins of California. On this general map the names have been indicated of the fields and

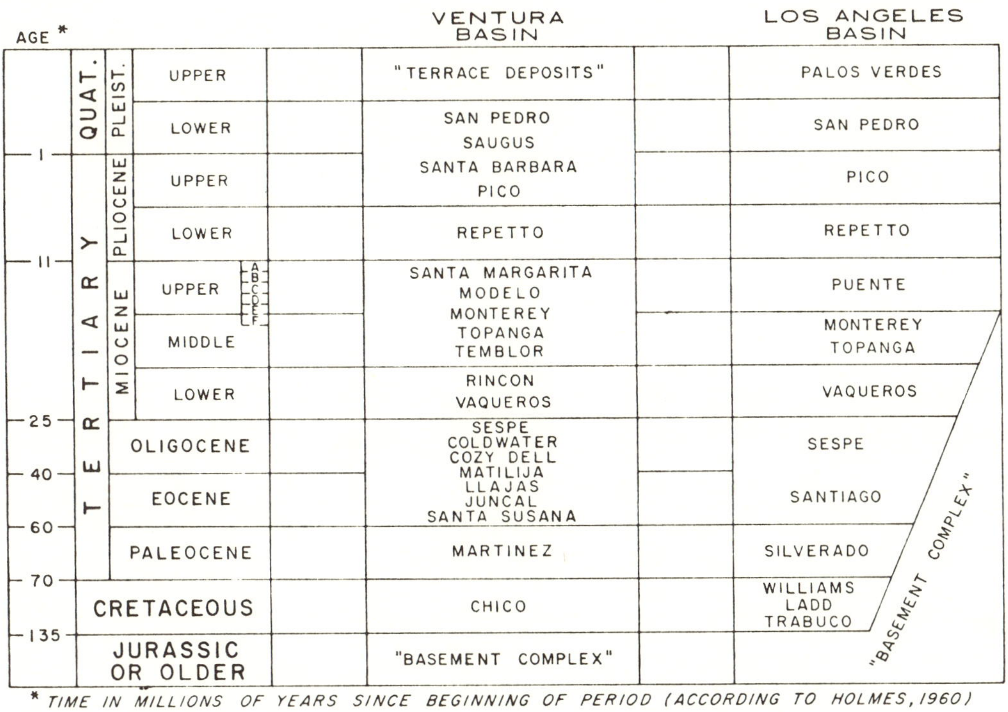

Fig. 3. Stratigraphic subdivisions and generalized correlation of formation names in the Los Angeles and Ventura basins, California.

wildcat wells which are discussed in the text of this paper. Figure 3 shows the formation names and the stratigraphic subdivisions used in Los Angeles and Ventura basin geology.

All subsurface samples discussed in this paper are conventional core samples which have been scraped and cleaned thoroughly before analysis. Only those samples are reported for which the indigenous nature of the hydrocarbons has been determined.

The graphs of Figs. 4 and 5 show the changes with increasing depth and age for indigenous hydrocarbons boiling above 325°C of (1) normal paraffin composition,

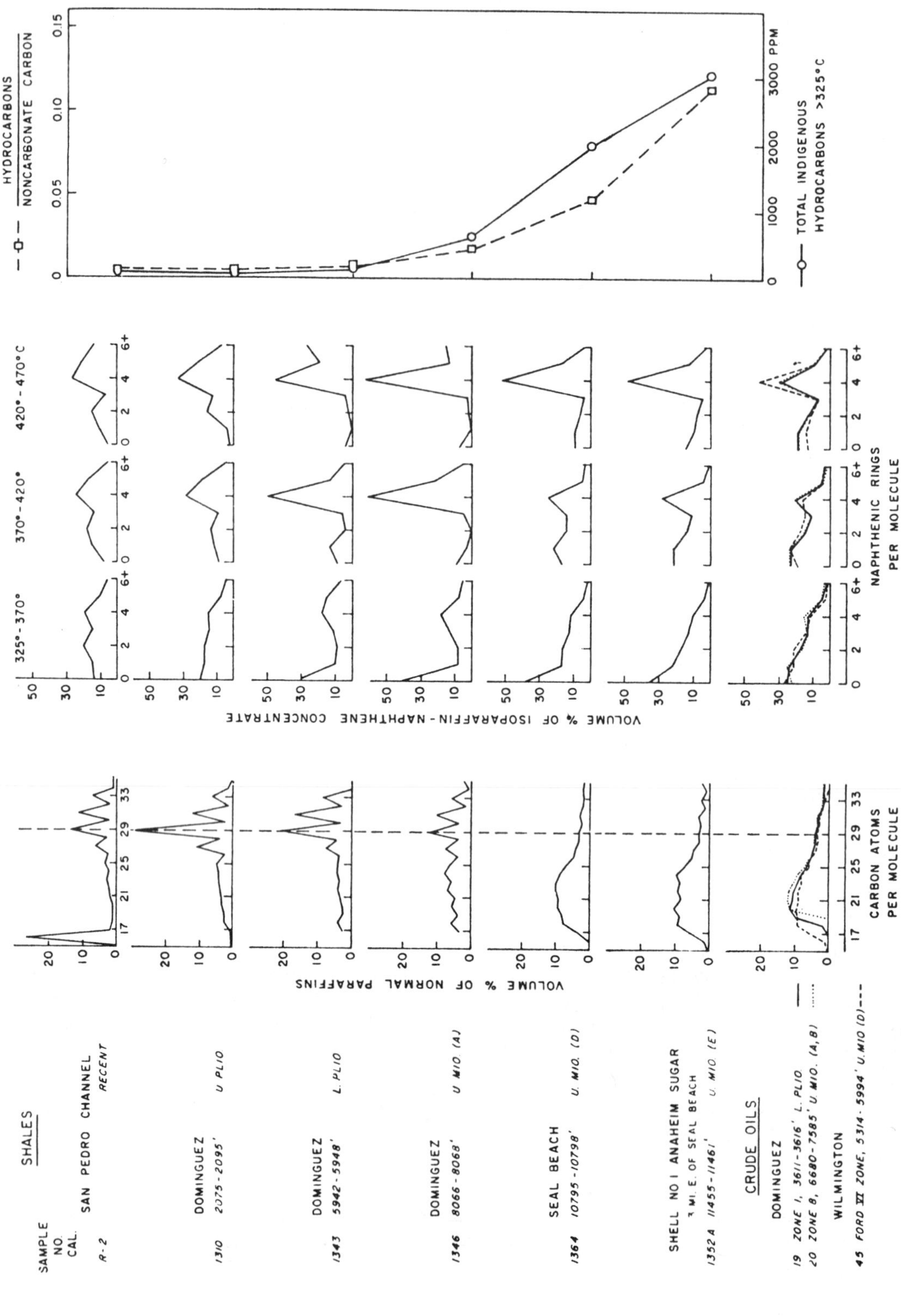

Fig. 4. Origin and development of petroleum paraffins and naphthenes from nonhydrocarbon organic matter as a function of depth of burial and time in the Los Angeles Basin, California.

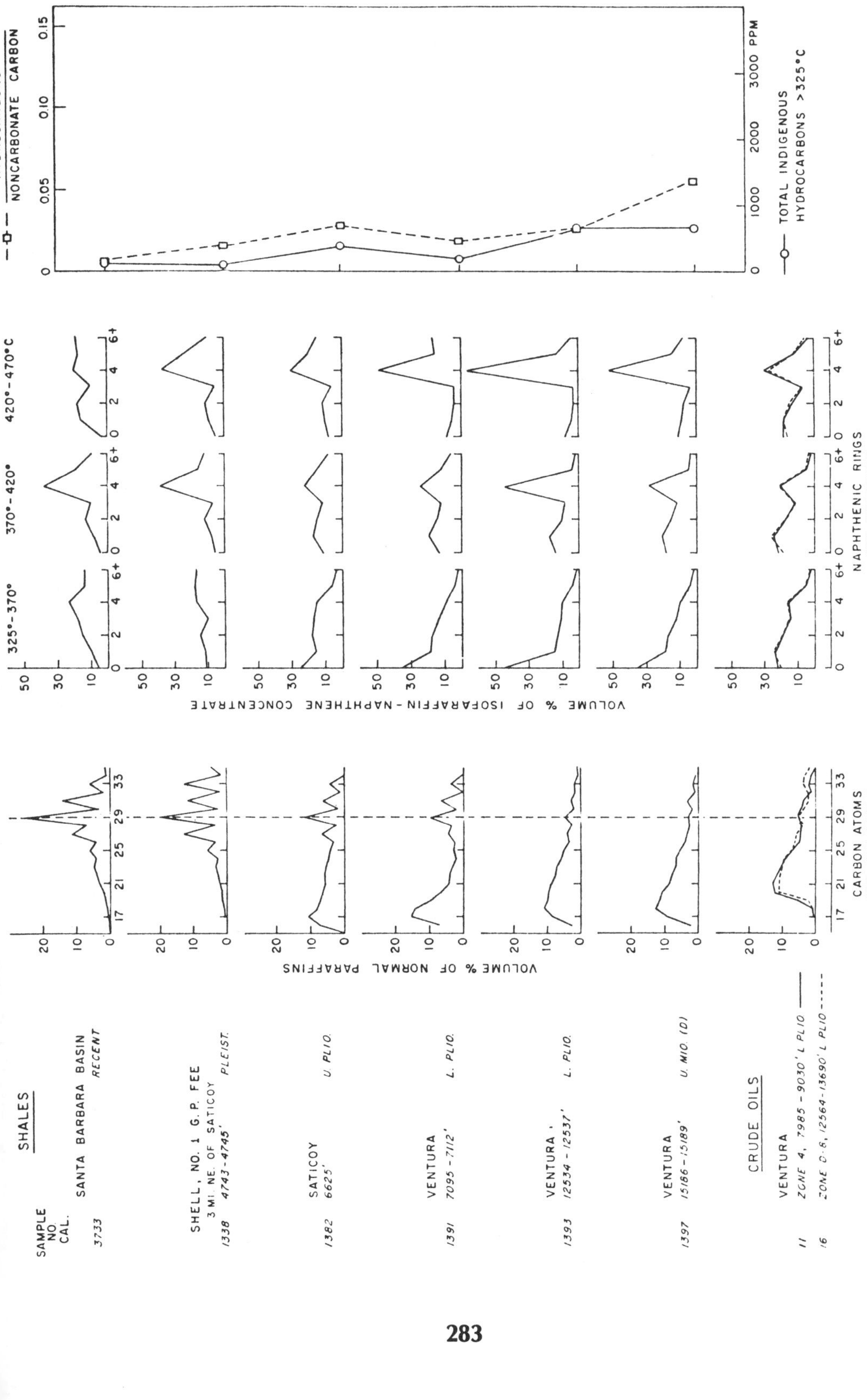

Fig. 5. Origin and development of petroleum paraffins and naphthenes from nonhydrocarbon organic matter as a function of depth of burial and time in the Ventura Basin, California.

(2) isoparaffin-naphthene composition, (3) hydrocarbon/noncarbonate carbon ratio, and (4) total hydrocarbon content. The subsurface samples analyzed are representative shales from the Los Angeles and Ventura basins and range in age from Pleistocene, through upper and lower Pliocene, down to upper Miocene division E. For comparison, at the tops of Figs. 4 and 5 data for Recent samples from offshore California have been included. Recent sample locations are shown on Fig. 2. Also for comparison, the normal paraffin and isoparaffin-naphthene distributions of some typical crude oils have been plotted at the bottoms of Figs. 4 and 5.

Change of the Hydrocarbon/Noncarbonate Carbon Ratio during Subsurface Oil Generation

The graphs shown at the extreme right of Figs. 4 and 5 demonstrate that both hydrocarbon content and hydrocarbon/noncarbonate carbon ratio increase strongly with depth of burial and age of the sediments. However, the increase is quite slow. It is not until well into the upper Miocene shales (divisions D and E, approximately 15 million years old according to the time scale of HOLMES, 1960) that enough hydrocarbons have been formed to raise the hydrocarbon/noncarbonate carbon ratio to the range of 0·030–0·120 which we have often observed in ancient oil source rocks (see Table 2).

In Fig. 6 gross hydrocarbon composition is plotted for each of the sediment samples and oil samples of Figs. 4 and 5. Gross hydrocarbon composition is defined as the percentage of total aromatics, the percentage of total isoparaffins plus naphthenes, and the percentage of paraffin wax (urea adduct, mostly normal paraffins). Figure 6 demonstrates that there is little, if any, clearcut change with depth and age of the gross hydrocarbon composition. This shows that during the subsurface oil generation process high-boiling aromatics, isoparaffins plus naphthenes, and paraffin wax are all formed and in approximately constant ratios. The combined information of Fig. 6 and of the graphs shown at the extreme right of Figs. 4 and 5 indicates that *the content of aromatics, of isoparaffins plus naphthenes, and of paraffin wax increases strongly with depth and age of the sediments.* This holds true for the Los Angeles basin as well as for the Ventura basin.

Change in Composition of Sediment Normal Paraffin Hydrocarbons during Subsurface Oil Generation

Although *gross* composition of the sediment hydrocarbons boiling above 325°C does not seem to vary appreciably during the oil generation process, a closer study of the *detailed* composition reveals very significant and systematic changes with age and depth. The very striking difference in composition of the normal paraffins from Recent sediments and from ancient sediments and crude oils has been demonstrated by STEVENS, BRAY and EVANS (1956) and BRAY and EVANS (1961) and has been discussed in a previous section. ROBINSON, CUMMINS and DINNEEN (1963) observed a decrease with depth of the odd/even ratio of the normal paraffins in a short, 900 ft section of Eocene Green River oil shale. In the present paper the gradual transition from one type of normal paraffin distribution to the other during the oil generation process will be discussed. At the top of Fig. 4, left, the normal paraffin distribution curve of a typical Recent sample from the San Pedro Channel, offshore from the Los Angeles basin, has been plotted. Similarly, at the top of Fig. 5, left, a normal paraffin distribution curve is shown of a Recent sample of the Santa Barbara basin, offshore from the Ventura basin. From Figs. 4 and 5, it appears that in the Ventura basin in marine lower Pleistocene shales at 4743–4745 ft and in the Los Angeles basin even in the lower Pliocene shales at 5942–5948 ft, hardly any change has taken place in the composition of the normal paraffins (BLUMER, 1958). However, as hydrocarbons begin to form, as indicated by the

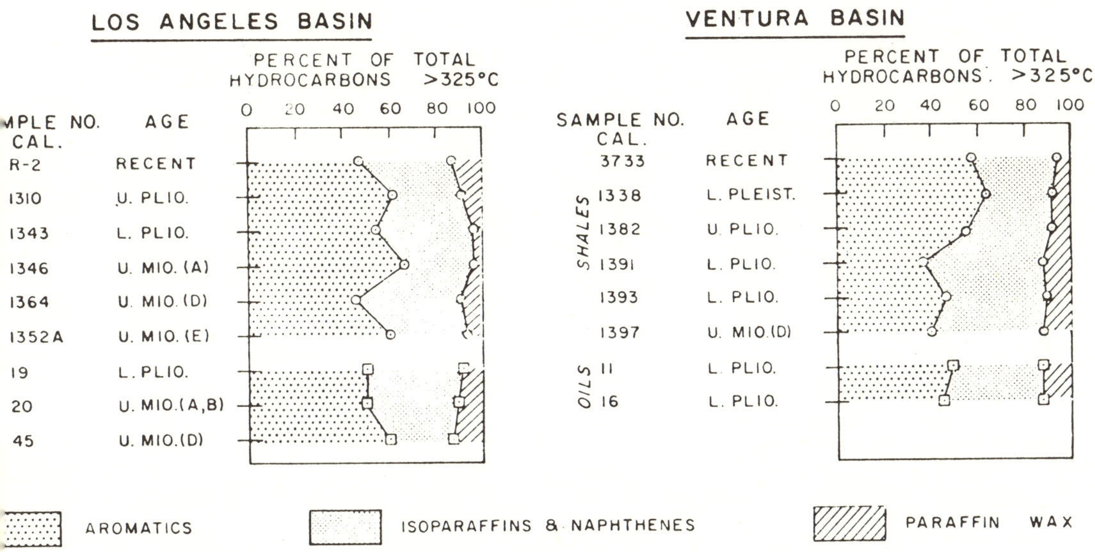

Fig. 6. Variation of gross composition of sediment hydrocarbons with depth and age and comparison with typical waxy crude oils of the same area.

increase of the hydrocarbon/noncarbonate carbon ratio, the composition of the normal paraffins begins to change. Examples are the upper and lower Pliocene samples from the Saticoy and Ventura fields of the Ventura basin (Fig. 5) and the upper Miocene division A sample from the Los Angeles basin. In these deeper and geologically older samples the oil generation process has started but is by no means completed. Normal paraffins in the C_{18}–C_{22} interval have formed, and the odd-carbon-number predominance in the C_{27}–C_{33} bracket has been reduced simultaneously but is still more pronounced than in crude oil.

As was pointed out in the discussion of Fig. 6, during the oil generation process the wax normal paraffins maintain approximately a constant proportion of the total hydrocarbons boiling above 325°C. Owing to the strong increase of the concentration of the total hydrocarbons boiling above 325°C, the concentration of the normal paraffins in the shales also increases strongly with depth and time. Because the subsurface oil generation process is thermal and nonbiological (see page 1042) and lacks the specificity of a biochemical process, the normal paraffins formed have neither an odd- nor an even-carbon-number predominance. Although about equal amounts of adjoining odd- and even-carbon-number normal paraffins are formed,

Table 3. Analysis of indigenous hydrocarbons boiling above 325°C from shales of

Sample No.	Field, Wildcat Well, or location	t Age	Depth (ft)	Non-carbonate carbon %	CO_2 as % $CaCO_3$	Hydrocarbons >325°C (ppm)	Hydrocarbons/non-carbonate carbon	Paraffin wax $\frac{2C_{29}}{C_{28} + C_3}$
Cal.			*Los Angeles Basin Shale Samples*					
R-2	San Pedro Channel	Recent	Water Depth 2800	1·91	10·8	91	0·0048	5·1
1310	Dominguez	U.Plio	2075–2095	1·59	2·5	67	0·0042	9·3
1378	Dominguez	L.Plio	3800–3802	1·71	10·9	201	0·012	7·5
1343	Dominguez	L.Plio	5942–5948	1·97	12·1	139	0·0071	6·1
1374	Dominguez	L.Plio	5980	2·11	9·9	181	0·0086	6·2
1375	Dominguez	U.Mio(A_2)	6650	2·98	10·2	276	0·0093	5·9
1376	Dominguez	U.Mio(A_5)	7137–7144	2·00	6·6	235	0·012	4·5
1346	Dominguez	U.Mio(A)	8066–8068	3·50	7·6	641	0·018	3·5
1347	Shell No. 1 Anaheim Sugar	L.Plio	9019–9025	4·06	6·4	707	0·017	2·8
1348	Shell No. 1 Anaheim Sugar	U.Mio(A)	9317–9320	4·72	5·7	832	0·018	2·7
1349	Shell No. 1 Anaheim Sugar	U.Mio(A)	9613–9630	3·38	4·7	1138	0·034	1·42
1364	Seal Beach	U.Mio(D)	10,795–10,798	4·26	33·2	1985	0·047	1·17
1351	Shell No. 1 Anaheim Sugar	U.Mio(D)	11,000–11,003	4·32	7·3	2697	0·062	0·84
1352A	Shell No. 1 Anaheim Sugar	U.Mio(E)	11,455–11,461	2·68	31·0	3032	0·113	1·00
			Ventura Basin Shale Samples					
3733	S. Barbara Basin	Recent	Water Depth 1889	2·27	—	133	0·0059	4·8
1338	Shell No. 1 G. P. Fee	Pleist	4743–4745	0·76	8·5	116	0·015	6·0
1382	Saticoy	U.Plio	6625	1·38	15·6	385	0·028	4·8
1385	Saticoy	U.Plio	7376–7378	1·54	12·0	464	0·030	5·1
1391	Ventura	L.Plio	7095–7112	1·05	13·4	202	0·019	3·8
1387	Saticoy	U.Plio	7982	0·98	11·0	220	0·022	4·6
1392	Ventura	L.Plio	10,813–10,817	3·11	14·3	472	0·015	2·4
1393	Ventura	L.Plio	12,534–12,537	2·58	14·8	665	0·026	2·2
1394	Ventura	L.Plio	13,301–13,304	4·14	9·9	964	0·023	2·5
1395	Ventura	L.Plio	14,647–14,656	3·67	14·4	1215	0·033	1·57
1396	Ventura	U.Mio(D)	14,934–14,937	2·33	13·3	1021	0·044	1·92
1397	Ventura	U.Mio(D)	15,186–15,189	1·21	8·6	666	0·055	1·35
1398	Ventura	U.Mio(D)	15,296–15,299	1·99	11·3	932	0·047	1·43
1399	Ventura	U.Mio(D)	15,407–15,410	2·71	10·5	922	0·034	1·48
			Miscellaneous Recent Sediment Samples					
LA R-11	Mississippi Delta	Recent	Lake Salvador, occasionally brackish	1·74	5·6	126	0·0072	6·3
LM 6A	Laguna Madre	Recent	Lagoon, brackish water	0·61	33·2	23	0·0038	3·8
GM 310	Gulf of Mexico	Recent	Shelf, water depth 360 ft	0·78	9·8	51	0·0065	5·3

* Mass spectrometric parent peak analyses, per cent of molecules with 1-ring, 2-rings etc., per molecule.
† 6-R is per cent of molecules with 6 or more rings per molecule.

the Los Angeles and Ventura Basins, California and from some Recent samples

Isoparaffins plus Naphthenes (per cent)																					
Fraction 2 (325–370°C)*							Fraction 3 (370–420°C)*							Fraction 4 (420–470°C)*							
0-R	1-R	2-R	3-R	4-R	5-R	6-R†	0-R	1-R	2-R	3-R	4-R	5-R	6-R†	0-R	1-R	2-R	1-R + 2-R	3-R	4-R	5-R	6-R†
Los Angeles Basin Shale Samples																					
14	15	20	15	20	10	6	8	15	18	14	24	16	5	6	11	15	26	7	26	21	14
20	18	18	15	16	8	5	10	12	14	10	29	19	6	3	4	17	21	13	34	21	8
38	21	17	11	8	4	1	20	17	13	7	14	21	8	14	11	8	19	3	18	30	16
32	10	9	11	17	15	6	9	14	4	6	50	13	4	4	0	2	2	4	45	19	26
42	17	14	12	8	5	2	14	13	6	8	21	28	10	7	1	0	1	2	49	22	19
56	11	10	9	8	4	2	16	11	4	3	32	27	7	8	0	0	0	1	59	17	15
43	14	13	12	11	5	2	15	12	7	3	41	16	6	8	0	0	0	1	67	15	9
42	8	7	13	18	7	5	8	2	0	4	61	21	4	7	Tr.	1	2	2	63	13	14
49	10	10	11	12	5	3	17	12	7	12	36	12	4	9	1	2	3	2	67	13	6
60	8	9	9	8	3	3	20	14	7	10	32	10	7	11	3	2	5	2	61	12	9
41	12	11	12	16	5	3	13	12	6	8	53	6	2	8	1	Tr.	1	2	70	16	3
39	17	16	12	11	4	1	17	22	14	14	25	5	3	9	9	6	15	4	52	17	3
35	18	16	13	13	4	1	17	18	11	11	36	4	3	11	7	4	11	4	60	12	2
36	21	17	12	10	3	1	21	21	14	11	28	4	1	14	10	8	18	5	48	13	2
Ventura Basin Shale Samples																					
6	10	15	18	23	14	14	4	8	13	10	37	19	9	3	15	17	32	10	20	17	18
11	12	15	10	17	18	17	6	8	12	7	39	16	12	5	9	11	20	5	37	23	10
25	16	18	17	15	6	3	11	17	15	12	22	15	8	8	10	11	21	6	30	20	15
23	19	20	15	14	6	3	12	22	19	16	17	11	3	11	16	14	30	7	25	16	11
36	19	17	13	9	4	2	13	19	14	12	24	12	6	8	5	4	9	4	48	15	16
35	17	17	12	11	5	3	16	19	14	10	17	17	7	9	10	9	19	5	31	23	13
48	15	14	10	8	3	2	17	18	14	9	31	8	3	8	4	8	12	4	62	10	4
45	15	13	11	10	4	2	14	17	10	8	44	5	2	8	4	3	7	3	66	12	4
25	19	16	12	21	5	2	9	5	7	6	65	6	2	8	3	6	9	3	68	10	2
25	18	17	16	15	6	3	13	20	15	15	26	7	4	9	12	10	22	7	39	17	6
45	14	15	13	9	3	1	25	20	17	14	17	5	2	11	12	10	22	7	46	11	3
36	19	17	12	10	4	2	18	20	15	11	28	5	3	10	8	7	15	3	51	14	7
31	19	18	13	13	4	2	14	20	19	12	25	6	4	6	6	12	18	4	42	27	3
34	17	16	14	12	5	2	12	18	14	12	34	7	3	6	5	5	10	3	48	29	4
Miscellaneous Recent Sediment Samples																					
16	19	21	16	15	8	5	13	19	18	13	16	12	9	10	16	13	29	6	10	12	33
31	42	13	3	6	4	1	13	10	4	2	29	31	11	16	13	6	19	3	21	26	15
19	19	19	12	17	10	4	12	20	17	11	17	12	11	2	26	21	47	19	17	9	6

Table 4. Analyses of hydrocarbons boiling above 325°C from crude oils from the

Sample No.	Field	Age	Zone	Depth (ft)	Gasoline <180°C (%)	Residue >325°C (%)	Hydrocarbons >325°C: Aromatics (%)	Iso-paraffins and naphthenes (%)	Paraffin wax (%)	Paraffin wax $\frac{2C_{29}}{C_{28}+C_{30}}$
CAL.				*Los Angeles Basin Oil Samples*						
19	Dominguez	L.Plio	1	3611–3616	19·0	54·0	50	41	9	1·15
20	Dominguez	U.Mio (A, B)	8	6680–7585	14·4	55·0	50	39	11	1·11
45	Wilmington	U.Mio (D)	Ford VI	5314–5994	13·6	64·8	60	27	13	1·11
36	Huntington Beach	U.Mio (D)	—	4410–4668	10·1	71·9	60	30	10	1·12
				Ventura Basin Oil Samples						
11	Ventura	L.Plio	4	7985–9030	16·0	64·0	49	39	12	1·19
16	Ventura	L.Plio	D-8	12,564–13,690	21·3	49·9	45	42	13	1·39
				Oil Samples from Louisiana, Wyoming and West Texas						
LA 13	Lake Washington	Mio	K_2	11,278–11,288	11·5	54·2	26	56	18	0·90
WYO 19	Hamilton Dome	Perm	Phosphoria	2300–2351	12·5	65·8	57	34	9	1·00
WT 51	Martin	Ord.	Ellenburger	8600–8900	28·2	40·8	51	37	12	0·98

* Mass spectrometric parent peak analyses, per cent of molecules with 1 ring, 2 rings etc., per molecule.
† 6-R is per cent of molecules with 6 or more rings per molecule.

the oil generation mechanism appears to favor the normal paraffins below about C_{25}, as indicated by the rise in the normal paraffin distribution curves of the older sediments and of the crude oils. The sharp dropoff of concentration below approximately C_{17} for the sediments and below about C_{19} for the oils is due exclusively to laboratory distillation.

Eventually, during the later stages of the oil generation process the concentration of normal paraffins formed in the subsurface (without odd predominance) is much greater than that which was originally present in the same sediments at the time of deposition. As a consequence, the strong odd-carbon-number predominance of the original normal paraffins, present from the time of deposition, is overshadowed and almost disappears. A small ripple usually remains, as indicated by values of the odd/even ratio $2C_{29}/(C_{28} + C_{30})$ of 1·1–1·3. The maximum value of this ratio which we have observed for crude oils is 1·6. This is very exceptional.

As is apparent from Figs. 4 and 5 and Tables 3 and 4, in the Los Angeles and Ventura basins, it is not until Miocene division D shales that the composition of the sediment normal paraffins begins to equal that of the oils. Tables 3 and 4 contain all the analytical data used in Figs. 4 and 5. The analytical data used in plotting the graphs of Figs. 7, 8 and 11 are also shown in Tables 3 and 4.

Los Angeles and Ventura Basins, California and from some other oil basins

Isoparaffins plus naphthenes (%)																					
Fraction 2 (325°–370°C)*							Fraction 3 (370°–420°C)*							Fraction 4 (420°–470°C)*							
0-R	1-R	2-R	3-R	4-R	5-R	6-R†	0-R	1-R	2-R	3-R	4-R	5-R	6-R†	0-R	1-R	2-R	1-R + 2-R	3-R	4-R	5-R	6-R†
Los Angeles Basin Oil Samples																					
25	24	18	13	13	5	2	23	23	15	11	20	5	3	19	19	13	32	7	29	10	3
22	25	17	13	15	5	3	23	24	15	11	20	5	2	19	19	13	32	6	30	11	2
27	22	21	14	12	3	1	19	24	18	15	16	5	3	13	14	12	26	7	41	10	3
25	24	18	13	14	4	2	23	25	14	12	20	4	2	13	10	7	17	5	51	11	3
Ventura Basin Oil Samples																					
22	23	19	14	15	5	2	21	24	17	11	20	5	2	18	18	13	31	7	29	12	3
20	24	19	14	16	5	2	19	25	17	11	20	5	3	16	18	14	32	7	28	12	5
Oil Samples from Louisiana, Wyoming and West Texas																					
26	23	17	13	10	7	4	23	25	17	14	11	6	4	19	25	18	43	11	16	8	3
41	23	18	10	6	2	0	35	24	18	12	7	3	1	29	24	18	42	9	13	6	1
25	45	18	7	3	1	1	26	38	20	11	4	1	0	19	34	19	53	9	9	8	2

Change in Composition of Sediment Isoparaffin and Naphthene Hydrocarbons during Subsurface Oil Generation

In the central portions of Figs. 4 and 5 the compositional change with depth and time of the high-boiling naphthenes and isoparaffins is shown for the Los Angeles and Ventura basins. The analyses were made in the same way as those of the Recent sediments of Fig. 1 discussed on pages 1024 and 1027.

Simultaneous with the increase of total hydrocarbon content and hydrocarbon/noncarbonate carbon ratio and the change of normal paraffin composition, the composition of the isoparaffin-naphthene concentrate also changes. Some of the differences in isoparaffin-naphthene composition may be due to differences in environment of deposition of the sediments of different ages considered. However, the very significant changes in composition which take place in the lower Pliocene and upper Miocene shales, where oil is formed at an appreciable rate, are undoubtedly due to the oil generation process. In lower Pliocene and in upper Miocene A, B, C and D shales, in fractions 3 and 4, the percentage of naphthenes with four rings per molecule decreases with increasing depth and age of the sediments, and the relative concentration of naphthenes with one and two rings per molecule increases. Figure 7 illustrates this interesting phenomenon on a larger scale than Figs. 4 and 5. The mass spectra of these and similar California samples,

2

according to a method developed by SCHISSLER *et al.* (1957), strongly suggest that the high four-ring-per-molecule peaks are due to a large extent to the presence of sterol derivatives. The presence of sterols in a variety of Recent sediments has been demonstrated by ERDMAN and SCHWENDINGER (1964).

The shale samples of Fig. 7 are all from one wildcat well, Shell No. 1 Anaheim Sugar, which is located east and downdip from the Seal Beach and Huntington

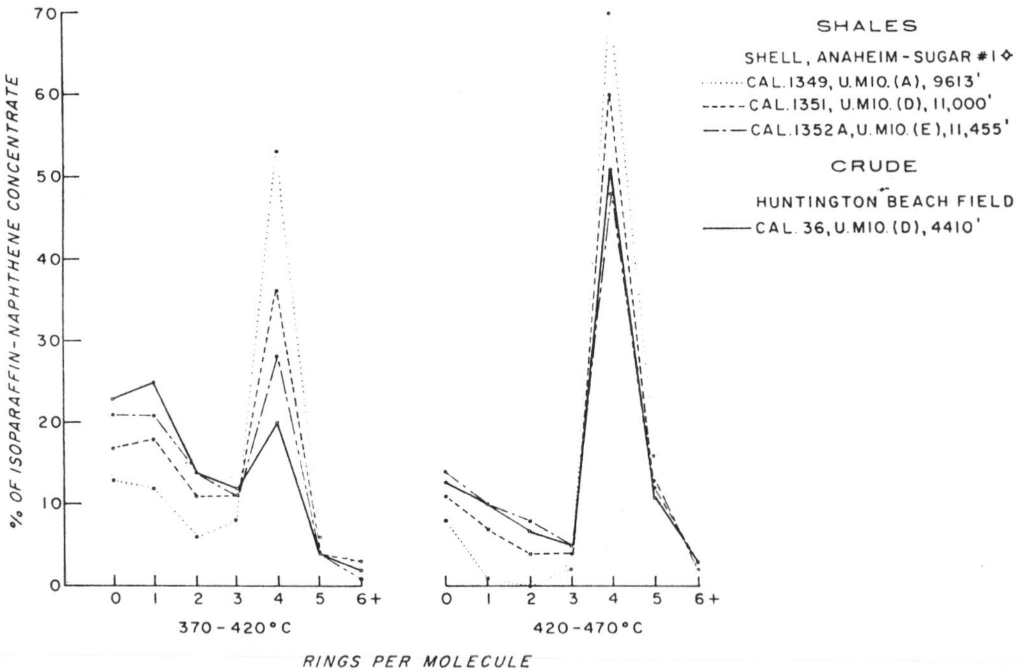

Fig. 7. Evolution of isoparaffins and naphthenes of the lube-oil range in shales of the Los Angeles Basin, California.

Beach oil fields (see Fig. 2). Also plotted in Fig. 7 is the analysis of the isoparaffin-naphthene concentrate from a typical waxy oil from nearby Huntington Beach field. As is apparent from Fig. 7, in upper Miocene E shales at 11,455 ft depth, the oil formation process has finally reached the point at which the composition of the sediment isoparaffin-naphthene concentrate is very similar to that of the waxy oils of the area. As has already been pointed out, in upper Miocene division D and E shales the shale normal paraffins have become identical with those of the oils. This suggests that upper Miocene D and E shales are the source of the oil of the Los Angeles basin. More supporting evidence for this conclusion is presented in the next section.

RELATIONSHIP BETWEEN OIL-SOURCE-ROCK HYDROCARBONS AND CRUDE-OIL HYDROCARBONS: SOURCE BEDS OF THE LOS ANGELES BASIN OILS

To facilitate the following discussion, some of the essential changes which, during the oil generation process, take place in the organic matter of subsurface

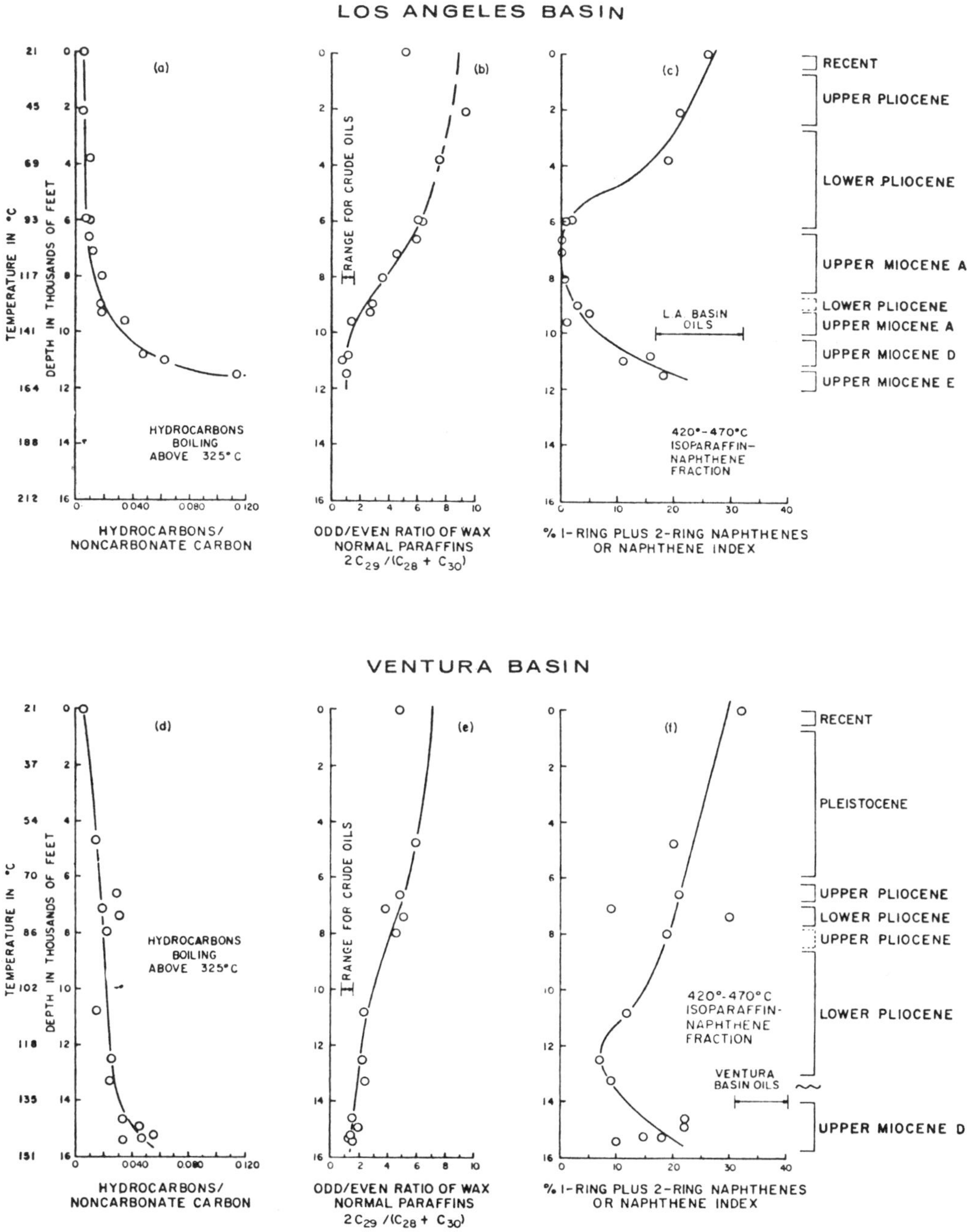

Fig. 8. Change of the hydrocarbon/noncarbonate carbon ratio, the normal paraffin odd/even ratio, and the naphthene index with depth and age in shales from the Los Angeles and Ventura basins, California.

shales are shown in Fig. 8. In Figs. 8(a), (b) and (c) data for the Los Angeles basin have been plotted, and in Figs. 8(d), (e) and (f) data for the Ventura basin have been plotted. In Fig. 8 analytical data of considerably more core samples have been used than in Figs. 4 and 5. The analytical data used are listed in Tables 3 and 4.

Figure 8(a) shows in detail the increase of the hydrocarbon/noncarbonate carbon ratio with depth and age of the Los Angeles basin shales. Figure 8(b) shows the continuous decrease of the normal paraffin odd/even ratio with depth and age. The end value of the odd/even ratio $2C_{29}/(C_{28} + C_{30})$ of shale normal paraffins is close to 1·0, the same as in the majority of crude oils. The change with depth and age of the "naphthene index," shown in Fig. 8(c), is even more interesting. "Naphthene index" is an abbreviated term for the sum per cent of one-ring and two-ring naphthenes in the 420–470°C isoparaffin-naphthene concentrate. The value of the naphthene index starts relatively high in Recent sediments, then drops essentially to zero, and later in the final stages of oil generation rises again to values identical with those for the crude oils of the same area. The range of values of the naphthene index as well as of the normal paraffin odd/even ratio for crude oils is shown on Fig. 8. In the Los Angeles basin, only in upper Miocene D and E shales which have been buried at least 10,500 ft do the naphthene index and the normal paraffin odd/even ratio values simultaneously equal the values of crude-oil hydrocarbons.

As is demonstrated in Fig. 8, the composition of shale hydrocarbons changes gradually with depth of burial and age. We have designated shale hydrocarbons *mature* when their composition is within the range valid for the oils of the same area. Both the normal paraffin odd/even ratio and the naphthene index can be used as a *maturity index* for shale hydrocarbons. In Fig. 9 these two maturity indices are plotted in a graph. Location, age, and depth of the samples, as well as numerical values of the maturity indices, are given in Table 5. Figure 9 demonstrates once more that only the hydrocarbons from some upper Miocene D and E shales have reached maturity and are similar to the crude-oil hydrocarbons. The fact that the hydrocarbons from other upper Miocene D shales and from all upper Miocene A to C shales have nor reached maturity will be explained in the next section.

From the results shown in Figs. 8 and 9, we conclude that upper Miocene divisions D and E shales are the major source of the Los Angeles basin oil. Middle Miocene shales, for instance, those of division F, may have contributed to the oil-generating process, but in the Los Angeles basin these strata are absent in many locations and, for this reason, are probably not important as a source of oil. In the Los Angeles basin, sediments older than middle Miocene consist of conglomerates, sandstones, red shales and schists, which are extremely unlikely to be oil source rocks. For these reasons the conclusion that in the Los Angeles basin the upper Miocene divisions D and E shales are the major source of the oil is well founded.

The results shown in Figs. 8(d), (e) and (f) and in Table 5 indicate that in the Ventura basin the upper Miocene D shales around 15,000 ft depth have not quite reached maturity. As in the Los Angeles basin, in the Ventura basin the upper Miocene D and E shales are undoubtedly a major source of petroleum in places where they are buried deeply enough to attain full maturity. However, in contrast with the Los Angeles basin, in the Ventura basin older Miocene shales as well as Oligocene and Eocene shales are present which may also have generated petroleum. The reason why the shale hydrocarbons in the Ventura basin mature at a deeper level than in the Los Angeles basin is explained in the next section.

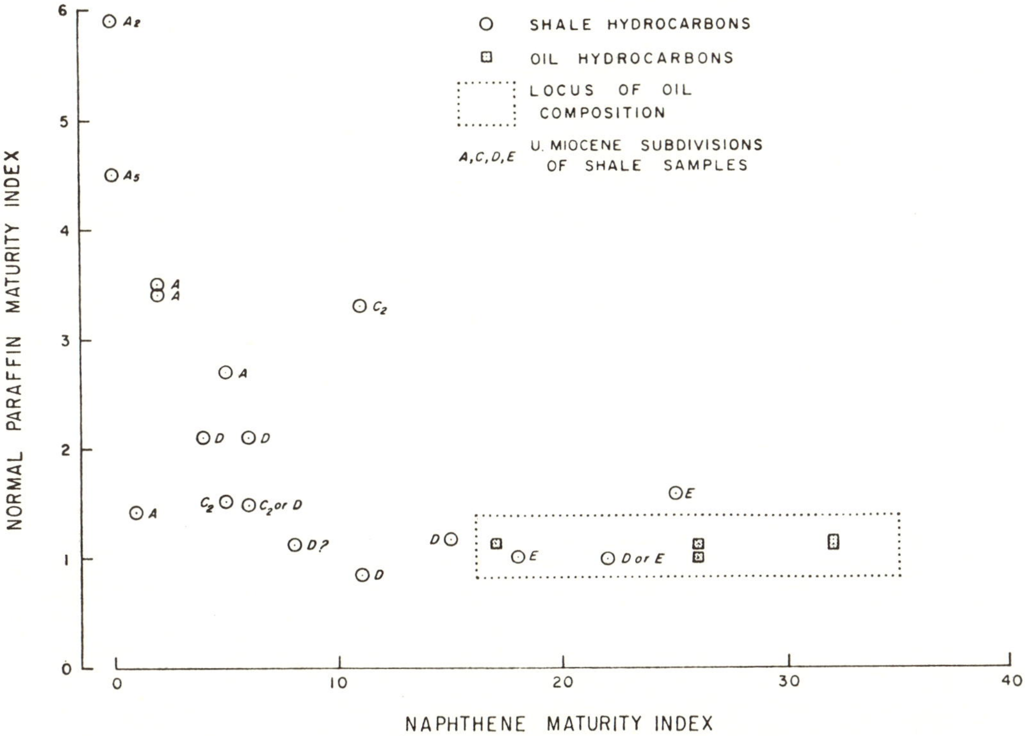

Fig. 9. Plot of the odd/even ratio $2C_{29}/(C_{28} + C_{30})$ of the normal paraffins (normal paraffin maturity index) versus the sum per cent of the one-ring and two-ring naphthenes in the 420–470°C isoparaffin-naphthene concentrate (naphthene maturity index). Los Angeles Basin data only.

Discussion of the Mechanism and the Influence of Temperature and Time on the Subsurface Generation of Petroleum

The increase of the temperature in the subsurface of the Los Angeles and Ventura basins will first be discussed. In Table 6 values are presented of the average temperature gradient in seven Los Angeles basin and six Ventura basin oil fields. The basic data for Table 6 are from papers by Van Orstrand (1934, 1939) and by Moses (1961). For the calculation of the temperature gradient values, an average surface temperature of 70°F (21°C) has been used. This is the average value which is obtained by extrapolation to zero depth of the Los Angeles basin subsurface temperature data of Van Orstrand (1934). As is demonstrated in Table 6, the average temperature gradient in the Los Angeles basin of 3·91°C/100 m (2·15°F/100 ft) is considerably greater than that in the Ventura basin, which is only

Table 5. Maturity indices of the hydrocarbons of Los Angeles and Ventura Basin Shale and crude oil samples

Sample No.	Field or Wildcat Well	Age	Depth (ft)	Noncarbonate carbon (%)	Hydrocarbons >325°C (ppm)	Hydrocarbons/ noncarbonate carbon	Normal paraffin maturity index $\frac{2C_{29}}{C_{28} + C_{30}}$	Naphthene maturity index*	Remarks
CAL				*Los Angeles Basin Shale Samples*					
1375	Dominguez	U.Mio(A_2)	6650	2·98	276	0·009	5·9	0	
1376	Dominguez	U.Mio(A_5)	7137–7144	2·00	235	0·021	4·5	0	
1346	Dominguez	U.Mio(A)	8066–8068	3·50	641	0·018	3·5	2	
1329	Shell No. 1 Lewis	U.Mio(A)	9236	3·18	1236	0·039	3·4	2	
1348	Shell No. 1 Anaheim Sugar	U.Mio(A)	9317–9320	4·72	832	0·018	2·7	5	
1349	Shell No. 1 Anaheim Sugar	U.Mio(A)	9613–9630	3·38	1138	0·034	1·42	1	
1361	Seal Beach	U.Mio(C_2)	9902–9903	4·02	1590	0·040	1·51	5	
1372	Dominguez	U.Mio(C_2 or D)	10,993–10,996	3·20	871	0·027	1·49	6	
1321	Brea-Olinda	U.Mio(C_2)	5132	1·77	265	0·015	3·3	11	Whittier Fault Trend
1358	Shell No. 1 Lewis	U.Mio(D ?)	10,007–10,010	2·86	1649	0·058	1·12	8	
1350	Shell No. 1 Anaheim Sugar	U.Mio(D)	10,663–10,670	5·80	2974	0·051	2·1	4	
1351	Shell No. 1 Anaheim Sugar	U.Mio(D)	11,000–11,003	4·32	2697	0·062	0·84	11	
1364	Seal Beach	U.Mio(D)	10,795–10,798	4·26	1985	0·047	1·17	15	
1323	Brea-Olinda	U.Mio(D)	5239	2·18	481	0·022	2·1	6	Whittier Fault Trend
1337	Brea-Olinda	U.Mio(D or E)	7957	4·18	937	0·022	0·98	22	Whittier Fault Trend
1336	Brea-Olinda	U.Mio(E)	1434	1·77	983	0·056	1·58	25	Whittier Fault Trend
1352A	Shell No. 1 Anaheim Sugar	U.Mio(E)	11,455–11,461	2·68	3032	0·113	1·00	18	
				Ventura Basin Shale Samples					
1396	Ventura	U.Mio(D)	14,934–14,937	2·33	1021	0·044	1·92	22	
1397	Ventura	U.Mio(D)	15,186–1,5189	1·21	666	0·055	1·35	15	
1399	Ventura	U.Mio(D)	15,407–15,410	2·71	922	0·034	1·48	10	
				Los Angeles Basin Oil Samples					
19	Dominguez	L.Plio(Z-1)	3611–3616				1·15	32	
20	Dominguez	U.Mio(A, B) (Z-8)	6680–7585				1·11	32	
45	Wilmington	U.Mio(D) (Z-Ford VI)	5314–5994				1·11	26	
36	Huntington Beach	U.Mio(D)	4410–4668				1·12	17	
24	Brea-Olinda	U.Mio(D)	4640–5400				0·99	26	Whittier Fault Trend
				Ventura Basin Oil Samples					
52	Saticoy	U.Plio(Z-F)	7243–7324				1·40	35	
53	Saticoy	U.Plio(Z-K)	7425–7640				1·09	35	
46	Ventura	U.Plio(Z-2b)	4795–5260				1·05	33	
11	Ventura	L.Plio(Z-4)	7985–9030				1·19	31	
16	Ventura	L.Plio(Z-D-8)	12,564–13,690				1·39	32	

* Sum per cent of the 1-ring and 2-ring naphthenes in the 420–470°C isoparaffin-naphthene concentrate.

Table 6. Temperature gradients in Los Angeles and Ventura Basin oil fields

Los Angeles Basin

Field	Depth (ft)	Reservoir temp. (°F)		Geothermal gradient °F/100 ft	Geothermal gradient °C/100 M*	Reference
Huntington Beach	1000	—		2·50	4·55	2
	2460	133		2·56	4·66	1
	3042	148		2·56	4·66	1
	3500	—		2·64	4·80	2
	4000	—	†(2)	2·68	4·88	2
	4230	175		2·48	4·51	1
	4795	170		2·09	3·80	1
			Average (7)	2·50	4·55	
Long Beach	1750	—	(2)	1·81	3·29	2
	2000	—	(3)	1·84	3·35	2
	2500	—	(5)	1·84	3·35	2
	3000	—	(7)	1·88	3·42	2
	3500	—	(5)	1·95	3·55	2
	4000	—	(3)	1·99	3·62	2
	5000	—		1·83	3·33	2
	9000	245		1·98	3·60	3
			Average (8)	1·89	3·44	
Santa Fe Springs	1000	—	(2)	2·03	3·69	2
	2000	—	(5)	1·89	3·44	2
	2500	—	(2)	1·99	3·62	2
	3000	—	(10)	1·97	3·59	2
	3500	—	(2)	1·94	3·53	2
	4000	—	(2)	1·95	3·55	2
	4250	—		1·90	3·46	2
	4525	—		1·92	3·49	2
			Average (8)	1·95	3·55	
Seal Beach	3500	—		2·26	4·11	2
West Coyote	2800	—		1·93	3·51	2
	3500	—		1·84	3·35	2
	3750	—		1·87	3·40	2
	4270	—		1·92	3·49	2
			Average (4)	1·89	3·44	
Whittier	2360	—		1·92	3·49	2
Wilmington	2760	139		2·50	4·55	1
	3192	156		2·69	4·90	1
	3396	150		2·36	4·30	1
	3530	168		2·78	5·06	1
	5585	226		2·79	5·08	1
			Average (5)	2·62	4·77	
Average for Los Angeles Basin (7 Fields)				2·15	3·91	

Ventura Basin

Field	Depth (ft)	Reservoir temp. (°F)		Geothermal gradient °F/100 ft	Geothermal gradient °C/100* M	Reference
Aliso Canyon	5230	144		1·41	2·57	1
Del Valle	7990	186		1·45	2·64	1
Newhall-Potrero	6140	170		1·63	2·97	1
	7048	170		1·42	2·58	1
	8680	198		1·47	2·68	1
	9833	205		1·37	2·49	1
	11,200	232		1·45	2·64	1
			Average (5)	1·47	2·68	
Rincon (Padre Canyon Area)	{7418	176		1·43	2·60	1
	{7576	179		1·44	2·62	1
	{8320	174		1·25	2·28	1
	9375	186		1·24	2·26	1
			Average (4)	1·34	2·44	
San Miguelito	6400	158		1·38	2·51	1
	7162	177		1·49	2·71	1
			Average (2)	1·44	2·62	
Ventura	3500	—		1·63	2·97	3
	7872	198		1·63	2·97	1
			Average (2)	1·63	2·97	
Average for Ventura Basin (6 Fields)				1·46	2·66	

* (°C/100 M) = 1·82 × (°F/100 ft)
† (Number of Values in Average)

References

1. MOSES P. L. (1961).
2. VAN ORSTRAND C. E. (1934).
3. VAN ORSTRAND C. E. (1939).

2·66°C/100 m (1·46°F/100 ft). The difference is probably caused by the much shallower depth of the basement in the Los Angeles basin. The considerable difference of the temperature gradients in the Los Angeles and Ventura basins is clearly shown by the lines in Fig. 10.

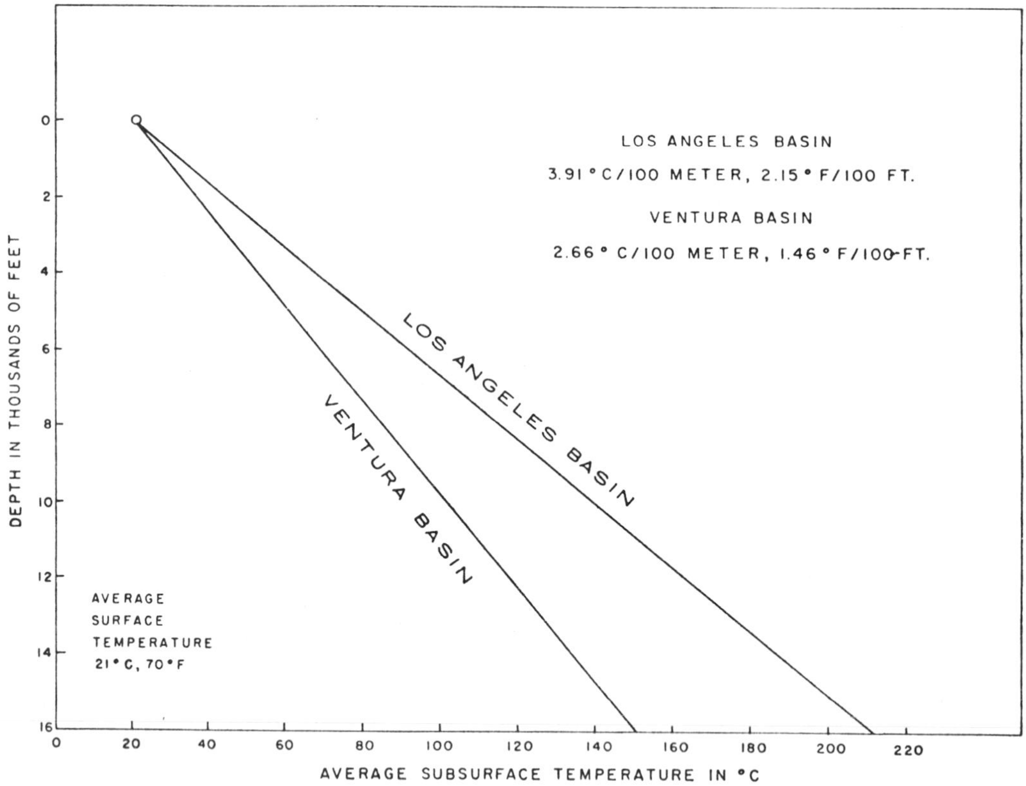

Fig. 10. Average temperature gradients in the Los Angeles and Ventura basins, California.

Figures 4, 5 and 8 demonstrate that in the Los Angeles basin the bulk of the oil generation takes place below 8000 ft depth and that in the Ventura basin it takes place below 12,000 ft. On the basis of the temperature depth relationships of Fig. 10, this means that in both the Los Angeles and Ventura basins the bulk of the oil generation takes place at temperatures above 115°C (239°F). Therefore, petroleum generation is a process occurring in an environment which has become sterilized because of the raised subsurface temperature. The conclusion that petroleum is formed essentially by nonbiological, thermal processes is inescapable. A shallow origin of petroleum in very young sediments, as has been considered by several investigators (SMITH, 1954; MEINSCHEIN, 1959; BAKER, 1960), is not compatible with the evidence presented.

The petroleum generation process, being chemical, consists of a multitude of subsurface reactions which convert part of the nonhydrocarbon components of the sediment organic matter both into hydrocarbons and into simple nonhydrocarbons, viz. compounds with a small percentage of sulfur or nitrogen which resemble hydrocarbons. As with other chemical reactions, the rates of these oil-forming reactions depend on both temperature and pressure. For the petroleum generation process, of these two factors, *temperature* is by far more important. The rate of different chemical reactions differs but, quite generally, increases exponentially with the absolute temperature. Because of this feature, the rate of most chemical reactions is increased by a factor of 2–3 for each rise in temperature of 10°C (or 18°F). Pressure, because of its effect on gas–liquid equilibria and component concentrations, influences the rate of part of the oil-forming reactions. However, the effect of increase of pressure on the reaction velocities involved is far less than exponential.

The strong increase of the hydrocarbon/noncarbonate ratio at greater depths and temperatures, shown in Figs. 4 and 5 and in Figs. 8(a) and (d), is supporting evidence for the chemical, temperature-controlled nature of the petroleum generation process. Additional and very convincing evidence for this conclusion is the much shallower depth of maximum oil generation and of equal composition of shale and oil hydrocarbons in the Los Angeles basin compared with the Ventura basin. This is undoubtedly caused by the considerably greater temperature gradient in the Los Angeles basin (Fig. 10). The present depth of equal composition of shale and oil hydrocarbons in normal positions in the Los Angeles basin, away from strongly uplifted fault blocks, is around 11,000 ft, where the temperature is about 150°C (302°F). As was pointed out in the previous section, even at 15,000 ft depth in the Ventura basin the upper Miocene D shale hydrocarbons are slightly immature. Therefore, in the Ventura basin the level at which upper Miocene D shale hydrocarbons mature is somewhat below 15,000 ft, or, in other words, at temperatures somewhat above 143°C (289°F). This suggests that the temperature at which hydrocarbons of oil source beds of the same age (upper Miocene D) mature is much the same in the Los Angeles and Ventura basins.

It should not be inferred that the high temperature and the great depths at which the bulk of the oil generation occurred in the Los Angeles and Ventura basins are general. This is not necessarily true for much older source rocks than those of upper Miocene age. In older source rocks more time has been available for the oil generation to proceed, and the same amount of petroleum may have been generated at a lower average temperature over a longer period of time. However, the amount of oil generated increases only linearly with time but exponentially with the absolute temperature. Because of this, *neither a fixed depth nor a fixed duration of the oil generation process exists.* The following factors influence the amount of petroleum which is formed in subsurface sediments:

(1) The amount and nature of the organic matter of the source beds.
(2) The temperature history of the source beds.
(3) The effect of catalysts, if present.

The importance of the temperature history of potential oil source sediments should be stressed. In the Los Angeles and Ventura basins oil generation is barely noticeable around 5000 ft depth. At the temperatures inherent to this depth the generation process is too slow to be of significance, even over a few epochs. The important thing is whether a potential source sediment has been exposed for a sufficiently long time to subsurface temperatures at which the rate of oil generation is sufficiently rapid. The temperature history of a potential source sediment is determined to a large extent by the tectonic history, i.e. how deeply the sediment has been buried, once or several times, and for how long. The shale samples from near the Whittier fault zone in the Los Angeles basin, shown in Table 5, are examples of rocks which have been uplifted. Shale sample CAL 1337, from upper Miocene division D or E in Brea Olinda field, is mature, despite its present relatively shallow depth of 7957 ft. For the location of Brea Olinda field, see Fig. 2. Upper Miocene E shale sample CAL 1336, also from Brea Olinda, has been uplifted even further. Its present depth is only 1434 ft; despite this shallow depth, its hydrocarbons are almost mature. Upper Miocene D and E samples CAL 1336 and CAL 1337 have been at much deeper levels during their tectonic history. Their present shallow depth is recent, and the uplifting took place mostly during Pleistocene time. Both shale samples must have been deep for a considerable period; otherwise, the indigenous hydrocarbons would not be mature. On the other hand, all upper Miocene A through C samples analyzed and several upper Miocene D samples are immature. They have not been deep and warm enough for a sufficiently long time. Temperature history of a potential oil source rock is a major factor in petroleum geology; it determines whether a potential oil source rock will be an oil-releasing source rock or not. The oil release process is discussed in the next section.

MECHANISM AND TIME OF OIL RELEASE FROM SOURCE BEDS

As demonstrated in the previous sections, shale hydrocarbon compositions identical with those of crude-oil hydrocarbons are not observed in sediments younger than upper Miocene divisions D and E in the Los Angeles and Ventura basins. We have analyzed oils from a variety of geologic provinces, and as yet, not a single oil residue has been encountered with a hydrocarbon composition similar to those of the young, immature hydrocarbons from Pliocene or upper Miocene A, B, or C shales, examples of which are illustrated in Figs. 4, 5, 8 and 9. If Pliocene or upper Miocene A, B or C shales *had* released any appreciable quantities of petroleum, the crude-oil analyses should show it. The lack of such evidence strongly suggests that the young shales have not released much oil. The question then arises, why not? Furthermore, what are the factors which govern the release of oil from source beds?

The following hypothesis explains the release of petroleum from its source beds. In source beds the oil is formed from, and within, the vastly more abundant nonhydrocarbon organic matter. This organic matter contains many hydrophobic groups in the derivatives of the lipids and proteins, which can dissolve and hold hydrocarbons. Hence, the organic matter has a certain sorption capacity for petroleum. On the other hand, clay minerals, in the presence of water, have no affinity for paraffins and naphthenes and little affinity for aromatics. In the initial stages of the oil generation process the amount of oil formed is far below the sorption capacity of the nonhydrocarbon organic matter; therefore, *the initial, immature petroleum will stay where it is formed,* viz. in the enclosing organic matter. Only in the later stages of the oil generation process, when enough oil is formed to exceed the sorption capacity of the organic matter, is oil expelled as a separate phase. Then, as a result of compaction, oil and water are squeezed through and out of the source beds in the direction of the hydrodynamic gradient.

Investigation of many compact ancient shales and argillaceous limestones has demonstrated that, for indigenous hydrocarbons boiling above 325°C, the maximum value of the hydrocarbon/noncarbonate carbon ratio is about 0·120. From the available data, it is inferred that, during the oil formation process, the sorption capacity is normally exceeded at hydrocarbon/noncarbonate carbon ratios in the range of about 0·030–0·120, and oil can then be released from the source formation.

The hypothesis presented here to elucidate the release of petroleum explains why young, immature oils, such as the hydrocarbon mixtures formed in the California lower Pliocene and upper Miocene A, B and C shales, are never found in reservoirs. It also explains why oil-releasing source rocks have a certain minimum age. In source sediments subjected to the normal geothermal intensity, a long time is required for enough oil to form to raise the hydrocarbon/noncarbonate carbon ratio to the critical release range of 0·030–0·120. Finally, the hypothesis presented explains why so many ancient shales, even of Ordovician age, have indigenous hydrocarbon contents of several hundred to several thousand parts per million. This is sorbed oil which has not been released because the amount is equal to, or less than, the sorption capacity of the rock. Determination of the amount of indigenous hydrocarbons and of the corresponding hydrocarbon/noncarbonate carbon ratio of shales and argillaceous limestones is the basis of an oil-source-rock evaluation method published several years ago (Philippi, 1956).

Vertical Migration of Petroleum in the Los Angeles and Ventura Basins of California

Most of the oil fields of the Los Angeles and Ventura basins produce from a multitude of horizons. The number of producing zones in Dominguez, for instance, is ten, in Long Beach eight, in Wilmington eight, in Ventura eight, and in Saticoy five. With the exception of the pre-Miocene production of the Ventura basin, the age of these producing zones ranges from upper Pliocene to upper Miocene D. In previous sections we have concluded that all of the Los Angeles basin oil originated deep in upper Miocene D and E shales and that the Ventura basin oil originated in the same zones and possibly also deeper. Because the majority of the producing zones of the Los Angeles and Ventura basins are considerably shallower than the oil source beds, vertical migration of petroleum must have been extremely common in both basins. Supporting evidence of a different nature for this conclusion follows from many crude-oil analyses. In Fig. 11(a) isoparaffin-naphthene curves are shown of a shallow zone 1 oil and a deep zone 8 oil from Dominguez. The two curves are so close that they can hardly be distinguished. Both are very different from the comparable curves of shale hydrocarbons from beds younger than upper Miocene D and E (see Fig. 4). In Fig. 11(b) isoparaffin-naphthene curves are compared of a shallow and deep Ventura field oil. Again, the two oils are the same within experimental error and are quite different from the hydrocarbons of the shales between the producing zones (see Fig. 5). These oil analytical results support

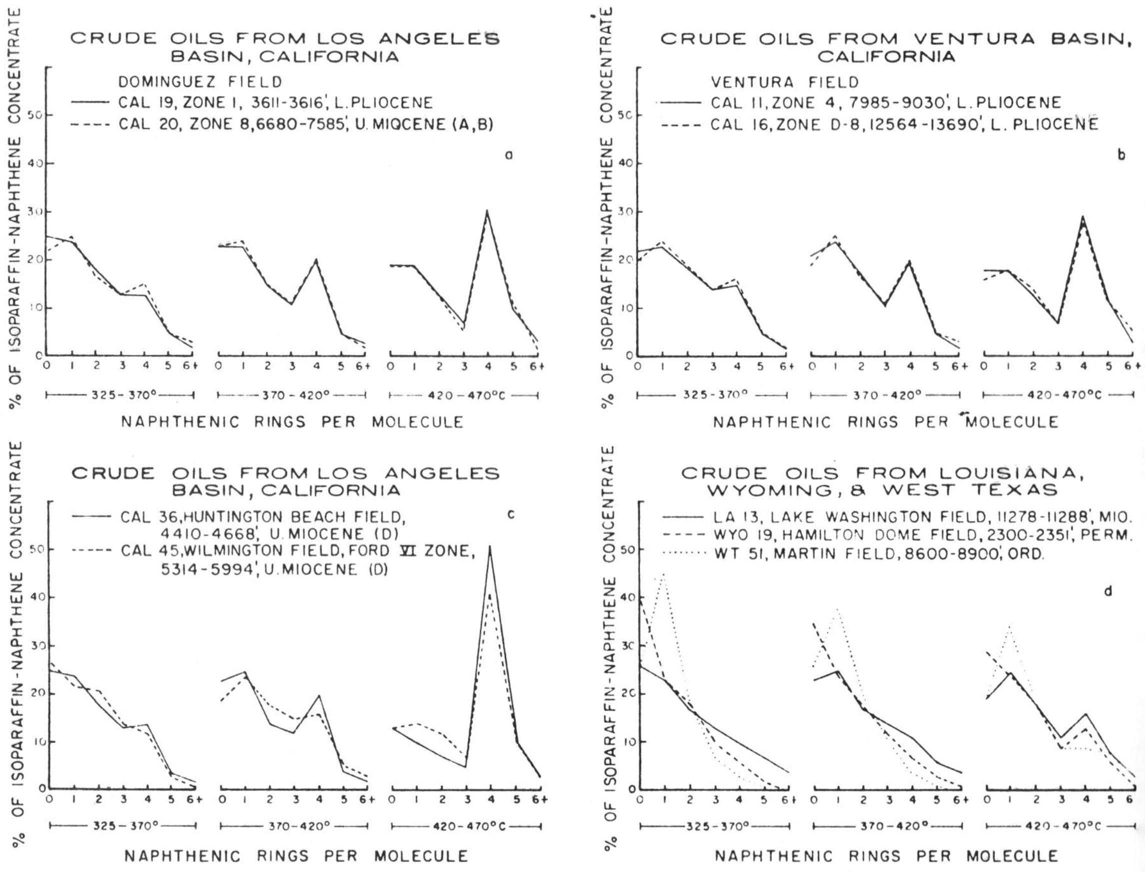

Fig. 11. Per cent of isoparaffins (zero rings) and of naphthenes with one ring, two rings, three rings etc., per molecule in isoparaffin-naphthene concentrates of crude oils from different oil basins.

the earlier conclusion that all the Dominguez field oils have the same deep origin and that all the Ventura field oils have the same deep origin, the oils having arrived in place by vertical migration. The evidence for the identical origin of the oils from all the producing zones in fields such as Dominguez and Ventura is more pertinent if Fig. 11(d) is considered. This figure shows the great differences in isoparaffin-naphthene composition between oils from other oil basins, which obviously originated in other source formations.

The paraffinic oils of the Los Angeles basin are not all the same. In Fig. 11(c) examples of oil analyses from the Huntington Beach and Wilmington fields are shown. In the 420–470°C isoparaffin-naphthene fractions these oils have a lower content of one- and two-ring-per-molecule naphthenes than the Dominguez oils of Fig. 11(a). The shale hydrocarbon analyses discussed suggest that the Huntington Beach and Wilmington oils have been released from source beds which are slightly less mature than the source beds of the Dominguez oils.

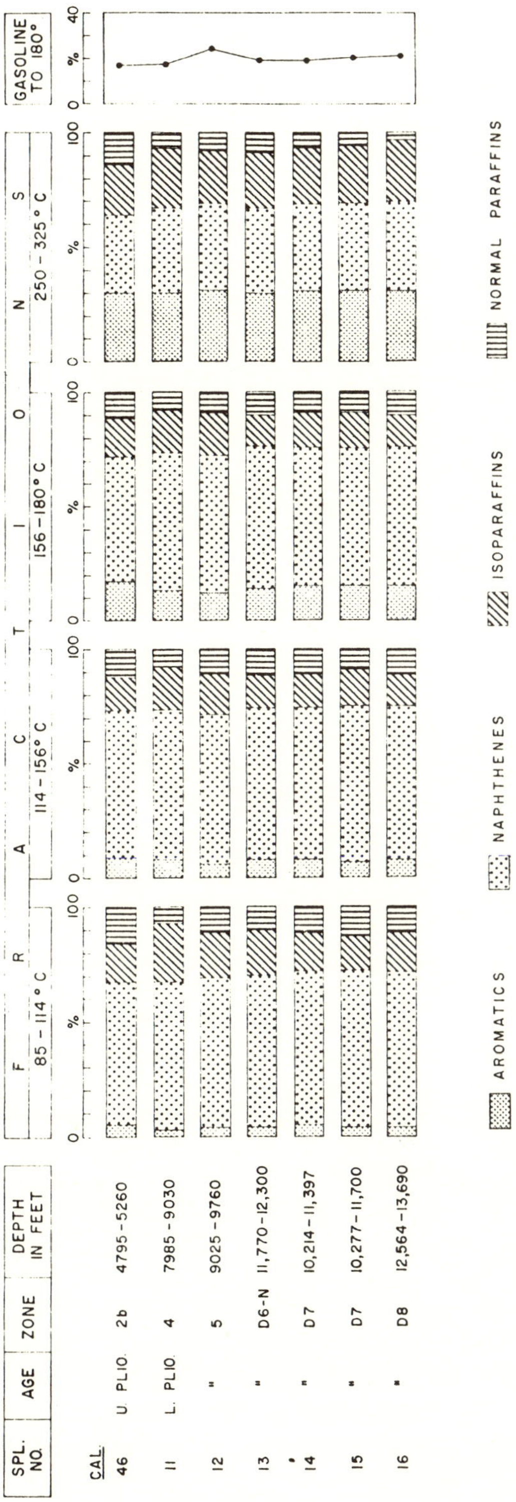

Fig. 12. Analyses of gasoline and midrange fractions from shallow and deep oils of the Ventura field, California.

Additional evidence for the identical origin of the oils of the various producing zones of the Ventura field is shown in Fig. 12. Depth of the oil samples analyzed ranges from 4795 to 13,690 ft. Despite this great depth interval, the composition of the gasoline and midrange fractions of the oils from the various zones of the Ventura field is very similar. The oil fractions were prepared by conventional high-efficiency distillation (SCHWARTZ, 1963) and were analyzed by mass spectrometry. Normal paraffin content of the oil fractions was determined by Linde 5A molecular sieves (SCHWARTZ and BRASSEAUX, 1957).

The oil analyses discussed strongly support the existence of extensive vertical migration of oil in California fields such as Dominguez and Ventura. This vertical migration of petroleum did not take place through the bulk of the shales occurring between the oil sands. The composition of the indigenous hydrocarbons from Pliocene shales present between the producing sands is very different from the composition of the oil hydrocarbons (see Fig. 8). Also, the indigenous hydrocarbon content of the Pliocene shales, especially of the shallower ones, is very small. If any appreciable amount of extraneous oil had migrated through the bulk of the Pliocene shales, it would show up in the analyses of the shale hydrocarbons and make these look "mature". *Evidently, the oil which originated deep in the upper Miocene D and E source rocks migrated upward through faults and fractures.* This result is in agreement with the geologic observation that oil seeps, all over the world, very often occur in fault zones.

REFERENCES

BAKER E. G. (1960) A hypothesis concerning the accumulation of sediment hydrocarbons to form crude oil. *Geochim. et Cosmochim. Acta* **19,** 309–317.

BLUMER M. (1962) An efficient still for milligram samples of high boiling materials. *Analyt. Chem.* **34,** 704–708.

BLUMER M. (1958) Unpublished manuscript.

BLUMER M., MULLIN M. M. and THOMAS D. W. (1963) Pristane in zooplankton. *Science* **140,** 974.

BRAY E. E. and EVANS E. D. (1961) Distribution of n-paraffins as a clue to recognition of source beds. *Geochim. et Cosmochim. Acta* **22,** 2–15.

ERDMAN J. G. and SCHWENDINGER R. B. (1964) Sterols in recent aquatic sediments. *Science* **144,** 1575–76.

HOLMES A. (1960) A revised geological time scale. *Trans. Edinb. Geol. Soc.* **17,** 183–215.

HOOD A. and O'NEAL M. J. (1959) Status of application of mass spectrometry to heavy oil analysis, in *Advances of Mass Spectrometry,* pp. 175–192. Pergamon Press, London.

HUNT J. M. and JAMIESON G. W. (1956) Oil and organic matter in source rocks of petroleum. *Bull. Amer. Ass. Petrol. Geol.* **40,** 477–488.

HUNT J. M. (1961) Distribution of hydrocarbons in sedimentary rocks. *Geochim. et Cosmochim. Acta* **22,** 37–49.

LARSEN B. and HAUG A. (1956) Carotene isomers in some red algae. *Acta Chem. Scand.* **10,** 470.

MEINSCHEIN W. G. (1959) Origin of petroleum. *Bull. Amer. Ass. Petrol. Geol.* **43,** 925–43.

MOSES P. L. (1961) Geothermal gradients, in *Drilling and Production Practice,* pp. 57–63. American Petroleum Institute.

O'NEAL M. J. (1954) Application of high molecular weight mass spectrometry to oil constitution, in *Applied Mass Spectrometry,* pp. 27–46. Institute of Petroleum.

O'NEAL M. J., HOOD A., CLERC R. J., ANDRE M. L. and HINES C. K. (1955) The determination of heavy oil composition by mass spectrometry. *Proc. Fourth World Petrol. Congr.,* Section V, 307, Carlo Columbo, Publ., Rome.

PHILIPPI G. T. (1956) Identification of oil source beds by chemical means. *Proc. Int. Geol. Cong., Mexico City*, Section III, Petrol. Geol., 25–38.

ROBINSON W. E., CUMMINS J. J. and DINNEEN G. U. (1963) Alteration of paraffinic compounds in Green River oil shale after deposition (Abstract). G.S.A. Meeting, November 17–20, New York, pp. 138A–139A.

SCHISSLER D. O., STEVENSON D. P., MOORE R. J., O'DONNELL G. J. and THORPE R. E. (1957) Steranes in Petroleum, A.S.T.M. Committee E-14, New York.

SCHWARTZ R. D. (1963) Abstract No. 48, p. 19J, 144th Meeting A.C.S., Los Angeles, California.

SCHWARTZ R. D. and BRASSEAUX D. J. (1957) Determination of normal paraffins in normal olefin free petroleum distillates by molecular sieve sorption and refractometry. *Analyt. Chem.* **29,** 1022–1026.

SIRAHANA K. (1938) Liquid unsaponifiable matter of the brown algae fat, VII, some new ingredients of brown algae. *J. Agr. Chem. Soc., Japan* **14,** 743.

SMITH P. V. (1954) Studies on origin of petroleum. Occurrence of hydrocarbons in Recent sediments. *Bull. Amer. Ass. Petrol. Geol.* **38,** 377–404.

STEVENS N. P., BRAY E. E. and EVANS E. D. (1956) Hydrocarbons in sediments of Gulf of Mexico. *Bull. Amer. Ass. Petrol. Geol.* **40,** 975–983.

TRASK P. D. and WU C. C. (1930) Does petroleum form at the time of deposition. *Bull. Amer. Ass. Petrol. Geol.* **14,** 1451–63.

VAN ORSTRAND C. E. (1934) Temperature gradients, in *Problems of Petroleum Geology*, pp. 989–1021, American Association of Petroleum Geology.

VAN ORSTRAND C. E. (1939) Temperature of the earth in relation to oil location, in *Temperature—Its Measurement and Control in Science and Industry*, pp. 1014–1033. Amer. Inst. Phys., Reinhold. New York.

WARTH A. H. (1956) *The Chemistry and Technology of Waxes*. Reinhold, New York.

37

Reprinted from *Am. Assoc. Petrol. Geol. Bull.* **58**:499–506 (1974)

Influence of Nature and Diagenesis of Organic Matter in Formation of Petroleum[1]

B. TISSOT,[2] B. DURAND,[2] J. ESPITALIÉ,[2] and A. COMBAZ[3]
Rueil-Malmaison, France, and Paris, France

Abstract Elevation of temperature and pressure during burial of sediments results in a physiochemical transformation of the kerogen, and particularly in changes of elementary composition, infrared spectra, and thermal properties of kerogen. The combined variation of these characters defines an evolution path for each kerogen type (organic matter from similar environments of deposition group along the same evolution path). The degree of evolution of individual samples can be evaluated by using these techniques.

An experimental evolution test, using thermogravimetric, infrared, and associated techniques, can reproduce the natural degradation of kerogen during its burial in a sedimentary basin. The test allows the petroleum geologist to make an evaluation of the petroleum potential of a given formation in the buried parts of a sedimentary basin, by using shallow or surface samples and simulating their degradation in the laboratory.

The major products generated during the three successive steps of kerogen evolution are carbon dioxide and water; oil; gas. The relative abundance of these products depends on the composition of the original kerogen. Kerogen from a "high" evolution path generates abundant oil, whereas that from a "low" evolution path produces methane at depth, but little or no oil.

General Scheme of Hydrocarbon Generation

Many studies have shown that most commercial hydrocarbons are generated by thermal transformation of kerogen during burial of the source rocks (Larskaya and Zhabrev, 1964; Louis, 1964; Philippi, 1965; Louis and Tissot, 1967; Albrecht and Ourisson, 1969; Vassoyevich *et al.*, 1969; Tissot *et al.*, 1971).

The general scheme of hydrocarbon generation is summarized on Figure 1, by gathering the results of several studies carried out in different sedimentary basins from France, north and west Africa, Canada, and South America. The diagram shows the relative amount and nature of hydrocarbons generated from sedimentary organic matter as a function of depth of burial. The depth scale is only indicative and may vary according to the nature of the original organic matter and to the geothermal gradient.

In young sediments, freshly deposited, little hydrocarbon is present. With the exception of methane of biochemical origin (marsh gas), these hydrocarbons are inherited more or less directly from the organisms incorporated in the sediment. They have very peculiar structures, characteristic of molecules synthesized by living organisms. Some are incorporated in the sediment without any change, like *n*-alkanes from higher plants or from algae; others, like steroids or pentacyclic triterpenoids, result from biogenic molecules through an early transformation (for example, loss of a functional group) that does not affect the hydrocarbon skeleton. These molecules can be considered as "geochemical fossils," in the same sense as

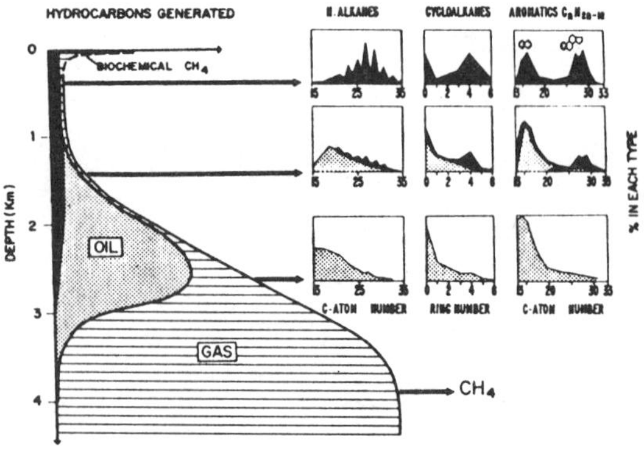

FIG. 1—General scheme of hydrocarbon generation. Depth scale represented is based on examples of Mesozoic and Paleozoic source rocks. It is only indicative and may vary according to nature of original organic matter, burial history, and geothermal gradient.

[1] Manuscript received, July 5, 1973; accepted, September 14, 1973. Read at the AAPG Research Symposium, Anaheim, California, May 15, 1973.

[2] Institut Francais du Petrole.

[3] Compagnie Francaise des Petroles.

The writers are indebted to Compagnie Francaise des Petroles and Compagnie ELF-Re for their assistance in supplying samples. The writers also thank B. Alpern, P. Robin, J. C. Roussel, C. Souron, A. Marchand, and M. F. Achard for their participation in reflectance, infrared, thermogravimetric, and ESR studies.

ammonites or foraminifers, are macro- or microfossils.

During a certain time and depth span, the hydrocarbon content remains low, as shown in Figure 1, to about 1,000 m. Then, beyond a certain threshold, new hydrocarbons are generated, first gradually, then more rapidly, by thermal breakdown of the insoluble organic matter ("kerogen"). These hydrocarbons have no characteristic structure, and the previous hydrocarbons (the geochemical fossils) are diluted progressively by the new ones. This is the principal phase of oil formation, as designated by Vassoyevich *et al.* (1969).

At greater depth, increased cracking of carbon-carbon bonds occurs and generates light hydrocarbons from kerogen and from previous oil as well. This is the phase of condensate and gas formation. Later, under epimetamorphic conditions, methane, carbon dioxide, and a carbon-rich residue would be the end point of organic matter.

Kerogen Evolution Paths

Although this is a general scheme of the evolution of organic matter during burial, the total amount of hydrocarbons that can be generated and the relative importance of oil and gas depend on the composition of the original organic matter; some formations may provide oil and gas, and others generate mostly gas. The chemical and physical study of the insoluble organic matter, or kerogen, is a way to characterize the various types of organic matter and thus to evaluate the oil and gas potential of the formations.

The elemental composition (atomic H/C and O/C ratios) of a number of kerogens is shown on Figure 2. This presentation allows a first approach to the classification of the kerogens. Kerogens taken at different depths from the same formation normally group along a curve that we propose to call the "evolution path," for example, the lower Toarcian shales of the Paris basin, the Silurian shales from the Sahara in Algeria and Libya, the Upper Cretaceous shales of the Douala basin, or the Mannville shales from the Lower Cretaceous of Alberta, which have been studied by McIver (1967).

Furthermore, rocks from closely related environments of deposition, like shales containing Tasmanites algae from the Silurian of the Sahara and from the lower Toarcian of the Paris basin, result in the same evolution path. The different paths start from rather different hydrogen and oxygen original contents, according to the nature of the organic material and to the conditions of deposition. All curves tend to join each other at great depths and approach pure carbon as shown in Figure 2 (Durand and Espitalié, 1973).

Kerogens rich in aliphatic structures, able to generate oil, have an original high H/C ratio and a relatively low O/C ratio (evolution path I on Fig. 2). This group includes algal-rich sediments, in particular those derived from lacustrine Botryococcus (like Autun and Campine bogheads), and their marine equivalents (like tasmanite from Tasmania). It also includes normal disseminated kerogen from the most prolific source rocks (including some source rocks from the Middle East).

Kerogens with a predominant aromatic structure result from abundant contributions from higher plants and terrestrial humic material accumulated in nonmarine or paralic environments. They have a high original oxygen content, with an O/C ratio that may reach 0.2 or 0.3 in some samples, and a relatively low H/C ratio (evolution path III on Fig. 2). This group includes the Mannville shales from the Lower Cretaceous of Alberta and the Upper Cretaceous of the Douala basin, Cameroon.

Most of the kerogens appear on Figure 2 between these two extreme types. Some have relatively high values of H/C ratio and low values of O/C ratio, like the Toarcian shales of the Paris basin, the Silurian shales of the Sahara, and the upper Paleozoic–Triassic shales of Spitzbergen (evolution path II on Fig. 2). Some others have a comparatively lower H/C ratio, like the Viking shales from the Cretaceous of Alberta (McIver, 1967, evolution path not represented here), resulting from the participation of plant material from the continent.

The global composition of the kerogen can be compared with the composition of the coal macerals by using the same diagram (Fig. 3). The evolution path of type I resembles the carbonization path of alginites. The kerogen type III, rich in material from higher plants, has an evolution path comparable to that of vitrinite. The other evolution paths, like type II, occupy an intermediate position, as exinite does, but the H/C ratio may be higher or lower according to the particular formation.

Transformation of Kerogen during Burial

The burial of sediments at depth results in a temperature and pressure increase that in-

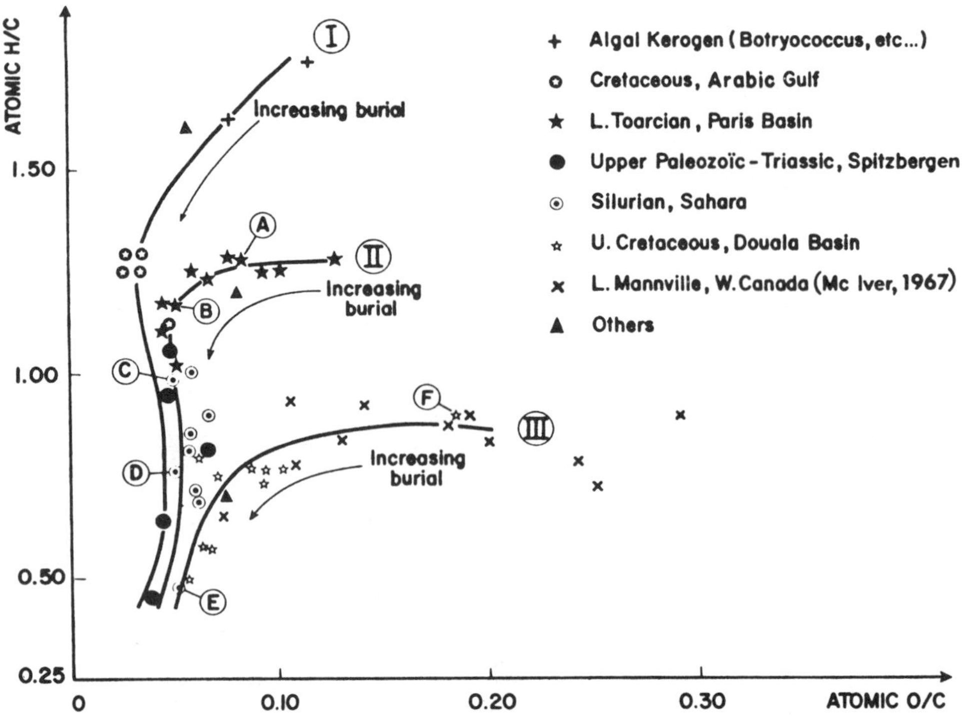

FIG. 2—Examples of kerogen evolution paths. Path I includes algal kerogen and excellent source rocks from Middle East; path II includes good source rocks from North Africa and other basins; path III corresponds to less oil-productive organic matter, but may include gas source rocks. Evolution of kerogen composition with depth is marked by arrow along each particular path.

duces a progressive transformation of kerogen (McIver, 1967). The successive steps of increasing thermal evolution can be characterized to locate them along an evolution path, like

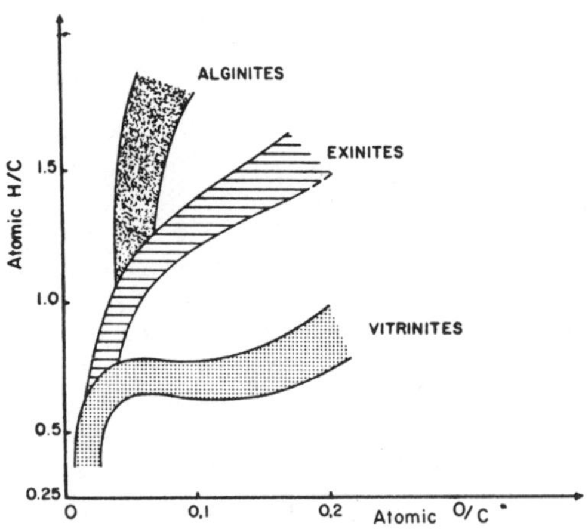

FIG. 3—Evolution of coal macerals (after Van Krevelen, 1961).

path II, that comprises the lower Toarcian shales of the Paris basin and the Silurian shales from the Sahara. In the same basin, for example the Paris basin, subsurface surveys enable geologists to reconstruct the burial history and arrange the samples in order of increasing burial depth (Tissot *et al.,* 1971). To compare samples from different basins, like the Paris basin and Sahara, burial depths cannot be used readily, as the age of the formations, the duration of burial, and the geothermal gradients may be different. However, the order of increasing evolution still can be evaluated by simultaneous use of vitrinite reflectance, microfossil preservation and color, and electron-spin resonance (ESR) measurements, together with available burial and geothermal data.

When the samples along an evolution path, like path II on Figure 2, are considered in this manner the shallow samples, with a low grade of evolution, are at the beginning of the curve with relatively high H/C and O/C ratios. A progressive evolution results first in a partial

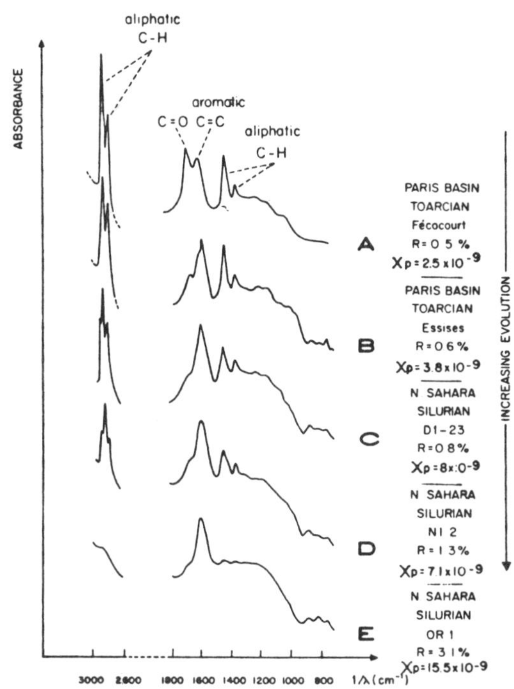

FIG. 4—Evolution of kerogen structure during burial. Samples are taken along evolution path II and marked *A-E* on Figure 2. Infrared spectra show progressive elimination of C = O (carbon dioxide and water formation) and aliphatic CH_2, CH_3 bands (hydrocarbons formation). Reflectance of vitrinite (*R*) and ESR data (paramagnetic susceptibility *Xp*, C G S electromagnetic units) are shown for comparison.

elimination of oxygen, marked by a decrease of the O/C ratio (sample *B*). Later the elimination of hydrogen, especially in the form of hydrocarbons, makes the H/C ratio decrease (samples *C, D*). In very deep samples, the H/C ratio is lowered to about 0.5, and the O/C ratio is only 0.05.

Infrared analysis of the same kerogen samples (Fig. 4) shows that the diminution of the oxygen content during the first step is related to the progressive elimination of the C = O group. The following stage, that of hydrocarbon generation, is linked with a decrease of the H/C ratio and of aliphatic bands. In very deep samples, aliphatic and carbonyl bands have nearly disappeared. The remaining oxygen probably is restricted to the nonlabile heterocyclic structures (Espitalié *et al.*, 1973).

The thermogravimetric analysis (TGA) of the same set of kerogen samples, all derived from a similar environment, allows a classification of these samples along their particular evolution path. On heating kerogens in an inert nitrogen atmosphere, the weight loss from shallow samples is important for this type of organic matter and may reach 70 percent. With increasing burial and increasing level of organic-matter transformation, the weight loss decreases progressively down to 10 percent in very deep samples from the western Sahara (Fig. 5).

Comparative Evaluation of Petroleum Potential by Simulated Evolution of Kerogen

In an attempt to simulate in the laboratory the natural evolution of kerogen during burial, a shallow and immature sample of the same sequence (type II) was used. Several aliquots of the sample were prepared and heated in a nitrogen atmosphere, with the same thermogravimetric equipment. The temperature was increased at a constant 4°C/minute; one sample was heated to 350°C, the second sample to 400°C, *etc.* . . . up to 600°C. The weight loss was recorded. The heated sample then was subjected to optical examination, reflectance measurement, elementary, and infrared analysis. The higher temperatures used are not found in a typical sedimentary basin, but they are used to achieve results in the short laboratory times that are comparable on a relative basis to the longer times and lower temperatures in nature.

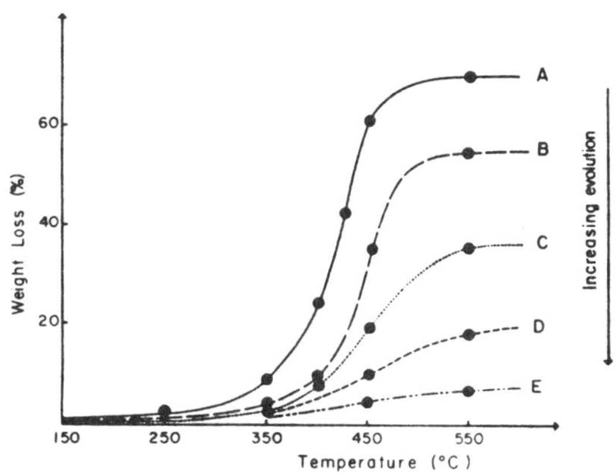

FIG. 5—Thermogravimetric analysis of kerogen. Samples are taken along evolution path II, and are marked *A-E* on Figure 2. Samples were heated at temperature increase of 4°C/min and weight loss was measured. Weight loss is highest in shallow samples taken along evolution paths I or II (here). Samples taken along path III would show least weight loss.

Heating to 350°C results in a rather small weight loss. Although the organic matter is somewhat darkened, its reflectance is still low (Fig. 6). From 350 to 470°C, kerogen shows a major change; the weight loss is maximal per time or temperature unit. The crystalline organization of carbon increases; reflectance goes up to 1.9 percent. Beyond 470°C the weight loss is small and very slow. But carbonization of kerogen goes on and is marked by a continuous increase of reflectance (2.9 percent at 600°C).

The elementary composition of the samples subjected to these experimental evolution tests varies continuously in relation to the temperature. It first is denoted by an oxygen loss, followed by a hydrogen loss. Furthermore, the succession of the different values of the H/C and O/C ratios reproduces exactly the natural evolution path that we have described previously (compare Figs. 7, 2).

The same observation applies to the infrared spectra; samples submitted to a thermal evolution in the laboratory show first a decrease of the carbonyl C = O bond followed by a progressive loss of both aliphatic and carbonyl groups. Again the succession of IR spectra resulting from the experimental evolution tests is nearly identical with the natural evolution with depth in a sedimentary basin that we have seen previously (compare Figs. 8, 4).

The step-by-step analysis of the chemical composition and the physical properties of the kerogen that has undergone either a simulated degradation in the laboratory or natural degradation during the geologic history of a sedimentary basin is an important comparison. Their coincidence allows us to make an evalua-

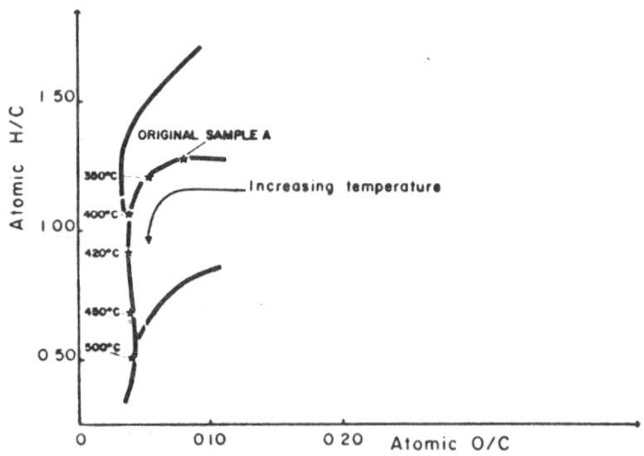

FIG. 7—Experimental evolution path of shallow kerogen. Aliquots of original sample (Fecocourt, lower Toarcian shales of Paris basin, marked *A* on Fig. 2) are respectively heated to temperatures shown on diagram (temperature increase 4°C/min). Succession of elementary compositions reproduces natural evolution path II shown on Figure 2.

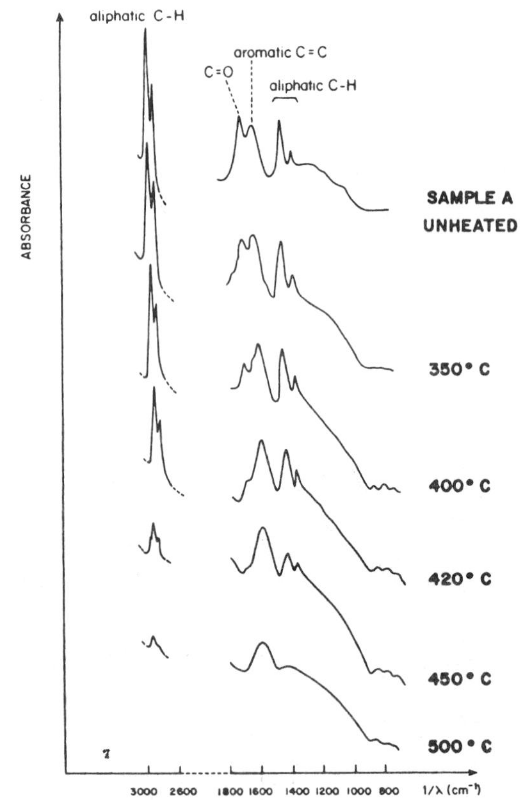

FIG. 8—Experimental evolution of kerogen (infrared analysis). Same aliquots of lower Toarcian shales from Fecocourt are heated to temperatures shown on diagram. With increasing temperature of experiments, succession of infrared spectra reproduces natural evolution during burial, described on Figure 4.

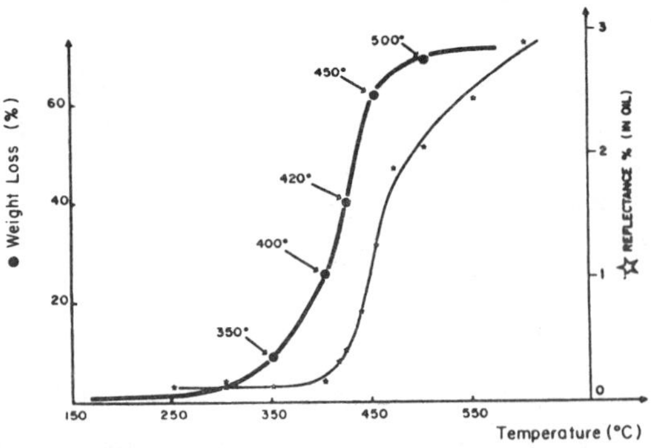

FIG. 6—Experimental evolution of kerogen *A* showing progressive weight loss and variation of reflectance of kerogen (temperature increase 4°C/min).

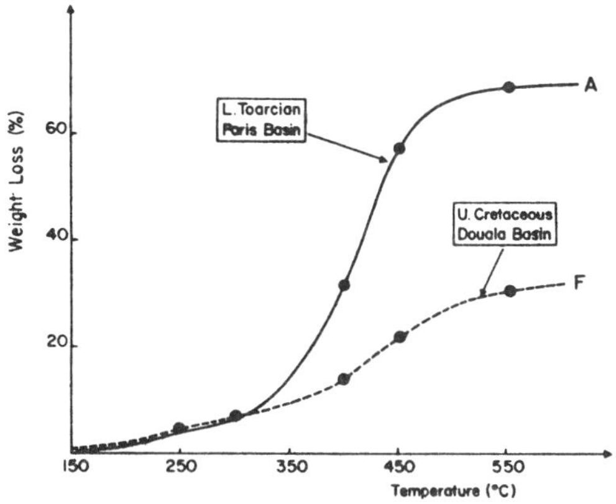

FIG. 9—Comparative experimental assay on two kerogens by using thermogravimetric analysis. Lower Toarcian shales of Paris basin (*A*) are two times more productive than Upper Cretaceous shales of Douala basin (*F*). Original samples *A* and *F* are marked on Figure 2.

tion of the petroleum potential of a given formation in the buried parts of a sedimentary basin, by using shallow or outcrop samples of the same formation and simulating their degradation in the laboratory.

As an example, two formations are compared here, one belonging to a relatively "high," hydrogen-rich, evolution path of type II (the lower Toarcian shales of the Paris basin), the other to a "low" evolution path of type III with a smaller hydrogen content (Upper Cretaceous shales of the Douala basin, studied by Albrecht and Ourisson, 1969). Two shallow kerogens, one from each of these formations, were heated gradually in the laboratory to simulate the evolutionary changes that would have occurred in nature if these shallow kerogens had been buried to greater depth.

On thermogravimetric analysis, the lower Toarcian shale kerogen yields twice as much volatile material as the Douala kerogen; the weight loss is between 60 and 70 percent, compared to 30 percent (Fig. 9). In addition, the infrared analysis of the two natural unheated samples shows that the lower Toarcian shales contain more aliphatic groups, capable of producing hydrocarbons, compared to the Upper Cretaceous of the Douala basin, where the aliphatic bands are much smaller and the C = O and aromatic bands are comparatively more important (Fig. 10). The relative abundances of these groups result in products of different composition from the thermal degradation of kerogen during the thermogravimetric analysis; CO_2 and H_2O amount to more than 50 percent of these products in the kerogen from the Douala basin, against only 25 percent in the kerogen of the lower Toarcian shales from the Paris basin. Altogether, the amount of petroleum resulting from the transformation of one gram of kerogen is much greater in the lower Toarcian shales. This is in agreement with the amount of hydrocarbons, resin, and asphaltenes effectively generated during burial at various depths, of these two types of kerogens, as measured directly by extraction of cores from the two basins. The amounts of natural petroleum, as extracted from cores, are plotted on Figure 11 as a function of the effective temperature corresponding to their natural burial depth. For a given temperature, the amount of petroleum generated in the lower Toarcian shales of the Paris basin is higher, as expected, than in the Upper Cretaceous from Douala basin.

CONCLUSION

The general structure usually proposed for kerogen is made of cyclic polycondensed nuclei, bearing alkyl chains and functions, and linked by heteroatomic bonds comprising in particular oxygen. As burial depth and tem-

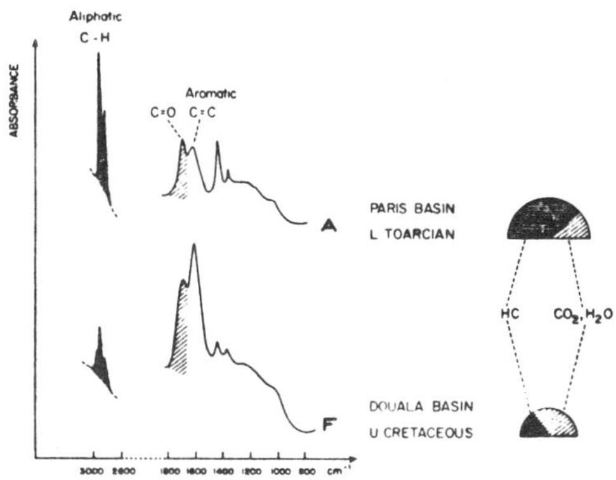

FIG. 10—Comparative infrared analysis of two shallow kerogens having maximum depth of burial of 800 m (left), and composition of products generated by simulated evolution (right). Lower Toarcian shales of Paris basin (*A*) are rich in aliphatic structures and generate mostly hydrocarbons upon simulated evolution. Upper Cretaceous shales of Douala basin (*F*) contain more C = O and aromatic C = C groups and generate much less hydrocarbons, but more carbon dioxide and water (surfaces on right are proportional to abundance of various products per gram of kerogen).

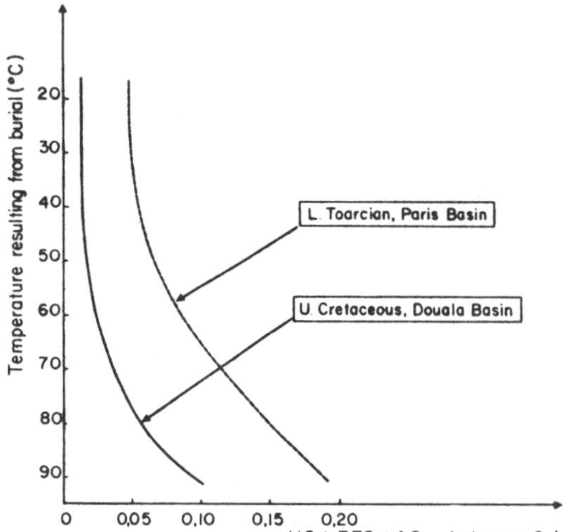

FIG. 11—Compared formation of petroleum in Paris and Douala basins, as function of temperature of source rocks. Transformation of organic matter into petroleum is measured on cores by ratio of chloroform extract (hydrocarbon + resins + asphaltenes) to total organic carbon (Tissot *et al.*, 1971). Comparison is restricted to interval 20–95°C, as deeper samples are available from Douala, but not from Paris basin.

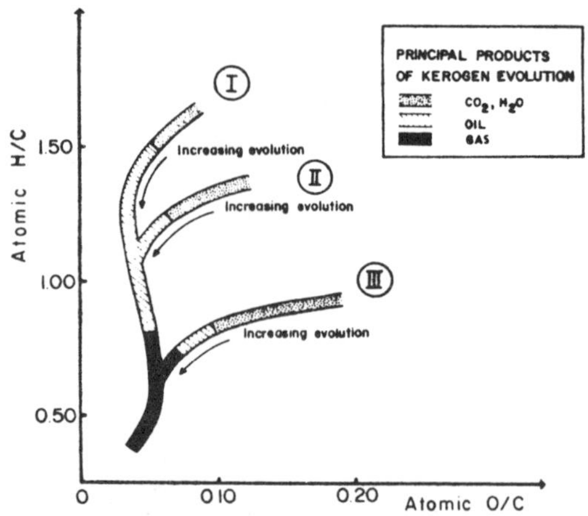

FIG. 12—Principal steps of kerogen evolution and hydrocarbon formation. Flat parts of curves indicate mostly oxygen loss as CO_2 and H_2O. Steep parts show loss of hydrogen and hydrocarbon generation. Evolution paths I or II correspond to kerogens able to generate abundant oil, whereas evolution path III is representative of hydrogen-poor kerogens, producing mostly methane at depth.

perature increase, heteroatomic bonds are broken successively, starting with some labile carbonyl and carboxyl groups, and roughly in order of ascending rupture energy. Heteroatoms, especially oxygen, are removed partly as CO_2 and H_2O; that step corresponds to a comparative greater decrease of O/C than H/C ratio, and is denoted on Figure 12 by a flattening of the evolution paths. As temperature continues to increase, hydrocarbon structures and particularly aliphatic groups are freed; this is the principal phase of oil formation. That step corresponds to a comparatively greater decrease of H/C than O/C ratio and is denoted on Figure 12 by a "steep" part of the evolution paths. In the final stage, breaking of C—C bonds becomes very important, and cracking generates gas from both kerogen and petroleum already formed (Fig. 12). All along the transformation the carbon content increases in the residual kerogen which becomes more condensed.

The relative importance of these three principal steps of the evolution is primarily dependent on the composition of the original kerogen; a "high" evolution path (like type I or II) corresponds to kerogen types that generate abundant oil, whereas a "low" evolution path (like type III) is representative of kerogen types that generate abundant CO_2 and H_2O, and produce methane at great depth, but little or no oil.

Chemical and physical analysis of kerogen, associated with simulated evolution in the laboratory, permits an evaluation of the comparative petroleum potential of the various kerogen types.

REFERENCES CITED

Albrecht, P., and G. Ourisson, 1969, Diagenese des hydrocarbures dans une serie sedimentaire epaisse (Douala, Cameroun): Geochim. et Cosmochim. Acta, v. 33, p. 138–142.

Durand, B., J. Espitalié, G. Nicaise, and A. Combaz, 1972, Etude de la matiere organique insoluble (kerogene) des argiles du Toarcien du Bassin de Paris: I, Etude par les procedes optiques, analyse elementaire, etude en microscopie et diffraction electroniques: Inst. Francais Petrole Rev., v. 27, no. 6, p. 865–884.

——— and J. Espitalié, 1973, Evolution de la matiere organique au cours de l'enfouissement des sediments: Acad. Sci. Comptes Rendus, t. 276, p. 2253–2256.

Espitalié, J., B. Durand, J. C. Roussel, C. Souron, 1973, Etude de la matiere organique insoluble (kerogene) des argiles du Toarcien du bassin de Paris: II, Etudes en spectroscopie infrarouge, en analyse thermique differentielle et en analyse thermogravimetrique: Inst. Francais Petrole Rev., v. 28, no. 1, p. 37–66.

Larskaya, Ye.S., and D. V. Zhabrev, 1964, Effect of stratal temperature and pressure on the composition of dispersed organic matter: Akad. Nauk SSSR

Doklady, v. 157, no. 4, p. 897–900 (in Russian); English translation: v. 157, no. 1–6, 1965, p. 135–138.
Louis, M., 1964, Etudes geochimiques sur les schistes cartons du Toarcien du bassin de Paris, *in* G. D. Hobson and M. C. Louis, eds., Advances in organic geochemistry 1964: Oxford, Pergamon Press, p. 85–94.
——— and B. Tissot, 1967, Influence de la temperature et de la pression sur la formation des hydrocarbures dans les argiles a kerogene: 7th World Petroleum Cong. Proc., v. 2, p. 47–60.
Marchand, A., P. Libert, and A. Combaz, 1969, Essai de caracterisation physico-chimique de la diagenese de quelques roches biologiquement homogenes: Inst. Francais Petrole Rev., v. 24, p. 3–20.
McIver, R. D., 1967, Composition of kerogen; clue to its role in the origin of petroleum: 7th World Petroleum Cong. Proc., v. 2, p. 25–36.
Philippi, G. T., 1965, On the depth, time, and mechanism of petroleum generation: Geochim. et Cosmochim. Acta, v. 29, p. 1021–1049.
Tissot, B., Y. Califet-Debyser, G. Deroo, and J. L. Oudin, 1971, Origin and evolution of hydrocarbons in early Toarcian shales, Paris basin, France: Am. Assoc. Petroleum Geologists Bull., v. 55, p. 2177–2193.
Van Krevelen, D. W., 1961, Coal: Amsterdam, Elsevier Publ. Co.
Vassoyevich, N. B., Yu. I. Korchagina, N. V. Lopatin, and V. V. Chernyshev, 1969, Principal phase of oil formation: Moskov. Univ. Vestnik, no. 6, p. 3–27 (in Russian); English translation: Internat. Geology Rev., 1970, v. 12, no. 11, p. 1276–1296.

38

Reprinted from *Preprint, Div. Petrol. Chem., Am. Chem. Soc.* **14**:E22–E31 (1969)

MICROBIOLOGICAL ALTERATION OF CRUDE OIL IN THE RESERVOIR

J. C. Winters and J. A. Williams

INTRODUCTION

Alteration of crude oil composition in the reservoir presents an imposing problem to anyone attempting to determine whether or not oils from different reservoirs are genetically related. To correlate an altered oil with unaltered oils, one must, by calculation, restore the composition and physical characteristics that the oil possessed before it was altered. Of course, it is first necessary to be able to recognize an altered oil, and to identify the nature of the alteration.

Crude oil may be altered in the reservoir by several natural phenomena. Maturation of an oil in a deep reservoir can change its properties to such an extent that it no longer resembles any of its genetically related counterparts which have remained at shallower depths. Occasionally, a shallow oil accumulation located near an outcrop or fault shows evidence of having undergone oxidation, either from atmospheric oxygen penetrating to the reservoir, or from oxygen dissolved in surface waters which contact the oils. Other types of alteration may result from reaction of the crude oil with sedimentary constituents, such as elemental sulfur.

The literature frequently refers to crude oils which seem to have lost their paraffin wax fraction, leaving them with "naphthenic" characteristics Some of the authors reporting this type of alteration have speculated that microorganisms have selectively destroyed the paraffins. Their speculation has been based on a belief that

a. there are aerobic microbes which selectively oxidize n-paraffins, or
b. the altered oils are usually found in relatively shallow reservoirs, which would be more easily reached by percolating surface waters bearing the oxygen necessary to sustain microbial oxidation processes.

In controlled laboratory experiments, Davis (1) has shown that crude oil can be attacked by soil bacteria with a resulting decrease in n-paraffin content.

Although the possibility of microbial attack on crude oil in the reservoir has been mentioned in the literature, there has not been reported any direct evidence that such type of crude oil alteration does occur. The possibility that the compositional anomalies resulted from some other unknown natural process could not be precluded.

DISCUSSION

Several types of data obtained on subsurface fluids are pertinent to the question. Figure 1 is a portion of a plot of correlation index vs. fraction number taken from a publication by Wenger and Reid of the Bureau of Mines (2) In their study of Powder River Basin crudes they found some anomalous correlation index curves for a few of the Cretaceous oils. These curves had generally high correlation index values compared to typical Cretaceous oils, indicating a lower paraffin content. They concluded that the anomalous oils had been altered by removal of n-paraffins, but did not suggest the possibility of alteration by microorganisms. H. M. Smith in a later publication (3) noted the same kinds of unusual correlation index curves and suggested that microbial action was responsible.

Normal-paraffin distributions of about 1,000 crude oils, obtained by gas chromatographic analysis at the Pan American Research Center, have confirmed that the n-paraffins are reduced in quantity or completely missing in some crude oils when compared with other oils from similar sources. Figure 2 shows the comparison of a typical chromatogram obtained on an altered oil compared to that of an unaltered one The gas chromatographic determinations were made on aliphatic fractions obtained by liquid chromatography on activated alumina. The heptane eluate was

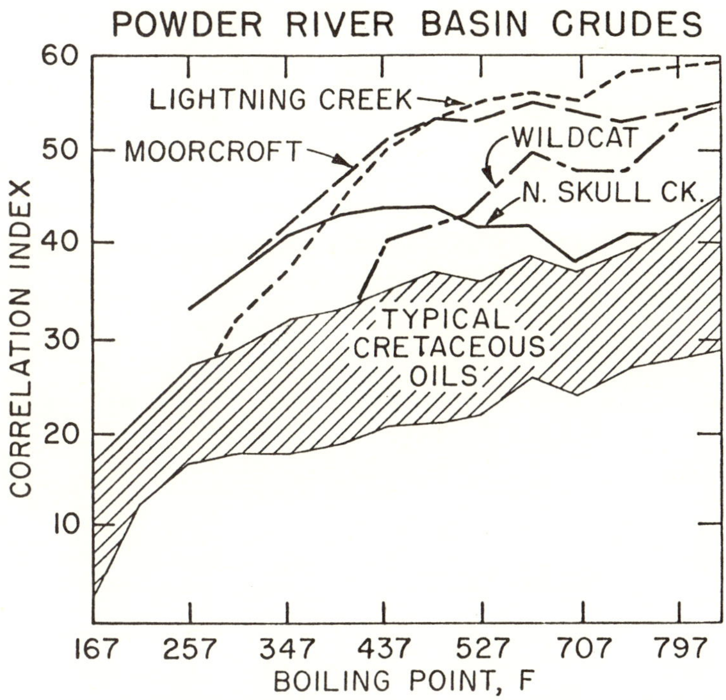

Figure 1. Correlation index curves for Cretaceous oils from the Powder River Basin.

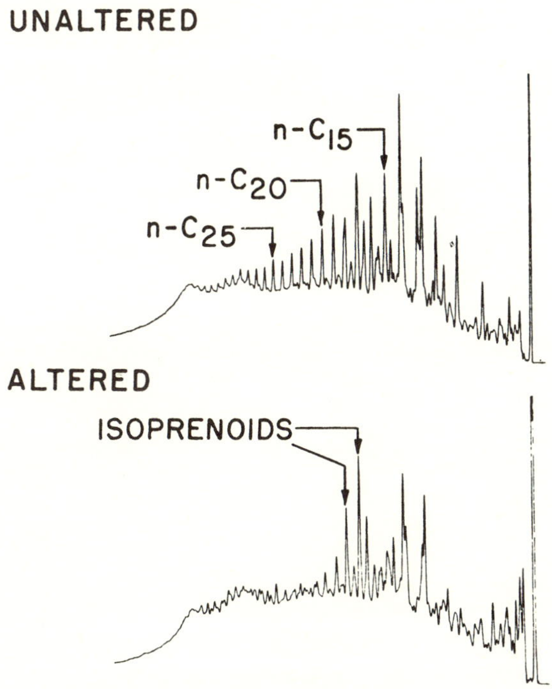

Figure 2. Typical gas chromatograms of aliphatic fractions of unaltered and altered crude oil.

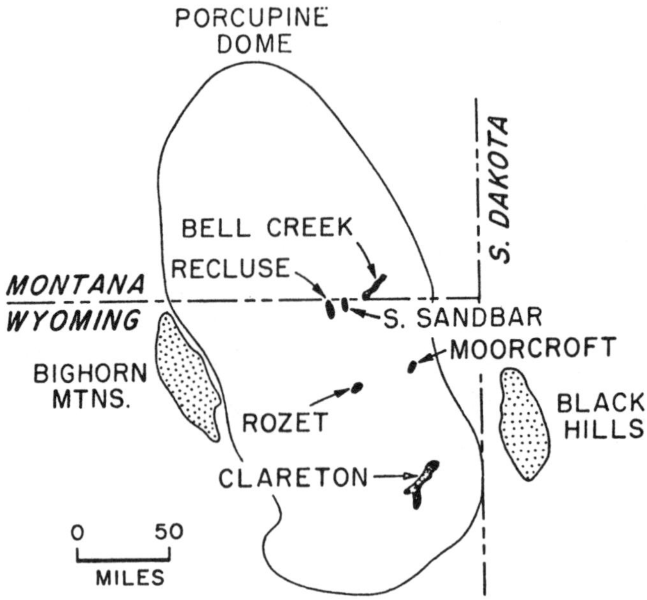

Figure 3. Index map of Powder River Basin, Wyoming and Montana, showing location of important [illegible] oil fields.

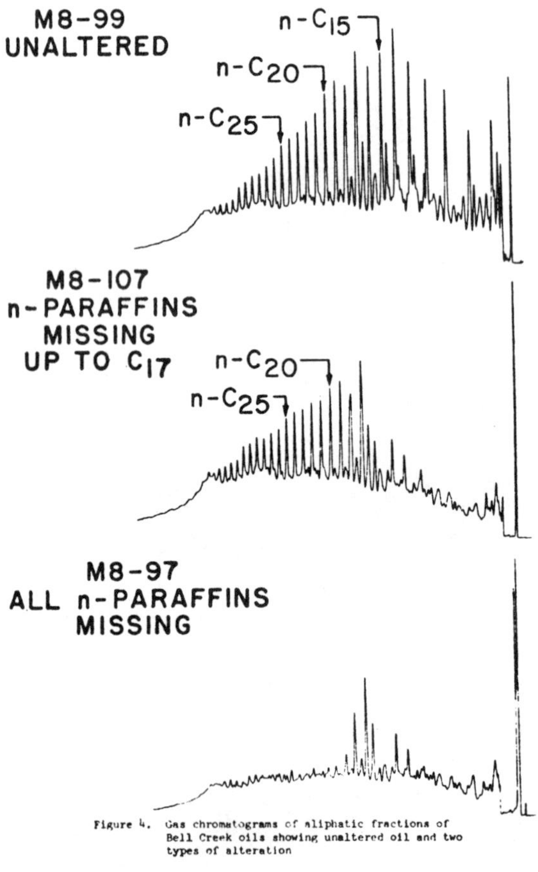

Figure 4. Gas chromatograms of aliphatic fractions of Bell Creek oils showing unaltered oil and two types of alteration

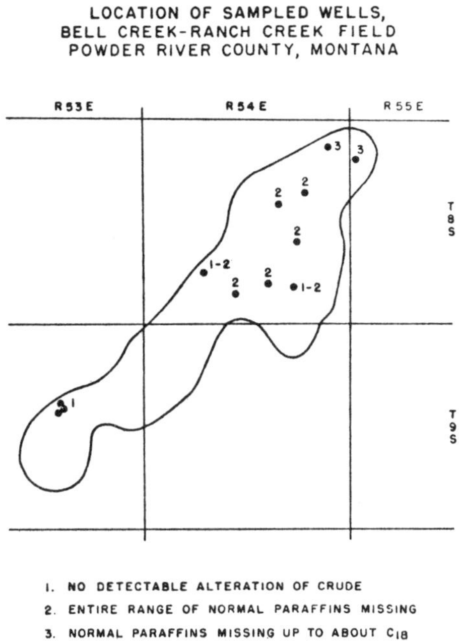

Figure 5. Location of sampled wells, Bell Creek-Ranch Creek Field, Montana

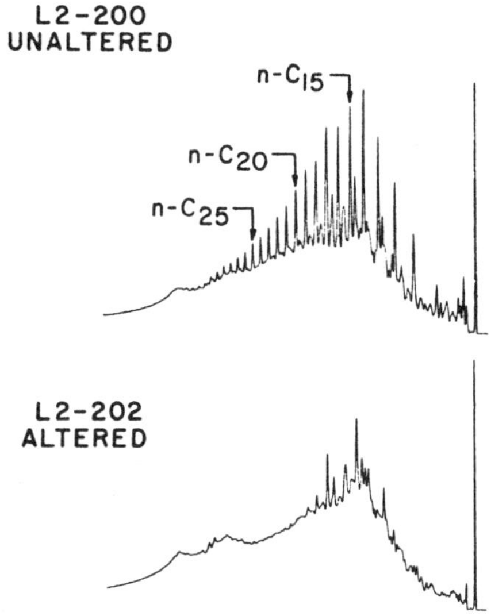

Figure 6. Gas chromatograms of aliphatic fractions of unaltered and altered oil from East and West Hackberry fields, Louisiana

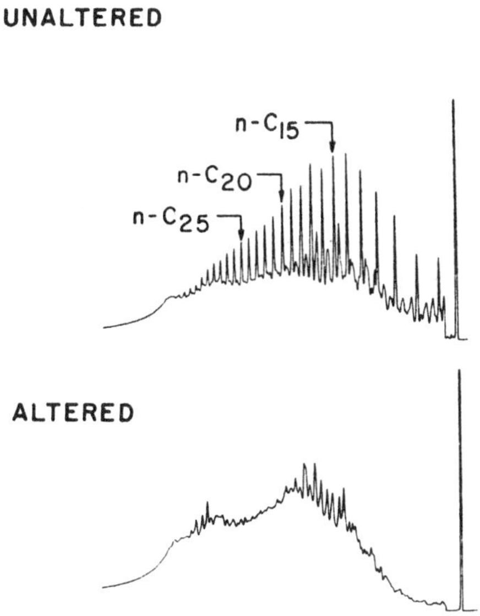

Figure 7. Gas chromatograms of aliphatic fractions of unaltered and altered oil from North Africa

concentrated by reduced pressure evaporation prior to injection into the chromatograph, so there is some loss of n-paraffins up to about C_{12} during sample preparation. Separation was made on a 4' x 1/8" thin wall stainless steel column packed with 3 percent OV-1 on 60-80 mesh Gas-chrom Q, using 40 ml/min nitrogen flow. Each run was made isothermally at 100°C for three minutes, then programmed at 13°C/min up to 350°C.

The chromatogram for an altered crude does not have the regularly spaced n-paraffin peaks which are quite prominent with the aliphatic fraction of most crudes. The three major iso-prenoids--C_{18}, C_{19}, and C_{20}--present in most crudes appear to be relatively unaffected and project well above the other non-normal hydrocarbons. On these chromatograms there is not enough detail below C_{10} to determine the extent of n-paraffin loss relative to non-normal constituents. However we have observed, without exception, that all crudes that have higher n-paraffins missing also have a marked deficiency of C_6-C_9 n-paraffins as shown by separate gas chromatographic analysis. The C_4 and C_5 n-paraffins also show some reduction relative to their non-normal isomers, but the deficiency is not as great as is observed among the higher n-paraffins. We expect that n-butane and n-pentane would be less affected by microorganisms than would higher homologs because the smaller molecules would be less accessible, and in addition, the energy requirements for microbial oxidation should be higher for the shorter-chain molecules.

Although numerous examples of n-paraffin removal from crude oils have been observed, this information alone is not conclusive evidence that microbial action on oil actually takes place in the reservoir. Some recent analyses of reservoir fluids from Bell Creek field in Montana provide important direct evidence on the question of subsurface alteration of crude oil by microorganisms. Bell Creek field, shown in Figure 3, is a recent discovery near the northern edge of the Powder River Basin, lying near the Montana-Wyoming border. It is already one of the largest fields in the basin. Production from sands in the Cretaceous Muddy formation is by natural pressure flow, and secondary recovery methods have not yet been initiated. There is therefore little chance of artificial contamination of the reservoir. Like other Muddy reservoirs in the Powder River Basin, Bell Creek is generally believed to have derived its oil from the organic-rich Cretaceous shales which essentially enclose the producing sands. Ranch Creek field, to the southwest of Bell Creek, was discovered before Bell Creek, but is now considered to be part of the same reservoir system. An important feature of the Bell Creek-Ranch Creek reservoir system is that it is located near the edge of the basin and is favorably situated to be reached by oxygen-containing surface waters. Average reservoir depth in the field is about 4,500 feet. Reservoir temperatures are about 95-105°F, which is sufficiently low for microorganisms to be active.

The n-paraffin distributions of 12 oil samples from various wells in Bell Creek field show some unusual characteristics. One might expect all oil samples from the same reservoir to have very similar composition, but those from Bell Creek show considerable variation. Figure 4 shows the chromatograms obtained from three different oil samples. The top chromatogram shows a n-paraffin distribution quite similar to that observed in most Powder River Basin Muddy oils. No significant alteration is indicated. The bottom chromatogram shows the entire range of n-paraffins, as signified by the sharp, prominent, regularly spaced peaks, to be missing. The crude oil composition shown in the middle chromatogram is quite unusual in that the n-paraffins are missing only up to about 17 or 18 carbon atoms, but the higher carbon number n-paraffins are all present in about the same proportion as they are found in unaltered Muddy oils. Some of the Bell Creek oils analyzed gave chromatograms which suggested intermediate stages of alteration.

Figure 5 shows an outline of Bell Creek field with the approximate location of each sampled well. At each sample location is a number signifying the type of n-paraffin chromatogram obtained. Number 1 is the unaltered type, 2 is the type with all n-paraffins missing, and 3 represents the oils with n-paraffins missing only up to C_{17}. There seems to be a progressive change from 1 to 2 to 3 going toward the northeast.

The reason for this observed variation in n-paraffin distributions of Bell Creek oils is not completely clear. The physical state of the oil at the time when presumed microbial attack was initiated may be an important factor. In the extreme northeast part of the field, the higher n-paraffins may not be in solution but rather in some form of solid solution in the oil. In such a case, they would not be free to diffuse to the oil-water interface, where microorganism activity must take place. The lower-molecular-weight paraffins would still be in the liquid phase and thus able to diffuse to the reaction interface.

The presence of a n-paraffin solid solution in an oil would depend primarily on

1. the ability of the oil to act as a solvent, and
2. reservoir temperature.

Oil originally emplaced in a reservoir should have nearly uniform composition, so the ability of the oil to retain n-paraffins in solution should be nearly the same throughout the field. Temperature should therefore play an important part in determining whether part or all of the n-paraffin range is affected. Bell Creek does not seem to have much temperature variation in the reservoir; the observed range is probably within the accuracy of bottom-hole temperature measurements.

Table I

NITROGEN CONTENT OF CRUDES FROM BELL CREEK
AND OTHER MUDDY RESERVOIRS

Sample	Field	Extent of n-Paraffin Loss	% Nitrogen
M8-97	Bell Creek	All	0.14
M8-98	Bell Creek	All	0.15
M8-104	Bell Creek	All	0.14
M8-107	Bell Creek	Below C_{18}	0.14
M8-108	Bell Creek	Below C_{18}	0.14
M8-102	Bell Creek	Partial loss, entire range	0.09
M8-99	Ranch Creek	None	0.05
W4-306	Recluse	None	0.04
W4-315	South Sandbar	None	0.04

Table II

OPTICAL ROTATION OF BELL CREEK OILS

Sample	Extent of n-Paraffin Loss	Specific Rotation Aliphatic Fraction
M8-97	All	1.05
M8-98	All	0.86
M8-103	All	0.95
M8-104	All	0.88
M8-105	All	0.86
M8-106	Almost all, entire range	0.89
M8-107	Below C_{18} only	0.80
M8-108	Below C_{18} only	0.78
M8-102	Partial loss, entire range	0.46
M8-96 *	None	0.46
M8-99 *	None	0.48
M8-101*	None	0.39

*Samples from Ranch Creek Field.

Table III

LIGHT ENDS ANALYSIS OF BELL CREEK OILS

Sample	Extent of n-Paraffin Loss	$n-C_7$/MCH	3MH/MCH	CH/MCP	2,4DMP/3MH
M8-97	All	0.15	0.50	0.61	0.08
M8-107	Below C_{18} only	0.02	0.63	0.54	0.08
M8-99*	None	0.83	0.55	0.84	0.09

*Oil from Ranch Creek

$n-C_7$.......normal heptane
MCH........methylcyclohexane
3MH........3-methylhexane
CH.........cyclohexane
MCP........methylcyclopentane
2,4DMP.....2,4-dimethylpentane

Table IV

LIGHT ENDS ANALYSIS OF EAST AND WEST HACKBERRY OILS

Sample	Producing Zone	Depth (ft.)	Extent of n-Paraffin Loss	$n-C_7$/MCH	3MH/MCH	CH/MCP	2,4DMP/3MH
L2-197	Camerina D	9,700	None	0.59	0.29	1.37	0.15
L2-200	Camerina 2	10,900	None	0.55	0.26	1.46	0.15
L2-203	Camerina A	4,800	None	0.45	0.21	1.51	0.13
L2-202	Amphestegina B	5,700	All	< 0.01	< 0.01	1.46	5.90

However, a very small temperature decrease may be sufficient to cause solid solution formation in the oil, with resulting immobilization of higher n-paraffins. This kind of effect would likely be observed only in crudes with high n-paraffin content, such as Bell Creek oil. Most crudes probably have enough solvent ability to retain all n-paraffins in solution even at the lowest subsurface temperatures, and any microbial oxidation would thus affect the entire range of n-paraffins.

The complete lack of alteration observed in the extreme southwest part of the reservoir is apparently due to factors other than temperature. Measured temperatures in the Ranch Creek end of the reservoir average about 5 to 10 degrees warmer than those found in Bell Creek, but are probably not sufficiently high to prevent microorganism activity. Presently available information is not adequate to determine whether the lack of alteration might be due to inability of paraffin-oxidizing microorganisms to reach this part of the reservoir, or to a relative inaccessibility to fresh oxygen-containing waters resulting in a microbial-oxidation rate so slow that alteration cannot yet be detected, or to other changes in reaction environment across the reservoir.

The importance of the data on Bell Creek oils is simply that we have failed to find any natural process, other than a microbiological one, that can account for the observed variations in n-paraffin distribution. From the sharp variations in composition, we conclude that if microbial activity has actually occurred, it is either still going on or has only recently stopped Otherwise, the oil should have had time to equilibrate and would be more uniform in composition throughout the reservoir.

More direct evidence of microorganism activity in Bell Creek field has also been found. Waters co-produced with some of the oil samples were cultured at our Whiting Research Laboratory. Cultures of some of the waters produced aerobic microorganisms that consume both n-butane and n-hexadecane at reservoir temperatures, while some did not. The results matched the wells in which gas chromatographic data indicated there is, and is not, composition change due to microorganisms.

Other evidence for microbial alteration of Bell Creek oil was obtained from nitrogen analyses and optical rotation measurements. Table I gives the nitrogen content of some of the Bell Creek oils. Altered oils from the middle and updip parts of the field contain three times as much nitrogen as the unaltered oil. Nitrogen content of two unaltered Muddy oils from other Powder River Basin fields is also included for comparison. Data published by Wenger and Reid (2) show that altered oils contain up to four to five times as much nitrogen as unaltered oils from the same producing formation. Some increase in nitrogen content is to be expected, of course, because nitrogen-containing compounds in the crude are concentrated by removal of paraffins However, the amount of paraffins lost should not exceed 25 percent of the original amount of oil, and the actual loss of material from the oil is probably less. Even this amount of concentration is far from sufficient to explain a threefold increase in nitrogen. A large portion of the nitrogen present in the altered oil must then be a product of microbiological activity, possibly amino acids derived from the proteinaceous material normally produced by paraffin-oxidizing microorganisms.

Table II shows the optical rotation of Bell Creek oils. The specific rotation of the aliphatic fraction of the altered oils is about twice that of unaltered oils. As is the case with nitrogen content, the optical rotation should be somewhat higher in altered oils because the n-paraffins do not contribute to optical activity. However, a twofold increase is more than would be expected from just the loss in n-paraffins. Again, it appears that compounds derived from the microorganism activity have been incorporated in the oil. The crudes which have lost only the lower-molecular-weight paraffins have rotations only slightly lower than the oils from which all n-paraffins have been removed. The optical rotation measurements were made on fractions from which most of the compounds through about 12 carbon atoms were lost by evaporation. The resulting C_{13}+ fractions for the partially altered and unaltered oils should then be different only in amount of C_{13}-C_{17} n-paraffins present, yet the rotation of partially altered oil is almost twice that of unaltered oil. These data support the possibility of optically active compounds produced by microorganisms being added to the original oil

Results of light ends analysis of three Bell Creek oils typical of each of the three types already discussed are given in Table III. The methylcyclohexane/n-heptane ratio in the altered oils is indicative of the loss of n-paraffins in the light ends. The methylcyclohexane/3-methylhexane, cyclohexane/methylcyclopentane, and 2,4-dimethylpentane/3-methylhexane ratios are relatively constant throughout the field, indicating that hydrocarbon types other than n-paraffins, but with similar volatilities, have not been significantly affected.

Although Bell Creek field has been the most intensively studied from the standpoint of possible microbiological activity, numerous other examples of the same type of alteration have been observed One of the more interesting examples was noted in the analysis of crudes from East and West Hackberry fields in Cameron Parish, Louisiana. These fields produce from several different Miocene and Oligocene reservoirs situated on the flanks of a salt dome. Oil in the youngest producing zone, of Upper Miocene age, has been severely altered, while the oil in all the older zones has apparently not been affected. Figure 6 shows typical n-paraffin distributions of both the altered and unaltered oils. Altogether, 5 altered oils, all from the youngest producing zone, and 21 unaltered oils from the two fields were analyzed. Actually, the shallowest oil analyzed, from a depth of 4,800 feet, did not come from the youngest zone. This oil was unaltered. Oxygen-containing waters may have been able to reach only the youngest zones even though production from the older zones is sometimes shallower and in the temperature range favorable for microorganism activity.

Light ends analysis of the East and West Hackberry field oils provides some unusual data, as shown in Table IV. In this case, the altered oil shows a marked loss of both n-paraffins and singly branched isoparaffins, as indicated by the four isomer ratios. The MCH/n-C_7 ratio shows an almost complete loss of n-paraffins from the altered oil. The MCH/3MH ratio is also much higher in the altered oil, suggesting that the isoparaffins have also been greatly affected. The CH/MCP is nearly constant for all four oils, so it does not appear that the light cycloparaffins have been affected. The sharp increase in the 2,4 DMP/3MH ratio for the altered oil is evidence that singly branched isoparaffins have been destroyed to a much greater extent than doubly branched ones, at least in the low carbon number range. Apparently, the microorganisms begin to consume the more difficult branched paraffins after the n-paraffins have been destroyed.

Another example of apparent microbiological alteration is shown on Figure 7. An oil from a wildcat well in North Africa shows extensive n-paraffin loss when compared to another oil believed to be from the same source. Light ends analysis also indicates some isoparaffin loss. In this example, the optical rotation is increased almost threefold, which is much too great an increase to result merely from the concentration of optically active compounds accompanying n-paraffin loss.

About 10 percent of all crude oils analyzed in our laboratories have compositional characteristics strongly suggestive of microbial alteration. The percentage of altered oils varies from one producing province to another, but at least a few examples can be found in almost any major oil-producing area of the world. Several giant fields are included among those which have apparently been altered. A consideration of the effects of microbiological activity on crude oil in the reservoir should therefore be of considerable importance to petroleum geochemists.

ACKNOWLEDGMENTS

We wish to acknowledge the important contributions made by some of our coworkers at the Pan American and American Oil Company Research Laboratories. J. A. Momper, who for the past several years has assimilated much information on microbiological processes in geologic environments, provided valuable assistance in development of the ideas presented in this paper. O. P. Funderburk specially prepared the gas chromatograms and aided in interpretation of optical rotation data. K. C. Chao cultured the water samples from Bell Creek and provided the necessary interpretation of the information he obtained. E. Field spent valuable time discussing the question of possible subsurface microbiological alteration, and offered several helpful suggestions.

LITERATURE CITED

(1) Davis, J. B., "Petroleum Microbiology", Elsevier (Amsterdam) 1967, p 110.

(2) Wenger, W. J., and Reid, B. W., Wyoming Geol. Assn. Guidebook, 13th Annual Field Conference, 148 (1958).

(3) Jones, T. S., and Smith, H. M., Fluids in Subsurface Environments--a Symposium, Amer. Assoc. of Petroleum Geologists, Memoir No. 4, 101 (1965).

(4) Berg, R. R., and Davies, D. K., Bull. Amer. Assoc. Petr. Geol. 52, No. 10, 1888 (1968).

39

Reprinted from *Science* **176**:1120–1122 (1972)

OIL POLLUTION: PERSISTENCE AND DEGRADATION OF SPILLED FUEL OIL

Max Blumer and Jeremy Sass

Abstract. *In September 1969, approximately 600 metric tons of number 2 fuel oil were spilled in Buzzards Bay, Massachusetts. Two years later, fuel oil hydrocarbons still persisted in the marsh and in offshore sediments. Hydrocarbon degradation is slow, especially below the immediate sediment surface and appears to proceed principally through microbial utilization of alkanes and through partial dissolution of the lower-boiling aromatic hydrocarbons. The boiling range of the spilled oil and the relative abundances of homologous hydrocarbons (for example, phytane and pristane) have been well preserved. The findings are in agreement with the known geochemical stability of hydrocarbons. Fuel oil is an appreciable fraction of whole crude oil. This fact suggests that oil products and crude oils have a considerable environmental persistence.*

Oil pollution of the sea and the public awareness of it have increased rapidly in recent years. In spite of this, little information exists on the persistence and long-term effect of hydrocarbon pollutants in the marine environment.

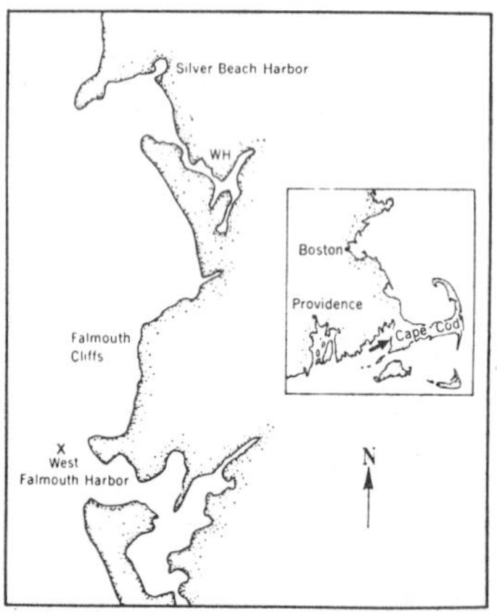

Fig. 1. Site of the September 1969 Buzzards Bay oil spill (×). Samples described here were taken at Silver Beach Harbor and Wild Harbor River (*WH*). For analyses from additional sites, see (*3*).

The fate of fuel oil from a spill in Buzzards Bay, off West Falmouth, Massachusetts, on 16 September 1969 (Fig. 1) has been under investigation for more than 2 years. The relatively small spill involved the discharge from a stranded barge of approximately 600 metric tons of number 2 fuel oil into the coastal waters and marshes. It led to persistent pollution of the sediments and of the fisheries resources (*1–3*).

Among the events that followed the 1969 spill, we can distinguish three distinct, although partly overlapping, series of events. Within the first few hours or days after the accident there was a heavy kill of organisms that came into contact with the oil; the effect extended over all phyla and over benthic and intertidal organisms (*2, 4*).

Next, within weeks or months after the spill, the oil spread to areas that had not been affected initially, and the kill extended, although in some cases more slowly than the spread of the oil, to outlying areas. For a considerable time after the spill, the oil prevented resettlement of the sediments by the original fauna. At present, degradation of the oil is evident; concurrently with the chemical changes in the oil, the immediate toxicity of the oil in the sediment has been reduced. This reduction in the toxicity has permitted resettlement of the polluted region, first by the most resistant faunal opportunists and later by a more varied and normal fauna (*4*).

Our analyses show the slow decrease of the oil content of the sediments from the high initial values in the marshes [Wild Harbor River (Fig. 2A), predominantly sand with clay layers] and offshore [station 31, Silver Beach Harbor (Fig. 2B), predominantly clay with sand; water depth, 1.5 m]. At both locations the oil content in the uppermost 2½ cm of sediment has remained well above the content of the indigenous sedimentary hydrocarbons, which average 5 to 7 mg per 100 g of dry sediment. The steep increase in the oil content at station 31 between December 1969 and March 1970 is attributed to the redistribution of oil from the same spill, since the gas chromatograms of the sediment extracts before and after this time agree in terms of boiling range, boiling point distribution, and characteristic detail and since no new spill occurred in that area. The degradation of the oil appears to involve both bacterial utilization and partial dissolution. Our experience with field samples confirms qualitatively those laboratory observations that indicated a decrease in the degradation rate from normal alkanes to iso- and cycloalkanes and to aromatic hydrocarbons. The ratio of *n*-heptadecane to pristane [*n*-C_{17}/pristane (Fig. 3, B and C)], is a sensitive indicator of incipient oil degradation (*2, 3*). It decreases more rapidly in the marshes and in lightly polluted offshore sediments than at the heavily polluted station 31; presumably, this reflects differences in the availability of oxygen. At station 31, the *n*-C_{17}/pristane ratio remained nearly constant for the first 3 months after the spill; a slight decrease in December 1969 was followed by a rise, which coincided with the steep rise in pollution level at that station in March 1970. A similar trend in the *n*-C_{17}/pristane ratio at more distant stations during the same period suggests the seaward spread of the fuel oil from the heavily polluted marshes and inshore regions during the first winter after the spill (*2, 3*).

The rate of bacterial degradation, even of the *n*-alkanes, is much lower in the environment than under laboratory conditions with the use of aeration, nutrients, agitation, and elevated temperatures. After 2 years small amounts of *n*-alkanes still persisted.

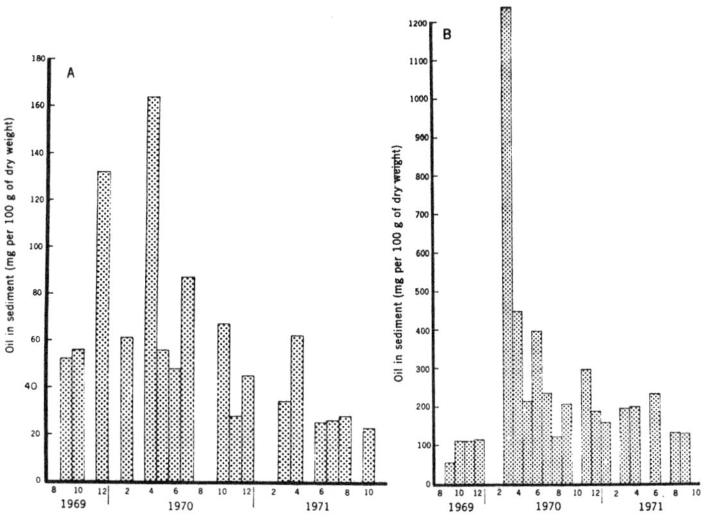

Fig. 2. Hydrocarbon content of marsh and offshore sediments, affected by the fuel oil spill in Buzzards Bay. (A) Wild Harbor River stations 2, 4, and 5 (average); (B) Silver Beach Harbor station 31. The weight of oil was determined according to the method of Blumer and Sass (*3*).

This finding is not surprising, however, in view of the well-documented geochemical stability of sedimentary hydrocarbons (*5*). Branched and cyclic hydrocarbons are attacked even more slowly than the *n*-alkanes; after 2 years, isoprenoids (phytane, pristane, and the C_{18} homolog are evident) and alicyclic and aromatic hydrocarbons remained prominent in the polluted sediments.

Mass spectral analyses of the aromatic hydrocarbons at station 31 from March 1970 to April 1971 show a relative increase in the more highly substituted benzenes, naphthalenes, and tetrahydronaphthalenes at the expense of the lower homologs, especially of the most soluble compounds, such as naphthalene and the C_1 to C_3 alkylnaphthalenes. This result suggests that dissolution is more important than bacterial utilization of the aromatic hydrocarbons; the latter process would not show the same molecular weight dependence.

The ratio of saturated to aromatic hydrocarbons in the fuel oil in the sediments at station 31 remained nearly constant from March 1970 to April 1971. Thus the overall rate of depletion of the saturated and aromatic hydrocarbons is the same, even if there is preferred bacterial degradation or dissolution of some components within each group.

Analyses of a core from station 31 in September 1971 show the presence of fuel oil to a depth of 7.5 cm below

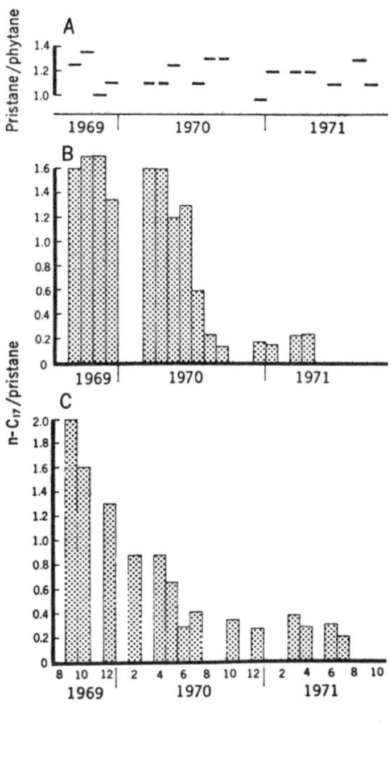

Fig. 3. The pristane/phytane ratio in sediments (A) at Silver Beach Harbor station 31 and the *n*-heptadecane/pristane ratio in sediments (B) at Silver Beach Harbor station 31 and (C) at Wild Harbor River station 4.

the sediment surface. On the basis of the n-C_{17}/pristane ratio, the oil below 2.5 cm is as fresh after 24 months as the oil at the surface was after 10 months (*3*). In the marshes penetration has been greater, extending to at least 60 cm, and bacterial degradation and dissolution are evident at that depth; this is probably the result of the greater permeability and aeration of the marsh sediments.

Some properties of the fuel oil have changed little, in spite of its gradual degradation. Initially the boiling range of the spilled oil extended from 170° to 370°C, and the normal alkanes ranged from decane to docosane. The boiling range of the fuel oil in the sediments is well preserved after 2 years; even at stations with low pollution level, C_{13} to C_{14} alkanes are still present. At station 31 and in the marshes dodecane can still be detected.

Adjacent members of the same or of closely related homologous series are affected by weathering to nearly the same degree; therefore, their concentration ratios change only slowly. Thus, the ratio of pristane to phytane at station 31 has remained 1.17 ± 0.09 during the 2-year period (Fig. 3A). This value differs characteristically from that encountered in unpolluted recent sediments (*6*). The pristane/phytane ratio and similar isomer ratios are distinguishing features that vary from one crude oil to another. That this ratio remains constant suggests that the identification of fossil fuel pollutants in the environment and their distinction from the natural hydrocarbon background may be possible over extended time periods.

Our continuing investigation will provide a framework for the consideration of the effect and fate of oil pollution in the coastal environment. Hydrocarbons in the boiling range of the oil spilled at West Falmouth are abundant in crude oil. Smith, in summarizing analyses of 6496 crude oils, showed that the "gas oil" content of the vast majority ranges from 20 to 35 percent, except for Tertiary oil, where higher values occur (*7*). Smith has defined gas oil as boiling between 200°C at 760 mm and 225°C at 40 mm (or 335°C at 760 mm). This range is comprised within that of this fuel oil (170° to 370°C, 760 mm). Thus, the "fuel oil" component, in the sense of the oil spilled at West Falmouth, constitutes one-quarter to one-third of the weight of a large majority of all crude oils. Therefore, we anticipate that the environmental effect and the persistence of a crude oil spill resembles that of a fuel oil spill qualitatively, and to a degree quantitatively. The presence in whole crude oils or residual oils of less rapidly degraded and less soluble hydrocarbons of higher molecular weight should lead to greater environmental persistence.

The preservation of hydrocarbons in marine sediments for geologically long time spans is one of the accepted key facts in the current thought on petroleum formation. Similarly, the uptake of dietary hydrocarbons in the food chain and their preservation in the lipids of organisms seem to be well documented. Our present findings do not contradict these background data.

References and Notes

1. M. Blumer, G. Souza, J. Sass, *Mar. Biol.* **5**, 195 (1970).
2. M. Blumer, J. Sass, G. Souza, H. Sanders, F. Grassle, G. Hampson, unpublished manuscript, reference 70-44, Woods Hole Oceanographic Institution, 1970.
3. M. Blumer and J. Sass, unpublished manuscript, reference 72-19, Woods Hole Oceanographic Institution, 1972.
4. H. Sanders, F. Grassle, G. Hampson, unpublished manuscript, reference 72-20, Woods Hole Oceanographic Institution, 1972.
5. G. T. Philippi, *Geochim. Cosmochim. Acta* **29**, 1021 (1965); N. P. Stevens, E. E. Bray, E. D. Evans, *Amer. Ass. Petrol. Geol. Bull.* **40**, 975 (1956).
6. M. Blumer and W. D. Snyder, *Science* **150**, 1588 (1965).
7. H. M. Smith, *U.S. Bur. Mines Bull. 642* (1968).

8. Contribution No. 2835 of the Woods Hole Oceanographic Institution. Supported by the Office of Naval Research (N00014-66, contract CO-241) and by the National Science Foundation (GA-19472). The work reported here is part of a continuing investigation; chemical and biological data available in November 1971 are compiled in (*3*) and (*4*), respectively.

Reprinted from *Proc. Natl. Acad. Sci.* **70**:1331–1334 (1973)

Racemization Reaction of Aspartic Acid and Its Use in Dating Fossil Bones

(Olduvai Gorge/5,000–70,000-years-old range/hominids)

JEFFREY L. BADA* AND REINER PROTSCH†

* Scripps Institution of Oceanography and Institute of Marine Resources, University of California, San Diego, La Jolla, Calif. 92037; and
† Department of Anthropology, Institute of Geophysics and Planetary Physics, University of California, Los Angeles, Los Angeles, Calif. 90024

Communicated by William A. Nierenberg, March 5, 1973

ABSTRACT In the time interval datable by radiocarbon, and at the temperatures of most archeological sites, a substantial amount of racemization of aspartic acid takes place. By determination of the amount of racemization of aspartic acid in bones from a particular location which have been dated by the radiocarbon technique, it is possible to calculate the *in situ* first-order rate constant for interconversion of the L- and D enantiomers of aspartic acid. Once this "calibration" has been calculated, the reaction can be used to date other bones from the deposit that are either too old to be dated by radiocarbon or that are too small for radiocarbon dating. The only assumption required with this approach is that the average temperature experienced by the "calibration" sample is representative of the average temperature experienced by older samples. This "calibration" technique is used herein to date bones from the Olduvai Gorge area in Tanzania, Africa.

Only L-amino acids are commonly found in living organisms. However, recent studies have shown the occurrence of D isomers in fossil materials. The amino acids in fossil shells (1–3), sediments (4–6), and bones (7, 8) are partially racemized, with the amount of racemization increasing with the age of the fossil. This racemization reaction has obvious applications in geochronology, and it appears that the reaction can be used to date deep-sea sediments (4–6) and fossil bones (7, 8) found in certain environments.

Most studies have concentrated on isoleucine. The epimerization of L-isoleucine produces the nonprotein amino-acid D-alloisoleucine. (For the reaction involving isoleucine, the process is more properly termed epimerization rather than racemization.) L-Isoleucine and D-alloisoleucine are directly separable on an automatic amino-acid analyzer. In contrast, in order to determine the amount of racemization of other amino acids, a suitable diastereomeric derivative must first be synthesized. The isoleucine epimerization reaction in bone has a half-life (i.e., the time required for the ratio of alloisoleucine to isoleucine to reach 0.345) at 20° in excess of 100,000 years (8), and evidence suggests that the reaction can be used to date fossil bones too old to be datable by radiocarbon. The rate of the amino-acid racemization reaction is dependent upon temperature. Therefore, in order to use the reaction to calculate ages which are reasonably accurate, some estimate of the temperature history of a fossil bone must be available. Due to the problem of temperature uncertainties, the isoleucine epimerization reaction can apparently only be used to date fossil bones found in certain environments (7, 8) such as caves where paleotemperatures might be estimated from $^{18}O/^{16}O$ ratios in stalagmites and stalactites.

Preliminary results obtained from modern bone fragments heated at elevated temperatures show that amino acids such as aspartic acid and alanine are racemized substantially faster than is isoleucine (Bada, J. L., Kvenvolden, K. A. and Peterson, E., in preparation). These results indicate the following sequence of racemization rates: aspartic acid > alanine ≅ glutamic acid > isoleucine ≅ leucine. This pattern was also found in several fossil bones.

Of particular interest is the racemization reaction of aspartic acid; this reaction can be written as

$$\text{L-aspartic acid} \underset{k_{asp}}{\overset{k_{asp}}{\rightleftharpoons}} \text{D-aspartic acid,} \qquad [1]$$

where k_{asp} is the first-order rate constant for interconversion of the L and D enantiomers of aspartic acid. In bone, this reaction has a half-life (i.e., the time required for the ratio of D- to L-aspartic acid to reach 0.333) at 20° of about 15,000–20,000 years (Bada, J. L., Kvenvolden, K. A. and Peterson, E., in preparation). Thus, in the time interval datable by radiocarbon, a substantial amount of racemization of aspartic acid will take place. This finding suggests that by determining the extent of racemization of aspartic acid in bones from a particular site which have been dated by the radiocarbon technique, it should be possible for one to calculate the *in situ* k_{asp} value for the deposit. Once this "calibration" has been done, the aspartic-acid racemization reaction can be used to date other samples from the area that are either too old to be dated by radiocarbon (i.e., older than 45,000 years), or for which insufficient amounts are available for radiocarbon dating (only a few grams are needed for amino-acid dating, compared to hundreds of grams needed for radiocarbon dating). This "calibration" procedure eliminates the need for evaluating the temperature history of a bone before it can be dated by the amino-acid racemization reaction. In this paper, we report how this approach was used to date fossil bones from the Olduvai Gorge area in Tanzania, Africa.

MATERIALS AND METHODS

The bones from the Naisiusiu Beds were sent to us by M. D. Leakey, Centre for Prehistory and Paleontology of the National Museum, Nairobi, Kenya. The actual location in Olduvai Gorge where the bones were collected is described elsewhere (9). About 10 g of bone were used for the amino-

* To whom reprint requests should be addressed.

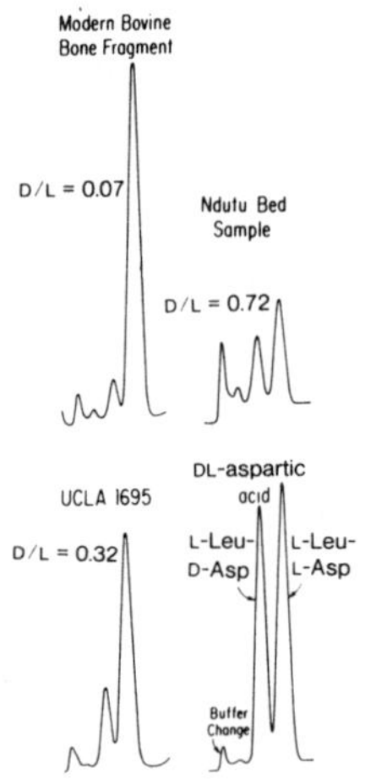

FIG. 1. Part of the amino-acid analyzer printouts for the various bone samples showing the diastereomeric dipeptides L-leucyl-D-aspartic acid and L-leucyl-L-aspartic acid. A 56 × 0.9-cm column filled with Beckman–Spinco AA-15 resin was used for the separation. The column was eluted with pH 3.24 buffer for 40 min, then the buffer was switched to pH 4.25 for the remainder of the run. The buffer change, L-Leu-D-Asp, and L-Leu-L-Asp peaks come at 72, 79, and 83 min, respectively, with this elution sequence. The small peak that elutes between the buffer peak and L-Leu-D-Asp is glycine that was not completely separated from aspartic acid on the Dowex 50 (H^+) column.

acid analysis; 412 g were used for the radiocarbon age determination.

The Ndutu Bed bone was provided by R. L. Hay, University of California, Berkeley. The sample was from a site located at the rim of the Gorge, between the Main and Side Gorges and 150 m south of archeological site FLK-NNI and 70 m west of the recently excavated site in the Masek Beds. At the sampled locality, the Ndutu Beds comprise 1.5 m of horizontal unstratified zeolitic eolian tuff, the lithology and mineral content of which indicate assignment to the upper unit of the Ndutu Beds. A piece of bone weighing about 10 g was used for the amino-acid analysis.

Several pieces of the Eyasi I Hominid bone amounting to 6.3 g were generously provided by H. Müller–Beck, University of Tübingen. Microanalysis of the pieces (determined at the Univ. of California, Los Angeles by the micro-Dumas method) gave 2.0% fluorine and 0.20% nitrogen, the same as those reported by Oakley and Campbell (10) for the skull Eyasi I. These measurements demonstrate the contemporaneity of the separate pieces and the skull. Before the amino-acid analysis, the Eyasi bones were refluxed in diethyl ether for 30 days in order to remove the preservative (shellac) that had been applied to the bones. After this treatment, the bones were processed in the manner described below.

Before amino-acid analyses, the fossil bones were carefully cleaned by repeated washings in dilute hydrochloric acid (0.2 M), using ultrasonication to loosen dirt and adhering soil. After cleansing, the bones were hydrolyzed in excess double-distilled 6 M HCl for 24 hr. Upon completion of the hydrolysis, the HCl solution was evaporated to dryness and the residue was dissolved in double-distilled water and desalted on unused Dowex 50W-X8 (100–200 mesh) resin, which had been cleaned with NaOH and regenerated with doubly-distilled HCl. The amino acids were eluted from the resin with 1.5 M NN_4OH prepared by dissolving NH_3 gas into double-distilled water. The effluent was concentrated in a rotary evaporator at 53° under reduced pressure.

The ratio of the D- to L- enantiomers of aspartic acid was determined by preparing diastereomeric dipeptides using the procedure described by Manning and Moore (11). Before the dipeptide synthesis, the aspartic acid was separated from the total amino-acid mixture by ion-exchange chromatography (12) on Dowex 50W-X8, using 1.5 M HCl to elute the aspartic acid. This separation step was necessary because the dipeptide synthesis of the total amino-acid mixture isolated from the bone resulted in a large number of peaks, many of which overlapped one another. After the ion-exchange separation, the isolated aspartic acid was taken up in borate buffer (pH = 10.4), the solution was cooled to 2–4°, and a small amount of L-leucine-*N*-carboxyanhydride (obtained from Cyclo Chemical Co.) was added to the solution in order to synthesize L-leucyl-D-aspartic acid and L-leucyl-L-aspartic acid. A Beckman–Spinco model 118 automatic amino-acid analyzer was used to separate the diastereomeric dipeptides.

The radiocarbon age of the Naisiusiu Bed bone was determined at the Univ. of California, Los Angeles Radiocarbon Laboratory by use of the method described by Protsch (13).

RESULTS

The Olduvai Gorge sequence consists of several stratigraphic sections, the oldest of which has an age of several million years. The uppermost part of the Gorge is upper Pleistocene (14, 15); this stratigraphic section is subdivided into a younger formation, the Naisiusiu Beds, and an older formation, the Ndutu Beds (9). Bones from the middle section of the Naisiusiu Beds were used to "calibrate" the aspartic-acid racemization reaction for Olduvai Gorge. Radiocarbon analysis of the collagen fraction isolated from the bones yielded an age of of 17,550 ± 1,000 yr (UCLA-1695). This age agrees closely with the age (9) of 17,000 yr (Lamont-no.?) determined on ostrich eggshell found in the same stratigraphic horizon and in close proximity to the bones.

Part of the amino-acid analyzer printout of the Naisiusiu Bed bone is shown in Fig. 1. Also included in Fig. 1 is the printout for a modern bovine-bone fragment and a DL-aspartic-acid mixture. The ninhydrin color yield is lower for the L-D dipeptide than for the L–L dipeptide (11). Thus, in order to obtain the actual ratio of D- to L- aspartic acid in a sample, the measured ratio must be multiplied by a conversion factor. The factor determined for our analyzer was 1.30.‡

The first-order rate constant (k_{asp}) for interconversion of the D and L enantiomers of aspartic acid can be calculated from the equation [see refs. (6) and (8) for derivation]

$$\ln\left[\frac{1 + D/L}{1 - D/L}\right] - \ln\left[\frac{1 + D/L}{1 - D/L}\right]_{t=0} = 2 \cdot k_{asp} \cdot t, \quad [2]$$

‡ We carefully tested the dipeptide method by preparing samples with D to L-aspartic-acid ratios ranging from 0.1 to 0.75, then converting these to the diastereomeric dipeptides. In all cases the D- to L-aspartic acid ratios estimated with the amino-acid analyzer were within 1–2% of the known values.

where t = time (in years) and D/L is the aspartic acid enantiomeric ratio in the bone. Kinetic studies at elevated temperatures have shown that the epimerization reaction of isoleucine in bones follows this equation essentially to equilibrium (8). The $t = 0$ term in Eq. [2] can be evaluated by setting t equal to zero and using D/L = 0.07, the aspartic acid enantiomeric ratio determined for a modern bovine bone carried through the same analytical steps as the fossil bones. The calculated value for the $t=0$ term is 0.14.

The D to L-aspartic-acid ratio determined for the Naisiusiu Bed bone was 0.32. Substitution of this ratio and the radiocarbon age of 17,550 ± 1000 yr into Eq. [2] yields $k_{asp} = 1.48 \pm 0.09 \times 10^{-5}\ yr^{-1}$; this value thus represents the average value of k_{asp} for Olduvai Gorge over about 18,000 yr.

It should be possible to use this k_{asp} value to estimate the age of other fossil bones from Olduvai Gorge. To test this proposal, amino-acid analyses were performed on a bone from the Ndutu Beds. The Ndutu Beds are older than the Naisiusiu Beds and appear to be beyond the range of radiocarbon dating (9). The Ndutu Bed bone was found to have a D to L aspartic-acid ratio of 0.72; part of the amino-acid analyzer printout for this bone is also shown in Fig. 1. Substitution of D/L = 0.72 and $k_{asp} = 1.48 \pm 0.09 \times 10^{-5}\ yr^{-1}$ into Eq. [2] gives an age of 56,000 ± 3,500 yr.

Analysis of fossil shells from the upper unit of the Ndutu Beds indicated a radiocarbon age of older than 29,000 yr (9). Based on mineral alterations, the upper Ndutu Beds have been estimated (9) to range in age from 30,000 to 70,000 years old. The artifacts associated with these beds appear to represent a Middle Stone Age culture (9). Thus, the age of 56,000 yr deduced for the Ndutu Bed bone using the aspartic-acid racemization reaction is in agreement with the general age range predicted from other information.

It should be emphasized that the only assumption concerning the temperature, and therefore the value of k_{asp}, made in this calculation is that the average temperature at Olduvai Gorge over the last 18,000 yr is representative of the average temperature over the last 56,000 yr. This assumption seems justified based on the estimated temperature fluctuations during the Pleistocene glacial cycles for the tropical Atlantic Ocean (16) and for a cave in New Zealand (17). The estimated average temperatures in these environments over the last 20,000 yr is only about 0.5° higher than the average temperature over much longer periods. If a similar situation existed in Olduvai Gorge, then the age of the Ndutu Bed bone would be increased from 56,000 yr to about 61,000 yr, assuming that the aspartic-acid racemization reaction has the same temperature dependence as the isoleucine reaction (8), which is the case in aqueous solution (18).

The value of k_{asp} determined above should be applicable not only to Olduvai Gorge, but also to bones from sites in the general vicinity. In fact, we have used the aspartic-acid racemization reaction "calibrated" by the Olduvai Gorge bones to estimate the age of some hominid bones found at Lake Eyasi, a site located about 65 k southwest of Olduvai Gorge. Fragmentary pieces of three hominid skulls were found on the northeast shore of Lake Eyasi in 1935 by a German expedition (19). A reconstruction of one of the skulls (Eyasi I) has been generally considered to display *Homo erectus* affinities (19, 20). However, the associated faunal and cultural material were upper Pleistocene (21, 22). More recent authors (23) have suggested that the Lake Eyasi hominids might belong to the African Neanderthals, and could thus be contemporaneous with other African hominids such as Broken Hill, Saldanha, and Florisbad. Fauna associated with the Saldanha and Florisbad hominids have recently been dated by radiocarbon (13); the ages are Saldanha, 40,680 ± 2000 yr (UCLA 1742) and Florisbad, 38,683 ± 2000 yr (UCLA 1745B). The Broken Hill skull has been roughly estimated (24) to have an age of about 40,000 yr, i.e., early Middle Stone Age. There have been no attempts to estimate the age of the Lake Eyasi hominids by radiocarbon because of the small amount of bone which was found, and because of the unavailability of any associated faunal matrial.

The ratio of D- to L-aspartic acid for the Eyasi I Hominid bone was found to be about 0.5. Based on the k_{asp} value determined for Olduvai Gorge, this ratio indicates an age of 34,000 yr. The only factor that could make this age greatly in error is if the temperatures at Lake Eyasi and Olduvai Gorge were substantially different. However, this is not the case: the present day climates in the two regions are essentially the same (R. L. Hay, personal communication). The 34,000-yr age does indeed make the Lake Eyasi hominids broadly contemporary with Saldanha and Florisbad, and possibly with Broken Hill also. A more detailed discussion of the archeological and anthropological implications of the Lake Eyasi results will be given elsewhere.

DISCUSSION

These results illustrate well the application of the amino-acid racemization reaction to the dating of fossil bones. By use of the approach described here, the aspartic-acid racemization reaction can be used to date fossil bones that have ages ranging from 5,000 yr up to 60,000 and 70,000 yr, the actual range dependent upon the temperature of the location. By use of a similar approach with other amino acids such as alanine, which racemize at rates slower than that of aspartic acid, it should be possible to date bones as old as 100,000–150,000 yr. With the isoleucine epimerization it might be possible to date bones as old as several hundred thousand years, but it will be difficult to "calibrate" this reaction since only small amounts of epimerization would take place during the time interval datable by radiocarbon. It may be possible to calibrate the isoleucine reaction by use of bones dated by the racemization of other amino acids.

The amino-acid dating method extends well beyond the time range applicable to radiocarbon. Furthermore, because of the small amounts of material needed, even bones thought to be datable by radiocarbon but which were available only in small quantities could be dated by use of the appropriate amino-acid racemization reaction. By using expendable faunal material to "calibrate" the racemization reaction, only a few grams of valuable hominid material would have to be used for amino-acid dating. Thus, the amino-acid racemization reaction should prove to be a valuable tool in paleoanthropology and provide a powerful chronological aid in unraveling the history of human evolution.

We are grateful to M. D. Leakey, R. Hay, and H. Müller-Beck for providing us with the bone samples. Also, we thank C. Lee, R. Berger, and S. Miller for reviewing the manuscript and R. A. Schroeder and Lydia Deems for technical assistance. Acknowledgement is made to the donors of the Petroleum Research Fund, administered by the American Chemical Society, and to the L.S.B Leakey Foundation for partial support of this research. The radiocarbon work was supported by NSF Grant GA-4349.

1. Hare, P. E. & Mitterer, R. M. (1967) *Carnegie Inst. Washington Yearb.* **65**, 362–364.
2. Hare, P. E. & Abelson, P. H. (1968) *Carnegie Inst. Washington Yearb.* **66**, 526–528.
3. Hare, P. E. & Mitterer, R. M. (1969) *Carnegie Inst. Washington Yearb.* **67**, 205–208.
4. Bada, J. L., Luyendyk, B. P. & Maynard, J. B. (1970) *Science* **170**, 730–732.
5. Wehmiller, J. F. & Hare, P. E. (1971) *Science* **173**, 907–911.
6. Bada, J. L. & Schroeder, R. A. (1972) *Earth Planet. Sci. Lett.* **15**, 1–11.
7. Turekian, K. K. & Bada, J. L. (1972) in *Calibration of Hominoid Evolution*, eds. Bishop, W. W. & Miller, J. A. (Scottish Academic Press, Edinburgh), pp. 171–185.
8. Bada, J. L. (1972) *Earth Planet. Sci. Lett.* **15**, 223–231.
9. Leakey, M. D., Hay, R. L., Thurber, D. L., Protsch, R. & Berger, R. (1972) *World Archeol.* **3**, 328–341.
10. Oakley, K. P. & Campbell, B. G. (1967) *Catalogue of Fossil Hominids. Part I: Africa* (British Museum, London), p. 106.
11. Manning, J. M. & Moore, S. (1968) *J. Biol. Chem.* **243**, 5591–5597.
12. Wall, J. S. (1953) *Anal. Chem.* **25**, 950–953.
13. Protsch, R. (1973) Ph.D. Dissertation, Department of Anthropology, University of California, Los Angeles, pp. 233.
14. Leakey, L. S. B., Protsch, R. & Berger, R. (1968) *Science* **162**, 559–560.
15. Leakey, L. S. B. (1969) *Science* **166**, 532.
16. Emiliani, C. (1971) *Science* **171**, 571–573.
17. Hendy, C. H. & Wilson, A. T. (1968) *Nature* **219**, 48–51.
18. Bada, J. L. (1971) *Advan. Chem. Ser.* **106**, 309–331.
19. Kohl-Larsen, L. & Reck, H. (1936) *Geol. Rundch.* **27**, 401–441.
20. Weinert, H. (1939) *Z. Morphol. Anthropol.* **38**, 252–307.
21. Leakey, L. S. B. (1936) *Nature* **138**, 1082–1084.
22. Leakey, L. S. B. (1946) *J. East Afr. Nat. His. Soc.* **19**, 40–43.
23. Tobias, P. V. (1968) in *Evolution and Hominisation*, ed. Kurth, G. (Fisher Verlag, Stuttgart), p. 187.
24. Oakley, K. P. (1968) in *Frameworks for Dating Fossil Man* (Aldine, Chicago), p. 317.

41

Reprinted from *Earth and Planetary Sci. Letters* **28**:275–282 (1975)

AGES AND DIAGENETIC TEMPERATURES OF PLEISTOCENE DEPOSITS OF FLORIDA BASED ON ISOLEUCINE EPIMERIZATION IN *MERCENARIA* *

RICHARD M. MITTERER

Institute for Geosciences, The University of Texas at Dallas, Richardson, Texas (USA)

Received August 4, 1975
Revised version received September 25, 1975

The epimerization reaction of isoleucine in skeletal carbonates provides a basis for estimating the age and the average diagenetic temperature experienced by fossils. The kinetics of this reaction in *Mercenaria*, a bivalve, are established by a combination of elevated temperature heating experiments and analysis of Holocene fossils of known age and diagenetic temperature. The Arrhenius activation energy for the reaction is 29.4 kcal/mole. Application of the kinetic relations to Pleistocene marine deposits of Florida yields ages of about 134,000, 180,000, 236,000 and 324,000 years for these deposits. These ages correspond to the times of eustatic high sea levels dated radiometrically throughout the world. Average diagenetic temperatures for the Late Pleistocene in Florida were about 12°C less than the present values; however, the average temperature gradient was close to the present gradient.

1. Introduction

After burial and fossilization the original protein in mollusk shells and other skeletal structures undergoes a series of diagenetic changes, which include hydrolysis to the constituent amino acids, partial conversion of these to non-protein amino acids or other compounds, and the racemization or epimerization of the original L-amino acids to D,L compounds. One basis for quantitating the extent of amino acid diagenesis in fossils is the reversible, first-order reaction in which isoleucine (ileu) interconverts to its non-protein diastereoisomer, alloisoleucine (alleu):

$$\text{L-isoleucine} \underset{k_2}{\overset{k_1}{\rightleftharpoons}} \text{D-alloisoleucine} \qquad (1)$$

Alloisoleucine is initially present at trace levels or absent in living organisms; after death and burial, alloisoleucine forms and slowly increases relative to isoleucine with increasing age of the fossil to an equilibrium ratio of alleu/ileu ≅ 1.30. The systematic interconversion of isoleucine and alloisoleucine in fossils and in elevated temperature experiments has been noted for mollusk shells [1–5], foraminifera [6–9], and bones [10]. Since the alleu/ileu ratio is a function of the age of a sample and the diagenetic temperature to which it has been exposed, it is feasible to deduce ages or paleotemperatures of fossils from a knowledge of the kinetic relationships of eq. 1. In addition, there are a number of other potential applications, including stratigraphic correlation, identification of reworked fossils, and verification of radiocarbon dates [4]. The latter application is particularly significant for carbonate fossils in the critical range of 25,000–35,000 years B.P. Many radiocarbon dates provide evidence of a high sea level in this time interval (e.g., [11,12]). Considering that a slight contamination of fossils beyond the range of the radiocarbon method will result in erroneous dates, an independent check on such dates would be useful.

In order to estimate ages and paleotemperatures, the kinetics of the reaction must be known. There are, however, several basic sources of uncertainty in apply-

* Contribution No. 292, Institute for Geosciences, The University of Texas at Dallas, P. O. Box 688, Richardson, Texas 75080.

ing results of laboratory heating experiments to temperature and age determinations. These are: (1) the uncertainty in the equilibrium value of alleu/ileu which varies from 1.25 in heated specimens to about 1.40 in fossil shells [1,2]; (2) the uncertainty in k_1 calculated by extrapolating rate data determined at elevated temperatures down to the temperature range of 10–25°C; and (3) the uncertainty that the elevated temperature rate law is the same as the low temperature rate law. Consequently, in the absence of any other confirming evidence, extrapolation of rate data from elevated temperatures to lower values may yield inaccurate estimates of paleotemperatures or ages. One kind of confirmatory evidence is obtained by empirically comparing independently determined ages or temperatures with those calculated from the kinetic data. Another approach is to use samples of known age as a "calibration" for dating unknown fossils [13].

The purpose of this study is twofold: (1) to obtain more detailed kinetic data for interconversion of isoleucine and alloisoleucine in *Mercenaria*; and (2) to make preliminary estimates of paleotemperatures and ages obtained by application of the rate equation to Late Pleistocene specimens.

2. Materials and methods

Mercenaria (sp.), a bivalve presently found along the eastern U.S. and Gulf of Mexico coasts, was chosen for this study because it is a common fossil in Late Cenozoic deposits of these regions and has been used in a number of previous investigations of amino acid diagenesis [1–5].

Mercenaria shells characteristically have three discrete aragonitic structural layers. Amino acid analyses of the individual layers reveal significant differences in composition [14]. To avoid the unknown effects these differences might introduce in the rate of isoleucine–alloisoleucine interconversion, the middle layer, which is easily separated with a dental tool, was sampled consistently for all analyses.

For the heating experiments, approximately 200-mg chips from a fresh *Mercenaria* shell, exclusive of muscle scars and beak area, were sealed under nitrogen in pyrex tubes containing 1 ml of deionized-distilled water and heated at various temperatures in refluxing organic liquids. At the end of the heating period the samples were processed for amino acid analyses according to previously described procedures [14].

Chips of the middle layer of fossil *Mercenaria* taken from the growth edge were cleaned successively by a dental tool and dilute HCl prior to hydrolysis with 6*N* HCl.

3. Results and discussion

3.1. Kinetics of isoleucine epimerization in Mercenaria

The integrated first-order rate equation for the reaction described in eq. 1 can be expressed as [15]:

$$\ln\left(\frac{A_e}{A_e - A_t}\right) - \text{constant} = \left(1 + \frac{k_2}{k_1}\right) k_1 t \qquad (2)$$

A_e and A_t are the mole fractions of alloisoleucine [alleu/(alleu + ileu)] at equilibrium and in the sample respectively; for an equilibrium value of alleu/ileu = 1.30, $A_e = 0.565$. The integration constant, determined from the analysis of several modern samples which were not heated, is 0.018; t is the age of the sample (or heating time) and $k_2/k_1 = 1/1.30$. By inserting known values and solving for k_1, the forward rate constant, eq. 2 reduces to:

$$k_1 = 0.565\ [\ln(0.565/0.565 - A_t) - 0.018]/t \qquad (3)$$

Preliminary heating experiments indicated that the isoleucine epimerization reaction follows the first-order kinetic expression (eq. 2) in *Mercenaria* [2,16].

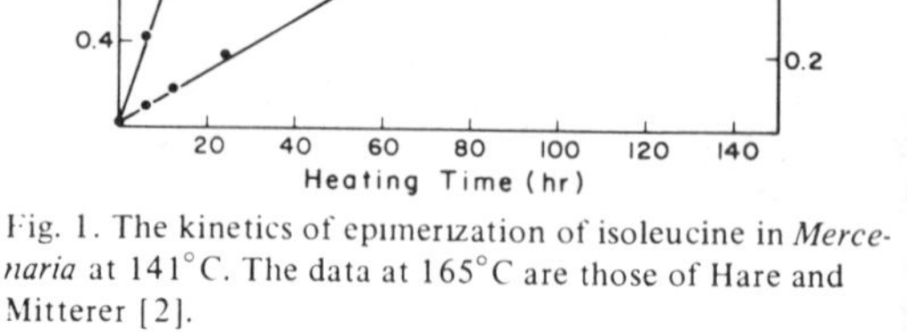

Fig. 1. The kinetics of epimerization of isoleucine in *Mercenaria* at 141°C. The data at 165°C are those of Hare and Mitterer [2].

TABLE 1

Kinetics of epimerization of isoleucine in heated *Mercenaria*

Temp. (°C)	Heating time (hr)	alleu/ileu	k_1 (hr^{-1})
165	10	0.575	5.77×10^{-2}
154	8	0.175	2.03×10^{-2}
	14	0.290	1.97×10^{-2}
144	47.5	0.479	9.91×10^{-3}
	67	0.595	8.95×10^{-3}
141	6	0.058	7.92×10^{-3}
	12	0.101	7.49×10^{-3}
	24	0.197	7.68×10^{-3}
	48	0.367	7.38×10^{-3}
	96	0.613	6.47×10^{-3}
	144	0.875	6.79×10^{-3}
132	97.5	0.287	2.80×10^{-3}
	197	0.552	2.80×10^{-3}
127.5	287	0.624	2.21×10^{-3}
	364	0.667	1.88×10^{-3}
117.5	308.5	0.333	1.04×10^{-3}
	373	0.340	8.76×10^{-4}

The results of the present study confirm the previous data and demonstrate that the reaction follows first-order kinetics to an alleu/ileu ratio of ~0.90 (Fig. 1). The kinetics for this reaction in *Mercenaria* thus differ somewhat from the kinetics in foraminiferal sediment, which is also a carbonate matrix. In the sediment, the reaction approximates first-order kinetics only up to an alleu/ileu ratio of ~0.35 [8].

The alleu/ileu ratios were measured for *Mercenaria* subjected to elevated temperatures from 117.5 to 165°C for various lengths of time. Values of k_1, calculated from the alleu/ileu ratio and eq. 3, are given in Table 1.

In order to estimate diagenetic temperatures or ages of fossils at the lower in-situ temperatures of natural environments, a large extrapolation of the Arrhenius plot is necessary. An empirical approach to obviate this extrapolation is to employ natural samples having precisely known ages and diagenetic temperatures to calculate values of k_1 at the lower temperatures. Although Pleistocene fossils as old as ~400,000 years can be dated by various radiometric methods, the thermal history of the earth has varied considerably during this time and it would be difficult to estimate an average diagenetic temperature for such specimens. Holocene fossils, particularly those with ages less than 5000 years from subtropical latitudes, have not been subjected to glacial-interglacial climatic extremes. Although there are known temperature fluctuations in historical times, these have been relatively minor and long-term averages of mean temperature in subtropical latitudes probably represent reasonable approximations of the average diagenetic temperature which prevailed for the Late Holocene. One possible problem with this approach is that ground temperatures are not as readily available as mean air temperatures and, depending on the albedo, the two values may differ slightly.

A series of seven radiocarbon-dated *Mercenaria* from southeastern United States and Mexico was employed to obtain an empirical relationship between in-situ diagenetic temperature (as approximated by mean air temperature) and k_1, the forward rate constant of eq. 3. Mean air temperatures in the collecting localities range from 19.5 to 23.5°C. Values for k_1 and mean temperatures are given in Table 2 and included in Fig. 2 together with the data from the elevated temperature heating experiments. If the line through the high temperature points is extended to lower temperature values it passes through the data points obtained from the natural samples. A least squares line fitted to all data has the equation:

$$\text{Log}\, k_1 (yr^{-1}) = 17.29 - 6417/T(°K) \qquad (4)$$

Based on this fit there is no significant difference in the temperature dependence of the rate of isoleucine epimerization as determined from heated samples and natural samples. In addition, the inclusion of the natural samples in the Arrhenius plot removes the considerable uncertainty inherent in the extrapolation of data obtained solely at higher experimental temperatures. Eqs. 3 and 4 may be used to calculate diagenetic temperatures from independently dated *Mercenaria*, or conversely, to determine approximate ages if independent estimates of diagenetic temperatures are available.

The Arrhenius activation energy (E_a = slope × 2.303R) for the isoleucine-alloisoleucine intercon-

TABLE 2

Kinetics of epimerization of isoleucine in fossil *Mercenaria*

Sample	Age (yr)	Mean temp. (°C)[e]	alleu/ileu	k_1 (yr^{-1})
FLA-05	1160 ± 120[a]	22.0	0.05	3.41×10^{-5}
FLA-08	1450 ± 100[b]	23.5	0.08	4.77×10^{-5}
MEX-10	2070[c]	23.0	0.10	4.30×10^{-5}
FLA-16	3440 ± 110[b]	23.5	0.16	4.30×10^{-5}
GA-09	3468[c]	19.6	0.09	2.28×10^{-5}
FLA-18	4500 ± 250[d]	22.0	0.18	3.73×10^{-5}
FLA-17	4550 ± 275[d]	22.0	0.17	3.47×10^{-5}

[a] Undated sample; age is that of associated shell (from Long and Mielke [17]).
[b] Isotopes-Teledyne (I-4027 and I-4028).
[c] Undated sample; age is average of three dates from deposit.
[d] Geochron Laboratories (GX 2921 and GX 2922).
[e] From U.S. Weather Bureau [18].

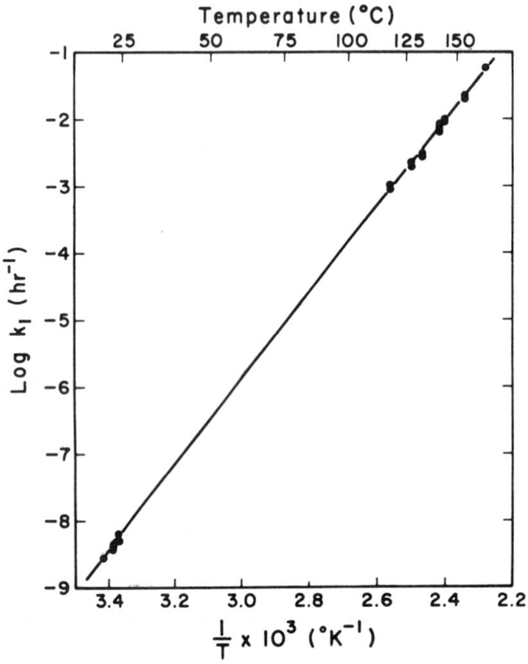

Fig. 2. Log k_1 vs. $1/T$ (°K) for temperatures from 19.6 to 165°C. The high temperature data are from laboratory experiments (Table 1); the low temperature data are from natural specimens (Table 2). The line is a least squares fit to all data. The least squares fit to the subsets alone are: (1) heated specimens: $\log k_1$ (yr^{-1}) = 17.29 – 6415/T(°K); (2) natural specimens: $\log k_1$ (yr^{-1}) = 17.99 – 6625/T(°K).

version in *Mercenaria*, calculated from eq. 4, is 29.4 kcal/mole. Other reported values (in kcal/mole) and their respective sample matrix are: deep-sea foraminiferal sediment – 27.5 [8]; buffered solution of free isoleucine (pH = 7.6) – 31.5 [19]; bone – 33.4 [20]. The activation energy for the two samples with a carbonate matrix (*Mercenaria* and forminifera) differ slightly suggesting that the overall mechanisms of interconversion differ. Because of the diversity of carbonate shell structures, their organic content and amino acid composition, there apparently is no single activation energy for all carbonate systems, so it is necessary to establish the kinetic relations for each carbonate structure independently.

3.2. Absolute chronology of Pleistocene deposits in Florida

Geochemical, sedimentological and paleontologica investigations of deep-sea sediments have expanded our knowledge of Pleistocene chronology, paleotemperatures and glacial-interglacial cycles. In tropical en vironments radiometric ages as old as about 500,000 years have been obtained from coral reefs deposited during high sea levels [21]. Other carbonates generall are not as well suited for radiometric age measurements in the time span >25,000 years B.P. Consequently, the history of Pleistocene deposits in tempe ate coastal areas – their chronology, number of units and correlation – is less complete because of the difficulty in correlating the frequently discontinuous and poorly exposed sediments. The amino acid method offers a new approach to decipher stratigraphic deposits generally unsuitable for radiometric dating and marginally suitable for physical and biostratigraphic techniques.

The age and correlation of the fossiliferous, post-Miocene, marine deposits of Florida have been subjects of considerable controversy and debate (see DuBar [22] for a recent summary). On the basis of alleu/ileu ratios, Mitterer [4] demonstrated the presence of at least five post-Miocene, probably Pleistocene, marine units apparently deposited during interglacial high sea levels in southern Florida and proposed a tentative correlation with similar deposits in central and northern Florida. A detailed stratigraphic study in southern Florida by Perkins [23] has corroborated the presence of five marine units deposited during five separate transgressive-regressive cycles. The absolute chronology and average diagenetic temperature of some of these Pleistocene units can be deduced from eq. 3 and 4. The only limitation is that the alleu/ileu ratio of datable units must fall on the linear portion of the curve in Fig. 1, i.e., alleu/ileu < ~0.90. The dating approach is the "calibration" technique of Bada and Protsch [13]; the age of unknown specimens can be estimated from a knowledge of the rate constant at the in situ diagenetic temperature, which is calculated from specimens of known age subjected to the same diagenetic temperature.

Several inferences strongly point to an age of ~125,000 years for the youngest marine Pleistocene deposit, the Coffee Mill Hammock Member of the Fort Thompson Formation (Unit V of Mitterer [4]):

(1) Radiometric ages of ~130,000 years B.P. are obtained for the Miami and Key Largo Formations of southernmost Florida [24,25], both of which are considered to be time-equivalents of the Coffee Mill Hammock Member [26].

(2) Radiometric dating of fossil corals throughout the world has yielded a number of dates in the 120,000–130,000-year range indicating a significant interglacial interval at this time [12,27].

(3) This interglacial, based on sea level curves for the past 250,000 years, constructed from a variety of data, represents the last time sea level was significantly higher than present and probably the last time the Florida platform was significantly submerged [12,28]. It is unlikely, therefore, that the southern Florida peninsula has received any younger marine deposits of significant extent. The evidence that Unit V (Coffee Mill Hammock Member) was deposited during the 120,000–130,000-year B.P. interglacial high stand of sea level is strong.

The underlying deposits also were apt to be deposited during higher-than-present sea level stands and the ages of these deposits should represent the times of these high sea levels. The ages of the underlying deposits can be deduced from the degree of isoleucine epimerization in *Mercenaria* by applying the rate constant calculated from fossils of the Coffee Mill Hammock Member with an assumed age of ~125,000 years. Although most dates in this time span range from 120,000 years to 130,000 years, those from southernmost Florida are slightly greater and bracket 130,000 years [24,25]. Accordingly the rate constant is calculated by assuming an age of 125,000 years for the lowest alleu/ileu ratio in Unit V. Because of the regional difference in temperature, the rate constant is calculated separately for each region. The calculated ages are given in Table 3.

The deduced ages for the Pleistocene units with alleu/ileu <0.90 are approximately: Unit V – 134,000, Unit IV – 180,000, Unit III – 236,000 and Unit II – 324,000 years old. The latter data is probably slightly low as the alleu/ileu ratio for this unit falls on the non-linear portion of Fig. 1. Several arguments reflect on the accuracy of these dates. First, the dates from each region are concordant. This correspondence supports the stratigraphic correlation previously proposed [4]. Second, the diagenetic temperature calculated for each region decreases systematically from south to north as expected. (This is discussed more fully in the next section.) Third, all ages correspond to the ages of high sea levels dated radiometrically at many localities worldwide [12,21,23]. This close correspondence between radiometrically dated and amino acid dated high sea levels is striking and gives confidence in the amino acid dating method.

The ages deduced above, however, are dependent on the original assumption that deposition of Unit V ceased ~125,000 years B.P. If Unit V was deposited during another high stand of the sea then the ages in Table 3 are in error; e.g., if Unit V is younger than 125,000 years the ages in Table 3 are too high. But,

TABLE 3

Ages of Pleistocene deposits in Florida deduced from the isoleucine epimerization reaction[a]

Unit	South $k_1 = 5.52 \times 10^{-6}$		Central $k_1 = 4.12 \times 10^{-6}$		North $k_1 = 3.10 \times 10^{-6}$	
	alleu/ileu	age (10^3 yrs)	alleu/ileu	age (10^3 yrs)	alleu/ileu	age (10^3 yrs)
V	0.67–0.75	125–144	0.52–0.60	125–147	0.40–0.45	125–142
IV	0.85–0.92	170–191	0.70–0.71	176–180	0.53–0.60	170–195
III	0.98–1.01	212–223[b]	0.92[c]	256	0.67–0.74	222–251
II					0.85–0.93	302–346

[a] Deduced ages are based on an assumed age of 125,000 years for the lowest alleu/ileu ratio obtained from Unit V in each region.
[b] Probably a minimum age; alleu/ileu exceeds ~0.90.
[c] One specimen.

with no other assumed age (e.g., the 82,000- or 105,000-year B.P. high sea levels [27]) can such a close correspondence be obtained between radiometric dates and amino acid dates for earlier high sea levels.

Since the discovery of vertebrate fossils in the Caloosahatchee Formation (Units I and II) there has been considerable controversy whether these deposits are Pleistocene or Pliocene in age [29]. The calculated ages of the deposits from Table 3 bear on this controversy. The deduced ages are in agreement with the mid-Pleistocene age assignment for the Upper Caloosahatchee Formation (Unit II) based on a variety of vertebrate remains [22]. However, this age conflicts with a long-established Pliocene age assignment based on invertebrate evidence (see Brooks [26] for a summary). The latter age is supported by He–U dates on corals [30]. Thus, the controversy continues; invertebrate fauna and He–U dates indicate a Pliocene age and vertebrate fauna and now amino acid dates support a mid-Pleistocene age. Were it not for the He–U dates, the amino acid dates would be taken as confirming a mid-Pleistocene age. Lesser reliance is placed on the invertebrate evidence because the detailed amino acid survey of these deposits has shown significant reworking of older fossils into younger units [4]. As the Caloosahatchee Formation overlies highly fossiliferous Miocene units there is a strong probability that many of the extinct species reported from the Caloosahatchee Formation were derived from the underlying deposits. In view of the agreement between radiometrically dated and amino acid dated high sea levels, the internal consistency of the amino acid dates for regions with different diagenetic temperatures, and the vertebrate data, the weight of evidence supports a mid-Pleistocene age for the Upper Caloosahatchee Formation.

3.3. *Late Pleistocene diagenetic temperatures in Florida*

Pleistocene fossils have experienced lower diagenetic temperatures for portions of their burial history; consequently, average diagenetic temperatures of dated Pleistocene samples calculated from eq. 4 will be lower than present mean temperatures for the region. A single calculated temperature cannot accurately express the diagenetic temperature history of a fossil which has been subjected to wide variations in thermal regimes. The calculated value will only depict the "average" thermal history of a fossil and thus will fall between the extremes of glacial and interglacial climate. In addition, the computed value is not a true average because k_1 does not vary linearly with temperature over an extended temperature range; consequently the "average" diagenetic temperature calculated from eq. 4 may be slightly higher than a true average. Approximate as it is, however, the calculated diagenetic temperature will represent an upper limit to average glacial temperatures and will yield a minimum amplitude for glacial-interglacial temperature variation in a particular region. The first-order rate constant for each region, given in Table 3, can be inserted in eq. 4 to calculate average diagenetic temperatures in these regions. The values are given in Table 4 together with

TABLE 4

"Average" diagenetic temperatures in Florida during the Late Pleistocene calculated from the isoleucine epimerization reaction[a]

	k_1	Present av. temp. (°C)	Av. Pleistocene temp. (°C)	ΔT(°C)[b]
North	3.10×10^{-6}	20.9	8.3	12.6
Central	4.12×10^{-6}	22	9.8	12.2
South	5.52×10^{-6}	23.5	11.4	12.1

[a] Deduced average Pleistocene temperatures are based on an assumed age of 125,000 years for the lowest alleu/ileu ratio obtained from Unit V in each region.

[b] Difference between present and Pleistocene "average" temperatures.

glacial-interglacial temperature amplitude. The average diagenetic temperatures are from 8.3 to 11.4°C which is a decrease of about 12°C from present mean temperatures. These average diagenetic temperatures are in accord with climatic estimates for the southeastern U.S. based on independent criteria. Pollen diagrams from Georgia and the Carolinas indicate a predominantly boreal character to the vegetation [31]. Similar flora is now found in northern New England nd the northern Great Lakes Region, a displacement f at least 1000 km northward. Emiliani [32] has summarized evidence from a number of sources indicating that the amplitude of glacial-interglacial temperature variations was as much as 10°C in continental mid-latitudes. Thus, the average diagenetic temperatures deduced from the isoleucine epimerization reaction are in general agreement with climatic estimates obtained from independent studies. More estimates are necessary, however, to firmly establish the thermal history of this region.

An "averaged" Pleistocene temperature gradient may also be calculated from the diagenetic temperatures. The average temperature difference over the last 125,000 years between northern and southern Florida is 3.1°C (Table 4). This value is not dependent on the absolute age of the deposits, as long as they are contemporaneous. Historical weather records give a gradient between these localities of about 2.6°C [18]. Therefore, during the Late Pleistocene, although average temperatures were significantly lower than present values, the average temperature gradient between northern and southern Florida was close to (perhaps slightly greater) than the present gradient.

4. Conclusions

Because of the temperature dependence of k_1, the forward rate constant, the time for alleu/ileu to reach ~0.90 increases with decreasing temperatures and thus the time range available for dating *Mercenaria* by the isoleucine epimerization reaction is greater at lower temperatures. At an average temperature of 5°C the time to reach alleu/ileu = 0.90 is about 0.6×10^6 years; at 0°C the time is 1.6×10^6 years. At the lower diagenetic temperatures which prevailed in the Atlantic Coastal Plain north of Florida it should be possible to date all Pleistocene deposits. Based on the present results, at least five Pleistocene marine cycles can be anticipated throughout the Atlantic Coastal Plain.

The amino acid dating method offers the potential to establish the absolute chronology of the entire Pleistocene sequence in the Atlantic Coastal Plain, and a precise stratigraphic correlation of these deposits from New England to Florida. In addition, an approximation to the average Pleistocene temperature for various parts of the U.S. east coast can be calculated.

Acknowledgements

I thank H.K. Brooks, V.J. Henry and E.W. Behrens for providing samples. This work was supported by Grant GA-30695 from the Oceanography Section of the National Science Foundation.

References

1 P.E. Hare and R.M. Mitterer, Non-protein amino acids in fossil shells, Carnegie Inst. Wash. Yearbook 65 (1967) 362.

2 P.E. Hare and R.M. Mitterer, Laboratory simulation of amino acid diagenesis in fossils, Carnegie Inst. Wash. Yearbook 67 (1969) 205.

3 R.M. Mitterer, Calcified proteins in the sedimentary environment, in: Advances in Organic Geochemistry 1971, H.R. von Gaertner and H. Wehner, eds. (Pergamon, Oxford, 1972) 441.

4 R.M. Mitterer, Pleistocene stratigraphy in southern Florida based on amino acid diagenesis in fossil *Mercenaria*, Geology 2 (1974) 425.

5 R.M. Mitterer, Amino acid and protein geochemistry in mollusk shells, Ph. D. Thesis, Florida State Univ., (1966) 151 pp.

6 J.L. Bada, B.P. Luyendyk and J.B. Maynard, Marine sediments; dating by the racemization of amino acids, Science 170 (1970) 730.

7 J. Wehmiller and P.E. Hare, Racemization of amino acids in marine sediments, Science 173 (1971) 907.

8 J.L. Bada and R.A. Schroeder, Racemization of isoleucine in calcareous marine sediments: kinetics and mechanism, Earth Planet. Sci. Lett. 15 (1972) 1.

9 K. King, Jr. and P.E. Hare, Species effects in the epimerization of L-isoleucine in fossil planktonic foraminifera, Carnegie Inst. Wash. Yearbook 71 (1972) 596.

10 J.L. Bada, The dating of fossil bones using the racemization of isoleucine, Earth Planet. Sci. Lett. 15 (1972) 223.

11 J.D. Milliman and K.O. Emery, Sea levels during the past 35,000 years, Science 162 (1968) 1121.

12 J. Chappell, Geology of coral terraces, Huon Peninsula, New Guinea: a study of Quaternary tectonic movements and sea-level changes, Geol. Soc. Am. Bull. 85 (1974) 553.

13 J.L. Bada and R. Protsch, Racemization reaction of aspartic acid and its use in dating fossil bones, Proc. Nat. Acad. Sci. U.S.A. 70 (1973) 1331.

14 P.E. Hare, Geochemistry of proteins, peptides, and amino acids, in: Organic Geochemistry, G. Eglinton and M.T.J. Murphy, eds. (Springer-Verlag, New York, 1969) 438 pp.

15 H.G. Bray and K. White, Kinetics and Thermodynamics in Biochemistry (Academic Press, New York, 1966) 2nd ed., 418 pp.

16 P.E. Hare, Effect of hydrolysis on the racemization rate of amino acids, Carnegie Inst. Wash. Yearbook 70 (1971) 256.

17 A. Long and J.E. Mielke, Smithsonian Institution radiocarbon measurements, IV, Radiocarbon 9 (1967) 380.

18 U.S. Weather Bureau, Climates of the States (Govt. Printing Office, Washington, D.C., 1962) 10.

19 J.L. Bada, Kinetics of the nonbiological decomposition and racemization of amino acids in natural waters, in: Nonequilibrium Systems in Natural Waters, J.D. Hem, ed., Adv. in Chemistry Series (American Chemical Society, Washington, D.C., 1971) 309 pp.

20 J.L. Bada, K.A. Kvenvolden and E. Peterson, Racemization of amino acids in bones, Nature 245 (1973) 308.

21 M.L. Bender, F.T. Taylor and R.K. Matthews, Helium-uranium dating of corals from Middle Pleistocene Barbados reef tracts, Quatern. Res. 3 (1973) 142.

22 J.R. DuBar, Summary of the Neogene stratigraphy of southern Florida, in: Post-Miocene Stratigraphy, Central and Southern Atlantic Coastal Plain, R.Q. Oaks, Jr. and J.R. DuBar, eds. (Utah State Univ. Press, Logan, 1974) 206.

23 R.D. Perkins, Discontinuity surfaces as a stratigraphic tool: the Pleistocene of South Florida, Abstracts Geol. Soc. Am. Annual Meet. 6 (1974) 908.

24 J.K. Osmond, J.R. Carpenter and H.L. Windom, ^{230}Th/^{234}U age of the Pleistocene corals and oolites of Florida, J. Geophy. Res. 70 (1965) 1843.

25 W.S. Broecker and D.L. Thurber, Uranium-series dating of corals and oolites from Bahamian and Florida Key limestones, Science 149 (1965) 58.

26 H.K. Brooks, The Plio-Pleistocene of Florida, with special reference to the strata outcropping on the Caloosahatchee River, in: Late Cenozoic Stratigraphy of Southern Florida – A Reappraisal, R.D. Perkins, ed. (Second Ann. Field Trip of Miami Geol. Soc. Guidebook, Miami 1968) 3.

27 K.J. Mesolella, R.K. Matthews, W.S. Broecker and D.L. Thurber, The astronomical theory of climatic change: Barbados data, J. Geol. 77 (1969) 250.

28 R.K. Matthews, Relative elevation of Late Pleistocene high sea level stands: Barbados uplift rates and their implications, Quatern. Res. 3 (1973) 147.

29 J.R. DuBar, Stratigraphy and paleontology of the late Neogene strata of the Caloosahatchee River of southern Florida, Fla. Geol. Surv. Bull. 40 (1958).

30 M.L. Bender, Helium-uranium dating of corals, Geochim. Cosmochim. Acta 37 (1973) 1229.

31 D.R. Whitehead, Late-Wisconsin vegetational changes in unglaciated eastern North America, Quatern. Res. 3 (1973) 621.

32 C. Emiliani, Pleistocene paleotemperatures, Science 168 (1970) 822.

Reprinted from *Proc. Natl. Acad. Sci.* **68**:486–490 (1971)

Nonprotein Amino Acids in the Murchison Meteorite

KEITH A. KVENVOLDEN, JAMES G. LAWLESS, AND CYRIL PONNAMPERUMA

Exobiology Division, Ames Research Center, NASA, Moffett Field, California 94305

Communicated by W. F. Libby, December 11, 1970

ABSTRACT Twelve nonprotein amino acids appear to be present in the Murchison meteorite. The identity of eight of them has been conclusively established as *N*-methylglycine, β-alanine, 2-methylalanine, α-amino-*n*-butyric acid, β-amino-*n*-butyric acid, γ-amino-*n*-butyric acid, isovaline, and pipecolic acid. Tentative evidence is presented for the presence of *N*-methylalanine, *N*-ethylglycine, β-aminoisobutyric acid, and norvaline. These amino acids appear to be extraterrestrial in origin and may provide new evidence for the hypothesis of chemical evolution.

The formation of organic compounds and their accumulation have been considered to be a necessary preamble to the appearance of life on the primordial earth (1). Various forces, such as ultraviolet light from the sun, heat from volcanos, electrical discharges in the form of lightning, and ionizing radiation from radionuclides acting upon the reducing atmosphere of the primitive earth may have produced a large number of organic compounds until the early oceans had the consistency of a "hot dilute soup" (2–5). Considerations of stellar and planetary evolution lead us to believe that the sequence of events that led to life on earth may have been duplicated elsewhere in innumerable planetary systems in the universe (6–9).

In attempting to substantiate this hypothesis, two avenues have generally been used. In the synthetic approach, the conditions of a primitive planet have been simulated in the laboratory. Many of the constituents of proteins and nucleic acids have been synthesized in this manner (10, 11). In analytical studies, investigations have centered around the interstellar medium, lunar samples, ancient sediments, and meteorites. Radioastronomical observations have clearly demonstrated the presence of the smaller molecules important to chemical evolution, such as water, ammonia, hydrogen cyanide, formaldehyde, cyanogen, carbon monoxide, and cyanoacetylene, as components of interstellar matter (12). The analysis, however, of lunar samples from the Apollo 11 and 12 missions (13–15) showed no conclusive evidence of any molecules of significance to chemical evolution. The oldest sediments available on the earth, dated at about 3.1–3.4 billion years, are reported to contain microscopic evidence of early life (16, 17). On account of the apparent absence of unmetamorphosed samples older than these rocks, and because any prebiotic terrestrial organic matter would probably have been consumed by microorganisms once life arose, the possibility of discovering any molecules predating the emergence of life appears to be remote.

The analysis of meteorites has provided some evidence for the presence of extraterrestrial carbon compounds (18). Carbonaceous chondrites, for example, are known to contain up to 5% carbon (19, 20). The finding of amino acids indigenous to such samples would constitute convincing evidence for extraterrestrial chemical evolution. In our analysis of the Murchison meteorite we have already reported (21) the presence of five amino acids commonly found in protein: glycine, alanine, valine, proline, and glutamic acid, and two not generally found in proteins: *N*-methylglycine (sarcosine) and α-aminoisobutyric acid (2-methylalanine). We have now identified aspartic acid and six additional nonprotein amino acids: β-alanine, α-amino-*n*-butyric acid, β-amino-*n*-butyric acid, γ-amino-*n*-butyric acid, isovaline, and pipecolic acid. Furthermore, there appears to be tentative evidence for at least four other nonprotein amino acids: β-aminoisobutyric acid, norvaline, *N*-methylalanine, and *N*-ethylglycine.

ANALYSIS

The amino acids were obtained from the meteorite by extraction of a pulverized sample with boiling distilled water, followed by acid hydrolysis of the aqueous extract (22). After the removal of inorganic salts from the acid hydrolysate, the resulting material was analyzed for amino acids by ion-exchange chromatography, gas chromatography, and gas chromatography combined with mass spectrometry. The residue left after the water extraction of the meteorite was hydrolyzed with acid and similarly analyzed. In a procedural blank, no amino acids could be detected.

A piece of the Murchison meteorite (23) that appeared to be massive in character, and showed signs of the least exterior contamination, was treated in the following manner. The ablation crust was first chiselled away. The interior pieces were then pulverized for 15 sec in an Angstrom Disc Mill (T100).

10 g of the pulverized meteorite were refluxed with 50 ml of triply-distilled water for 20 hr at about 110°C. The extract was recovered, after centrifugation, by decantation. The residue was twice rinsed with 50 ml of water. The extract and rinses were evaporated to dryness, and hydrolyzed with 50 ml of 6 N HCl for 20 hr at about 110°C. The resulting hydrolysate was again evaporated to dryness.

The dried hydrolysate was dissolved in about 10 ml of water and charged on a 50-ml column packed with Dowex 50 (Bio-Rad) ion-exchange resin, prepared in the H^+ form. The column was successively eluted with two bed volumes of water and two bed volumes of 2 N NH_4OH. The NH_4OH eluate was collected, evaporated to dryness, and redissolved in 5 ml of water; 0.5 ml was removed for analysis by ion-exchange chromatography using a conventional amino acid analyzer. In a subsequent experiment on a different sample, the procedure included use of Dowex 50 (H^+) followed by Dowex 1 (OH^-).

Ion-exchange chromatography

The pH of the 0.5-ml solution was adjusted to about 3 and 1 ml of pH 2.2 buffer was added. This mixture was charged on the column of an automatic amino acid analyzer (Spinco Model 120) for determination of neutral and acidic amino acids.

Preparation of diastereoisomeric derivatives for gas chromatography

The remaining 4.5 ml of solution were evaporated to dryness. *N*-trifluoroacetyl-D-2-butyl esters of amino acids were prepared (24). The dried residue from the aqueous solution was refluxed for 3 hr at 110°C under anhydrous conditions with 0.5 ml of D-2-butanol–8 N HCl (Kroon Research) which was 96% enriched in the D isomer. Excess D-2-butanol was removed by evaporation. To the product was added 0.5 ml of dichloromethane and 0.05 ml of trifluoroacetic anhydride (Eastman). The mixture reacted at room temperature for 16 hr and was evaporated at room temperature. Dichloromethane was added and portions of this solution were analyzed by gas chromatography and gas chromatography–mass spectrometry. Diastereoisomeric derivatives of amino acid standards were prepared by using DL-2-butanol instead of D-2-butanol.

Gas chromatography

Gas chromatographic analyses were performed on Perkin Elmer 881 and 900 gas chromatographs (flame-ionization detection) on two different capillary columns (150 ft × 0.02 in.); one was coated with UCON 75 H 90,000 and the other with XE 60 (both phases available through Applied Science).

Gas chromatography–mass spectrometry

The effluent from the Perkin Elmer 881 gas chromatograph (with the UCON 75 H 90,000 column installed) was introduced into a CEC 21-491 mass spectrometer through a membrane separator and the mass spectra of the eluted compounds were obtained.

Hydrolysis of meteorite residue

The meteorite residue, after water extraction, was hydrolyzed with 50 ml of 1 HCl under reflux for 20 hr at about 110°C. This hydrolysate was treated in a manner similar to that described for the hydrolysate of the water extract.

IDENTIFICATION

Ion-exchange chromatography with an automated amino acid analyzer revealed the presence of several common amino acids in the hydrolyzed extract as well as in the hydrolysate of the residue. The concentration of compounds in the latter was about four times smaller than in the hydrolyzed aqueous extract. In this paper we have, therefore, directed our attention to the amino acids in the hydrolyzed aqueous extract.

We have previously reported (21) that the hydrolyzed aqueous extract of this meteorite contained the amino acids glycine (6 μg/g), alanine (3 μg/g), valine (2 μg/g), proline (1 μg/g), and glutamic acid (3 μg/g). We had also found the nonprotein amino acids *N*-methylglycine and α-aminoisobutyric acid. The identity of these compounds was established by gas chromatography and gas chromatography combined with mass spectrometry. It was further shown that the enantiomers of each of the amino acids with asymmetric centers, namely alanine, valine, proline, and glutamic acid, were present in about equal amounts.

By a detailed examination of the gas chromatograms and mass spectra of this meteorite extract, we have found ten additional nonprotein amino acids. The identities of six of these have been definitely established. The identification of the four remaining may be considered tentative. Of the amino acids found in protein, we have also identified D- and L-aspartic acid, and confirmed the presence of L-glutamic acid, the mass spectrum of which we had not obtained previously (21).

The identification of the amino acids was based on the following criteria.

(*1*) Equivalence of the mass spectrum of the isolated amino acid with that of an authentic standard. The compounds compared were *N*-trifluoroacetyl-2-butyl esters of amino acids.

(*2*) Coincidence of retention times resulting from the cochromatography of the *N*-trifluoroacetyl-2-butyl ester of the isolated amino acid with the derivative of the corresponding standard.

Table 1 lists the nonprotein amino acids whose presence was established by these criteria. Fig. 1 illustrates the gas-chromatographic separation of the derivatives of these compounds. In four cases where the spectra obtained from the sample resembled the standard, but were weak, the identifications were considered tentative. Cochromatography of standards supported the tentative identifications.

The most abundant ion fragments for the nonprotein amino acids identified in this meteorite sample are shown in Table 2. Mass spectra of the *N*-trifluoroacetyl-2-butyl esters of authentic amino acid standards resemble closely, with a few exceptions, the mass spectra of *N*-trifluoroacetyl-*n*-butyl

TABLE 1. *Nonprotein amino acids in the Murchison meteorite*

Chromatographic peak number (Fig. 1)	Name	Supplier of authentic amino acid standard†
10	*N*-methylglycine (sarcosine)*	K & K
20	β-alanine	K & K
5	*N*-methylalanine (DL?) (tentative identification)	Cyclo
11	*N*-ethylglycine (tentative identification)	Cyclo
2	α-aminoisobutyric acid (2-methylalanine)*	Eastman
6 and 8	α-amino-*n*-butyric acid (D and L)	K & K
14 and 15	β-aminoisobutyric acid (D and L) (tentative identification)	Calbiochem
16	β-amino-*n*-butyric acid (DL?)	Mann
23	γ-amino-*n*-butyric acid	Mann
1	Isovaline (DL?)	K & K
12 and 13	Norvaline (D and L) (tentative identification)	Mann
17 and 18	Pipecolic acid (D and L)	Calbiochem

* Previously identified (21).

† The identities of the standards were supported by mass spectra obtained for *N*-trifluoroacetyl-*n*-butyl and *N*-trifluoroacetyl-DL-2-butyl ester derivatives.

TABLE 2. *Most abundant ion fragments of the nonprotein amino acids in the Murchison meteorite*

m/e	*N*-methyl-glycine	β-alanine	α-aminoiso-butyric acid	α-amino-*n*-butyric acid	β-amino-*n*-butyric acid	γ-amino-*n*-butyric acid	Isovaline	Pipecolic acid
100							6	
101							3	
102			3	4				
110	2							
112	5				7	8		
114			10		9	4	8	
116		7			2	3	2	
126	4	28		9		26	3	6
127		6				7		
130				2	8			
139		12			4	9		
140	100	5		15	80	12	1	
141	43	5			7			
153			4	5	18	7		
154			100	100	4	28		
155			16	46	9	4		
166	4	3	2		7		10	
167	1	4					6	
168	27	100					100	
169	2	7					12	
180			2	4	5	2		100
181					5	14		10
182			12	16	100	100		1
184							12	
185	15	8						
186	2	10						
194							2	
195							1	
196							6	
199					11	15		
200					18	5		
208								7
225								4
240							1	
241	2							
281								5

esters (Lawless and Chadha, manuscript in preparation). The mass spectra taken during the gas-chromatographic separation suggest the presence of at least two other isomers of valine (M^+ 269), and two other homologs of proline (M^+ 281) (isomers of pipecolic acid). A major problem, however, in the identification of these compounds is the unavailability of standards, without which identification is equivocal.

Three nonprotein amino acids, isovaline, β-amino-*n*-butyric acid, and *N*-methylalanine, all of which have asymmetric carbon atoms, did not yield two peaks, corresponding to each diastereoisomeric derivative, when the *N*-trifluoroacetyl-DL-butyl esters of these amino acid standards were chromatographed both on UCON 75 H 90,000 and on XE 60. For this reason, it was impossible to estimate the ratio of D and L isomers for these compounds as found in this meteorite. Consequently, the expression (DL?) is used in Table 1 in the case of these amino acids.

CONCLUSION

The analytical procedures of ion-exchange chromatography, gas chromatography, and gas chromatography combined with mass spectrometry have established unequivocally that six amino acids commonly found in protein, and eight which do not occur in natural protein, are present in the Murchison meteorite. All the amino acids that have asymmetric carbon atoms and whose diastereoisomeric derivatives could be separated by the gas chromatographic method we used appeared to consist of approximately equal amounts of the D and L isomers.

The possibility of terrestrial contamination is inherent in meteorite analysis. Recent biological contamination would, however, invariably lead to the predominance of L isomers of the amino acids commonly found in protein (25, 26). The identification of nonprotein amino acids with an almost equal abundance of D and L isomers argues strongly against this possibility.

If the amino acids are indigenous to the meteorite and therefore extraterrestrial, the question immediately arises as to their origin. Two possibilities present themselves: these amino acids were present at some period of time in the meteorite in one stereoisomeric form and were then racemized in the course of time (27, 28), or the two forms were always present in nearly equal amounts. The possibility that an extraterrestrial biota was responsible for D or L amino acids initially cannot be dis-

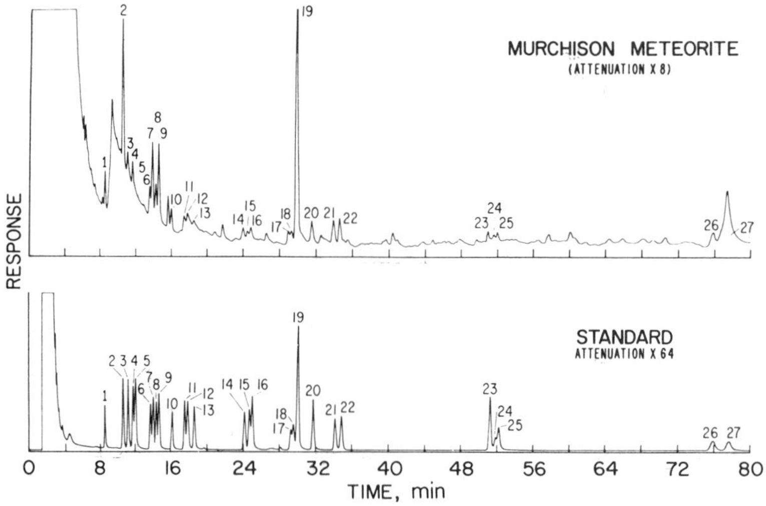

FIG. 1. Gas chromatogram of *N*-trifluoroacetyl-D-2-butyl esters of amino acids in acid-hydrolyzed aqueous extract of Murchison meteorite compared with a gas chromatogram of *N*-trifluoroacetyl-DL-2-butyl esters of standard amino acids. Gas chromatography: 0.02 inch × 150 foot capillary column coated with UCON 75 H 90,000; temperature programmed from 100 to 150°C at 1°C/min. Identification for the sample is as follows: (1) (DL?)-isovaline; (2) α-aminoisobutyric acid; (3) D-valine; (4) L-valine; (5) (DL?)-*N*-methylalanine (?); (6) D-α-amino-*n*-butyric acid; (7) D-α-alanine; (8) L-α-amino-*n*-butyric acid; (9) L-α-alanine; (10) *N*-methylglycine; (11) *N*-ethylglycine (?); (12) D-norvaline (?); (13) L-norvaline (?); (14) D-β-aminoisobutyric acid (?); (15) L-β-aminoisobutyric acid (?); (16) (DL?)-β-amino-*n*-butyric acid; (17) D-pipecolic acid; (18) L-pipecolic acid; (19) glycine; (20) β-alanine; (21) D-proline; (22) L-proline; (23) γ-amino-*n*-butyric acid; (24) D-aspartic acid; (25) L-aspartic acid; (26) D-glutamic acid; (27) L-glutamic acid (masked by an unidentified peak in Murchison analysis). The same identifications hold for the chromatogram of the standard, except for the compounds represented by peaks (1), (5), and (16). These compounds are known to be DL.

counted entirely. However, by analogy with terrestrial life, it would be difficult to explain the presence of several nonprotein amino acids. The production of either the D or the L form by an abiogenic process and subsequent racemization may also be considered unlikely, since no well-defined evidence is available for abiotic processes that would produce one form rather than the other, in spite of extensive experimentation (29). We are therefore led to believe that the amino acids in the Murchison meteorite were most likely produced in both forms (DL) by an abiotic process. Of the twelve nonprotein amino acids positively and tentatively identified by us, five have been identified in laboratory experiments simulating primitive earth or planetary conditions (30, 31). The presence of all the isomers of the amino acids of two and three carbon atoms and all but two of the isomers of the amino acids with four carbon atoms, suggests a random synthesis. In addition, the previously reported random distribution of isomers of aliphatic hydrocarbons and heavy (^{13}C) isotopic values of the extractable carbon (21), point to an abiogenic origin of the organic matter in this meteorite. We conclude that our analysis of the Murchison meteorite provides a new basis for the study of chemical evolution and the search for extraterrestrial life (9).

We thank Mr. G. E. Pollock, Life Detection Branch, Exobiology Division, Ames Research Center, NASA, for providing the gas-chromatographic columns used in this study.

1. Oparin, A. I., *Proiskhogden. Zhizni*, Moscovsky Robotchii, Moscow (1924).
2. Haldane, J. B. S., *Rationalist Annual*, **148**, 3 (1928).
3. Oparin, A. I., *The Origin of Life on the Earth*, Proc. of the First International Symposium, Moscow, 1957 (Pergamon Press, London, 1959).
4. Fox, S. W., ed., *The Origins of Prebiological Systems and of Their Molecular Matrices*, Proc. of a conference at Wakulla Springs, Fla., Oct. 27–30, 1963 (Academic Press, New York, 1965).
5. Buvet, R., and C. Ponnamperuma, eds., *Proc. of the Third International Conference on the Origin of Life*, Pont-a-Mousson, France, April 20–25, 1970 (North Holland, Amsterdam, in press, 1971).
6. Shklovskii, I. S., and C. Sagan, *Intelligent Life in the Universe* (Holden-Day, San Francisco, 1966).
7. Brown, H., *Science*, **145**, 1117 (1964).
8. Huang, S. S., *Amer. Sci.*, **47**, 397 (1959).
9. Ponnamperuma, C., and H. P. Klein, *Quart. Rev. Biol.*, **45**, 235 (1970).
10. Ponnamperuma, C., and N. W. Gabel, *Space Life Sci.*, **1**, 64 (1968).
11. Lemmon, R. M., *Chem. Rev.*, **70**, 95 (1970).
12. Snyder, L. E., and D. Buhl, *Sky and Telescope*, **40**, 267 (1970).
13. Lunar Science Conference, *Science*, **167**, 751 (1970).
14. Chang, S., K. Kvenvolden, J. Lawless, C. Ponnamperuma, and I. R. Kaplan, *Science*, in press (1970).
15. Chang, S., and K. Kvenvolden, *Exobiology*, ed. C. Ponnamperuma (North Holland, Amsterdam, in press, 1971).
16. Barghoorn, E. S., and J. W. Schopf, *Science*, **152**, 758 (1966).
17. Engel, A. E. J., B. Nagy, L. A. Nagy, C. G. Engel, G. O. W. Kremp, and C. M. Drew, *Science*, **161**, 1005 (1968).
18. Hayes, J. M., *Geochim. Cosmochim. Acta*, **31**, 1395 (1967).

19. Nagy, B., *Geol. Foeren. Stockholm Foerh.*, **88**, 235 (1966).
20. Studier, M. H., R. Hayatsu, and E. Anders, *Science*, **149**, 1455 (1965).
21. Kvenvolden, K. A., J. Lawless, K. Pering, E. Peterson, J. Flores, C. Ponnamperuma, I. R. Kaplan, and C. Moore, *Nature*, **228**, 923 (1970).
22. Fox, S. W., K. Harada, P. E. Hare, G. Hinsch, and G. Mueller, *Science*, **167**, 767 (1970); Hare, P. E., K. Harada, and S. W. Fox, *Geochim. Cosmochim. Acta*, Suppl. 2, 2, 1799 (1970).
23. Report 779, Center for Shortlived Phenomena, Smithsonian Institution, Cambridge, Mass., Oct. 7, 1969.
24. Pollock, G. E., V. I. Oyama, and R. D. Johnson, *J. Gas Chrom.*, **3**, 174 (1965); Pollock, G. E., and V. I. Oyama, *J. Gas Chrom.*, **4**, 126 (1966).
25. Wald, G., *Ann. N.Y. Acad. Sci.*, **69**, 352 (1957).
26. Corrigan, J. J., *Science*, **164**, 142 (1969).
27. Kvenvolden, K. A., E. Peterson, and F. S. Brown, *Science*, **169**, 1079 (1970).
28. Bada, J. L., B. P. Luyendyk, and J. B. Maynard, *Science*, **170**, 730 (1970).
29. Bonner, W. A., *Exobiology*, ed. C. Ponnamperuma (North-Holland, Amsterdam, in press).
30. Miller, S. L., *J. Amer. Chem. Soc.*, **77**, 2351 (1955).
31. Chadha, M., S., J. Lawless, J. Flores, and C. Ponnamperuma, in *Proc. Third International Conference on the Origin of Life*, Pont-a-Mousson, France, April 20–25, 1970, eds. R. Buvet and C. Ponnamperuma (North-Holland, Amsterdam, in press, 1971).

AUTHOR CITATION INDEX

SUBJECT INDEX

About the Editor

KEITH A. KVENVOLDEN is a geologist with the U.S. Geological Survey in Menlo Park, California, and specializes in organic geochemistry. When he joined the Survey in 1975, Dr. Kvenvolden had held research and administrative positions since 1966 at the National Aeronautics and Space Administration's Ames Research Center. He was a senior research technologist from 1961–66 in petroleum geochemistry with Mobil Oil Corporation. He is presently a consulting professor of Geology at Stanford University and adjunct professor at California State University, Hayward.

A graduate in geophysical engineering from the Colorado School of Mines, Dr. Kvenvolden received the M.S. and Ph.D. degrees in geology from Stanford University. He is currently chairman of the JOIDES Advisory Panel on Organic Geochemistry and is a member of many professional groups, including the Geochemical Society (Organic Geochemistry Division), International Association of Geochemistry and Cosmochemistry, Geological Society of America (fellow), American Association of Petroleum Geologists, Society of Economic Paleontologists and Mineralogists, American Chemical Society, and American Geophysical Union. He has also served as chairman of the Organic Geochemistry Division of the Geochemical Society and as associate editor of *Geochimica et Cosmochimica Acta* and *Biosystems.*

Dr. Kvenvolden is the author of more than one hundred papers and abstracts concerned with various aspects of organic geochemistry. He is the editor of the Benchmark Papers in Geology, Volume 14, *Geochemistry and the Origin of Life.*